Developmental Biology: Essential Concepts and Applied Principles

Developmental Biology: Essential Concepts and Applied Principles

Editor: Leonardo Reeves

MURPHY & MOORE
www.murphy-moorepublishing.com

Murphy & Moore Publishing,
1 Rockefeller Plaza,
New York City, NY 10020, USA

Visit us on the World Wide Web at:
www.murphy-moorepublishing.com

ISBN: 978-1-63987-156-8 (Hardback)

Cataloging-in-Publication Data

Developmental biology : essential concepts and applied principles / edited by Leonardo Reeves.
p. cm.
Includes bibliographical references and index.
ISBN 978-1-63987-156-8
1. Developmental biology. 2. Growth. 3. Biology. 4. Ontogeny. I. Reeves, Leonardo.
QH491 .D48 2022
571.8--dc23

Table of Contents

Preface

Developmental biology is concerned with the processes that helps animals and plants to grow and develop. Some of the major processes related to this field are cell differentiation, regional specification, morphogenesis and tissue growth. This field of study also includes the biology of metamorphosis, regeneration and asexual reproduction. The major developmental processes of the organisms associated with this field are divided into two categories. These are cell differentiation and regeneration. Cell differentiation studies the cell types that arise during development whereas regeneration is the ability to grow a missing part. This book traces the progress of this field and highlights some of its key concepts and applications. It presents researches and studies performed by experts across the globe. Coherent flow of topics, student friendly language and extensive use of examples makes this book an invaluable source of knowledge.

This book has been the outcome of endless efforts put in by authors and researchers on various issues and topics within the field. The book is a comprehensive collection of significant researches that are addressed in a variety of chapters. It will surely enhance the knowledge of the field among readers across the globe.

It gives us an immense pleasure to thank our researchers and authors for their efforts to submit their piece of writing before the deadlines. Finally in the end, I would like to thank my family and colleagues who have been a great source of inspiration and support.

Editor

1

Sonic Hedgehog Signaling and Development of the Dentition

Maisa Seppala [1,2], Gareth J. Fraser [3], Anahid A. Birjandi [1], Guilherme M. Xavier [1,2] and Martyn T. Cobourne [1,2,*]

1 Centre for Craniofacial and Regenerative Biology, King's College London Dental Institute, Floor 27, Guy's Hospital, London SE1 9RT, UK; maisa.seppala@kcl.ac.uk (M.S.); anahid.ahmadi_birjandi@kcl.ac.uk (A.A.B.); guivier@hotmail.com (G.M.X.)

2 Department of Orthodontics, King's College London Dental Institute, Floor 22, Guy's and St Thomas' NHS Foundation Trust, London SE1 9RT, UK

3 Department of Animal and Plant Sciences, Alfred Denny Building, University of Sheffield, Sheffield S10 2TN, UK; g.fraser@sheffield.ac.uk

* Correspondence: martyn.cobourne@kcl.ac.uk

Academic Editors: Henk Roelink and Simon J. Conway

Abstract: Sonic hedgehog (Shh) is an essential signaling peptide required for normal embryonic development. It represents a highly-conserved marker of odontogenesis amongst the toothed vertebrates. Signal transduction is involved in early specification of the tooth-forming epithelium in the oral cavity, and, ultimately, in defining tooth number within the established dentition. Shh also promotes the morphogenetic movement of epithelial cells in the early tooth bud, and influences cell cycle regulation, morphogenesis, and differentiation in the tooth germ. More recently, Shh has been identified as a stem cell regulator in the continuously erupting incisors of mice. Here, we review contemporary data relating to the role of Shh in odontogenesis, focusing on tooth development in mammals and cartilaginous fishes. We also describe the multiple actions of this signaling protein at the cellular level.

Keywords: odontogenesis; initiation; patterning; dental mesenchymal stem cells; tooth number

---

1. Introduction

Sonic hedgehog (Shh) signaling plays a key role in multiple developmental processes in vertebrates [1–3]. Shh is a highly-conserved marker for tooth development amongst all the toothed vertebrates [4–9]. The mammalian tooth is a site of dynamic Shh signaling activity, which begins in the dental lamina during initiation, and continues in well-defined regions of the tooth germ epithelium during subsequent coronal and root morphogenesis [8–13]. Several early investigations highlighted the importance of Shh signal transduction during mammalian tooth development, which includes the specification of boundaries between odontogenic and non-odontogenic epithelium during initiation, as well as cell cycle regulation, differentiation, and morphogenesis in the established tooth germ [9,14–17]. More recently, studies have identified a role for Shh in promoting morphogenetic movement in the early tooth bud through cell polarization, influencing both cuspal morphogenesis and root development during the later stages of development, and playing a key role in defining tooth number within the mature dentition [18–20]. Here, we review contemporary data relating to the role of Shh during odontogenesis in jawed vertebrates, describe the multiple actions of this signaling protein at the cellular level, and highlight the more recent identification of Shh as a potential primitive stem cell regulator in the murine incisor [21].

2. Shh Pathway Components and Signaling Activity

Shh transcription is consistently restricted to the epithelial component of the developing tooth germ, from initiation through to root development (Figure 1). However, the secreted signaling protein has been shown to diffuse within the surrounding epithelium and mesenchyme of the tooth, influencing reciprocal epithelial–mesenchymal signaling between these two embryonic tissue components in both mice and humans [9,13,16]. Shh signals through a complex pathway, with the availability and activity of secreted ligand modified at multiple levels in both producing and receiving cells in a tissue-dependent manner. The post-translational modification of Shh involves the addition of cholesterol [22,23] and palmitate [24] groups to the newly synthesized pre-protein. Cholesterol limits the movement of the processed ligand, and creates a higher concentration gradient close to the signal source [25]. Palmitate is required for the formation of a multimeric complex that facilitates Shh diffusion over multiple cell layers [26]. In addition, Shh creates a self-regulatory feedback loop by inducing the transcription of signaling components, such as the receptor *Patched1* (*Ptch1*), Glioma-associated oncogene family members *Gli1-2*, and the negative regulator Hedgehog-interacting protein (*Hhip1*). The target genes can be found in both Shh-expressing cells and surrounding tissues, demarcating regions where autocrine and paracrine signaling is active [9,27,28]. This multifaceted regulation of Shh is an essential part of defining the correct tissue responses, as it acts as a morphogen whose concentration gradient elicits different tissue responses and cell fates based on their distance from the signal source [28–30].

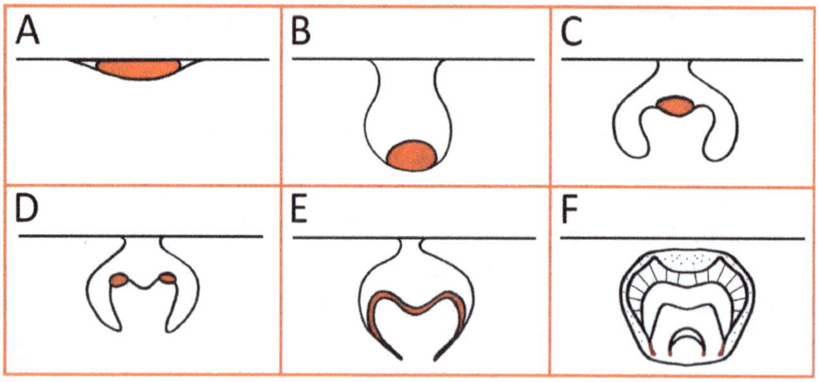

Figure 1. Schematic representation of *Shh* transcription in the developing tooth (red). (**A**) Initiation; (**B**) Early bud stage (at the tip of the bud in enamel knot precursor cells); (**C**) Early cap stage (in the primary enamel knot); (**D**) Late cap stage (in the secondary enamel knots); (**E**) Bell stage (in the internal enamel epithelium, pre-ameloblasts and stratum intermedium); (**F**) During root development (in Hertwig's epithelial root sheath).

In recent years, the primary cilium has been identified as an important region of the cell that coordinates a number of molecular signaling pathways, including Hedgehog and Wingless-type MMTV integration site family (Wnt) [31]. During signaling, pathway activation occurs following the binding of Shh to the primary receptor Ptch1 [32–34] (Figure 2). In the resting state, Ptch1 accumulates at the cilium, and inhibits transduction through the indirect suppression of Smoothened (Smo), a G-protein coupled receptor-like protein essential for transduction [35,36]. Activated Smo then accumulates at the ciliary membrane [37], and effects downstream intracellular signaling through the processing of Gli1-3 transcription factors [33,38,39]. Following pathway activation, Ptch1 rapidly accumulates on the cell surface to buffer signal activity in a ligand-dependent manner [40,41]. The relative availability of bound and unbound receptor is further regulated through the Shh-mediated transcriptional repression of its own co-receptors. These include the GPI-linked membrane glycoprotein Growth arrest-specific 1 (Gas1) [42], as well as the Ig/fibronectin single-pass membrane-spanning cell adhesion proteins Cdon (cell adhesion associated, oncogene regulated) and Boc (cell adhesion associated, oncogene regulated) [43,44]. Although a vast amount of previous and ongoing research has identified multiple intriguing features of this pathway, a complete understanding of the biochemistry remains elusive.

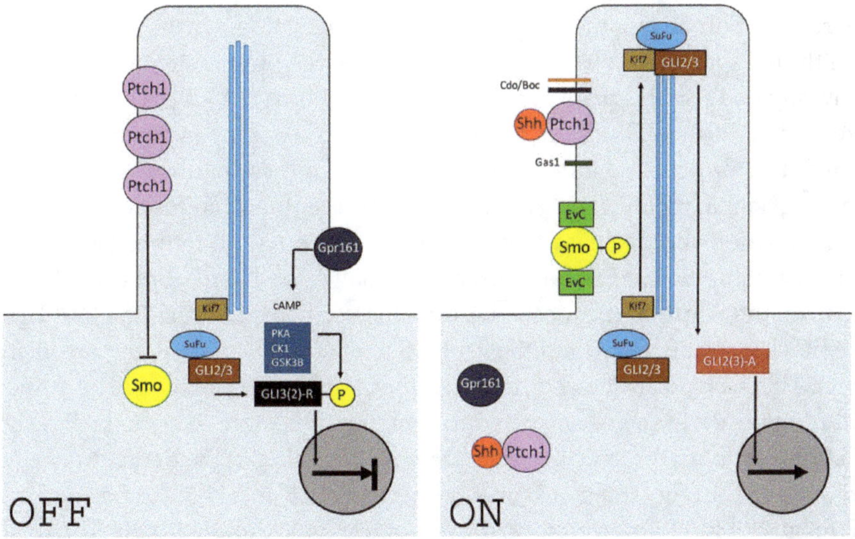

Figure 2. A simple diagrammatic representation of the (known) key events in Shh signaling at the primary cilium. In the absence of signal (OFF), Ptch1 accumulates at the ciliary membrane and represses Smo. GLI proteins are sequestered by Kif7 and Suppressor-of-fused (SuFu), and phosphorylated through the activity of Gpr161, PKA, CK1 and GSK3B. The phosphorylated GLI proteins are proteolytically processed into truncated repressor forms, which act to inhibit transcriptional activation of the pathway. The signaling pathway is activated (ON) through the binding of Shh to Ptch1, which is facilitated by Cdo, Boc and Gas1 co-receptors. The receptor complex is internalized and degraded, which leads to the accumulation of Smo in the ciliary membrane, where it is phosphorylated and interacts with Ellis-van Creveld (EvC) proteins in the basal component of the cilium. The Kif7, SuFu, and GLI complex moves to the tip of the cilium where GLI is concentrated and then transported back to the cytosol where it enters the nucleus as a full length activator of transcriptional targets.

3. Shh Induces Invagination of the Early Dental Lamina

During mouse embryonic day (E)11 *Shh* is expressed in the early dental placodes [8] and has an inverse relationship with *Wnt7b*, which is expressed in the non-odontogenic epithelium and inhibits *Shh*. These expression domains define the sites where teeth will and will-not form [17]. Subsequently, the *Shh*-expressing dental placode undergoes strong vertically-orientated cell division and becomes histologically visible. The placode further stratifies into basal (deeper) and multiple suprabasal (superficial) layers to produce an epithelial thickening, which invaginates into the underlying mesenchyme at around E11.5 and progresses to an early bud by E12.5 [45,46]. After a short period of quiescence, *Shh* expression becomes localized at the tip of the established tooth bud [9,15].

The repeated expression of *Shh* in epithelium of the developing tooth prompted early studies investigating the role of this signal pathway during tooth development [9,14–16,47]. One of the first investigations demonstrated that the application of ectopic Shh was capable of stimulating crude invagination of both odontogenic and non-odontogenic oral epithelium [9]. These data proved consistent with findings from mandibular explant culture experiments, where the opposite effect was achieved following the inhibition of Shh using the 5E1 blocking antibody [14]. Specifically, a loss of Shh function from E10.5 resulted in the early arrest of tooth development at the epithelial thickening stage, associated with the down-regulation of *Ptch1* and a lack of proliferation in deeper regions of the epithelial thickening. *Shh* mutant mice display severe defects associated with the first branchial arch [48,49], which makes them redundant as a model to study tooth development. To overcome this, mice have been generated using Cre-LoxP-mediated recombination to remove Shh function in the early oral epithelium from around E11.75 using a Keratin-14 (K14) promoter. Interestingly, these mice did form epithelial buds at E12.5, but they were associated with wider and shallower invaginations when compared to control teeth [15]. Collectively, these studies provided strong evidence of an inductive and mitogenic role for Shh during the early stages of tooth development.

More recently, the specific requirements for Shh during the progression from placode to bud have been further investigated using explant culture and confocal imaging [20]. A temporal 24-h inhibition and over-expression of Shh and Fibroblast growth factor (Fgf) signaling in mandibular cultures at E10.5 and 11.5 have revealed distinct and independent functions for these pathways in the early tooth germ. The inhibition of Fgf signaling at E11.5 using the pan-Fgf receptor inhibitor SU5402 results in a failure of epithelial stratification and decreased levels of proliferation, with reduced depth in the placodal region. Similarly, Fgf inhibition at E12.5 also produces smaller tooth buds, but even though there is less downward growth, the width of the tooth bud remains unaffected, indicating that Fgf signals drive proliferation, epithelial stratification, and the generation of suprabasal cells, but are likely dispensable for invagination. This study also showed that even though Shh has previously been shown to induce proliferation during the early stages of tooth development, it is more likely to be Fgf signaling that is required and sufficient to drive vertically-orientated cell division at the dental thickening stage. The inhibition of Shh using the Smo-antagonist cyclopamine at E10.5 and 11.5 had no effect on stratification or proliferation, but the tooth buds were shallower and wider. Cellular analysis revealed that these tooth germs lacked columnar-shaped cells and elongated nuclei in basal and suprabasal regions, with the difference most significant in shoulder regions of the invaginating dental lamina. When Shh activity was amplified using the Shh-agonist purmorphamine, the results were narrowed tooth bud necks that invaginated to a much deeper level in comparison to controls. Changes in cellular and nuclear shape are indicative of motile cell behavior, and Shh appears to coordinate the morphogenetic movement of cells required for normal invagination of the tooth germ. Significantly, this is contradictory to previous studies—as no changes in proliferation levels were noted—and it suggests an important role for Shh specifically influencing the cellular rearrangements required for driving the invagination of the early tooth germ [34].

4. Shh Governs Early Molar Morphogenesis

After the initial thickening stage, *Shh* is upregulated at E13.5 in the epithelial bud tip. The intensity of *Shh* expression subsequently increases a day later at the cap stage, when the primary enamel knot appears [9,12]. At the same time, *Ptc1*, *Gli1-2*, and *Smo* are all expressed in partially overlapping domains within both the dental epithelium and the surrounding mesenchyme, but their expression is excluded from the enamel knot [9].

The primary and secondary enamel knots are transient signaling centers [12] that express a number of signaling molecules, including *Fgf3*, *4*, and *9*; Bone morphogenetic protein (*Bmp*) *2*, *4*, and *7*; *Wnt10a* and *10b*; and *Shh* [12,50–53] and act to define tooth shape [12,54–57]. At E14.5, K14-Cre; *Shh* conditional knock-out molars show continued dental development, but morphogenesis is significantly disrupted. At this stage, the molars are around 25 per cent smaller than wild type and are dysmorphic, as they fail to obtain a normal cap stage form. Interestingly, these molars also show signs of disrupted bucco–lingual polarisation, as only the remnants of a lingual protrusion are observed. However, *Fgf4*, *Bmp2* and *Wnt10b* are all expressed in the enamel knot of these teeth, whilst ectopic *Wnt10b* and *Lef1* expression extends lingually. This suggests that the loss of Shh results in a partial change of cell fate, from proliferative to a less proliferative enamel knot-like population on the lingual side. In addition, E14.5 tooth germs are fused with the oral epithelium and lack any clear dental cord connecting the oral epithelium with the invaginating tooth germ. Thereafter, Shh appears to drive polarisation of the tooth in a bucco–lingual axis and induce proliferation in a differential manner, taking part in determining tooth shape [15]. This is supported by the stronger expression of *Ptch1* on the lingual side of the cap stage outer enamel epithelium (OEE) [9,15]. Furthermore, a molecular investigation of conditional knockouts shows the intact expression of *Bmp4*, *Msx1*, and *Pax9*, providing evidence that during the start of morphogenesis, Shh is dispensable for their induction. In contrast, *Pitx1* is downregulated and appears to be under the regulation of Shh [15].

The loss of *Ellis-van Creveld (EvC)* function also results in fusion of the first and second molars, reduced size of the first molar in comparison to the second and single root formations. In addition, *EvC* appears to be involved in a symmetrical bucco–lingual response to Shh during molar development. Intriguingly, *EvC* mutant mice show more severe defects on the buccal side rather than the lingual, distinguishing them from K14-Cre; *Shh* mice. This may be due to the different origins of the buccal and lingual tissues from the ectoderm and endoderm, respectively; as patients with EvC syndrome do not present with any known anomalies associated with endoderm-derived tissues [58,59]. At E14.5, the asymmetric molar phenotype is accompanied by a buccal shift in expression of *Shh, Wnt10b*, and *Lef1*-expressing enamel knot cells and a lingual shift of *Fgf4*. Although *Shh* expression remains robust in the epithelium, *Ptch1* and *Gli1* are absent from the surrounding mesenchyme, underlining that Shh signaling is disrupted in *EvC* mutants [58]. When EvC forms a complex with EvC2 it localises to the cilium, where it interacts with Smo and controls dissociation of the negative Hedgehog regulator Suppressor-of-fused (SuFu) from Gli3, and thus Gli3 movement into the cilia (see Figure 2) [60].

5. A Lack of Shh Results in a Failure of Ameloblast Polarization

The process of cytodifferentiation begins in the tooth germ at the bell stage, around E18 in a mouse. At this stage, the future tooth consists of an epithelial enamel organ (EEO) and the dental mesenchyme. The EEO is a complex structure consisting of the inner enamel epithelium (IEE), secondary enamel knots, and the OEE, which is separated from the IEE by an internally-located stratum intermedium (SI) and stellate reticulum (SR). Polarized pre-odontoblasts situated adjacent to the IEE within the mesenchymal compartment differentiate and secrete predentin, which induces the IEE to differentiate into pre-ameloblasts [61,62]. *Shh* is strongly expressed in the IEE with *Ptch1* in the underlying mesenchyme just prior to the differentiation of the pre-odontoblasts [16,63]. The inductive signal for odontoblast differentiation is known to originate from the IEE, and thereafter the spatio-temporal expression pattern of *Shh* and *Ptch1* implicates Shh as having a role in transmitting the required signal between these two tissue components, as well as being a marker for pre-ameloblasts.

Regardless of the strong expression of *Shh* in the IEE, K14-Cre; *Shh* mice do have differentiated pre-odontoblasts and pre-ameloblasts in the developing tooth at E19.5, with these cells expressing known markers that demonstrate functionality. Furthermore, a secreted enamel and dentin matrix is evident between the two cell populations. However, the cellular architecture is disrupted, as the odontoblasts fail to form a continuous monolayer, and only a proportion of these cells polarise and elongate. Similarly, the differentiating ameloblasts remain small and cuboidal, in contrast to the elongated and polarised cell shapes seen in control tooth germs [15]. The conditional removal of Shh responsiveness exclusively from the oral epithelium by inactivation of *Smo* under the K14 promoter (K14-Cre; *Smo*) results in the normal polarisation of the mesenchymally-derived pre-odontoblasts, but disrupted polarisation of the pre-ameloblasts, and hypocellularity of the SR and SI. An abnormal cytoskeletal organisation of the pre-ameloblasts is seen, associated with a lack of *Ptch2* and *Gli1* accumulation in the intracellular space on the basal side. This implicates a possible role for the stratum intermedium in providing a guiding signal for the subcellular trafficking of proteins that takes place during polarisation. In addition, K14-Cre; *Smo* mice show an absence of β-tubulin accumulation, and a polarized cytoplasmic localisation of E-Cadherin and Zonula Occludens-1 (ZO-1). This provides evidence of a lack of specialised junctional complexes that, together with microtubules, are important for establishing the cell polarity and the apical and basolateral domains. Further analysis has demonstrated reduced proliferation as well as early loss of the phosphorylated histone H3 (PH-H3) protein (a known marker of mitosis) and the Shh-induced *cyclinD1*. This indicates that decreased proliferation levels could be a consequence of premature exit from the cell cycle, and cytodifferentiation of ameloblast progenitors before predentin secretion. This suggests that Shh regulates the cell cycle and exerts its mitogenic role by promoting *cyclin D1* transcription to control the G1/S transition [16]. These studies have highlighted how Shh guides cytoskeletal organisation and promotes the arrangement of cell–cell junctional complexes, not only at sites of mesenchymal–epithelial interaction, but also within epithelial cell populations.

6. Disrupted Shh Signaling Causes Molar Fusion and Altered Cusp Morphology

The signaling molecules expressed in the enamel knot regulate morphogenesis by diffusing through the surrounding epithelial and mesenchymal tissues, where they are responsible for defining appropriate molecular and cellular responses. These molecular signals can positively and negatively interact at multiple levels to polarise and pattern tissues.

An elaborate model, based on Alan Turing's reaction–diffusion principles guiding pattern formation, has been redrawn to potentially explain cuspal patterning by Shh and Wnt signaling [18]. At E14.5, *Shh, Wnt10a,* and *Wnt10b* are similarly expressed in the enamel knot and secreted laterally to the surrounding tissues [50]. Whilst Wnt induces its own transcription factor *Lef1* in immediate proximity to its source and positively regulates *Shh* via *Fgf3 and Fgf4,* Shh ligand has the ability to diffuse faster than Wnt, and travels outside of the Wnt-created *Lef1* zone, where it has an opposing effect. Shh induces the expression of the Wnt inhibitor *Wise,* indirectly limiting the activity of Wnt and simultaneously creating the odontogenic *Lef1* and non-odontogenic *Wise* zones [16,18,64–66]. In this Shh–Wise–Wnt feedback model, Wnt acts as an activator; Shh a mediator; and Wise an inhibitor [18]. Consistently, a loss of Wise function [54] and multiple mouse models with reduced Shh signaling activity [15,16,18] have severely disrupted cuspal patterning in the molar dentition, manifesting as enlarged and fused molar teeth with multiple smaller cusps.

The importance of temporal Shh signaling in molar development has more recently been investigated by blocking Shh activity in utero at different time-points between E10–18 with 5E1 intra-peritoneal injection. When injections are carried out at E10 or 12, the non-odontogenic field is absent only between first and second molars, which are fused together. If injections are carried out at E16 the same effect moves further distally in the dental arch, with the second and third molars fused. This correlates with the sequential development of these teeth, progressing along the mesio–distal axis. This phenotype is associated with altered Wnt signaling domains between the fusing molars, as demonstrated by an expanded *Sp5* expression across the first and second molars, and in contrast to controls that demonstrated the separate molar domains of *Sp5.* Interestingly, reduced levels of Shh and increased levels of Wnt were associated with the accelerated development of the second molars in comparison to controls [18].

7. Shh Influences Tooth Root Development

Shh and constituent members of the signalling pathway are also expressed during early development of the molar tooth root in the mouse. Specifically, *Shh* transcripts localise to the epithelial cells of Hertwig's epithelial root sheath, whilst *Smo, Ptch1,* and *Gli* are expressed in the apical mesenchyme of the associated dental papilla and follicle [11]. The analysis of mesenchymal dysplasia (*mes*) mice, which harbour an abnormal C-terminus associated with Ptch1, has shown repressed proliferation, shorter roots and a disturbed molar eruption [67]. More recently, it has been shown that regulating appropriate levels of Shh signaling is crucial for the normal growth and development of the root, with both reduced and increased signal levels producing reduced levels of proliferation, and ultimately, root length. Moreover, levels of Hedgehog activity are regulated by nuclear factor I/C (*Nfic*), which binds Hhip1 in the apical papilla [68]. Signifcantly, *Nfic* mutant mice also develop short molar roots [69].

8. Shh and Its Role in Defining the Tooth Formula

In contrast to humans, mice have only a single incisor and three molars in each dental quadrant, separated by a non-odontogenic diastema. However, at the earliest stages of odontogenesis, multiple epithelial thickenings accompanied by the expression of *Shh* are identifiable in the diastema region.

These vestigial swellings only progress to a thickening or rudimentary bud stage, and thereafter regress via apoptosis with the simultaneous disappearance of *Shh* expression [70–74]. Interestingly, several mouse models with disrupted Shh, Wnt, Fgf, and Bmp signaling levels develop supernumerary teeth. These teeth typically arise from the most posterior rudimentary swelling, referred to as R2, and which during normal mouse development fuse with the main cap stage's first molar [71,73]. Although these premolar-like supernumeraries have been described as evolutionary remnants (the non-odontogenic diastema has been a prevailing feature of mice for over 50 million years) [19,55,73], they provide an excellent model to study how different signaling cascades interact to control tooth formula, by either inducing or suppressing the progression of tooth development, and defining the boundaries between odontogenic and non-odontogenic fields.

Whereas a reduction or loss of Shh signaling activity has been associated with molar fusion [15,16,18], increased activity in the diastema can result in altered molar field patterning and supernumerary tooth formation (Figure 3) [19,55]. Indeed, multiple knockout mouse models have identified the upregulation of *Shh* in the diastema, which as a primary or secondary response results in supernumerary tooth formation [38,47,54], and places *Shh* as a downstream target [54,55]. These studies demonstrate how a complex regulatory network—consisting of its own as well as other molecular signaling cascade members—is responsible for defining the pivotal threshold levels of Shh for the normal progression of tooth development.

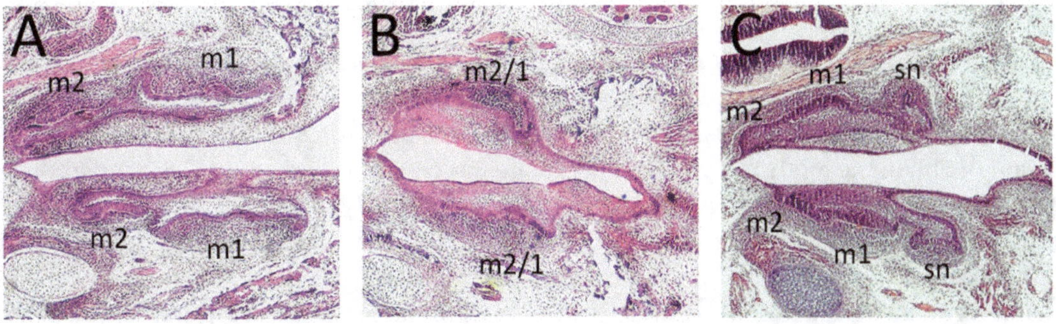

Figure 3. Disrupted Shh signaling and early tooth development. (**A**–**C**) Para-sagittal sections of the developing maxillary and mandibular molar dentition in (**A**) E16.5 Wild type; (**B**) E16.5 CreER™; *Shh* mutant; (**C**) E15.5 *Gas1* mutant embryos. In the wild type embryo, the first molars (m1) are at the late cap/early bell stage of development and the second molars (m2) are at the early cap stage. In CreER™; *Shh* mutant embryos, Shh signaling has been abrogated from E10.5 following the administration of Tamoxifen, with fusion between the first and second mandibular molars (m1/m2). In the *Gas 1* mutant, m1 are at the late cap stage and the m2 tooth germs are just beginning to form an early cap. However, a supernumerary premolar tooth (sn) is present in mesial to m1 in both quadrants.

Shh also negatively regulates transcription of the co-receptors *Gas1*, *Cdon*, and *Boc*, which are typically expressed at some distance from the signal source, and act in a reciprocal manner to *Ptch1* [42,75]. In peripheral regions of the Shh signaling domain where protein levels are low, the increased transcription of these co-receptors is thought to promote signaling in receiving cells, and help achieve appropriate threshold levels for signal transduction in the peripheral regions of the signaling domain [42,76–79]. Interestingly, Gas1 seems to influence Shh in a context-dependent manner: it positively regulates signaling in the facial midline and neural tube [42,76,78], whilst seemingly having an opposite inhibitory role in the early jaw [19,47]. During early tooth development, *Gas1* is abundant in the surrounding mesenchyme of early tooth germs and in the non-odontogenic diastema, expressed mainly in a reciprocal manner to *Ptch1*, and sharing only small areas of overlapping expression in the peripheral regions of their respective domains. When *Gas1* is over-expressed in the non-odontogenic diastema it downregulates *Ptch1*, which suggests an important early regulatory mechanism that restricts Shh activity to the tooth-forming regions of the mandible [47]. This is consistent with the molar phenotype of *Gas1* mutant mice, which demonstrates how an absence of Gas1 allows the diffusion of Shh protein further away from its source. This results at E12.5 in the presence of ectopic protein in the diastema, and an accompanied increased expression of *Ptch1*. Significantly, *Gas1* mutant mice present with single premolar-like supernumeraries located mesial to the first molars, which further supports the concept of Gas1 acting as an additional inhibitor required for limiting Shh activity to the tooth-forming regions [19]. The high-level regulation of Shh signaling in the tooth is further demonstrated by the strong expression of *Hhip1* in the peripheral odontogenic mesenchyme [27]. *Hhip1* encodes a membrane glycoprotein that can also bind and sequester Shh [80], although to date it only seems to retain an essential function in root development [68]. Primary cilia protein components also generally exert a negative regulatory effect on Shh activity, and seem to play a role in repressing tooth formation. Mice mutants for the cilia intraflagellar transport (IFT) protein (*IFT88* or *Polaris*) show an upregulation of Shh signaling activity, and form ectopic premolar-like teeth in the diastema [19]. However, the phenotypic consequences of a disrupted ciliary function have to be interpreted within the wider context of their potential effect on other signal pathways, including Wnt transduction.

The Sprouty (Spr) gene family encode universal receptor tyrosine kinase signaling (RTK) inhibitors that are induced by Fgfs, but which inhibit Fgf signaling to provide an important negative feedback regulatory mechanism [81]. *Fgf*s are found at multiple sites in the developing tooth and at multiple stages, with expression domains extending into the dental epithelium. There, they become localised to the enamel knot (*Fgf4* and *Fgf9*) and surrounding mesenchyme (*Fgf3* and *Fgf10*), including the diastema (*Fgf10*). Interestingly, *Spr2* and *Spr4* mutant mice form premolar-like diastema supernumeraries similar to those seen in *Gas1* and *Polaris* mutants [55]. Predictably, *Spr2* and *Spr4* mutants demonstrate ectopic Fgf activity, which spatio-temporally correlates with the formation of these supernumeraries. There is also the continued expression of *Shh* in R2 at E14.5, providing further evidence that Fgfs act upstream of Shh in the tooth and positively regulate its activity. Whereas *Spr4* is found exclusively in the surrounding mesenchyme, *Spr2* localizes more prominently to the bud epithelium. However, some activity is also present in the mesenchyme surrounding the developing first molar. Thereafter, Spr2 is considered to be required for the indirect inhibition of Fgf-induced *Shh* expression, and in normal development acts to suppress the development of the first molar [55]. In addition, Fgf and Wnt pathways form an important reciprocal signaling network between the epithelium and the mesenchyme at this stage, because *Fgf4* transcription relies on Wnt-induced *Lef1* [65].

Interestingly, even though diastema teeth develop in *Spr2* and *Spr4* mutants, they do not significantly affect first molar cusp morphology, indicating that Sprouty gene activity is more notably involved in defining tooth number rather than cusp patterning [55]. This is distinct from Wise-mediated loss of Wnt inhibition, which results in both the development of diastema supernumerary teeth with variable penetrance and altered cusp morphology [54]. The relationship between Shh and Wnt signaling during supernumerary tooth formation has been further investigated by generating *Wise+/−; Shh/GFPCre/+* compound mice, which exhibit reduced Shh dosage. These mice also display a supernumerary tooth phenotype that can be rescued by increasing Shh levels through a partial loss of *Ptch1* function, or reducing Wnt signaling levels by reducing *Lrp6*. In turn, when Shh is downregulated, Wnt signaling is upregulated, indicative of a Shh negative–feedback loop required for controlling the levels of Wnt signaling. Altogether, there is strong evidence available to suggest that the relative expression levels of Shh and Wnt play a major role in coordinating multiple patterning processes, and that an appropriate balance between these pathways is crucial for the normal progression of tooth development.

9. Shh as a Regulator of Dental Stem Cells

Mouse incisors grow continuously throughout life, and have provided a unique model to study the potential surface markers, anatomical location and role of stem cells in tooth development. The cervical loop is a component of the enamel organ that disappears in molars, but prevails in incisors and enables enamel formation to continue in adult mice. Actively proliferating transit amplifying cells

(TACs) are located close to the cervical loop, and are thought to represent the location of an epithelial stem cell reservoir [82,83]. Murine incisor epithelial stem cells can be identified in the incisor based on the expression of specific surface markers, which include *Gli1*, *Bmi1* and *Sox2* [83,84]. Investigations in the murine incisor using *Gli1*-LacZ reporter mice have found such expression (in addition to dental epithelial cells and labial post-mitotic odontoblasts) specifically in the mesenchyme surrounding the neurovascular bundle (NVB) in both arterioles and their surrounding nerves [84]. Interestingly, this expression was not induced by Shh originating from the dental epithelium, but Shh secreted by sensory neurons of the trigeminal ganglion, whose axons also innervate the inferior alveolar nerve and adult lower incisors. In the incisor mesenchyme, *Gli1*+ cells strongly overlap with H2BGFP-labelled quiescent stem cells, supporting *Gli1*+ as being a stem cell marker. The *Gli1*+ and H2BGFP double-labelled cells do not express many of the common stem cell-related surface markers such as *CD105, CD73, NG2, CD146, CD44*, or *Sca1*. However, culture-based studies using *Gli1-CE; Tdtomato* mice have shown that these cells rapidly lose *Gli1* expression when communication with the NVB is lost, and can subsequently differentiate into different MSC subpopulations that express *CD146, CD105, Sca1*, and *CD73*. In addition, upon hard tissue injury, *Gli1*+ cells can give rise to odontoblasts [21]. Similarly to Shh being critical for an osteoblast lineage commitment [85], it may also be essential for an odontogenic commitment of incisor stem cells through primitive *Gli1*+ cells giving rise to pericytes, which consecutively upon injury are stimulated to differentiate into odontoblasts. Although the inhibition of Shh signaling in mice fed for one month with a Hedgehog-antagonist did reduce dentin formation, it did not have any significant impact on mesenchymal cell proliferation, apoptosis, or the number of label-retaining slow cycling stem cells [84]. Therefore, Shh derived from sensory neurons has a key role in both murine adult incisor homeostasis and hard tissue tissue repair, but is not required for stem cell proliferation and maintenance. Significantly, *Gli1*+ cells are also found elsewhere in the body surrounding the arteries, and raise the possibility that they could also mark stem cells in other organs [86].

10. Shh as a Universal Marker of Vertebrate Tooth Development

Although the majority of tooth development research has focused on the murine dentition, it is clear that Shh is also a conserved marker of tooth development, from initiation through to morphogenesis, amongst all toothed vertebrates including fish, reptiles, and humans [4–9]. Even in the phylogenetically basal cartilaginous fishes, the Chondrichthyes (which include sharks, rays and holocephalans), shh is a key epithelial marker of tooth development and morphogenesis [6,7]. In the embryonic shark (*Scyliorhinus canicula*), *shh* is expressed in a restricted band along the oral jaw prior to tooth initiation [6] (Figure 4). This expression is an indicator of emerging dental competence in the pre-lamina epithelium or odontogenic band [6,87]. This early *shh* expression in the odontogenic band is obvious across vertebrate taxa, and demarcates the spatial restriction of tooth territories. The comparative appearance of *shh* expression in this region also confirms a variability that highlights the early diversity of tooth patterning among vertebrate clades [88].

Following initial establishment of the odontogenic band and the first appearance of *shh* expression, the first teeth in the shark jaw manifest as superficial thickened placodes in a similar pattern of emergence to the teeth of other vertebrates, such as reptiles and mammals. During this process of shark tooth placode maturation, *shh* expression persists within the epithelial placode (the early tooth bud) as the underlying mesenchyme begins to condense and trigger the cap stage of tooth morphogenesis (Figure 4). Here, the epithelial component surrounds the mesenchymal core or dental papilla in a characteristic form that appears to be highly conserved among all vertebrates. At this stage, *shh* expression becomes clearly localized to the apical cells of the tooth bud (Stage 32; Figure 4), which is indicative of the regions attributed to the enamel knot signaling centres that govern tooth shape in mammals [6]. In the shark, defined regions of *shh* expression (and the expression of other known enamel knot markers) in the presumptive primary enamel knot cells show a uniformity, and therefore a conservation of primary enamel knot signaling among vertebrates. This indicates that not only are the early phases of vertebrate tooth initiation conserved, but also that the later stages of tooth morphogenesis, and therefore the genetic programme (including Shh signaling) for general vertebrate tooth development, is an evolutionarily stable process.

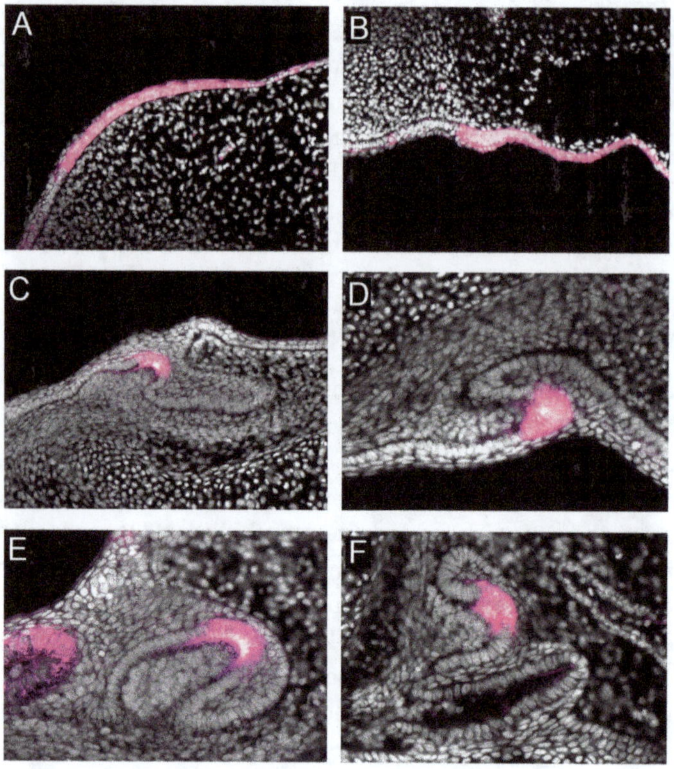

Figure 4. Conserved *shh* expression in shark tooth initiation and successional morphogenesis. Expression of *shh* in the odontogenic band marks the onset of dental competence in the shark (*Scyliorhinus stellaris*) lower jaw (**A**) and upper jaw (**B**). Development of the first tooth and subsequent morphogenesis in *Scyliorhinus canicula* is superficial with *shh* expression restricted to the dental epithelium at the apical cusp, lower jaw (**C**) and upper jaw (**D**). Dental lamina epithelium continues to invaginate lingual to the first tooth, from which new successional teeth will develop in the lower jaw (**E**) and upper jaw (**F**). New successional teeth in the shark continue to express *shh* during morphogenesis (**E**,**F**). Note *shh* is not expressed in the epithelial dental lamina away from developing teeth, and its expression is restricted to the enamel knot-like cells at the apex of each new tooth throughout morphogenesis. False colour, (magenta) gene expression; nuclear counterstain DAPI (white).

Following the establishment of the first tooth and the subsequent territories of the elasmobranch first generation dentition, the dental epithelial cells lingual to the first tooth continue to proliferate and invaginate lingually into the mesenchyme. This invagination oral to the jaw cartilages produces the early dental lamina, which is the continuous jaw-wide epithelial unit that plays a crucial role in the production of successional teeth, and houses progenitor cells for successional tooth development [89]. Sharks, and their cartilaginous relatives (the elasmobranchs), offer a novel tooth and jaw complex to study the locations where teeth develop and attach superficially to jaw cartilages via connective tissues rather than being anchored in a bony jaw as seen in bony vertebrates (Osteichthyans). This condition in elasmobranchs permits a unique flexibility during tooth development, and importantly during the production of multiple tooth generations, where each tooth position along the jaw has a related lingual succession of teeth (family) at various stages of development housed within a continuous dental lamina. Sharks are polyphyodont vertebrates with continuous tooth regeneration occurring throughout life [89]; therefore, several generations of teeth are present ahead of functions within the epithelial dental lamina (Figure 4). As each new tooth generation is initiated within the deeply invaginated distal tip of the dental lamina (successional lamina), *shh* is expressed within the dental epithelium, where it becomes more restricted to the cells of the primary enamel-like cell clusters as the tooth matures (Figure 4). Intriguingly, although *shh* is expressed in the odontogenic band, and therefore the early dental epithelium essential for first generation tooth formation, in non-mammalian polyphyodont species it is not expressed within the dental lamina (aside from the cells associated with the cusps of developing teeth). Importantly, it is absent from regions of the dental lamina that are actively initiating new tooth generation, the successional lamina. This characteristic of polyphyodonty is observed in teleost fishes [90], chondrichthyan fishes [6], and in reptilian models [5]. This suggests that although shh—and more generally, Hedgehog signaling—is a requirement for the normal development of first generation teeth, it is not essential to the subsequent development of successional tooth families in non-mammalian polyphyodont animals. This indicates a crucial role for Shh signaling during first tooth initiation. However, after the dental epithelium becomes competent for tooth formation, the wider dental epithelium (dental lamina, and successional lamina) is no longer dependent on shh for tooth production.

11. Conclusions

The current data suggests a key role for Shh signaling in tooth development that is complex, acts at multiple stages, coordinates various cellular processes and is relevant for multiple species. These processes are independently regulated and are involved in inducing Shh-mediated proliferation, cell polarisation, differentiation, morphogenesis and patterning. Shh interacts with several other signaling pathways at the molecular level, particularly Fgf and Wnt, to ensure the normal progression of tooth development up to and beyond the bud stage. In addition, signal transduction not only activates downstream target genes, but also inhibits pathway activity. This provides an important regulatory mechanism to ensure that signal levels achieve the appropriate cellular response.

Acknowledgments: M.S. and G.M.X. in the laboratory of M.T.C. are supported by the European Orthodontic Society and Academy of Medical Sciences (M.S.) and Academy of Medical Sciences (G.M.X.). The laboratory of G.J.F. is supported by the Natural Environment Research Council (NERC) Standard Grant NE/K014595/1 and Leverhulme Trust Research Grant RPG-211. We thank Liam Rasch for performing the in situ hybridization shown in Figure 4.

Conflicts of Interest: The authors declare no conflict of interest.

References

1. Briscoe, J.; Therond, P.P. The mechanisms of hedgehog signalling and its roles in development and disease. *Nat. Rev. Mol. Cell Biol.* **2013**, *14*, 418–431. [CrossRef] [PubMed]
2. Ingham, P.W.; McMahon, A.P. Hedgehog signaling in animal development: Paradigms and principles. *Genes Dev.* **2001**, *15*, 3059–3087. [CrossRef] [PubMed]
3. Xavier, G.M.; Seppala, M.; Barrell, W.; Birjandi, A.A.; Geoghegan, F.; Cobourne, M.T. Hedgehog receptor function during craniofacial development. *Dev. Biol.* **2016**, *415*, 198–215. [CrossRef] [PubMed]
4. Buchtova, M.; Handrigan, G.R.; Tucker, A.S.; Lozanoff, S.; Town, L.; Fu, K.; Diewert, V.M.; Wicking, C.; Richman, J.M. Initiation and patterning of the snake dentition are dependent on sonic hedgehog signaling. *Dev. Biol.* **2008**, *319*, 132–145. [CrossRef] [PubMed]
5. Handrigan, G.R.; Richman, J.M. Autocrine and paracrine shh signaling are necessary for tooth morphogenesis, but not tooth replacement in snakes and lizards (squamata). *Dev. Biol.* **2010**, *337*, 171–186. [CrossRef] [PubMed]
6. Rasch, L.J.; Martin, K.J.; Cooper, R.L.; Metscher, B.D.; Underwood, C.J.; Fraser, G.J. An ancient dental gene set governs development and continuous regeneration of teeth in sharks. *Dev. Biol.* **2016**, *415*, 347–370. [CrossRef] [PubMed]
7. Smith, M.M.; Fraser, G.J.; Chaplin, N.; Hobbs, C.; Graham, A. Reiterative pattern of sonic hedgehog expression in the catshark dentition reveals a phylogenetic template for jawed vertebrates. *Proc. Biol. Sci.* **2009**, *276*, 1225–1233. [CrossRef] [PubMed]

8. Bitgood, M.J.; McMahon, A.P. Hedgehog and bmp genes are coexpressed at many diverse sites of cell-cell interaction in the mouse embryo. *Dev. Biol.* **1995**, *172*, 126–138. [CrossRef] [PubMed]
9. Hardcastle, Z.; Mo, R.; Hui, C.C.; Sharpe, P.T. The shh signalling pathway in tooth development: Defects in gli2 and gli3 mutants. *Development* **1998**, *125*, 2803–2811. [PubMed]
10. Iseki, S.; Araga, A.; Ohuchi, H.; Nohno, T.; Yoshioka, H.; Hayashi, F.; Noji, S. Sonic hedgehog is expressed in epithelial cells during development of whisker, hair, and tooth. *Biochem. Biophys. Res. Commun.* **1996**, *218*, 688–693. [CrossRef] [PubMed]
11. Khan, M.; Seppala, M.; Zoupa, M.; Cobourne, M.T. Hedgehog pathway gene expression during early development of the molar tooth root in the mouse. *Gene Expr. Patterns* **2007**, *7*, 239–243. [CrossRef] [PubMed]
12. Vaahtokari, A.; Åberg, T.; Jernvall, J.; Keranen, S.; Thesleff, I. The enamel knot as a signaling center in the developing mouse tooth. *Mech. Dev.* **1996**, *54*, 39–43. [CrossRef]
13. Hu, X.; Zhang, S.; Chen, G.; Lin, C.; Huang, Z.; Chen, Y.; Zhang, Y. Expression of shh signaling molecules in the developing human primary dentition. *BMC Dev. Biol.* **2013**, *13*, 11. [CrossRef] [PubMed]
14. Cobourne, M.T.; Hardcastle, Z.; Sharpe, P.T. Sonic hedgehog regulates epithelial proliferation and cell survival in the developing tooth germ. *J. Dent. Res.* **2001**, *80*, 1974–1979. [CrossRef] [PubMed]
15. Dassule, H.R.; Lewis, P.; Bei, M.; Maas, R.; McMahon, A.P. Sonic hedgehog regulates growth and morphogenesis of the tooth. *Development* **2000**, *127*, 4775–4785. [PubMed]
16. Gritli-Linde, A.; Bei, M.; Maas, R.; Zhang, X.M.; Linde, A.; McMahon, A.P. Shh signaling within the dental epithelium is necessary for cell proliferation, growth and polarization. *Development* **2002**, *129*, 5323–5337. [CrossRef] [PubMed]
17. Sarkar, L.; Cobourne, M.; Naylor, S.; Smalley, M.; Dale, T.; Sharpe, P.T. Wnt/shh interactions regulate ectodermal boundary formation during mammalian tooth development. *Proc. Natl. Acad. Sci. USA* **2000**, *97*, 4520–4524. [CrossRef] [PubMed]
18. Cho, S.W.; Kwak, S.; Woolley, T.E.; Lee, M.J.; Kim, E.J.; Baker, R.E.; Kim, H.J.; Shin, J.S.; Tickle, C.; Maini, P.K.; et al. Interactions between shh, sostdc1 and wnt signaling and a new feedback loop for spatial patterning of the teeth. *Development* **2011**, *138*, 1807–1816. [CrossRef] [PubMed]
19. Ohazama, A.; Haycraft, C.J.; Seppala, M.; Blackburn, J.; Ghafoor, S.; Cobourne, M.; Martinelli, D.C.; Fan, C.M.; Peterkova, R.; Lesot, H.; et al. Primary cilia regulate shh activity in the control of molar tooth number. *Development* **2009**, *136*, 897–903. [CrossRef] [PubMed]
20. Li, J.; Chatzeli, L.; Panousopoulou, E.; Tucker, A.S.; Green, J.B. Epithelial stratification and placode invagination are separable functions in early morphogenesis of the molar tooth. *Development* **2016**, *143*, 670–681. [CrossRef] [PubMed]
21. Zhao, H.; Feng, J.; Seidel, K.; Shi, S.; Klein, O.; Sharpe, P.; Chai, Y. Secretion of shh by a neurovascular bundle niche supports mesenchymal stem cell homeostasis in the adult mouse incisor. *Cell Stem Cell* **2014**, *14*, 160–173. [CrossRef] [PubMed]
22. Porter, J.A.; Ekker, S.C.; Park, W.J.; von Kessler, D.P.; Young, K.E.; Chen, C.H.; Ma, Y.; Woods, A.S.; Cotter, R.J.; Koonin, E.V.; et al. Hedgehog patterning activity: Role of a lipophilic modification mediated by the carboxy-terminal autoprocessing domain. *Cell* **1996**, *86*, 21–34. [CrossRef]
23. Porter, J.A.; von Kessler, D.P.; Ekker, S.C.; Young, K.E.; Lee, J.J.; Moses, K.; Beachy, P.A. The product of hedgehog autoproteolytic cleavage active in local and long-range signalling. *Nature* **1995**, *374*, 363–366. [CrossRef] [PubMed]
24. Pepinsky, R.B.; Zeng, C.; Wen, D.; Rayhorn, P.; Baker, D.P.; Williams, K.P.; Bixler, S.A.; Ambrose, C.M.; Garber, E.A.; Miatkowski, K.; et al. Identification of a palmitic acid-modified form of human sonic hedgehog. *J. Biol. Chem.* **1998**, *273*, 14037–14045. [CrossRef] [PubMed]
25. Saha, K.; Schaffer, D.V. Signal dynamics in sonic hedgehog tissue patterning. *Development* **2006**, *133*, 889–900. [CrossRef] [PubMed]
26. Zeng, X.; Goetz, J.A.; Suber, L.M.; Scott, W.J., Jr.; Schreiner, C.M.; Robbins, D.J. A freely diffusible form of sonic hedgehog mediates long-range signalling. *Nature* **2001**, *411*, 716–720. [CrossRef] [PubMed]
27. Cobourne, M.T.; Sharpe, P.T. Expression and regulation of hedgehog-interacting protein during early tooth development. *Connect. Tissue Res.* **2002**, *43*, 143–147. [CrossRef] [PubMed]
28. Cohen, M.; Kicheva, A.; Ribeiro, A.; Blassberg, R.; Page, K.M.; Barnes, C.P.; Briscoe, J. Ptch1 and gli regulate shh signalling dynamics via multiple mechanisms. *Nat. Commun.* **2015**, *6*, 6709. [CrossRef] [PubMed]

29. Fan, C.M.; Porter, J.A.; Chiang, C.; Chang, D.T.; Beachy, P.A.; Tessier-Lavigne, M. Long-range sclerotome induction by sonic hedgehog: Direct role of the amino-terminal cleavage product and modulation by the cyclic amp signaling pathway. *Cell* **1995**, *81*, 457–465. [CrossRef]
30. Harfe, B.D.; Scherz, P.J.; Nissim, S.; Tian, H.; McMahon, A.P.; Tabin, C.J. Evidence for an expansion-based temporal shh gradient in specifying vertebrate digit identities. *Cell* **2004**, *118*, 517–528. [CrossRef] [PubMed]
31. Bangs, F.; Anderson, K.V. Primary cilia and mammalian hedgehog signaling. *Cold Spring Harb. Perspect. Biol.* **2017**, *9*. [CrossRef] [PubMed]
32. Goodrich, L.V.; Milenković, L.; Higgins, K.M.; Scott, M.P. Altered neural cell fates and medulloblastoma in mouse patched mutants. *Science* **1997**, *277*, 1109–1113. [CrossRef] [PubMed]
33. Haycraft, C.J.; Banizs, B.; Aydin-Son, Y.; Zhang, Q.; Michaud, E.J.; Yoder, B.K. Gli2 and gli3 localize to cilia and require the intraflagellar transport protein polaris for processing and function. *PLoS Genet.* **2005**, *1*, e53. [CrossRef] [PubMed]
34. Tukachinsky, H.; Kuzmickas, R.P.; Jao, C.Y.; Liu, J.; Salic, A. Dispatched and scube mediate the efficient secretion of the cholesterol-modified hedgehog ligand. *Cell Rep.* **2012**, *2*, 308–320. [CrossRef] [PubMed]
35. Taipale, J.; Cooper, M.K.; Maiti, T.; Beachy, P.A. Patched acts catalytically to suppress the activity of smoothened. *Nature* **2002**, *418*, 892–897. [CrossRef] [PubMed]
36. Zhang, X.M.; Ramalho-Santos, M.; McMahon, A.P. Smoothened mutants reveal redundant roles for shh and ihh signaling including regulation of l/r asymmetry by the mouse node. *Cell* **2001**, *105*, 781–792. [CrossRef]
37. Corbit, K.C.; Aanstad, P.; Singla, V.; Norman, A.R.; Stainier, D.Y.; Reiter, J.F. Vertebrate smoothened functions at the primary cilium. *Nature* **2005**, *437*, 1018–1021. [CrossRef] [PubMed]
38. Aberger, F.; Ruiz, I.A.A. Context-dependent signal integration by the gli code: The oncogenic load, pathways, modifiers and implications for cancer therapy. *Semin. Cell Dev. Biol.* **2014**, *33*, 93–104. [CrossRef] [PubMed]
39. Hui, C.C.; Angers, S. Gli proteins in development and disease. *Annu. Rev. Cell. Dev. Biol.* **2011**, *27*, 513–537. [CrossRef] [PubMed]
40. Casali, A.; Struhl, G. Reading the hedgehog morphogen gradient by measuring the ratio of bound to unbound patched protein. *Nature* **2004**, *431*, 76–80. [CrossRef] [PubMed]
41. Chen, Y.; Struhl, G. Dual roles for patched in sequestering and transducing hedgehog. *Cell* **1996**, *87*, 553–563. [CrossRef]
42. Martinelli, D.C.; Fan, C.M. Gas1 extends the range of hedgehog action by facilitating its signaling. *Genes Dev.* **2007**, *21*, 1231–1243. [CrossRef] [PubMed]
43. Kang, J.S.; Gao, M.; Feinleib, J.L.; Cotter, P.D.; Guadagno, S.N.; Krauss, R.S. Cdo: An oncogene-, serum-, and anchorage-regulated member of the ig/fibronectin type iii repeat family. *J. Cell Biol.* **1997**, *138*, 203–213. [CrossRef] [PubMed]
44. Kang, J.S.; Mulieri, P.J.; Hu, Y.; Taliana, L.; Krauss, R.S. Boc, an ig superfamily member, associates with cdo to positively regulate myogenic differentiation. *EMBO J.* **2002**, *21*, 114–124. [CrossRef] [PubMed]
45. Biggs, L.C.; Mikkola, M.L. Early inductive events in ectodermal appendage morphogenesis. *Semin. Cell Dev. Biol.* **2014**, *25–26*, 11–21. [CrossRef] [PubMed]
46. Pispa, J.; Thesleff, I. Mechanisms of ectodermal organogenesis. *Dev. Biol.* **2003**, *262*, 195–205. [CrossRef]
47. Cobourne, M.T.; Miletich, I.; Sharpe, P.T. Restriction of sonic hedgehog signalling during early tooth development. *Development* **2004**, *131*, 2875–2885. [CrossRef] [PubMed]
48. Chiang, C.; Litingtung, Y.; Lee, E.; Young, K.E.; Corden, J.L.; Westphal, H.; Beachy, P.A. Cyclopia and defective axial patterning in mice lacking sonic hedgehog gene function. *Nature* **1996**, *383*, 407–413. [CrossRef] [PubMed]
49. Yamagishi, C.; Yamagishi, H.; Maeda, J.; Tsuchihashi, T.; Ivey, K.; Hu, T.; Srivastava, D. Sonic hedgehog is essential for first pharyngeal arch development. *Pediatr. Res.* **2006**, *59*, 349–354. [CrossRef] [PubMed]
50. Dassule, H.R.; McMahon, A.P. Analysis of epithelial-mesenchymal interactions in the initial morphogenesis of the mammalian tooth. *Dev. Biol.* **1998**, *202*, 215–227. [CrossRef] [PubMed]
51. Kettunen, P.; Karavanova, I.; Thesleff, I. Responsiveness of developing dental tissues to fibroblast growth factors: Expression of splicing alternatives of fgfr1, -2, -3, and of fgfr4; and stimulation of cell proliferation by fgf-2, -4, -8, and -9. *Dev. Genet.* **1998**, *22*, 374–385. [CrossRef]
52. Kettunen, P.; Laurikkala, J.; Itaranta, P.; Vainio, S.; Itoh, N.; Thesleff, I. Associations of fgf-3 and fgf-10 with signaling networks regulating tooth morphogenesis. *Dev. Dyn.* **2000**, *219*, 322–332. [CrossRef]

53. Sarkar, L. The Wnt Signalling Pathway in Tooth Development. Ph.D. Thesis, University of London, London, UK, 1999.
54. Ahn, Y.; Sanderson, B.W.; Klein, O.D.; Krumlauf, R. Inhibition of wnt signaling by wise (sostdc1) and negative feedback from shh controls tooth number and patterning. *Development* **2010**, *137*, 3221–3231. [CrossRef] [PubMed]
55. Klein, O.D.; Minowada, G.; Peterkova, R.; Kangas, A.; Yu, B.D.; Lesot, H.; Peterka, M.; Jernvall, J.; Martin, G.R. Sprouty genes control diastema tooth development via bidirectional antagonism of epithelial-mesenchymal fgf signaling. *Dev. Cell* **2006**, *11*, 181–190. [CrossRef] [PubMed]
56. Pummila, M.; Fliniaux, I.; Jaatinen, R.; James, M.J.; Laurikkala, J.; Schneider, P.; Thesleff, I.; Mikkola, M.L. Ectodysplasin has a dual role in ectodermal organogenesis: Inhibition of bmp activity and induction of shh expression. *Development* **2007**, *134*, 117–125. [CrossRef] [PubMed]
57. Murashima-Suginami, A.; Takahashi, K.; Sakata, T.; Tsukamoto, H.; Sugai, M.; Yanagita, M.; Shimizu, A.; Sakurai, T.; Slavkin, H.C.; Bessho, K. Enhanced bmp signaling results in supernumerary tooth formation in usag-1 deficient mouse. *Biochem. Biophys. Res. Commun.* **2008**, *369*, 1012–1016. [CrossRef] [PubMed]
58. Nakatomi, M.; Hovorakova, M.; Gritli-Linde, A.; Blair, H.J.; MacArthur, K.; Peterka, M.; Lesot, H.; Peterkova, R.; Ruiz-Perez, V.L.; Goodship, J.A.; et al. Evc regulates a symmetrical response to shh signaling in molar development. *J. Dent. Res.* **2013**, *92*, 222–228. [CrossRef] [PubMed]
59. Ruiz-Perez, V.L.; Blair, H.J.; Rodriguez-Andres, M.E.; Blanco, M.J.; Wilson, A.; Liu, Y.N.; Miles, C.; Peters, H.; Goodship, J.A. Evc is a positive mediator of ihh-regulated bone growth that localises at the base of chondrocyte cilia. *Development* **2007**, *134*, 2903–2912. [CrossRef] [PubMed]
60. Yang, C.; Chen, W.; Chen, Y.; Jiang, J. Smoothened transduces hedgehog signal by forming a complex with evc/evc2. *Cell Res.* **2012**, *22*, 1593–1604. [CrossRef] [PubMed]
61. Sharpe, P.T. Neural crest and tooth morphogenesis. *Adv. Dent. Res.* **2001**, *15*, 4–7. [CrossRef] [PubMed]
62. Thesleff, I.; Hurmerinta, K. Tissue interactions in tooth development. *Differentiation* **1981**, *18*, 75–88. [CrossRef] [PubMed]
63. Koyama, E.; Yamaai, T.; Iseki, S.; Ohuchi, H.; Nohno, T.; Yoshioka, H.; Hayashi, Y.; Leatherman, J.L.; Golden, E.B.; Noji, S.; et al. Polarizing activity, sonic hedgehog, and tooth development in embryonic and postnatal mouse. *Dev. Dyn.* **1996**, *206*, 59–72. [CrossRef]
64. Gritli-Linde, A.; Lewis, P.; McMahon, A.P.; Linde, A. The whereabouts of a morphogen: Direct evidence for short- and graded long-range activity of hedgehog signaling peptides. *Dev. Biol.* **2001**, *236*, 364–386. [CrossRef] [PubMed]
65. Kratochwil, K.; Galceran, J.; Tontsch, S.; Roth, W.; Grosschedl, R. Fgf4, a direct target of lef1 and wnt signaling, can rescue the arrest of tooth organogenesis in lef1(-/-) mice. *Genes Dev.* **2002**, *16*, 3173–3185. [CrossRef] [PubMed]
66. Liu, F.; Chu, E.Y.; Watt, B.; Zhang, Y.; Gallant, N.M.; Andl, T.; Yang, S.H.; Lu, M.M.; Piccolo, S.; Schmidt-Ullrich, R.; et al. Wnt/beta-catenin signaling directs multiple stages of tooth morphogenesis. *Dev. Biol.* **2008**, *313*, 210–224. [CrossRef] [PubMed]
67. Nakatomi, M.; Morita, I.; Eto, K.; Ota, M.S. Sonic hedgehog signaling is important in tooth root development. *J. Dent. Res.* **2006**, *85*, 427–431. [CrossRef] [PubMed]
68. Liu, Y.; Feng, J.; Li, J.; Zhao, H.; Ho, T.V.; Chai, Y. An nfic-hedgehog signaling cascade regulates tooth root development. *Development* **2015**, *142*, 3374–3382. [CrossRef] [PubMed]
69. Steele-Perkins, G.; Butz, K.G.; Lyons, G.E.; Zeichner-David, M.; Kim, H.J.; Cho, M.I.; Gronostajski, R.M. Essential role for nfi-c/ctf transcription-replication factor in tooth root development. *Mol. Cell. Biol.* **2003**, *23*, 1075–1084. [CrossRef] [PubMed]
70. Lesot, H.; Peterkova, R.; Viriot, L.; Vonesch, J.L.; Tureckova, J.; Peterka, M.; Ruch, J.V. Early stages of tooth morphogenesis in mouse analyzed by 3d reconstructions. *Eur. J. Oral Sci.* **1998**, *106* (Suppl. S1), 64–70. [CrossRef] [PubMed]
71. Peterkova, R.; Hovorakova, M.; Peterka, M.; Lesot, H. Three-dimensional analysis of the early development of the dentition. *Aust. Dent. J.* **2014**, *59* (Suppl. S1), 55–80. [CrossRef] [PubMed]
72. Peterkova, R.; Peterka, M.; Viriot, L.; Lesot, H. Dentition development and budding morphogenesis. *J. Craniofac. Genet. Dev. Biol.* **2000**, *20*, 158–172. [PubMed]
73. Peterková, R.; Peterka, M.; Viriot, L.; Lesot, H. Development of the vestigial tooth primordia as part of mouse odontogenesis. *Connect. Tissue Res.* **2002**, *43*, 120–128. [CrossRef] [PubMed]

74. Viriot, L.; Lesot, H.; Vonesch, J.L.; Ruch, J.V.; Peterka, M.; Peterkova, R. The presence of rudimentary odontogenic structures in the mouse embryonic mandible requires reinterpretation of developmental control of first lower molar histomorphogenesis. *Int. J. Dev. Biol.* **2000**, *44*, 233–240. [PubMed]
75. Tenzen, T.; Allen, B.L.; Cole, F.; Kang, J.S.; Krauss, R.S.; McMahon, A.P. The cell surface membrane proteins cdo and boc are components and targets of the hedgehog signaling pathway and feedback network in mice. *Dev. Cell.* **2006**, *10*, 647–656. [CrossRef] [PubMed]
76. Allen, B.L.; Tenzen, T.; McMahon, A.P. The hedgehog-binding proteins gas1 and cdo cooperate to positively regulate shh signaling during mouse development. *Genes Dev.* **2007**, *21*, 1244–1257. [CrossRef] [PubMed]
77. Okada, A.; Charron, F.; Morin, S.; Shin, D.S.; Wong, K.; Fabre, P.J.; Tessier-Lavigne, M.; McConnell, S.K. Boc is a receptor for sonic hedgehog in the guidance of commissural axons. *Nature* **2006**, *444*, 369–373. [CrossRef] [PubMed]
78. Seppala, M.; Depew, M.J.; Martinelli, D.C.; Fan, C.M.; Sharpe, P.T.; Cobourne, M.T. Gas1 is a modifier for holoprosencephaly and genetically interacts with sonic hedgehog. *J. Clin. Investig.* **2007**, *117*, 1575–1584. [CrossRef] [PubMed]
79. Zhang, W.; Kang, J.S.; Cole, F.; Yi, M.J.; Krauss, R.S. Cdo functions at multiple points in the sonic hedgehog pathway, and cdo-deficient mice accurately model human holoprosencephaly. *Dev. Cell* **2006**, *10*, 657–665. [CrossRef] [PubMed]
80. Chuang, P.T.; McMahon, A.P. Vertebrate hedgehog signalling modulated by induction of a hedgehog-binding protein. *Nature* **1999**, *397*, 617–621. [CrossRef] [PubMed]
81. Reich, A.; Sapir, A.; Shilo, B. Sprouty is a general inhibitor of receptor tyrosine kinase signaling. *Development* **1999**, *126*, 4139–4147. [PubMed]
82. Harada, H.; Kettunen, P.; Jung, H.S.; Mustonen, T.; Wang, Y.A.; Thesleff, I. Localization of putative stem cells in dental epithelium and their association with notch and fgf signaling. *J. Cell Biol.* **1999**, *147*, 105–120. [CrossRef] [PubMed]
83. Juuri, E.; Saito, K.; Ahtiainen, L.; Seidel, K.; Tummers, M.; Hochedlinger, K.; Klein, O.D.; Thesleff, I.; Michon, F. Sox2+ stem cells contribute to all epithelial lineages of the tooth via sfrp5+ progenitors. *Dev. Cell* **2012**, *23*, 317–328. [CrossRef] [PubMed]
84. Seidel, K.; Ahn, C.P.; Lyons, D.; Nee, A.; Ting, K.; Brownell, I.; Cao, T.; Carano, R.A.; Curran, T.; Schober, M.; et al. Hedgehog signaling regulates the generation of ameloblast progenitors in the continuously growing mouse incisor. *Development* **2010**, *137*, 3753–3761. [CrossRef] [PubMed]
85. Rodda, S.J.; McMahon, A.P. Distinct roles for hedgehog and canonical wnt signaling in specification, differentiation and maintenance of osteoblast progenitors. *Development* **2006**, *133*, 3231–3244. [CrossRef] [PubMed]
86. Passman, J.N.; Dong, X.R.; Wu, S.P.; Maguire, C.T.; Hogan, K.A.; Bautch, V.L.; Majesky, M.W. A sonic hedgehog signaling domain in the arterial adventitia supports resident sca1+ smooth muscle progenitor cells. *Proc. Natl. Acad. Sci. USA* **2008**, *105*, 9349–9354. [CrossRef] [PubMed]
87. Fraser, G.J.; Graham, A.; Smith, M.M. Conserved deployment of genes during odontogenesis across osteichthyans. *Proc. Biol. Sci.* **2004**, *271*, 2311–2317. [CrossRef] [PubMed]
88. Fraser, G.J.; Bloomquist, R.F.; Streelman, J.T. A periodic pattern generator for dental diversity. *BMC Biol.* **2008**, *6*, 32. [CrossRef] [PubMed]
89. Martin, K.J.; Rasch, L.J.; Cooper, R.L.; Metscher, B.D.; Johanson, Z.; Fraser, G.J. Sox2+ progenitors in sharks link taste development with the evolution of regenerative teeth from denticles. *Proc. Natl. Acad. Sci. USA* **2016**, *113*, 14769–14774. [CrossRef] [PubMed]
90. Fraser, G.J.; Bloomquist, R.F.; Streelman, J.T. Common developmental pathways link tooth shape to regeneration. *Dev. Biol.* **2013**, *377*, 399–414. [CrossRef] [PubMed]

2

Engineering the *Drosophila* Genome for Developmental Biology

Dagmara Korona, Stefan A. Koestler and Steven Russell *

Department of Genetics, University of Cambridge, Downing Street, Cambridge CB2 3EH, UK; dk500@cam.ac.uk (D.K.); sak76@cam.ac.uk (S.A.K.)
* Correspondence: s.russell@gen.cam.ac.uk

Abstract: The recent development of transposon and CRISPR-Cas9-based tools for manipulating the fly genome in vivo promises tremendous progress in our ability to study developmental processes. Tools for introducing tags into genes at their endogenous genomic loci facilitate imaging or biochemistry approaches at the cellular or subcellular levels. Similarly, the ability to make specific alterations to the genome sequence allows much more precise genetic control to address questions of gene function.

Keywords: *Drosophila*; genome engineering; CRISPR-Cas9; protein tagging

---

1. Introduction

It is over a century since Thomas Hunt Morgan and his students established *Drosophila melanogaster* as a powerful system for exploring the basis of heredity [1,2]. By the latter period of the 20th century, due in part to the pioneering work of Ed Lewis, along with the large scale genetics screens by Christiane Nüsslein-Volhard and Eric Wieschaus, and coupled with the tremendous progress in molecular biology, the fly became well established as a model for developmental biology [2]. With an increasingly sophisticated range of genetic tools, very efficient transgenesis and the well-described anatomy available, considerable progress was made in characterising conserved developmental pathways [3]. The advent of genome sequencing and the post-genomic era inevitably led to new approaches in developmental biology, in particular, functional analysis of genes identified in high-throughput genomics screens or fly orthologues of mammalian genes identified in such studies [4]. Despite its sophistication, the major drawback of the fly as an experimental model was the lack of an easy system for targeting genomic changes in the organism via the type of homologous recombination used in yeast or vertebrate cells. While the development of gene targeting approaches [5,6] by the Golic laboratory addressed this deficiency in part, the systems are cumbersome and success tends to be highly locus specific. More recently, a range of transposon-based methods has allowed much greater flexibility in precisely manipulating genes in situ [7,8], however, the very recent advent of Clustered, Regularly Interspaced, Short Palindromic Repeat (CRISPR)-based genome engineering now offers the prospect of making virtually any desired change to the genome sequence much easier than previously achievable. Together, these methods are opening new avenues in developmental biology research, facilitating precise questions about gene/protein function, allowing the targeted expression or knock out of any gene and enabling high-resolution imaging of proteins, often in live tissues. Here we overview the most recent methods for manipulating the fly genome with a focus on approaches tackling gene manipulation and in vivo localisation for developmental studies.

2. Classical Methods

One of the major landmarks in the progression of the fly as an experimental tool for the study of development was the establishment of a robust method for generating transgenic animals via the *P* element transposon [9]. Since its introduction, a range of sophisticated *P* element-based tools have been constructed; including the Gal4-*UAS* system for precise spatial and temporal control of gene expression [10]; enhancer trapping for the genome-wide identification of developmentally regulated genes [11,12] and methods for systematic gene misexpression with engineered *P* transposable elements (EP) [13]. The utility and use of these by now classic tools have been well described in many reviews, and we here focus on some of the more modern applications of transposons. While the *P* element has been the workhorse of *Drosophila* transgenesis for over 30 years, it does exhibit an insertion bias [14], particularly towards gene promoters, making it less useful for targeting other genic regions such as introns. The establishment of other transposon systems, particularly *Minos* and *piggyBac* [15,16], along with the site-specific integrase PhiC31 [17], have opened more of the fly genome to engineering approaches. Considerable detail on newer engineering methods have recently been well reviewed by Venken and colleagues [18]; here we briefly touch on the most important systems, with a focus on those facilitating imaging studies or introducing precise changes to the genome sequence.

The introduction of a variety of inducible site-specific recombination systems, principally Flp-*FRT* [19] and Cre-*Lox* [20], have enhanced our ability to manipulate gene expression in space and time or to precisely engineer the genome. For example, introducing transgenic constructs containing *FRT* sites into the genome has considerable utility for functional and imaging analyses, facilitating spatial or temporal control of gene activation [21]. Generating mitotic or germ-line clones, techniques that are necessary to overcome pleiotropic or maternal effects of mutations and study gene functions during later stages of development, are simplified by the use of chromosomes with *FRT* sites close to the centromere. Similarly, the use of Flp-mediated recombination sees widespread use in methods aimed at marking specific populations of cells for lineage tracing or functional studies, particularly in the nervous system, and these have recently been well reviewed by others [22–27].

Projects by Exelixis and the DrosDel consortium have generated several thousand insertions with transposons carrying *FRT* sites that allow precise chromosome engineering [28,29]. For example, using the approach developed by Golic and Golic [30] (Figure 1a–c), DrosDel generated over 3000 insertions of re-arrangement screen 5 and 3 (RS5 and RS3) *P* elements, using these to build deletion coverage for almost 80% of the genome [29]. More importantly, combining the Exelixis and the DrosDel collections allows the construction of over 500,000 deletions ranging in size from 1 bp to 1 Mb that are precisely mapped at base pair resolution. These tools can be used to construct precise deletions in a homogeneous genetic background, for example, facilitating the analysis of complex regulatory regions [31], or genome-wide analysis of gene dosage effects to provide insights into gene regulatory network robustness [32]. Finally, *FRT* elements may be combined to generate other types of chromosomal aberrations, including translocations and inversions [30], and we have used such an approach to examine the consequences of disrupting developmental gene regulatory neighbourhoods [33]. The ability to precisely manipulate genomes at the chromosome level remains a powerful tool in the research armoury and the Bloomington Deficiency kit, rebuilt using the Flp-*FRT* approach, remains a widely used resource for the type of gene mapping and pathway expansion studies that have been important in building developmental networks [34].

The phiC31 system is a particularly useful tool for generating transgenic flies since a range of different transgenes may be inserted at an identical site in the genome [17,35], ameliorating concerns about position effects or other insertion site artefacts when comparing the effects of different constructs. PhiC31 integrase is directional and highly specific, mediating recombination between *attP* and *attB* sites to generate hybrid *attL* and *attR* sites that are not substrates for the enzyme (Figure 1d). A set of lines containing landing sites across the genome are available from stock centres and at least some of these have been functionally characterised in terms of expression properties [36]. Recently, a mutated integrase has been developed that allows reversal of the integration reaction [37], which is useful for building arrays of different transgenes at single sites in the genome. A second very useful approach available with phiC31 is recombinase-mediated cassette exchange (RMCE) [38]: in this case a genomic landing site containing a marker gene flanked by *attP* sites may be replaced by any other DNA sequence via a plasmid containing a gene or sequence of interest flanked by *attB* sites (Figure 1e). The RMCE system is incredibly powerful and since its introduction has been deployed as a component in a wide variety of genome engineering strategies (i.e., see Minos-mediated integration cassette (MiMIC) below).

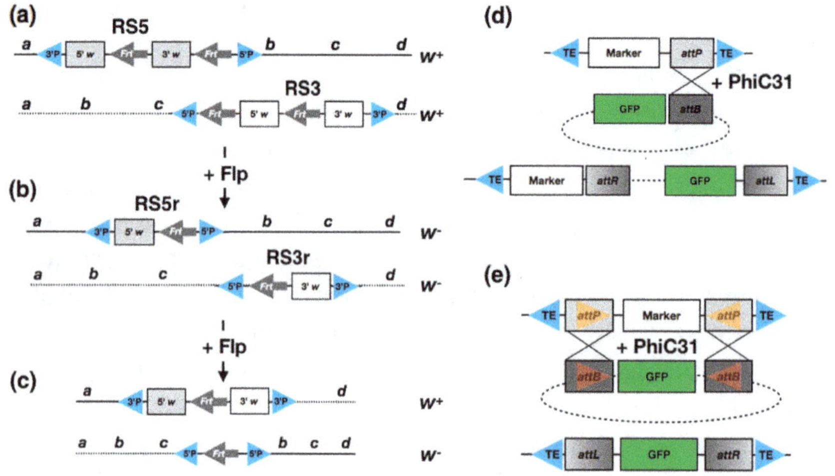

Figure 1. Genome engineering. (**a**–**c**) **The Golic method for generating precise chromosomal deletions.** (**a**) Two *P* elements, designated RS3 and RS5, are inserted at different locations (designated by a–d) on two homologous chromosomes and kept in separate fly lines. The elements contain a functional mini-*white* gene composed of multiple exons that for simplicity are drawn as grey or white boxes, representing 5′ and 3′ portions of the gene. There are two *Frt* sites (grey arrows) in each element, one of which is located within a mini-*white* exon. The elements differ in the position of the second *Frt* site and the orientation of the construct with respect to the *P* element ends (blue triangles). (**b**) Internal Flp-driven recombination between the *Frt* sites produces remnant forms of the *white* genes such that RS5r contains the 5′ end and RS3r the 3′ end with the intronic *FRT* site remaining. Each of these remnant elements are generated in separate fly lines that are phenotypically white eyed. (**c**) RS5r and RS3r elements are brought together in trans in a fly along with a source of Flp recombinase. FLP-mediated recombination between the elements produces a reconstituted functional *white* gene and the intervening genomic DNA is deleted. The reciprocal event creates a tandem duplication of the deleted segment, separated by an *FRT* site, but no white gene. (**d**) **The PhiC31 system.** A transposon (blue triangles mark the transposon ends) carrying a marker gene for genetic tracking and an *attP* site is inserted into the genome. Providing an *attB* containing plasmid with a gene or sequence of interest, in this case GFP, and a source of PhC31 integrase results in high efficiency integration of the plasmid into the genomic location. (**e**) **Recombinase-mediated cassette exchange (RMCE).** A transposon (blue triangles mark the transposon ends) carrying a marker gene flanked by *attP* site is inserted into the genome. Providing a plasmid with a gene or sequence of interest, in this case GFP, flanked by *attB* sites and a source of PhC31 recombinase results in high efficiency replacement of the genomic marker with the sequence of interest. With RMCE the inverted orientation of the *attP* and *attB* sites is critical for producing the desired exchange.

A key requirement for understanding gene function during development is the ability to localise gene products in space and time at the organismal, cellular and subcellular levels. For many years immunohistochemistry or immunofluorescence using specific antibodies was the primary method for collecting these data. Where antibodies are not available or there is a requirement for live imaging, transgenes containing tagged versions of a protein of interest have been useful. It is most common to express genes encoding tagged proteins under the control of the Gal4-*UAS* system, however, there can be concerns that over or misexpression may affect function. While it is sometimes possible to generate tagged transgenes that rescue null mutants, endogenous regulatory sequences are often too large or poorly characterised to make such an approach generally applicable. To try and address this P[acman] bacterial artificial chromosome (BAC) libraries for *D. melanogaster* [39] and Fosmid libraries for both *D. melanogaster* and *D. pseudoobscura* have been generated, which allow recombineering approaches in *E. coli* to generate tagged transgenes that are then integrated into the fly genome [40,41]. It would however clearly be preferable to tag endogenous loci in situ. Two transposon based approaches have been developed to facilitate in vivo protein tagging: the MiMIC method described below and various protein trap strategies. In the latter, a transposon containing an artificial exon, usually a fluorescent protein flanked by splice acceptor and donor sequences, is mobilised in the germline and progeny inheriting fluorescent reporter expression are recovered. A comprehensive screen requires a set of three vectors, each with the reporter in a different open reading frame. When the transposon inserts into an intron of a protein coding gene in the correct orientation, splicing will incorporate the artificial exon into the primary transcript and hence the fluorescent reporter into the protein [42] (Figure 2a–c). Due to the bias observed with *P* elements, particularly insertion hotspots in promoter and 5′ UTR regions [43], the use of *piggyback*-based transposons was found to generate a much higher frequency of bona fide protein traps. Together, three large-scale studies screened approximately 150 million embryos, recovering a little over 600 verified protein traps between them. Reassuringly, the majority of protein traps do not appear to substantially affect protein function [44–46]. While this may seem a relatively poor yield for such an effort, the lines generated have been incredibly useful for imaging studies down to the subcellular level [47] and have also facilitated mass spectrometry-based analysis of in vivo protein complexes that are purified via the tag [48]. Providing the ability to characterise protein localisation and interactions, protein trap lines have been used to study diverse developmental processes, including live imaging of anterior-posterior (A-P) polarity formation during oogenesis, the cell biology of epithelial morphogenesis and signalling pathway interactions [49–51]. Trapped lines have also facilitated the characterisation of genome wide transcription factor binding via chromatin immunoprecipitation (ChIP)-based studies [52].

One of the most useful and flexible transposon-based strategies available to the fly biologist is the MiMIC system [53] (Minos-mediated integration cassette): a collection of over 7000 *Minos*-based insertions that provide a platform for a variety of targeted engineering approaches. The core MiMIC transposon contains inverted phiC31 *attP* target sites within the *Minos* ends, a splice acceptor site followed by stop codons in all three reading frames. The element also contains a transcriptional/translational cassette encoding EGFP followed by a polyadenylation signal and a phenotypic marker (*yellow*). Insertion of a MiMIC element in the correct orientation into an intron of a coding gene will generate a truncated protein due to the splice acceptor and stop codons, thus acting as a gene trap. The true utility of MiMIC is the ability to use the phiC31 *attP* sites to replace the contents of the transposon with any piece of DNA via the RMCE system (Figure 2d). For example, regulatory sequences may be added to genes, functional reporters such as Gal4 or Flp may be inserted within 5′ UTRs or, in cases when inserts are within coding introns, a wide range of protein tags may be introduced. Around 30 different cassettes have been generated by the MiMIC team (Figure 2e), but any desired sequence can be easily cloned and introduced into the genome using the system. More recently, the team developed a genetic approach to circumvent the requirement for embryo injection [54], speeding up the generation of cassette replacement. Approximately 2000 genes currently have MiMIC insertions within introns, but the possibility of using CRISPR (see below) to place MiMIC insertions into the genome opens up the possibility that every intron containing gene in the genome could be amendable to protein trapping. The utility of the MiMIC system is attested to by over 100 research papers citing its use in a little over six years.

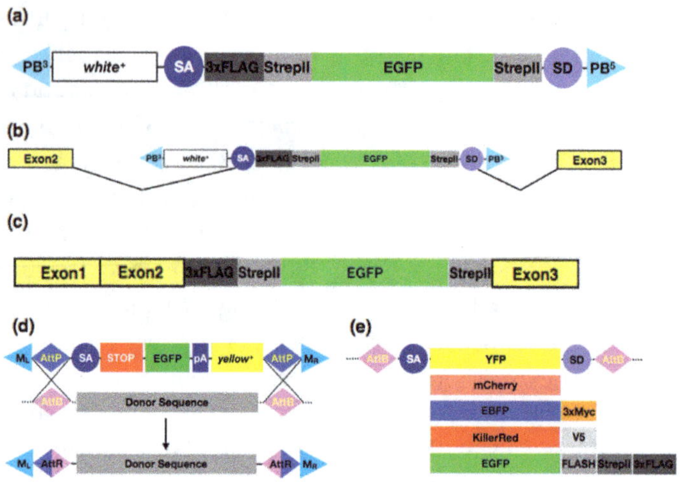

Figure 2. Protein Trapping. (**a**–**c**) ***piggyBac* transposon-based protein trapping**. (**a**) The *pigP* protein trap element used in [47]: the transposon ends (blue triangles) flank a genetic marker gene (*white*) and an artificial exon, which in this case contains the coding sequence for enhanced green fluorescent protein (EGFP) along with StrepII and 3XFLAG tags between splice donor and splice acceptor sites (blue circles). (**b**) Insertion of the transposon into an intron of a protein coding gene (represented by lines separating the yellow boxes) allows the possibility of splicing the artificial exon into the gene transcript. (**c**) If the transcript carrying the artificial exon is translated a tagged protein is generated. (**d**,**e**) **The** Minos-mediated integration cassette **(MiMIC) System.** (**d**) A *Minos*-based transposon with transposable element (TE) ends indicated by blue triangles, contains two *attP* sites in inverted orientation (blue diamonds) flanking a gene trap cassette with a splice acceptor (blue circle), stop codons in all three reading frames (red box), a fluorescent marker (EGFP) followed by a polyadenylation signal (blue box) and a genetic marker (*yellow*). The sequences internal to the *attP* sites may be replaced via a RMCE reaction by providing a donor sequence flanked by *attB* sites and a source of phiC31 integrase. (**e**) A variety of different fluorescent reporters have been developed that can be used to introduce tags into genes with MiMIC insertions in coding introns via RMCE.

Whether protein trap or engineered MiMIC strategies are used to generate tagged proteins, there are some useful approaches to knock down their function [55] in a tissue or temporally specific manner. Transcripts containing inserted tags may be targeted by RNA-mediated interference (RNAi) constructs directed against the tag, and lines containing *UAS*-driven RNAi recognising green-fluorescent protein/enhanced green fluorescent protein (GFP/EGFP) are available from stock centres. A second approach is via the use of an anti-GFP nanobody, which targets GFP-tagged proteins for degradation via the ubiquitin pathway [56,57] and *UAS*-driven lines are available. Both these methods have been shown to be effective with in vivo tagged proteins and offer a level of functional control over gene products that can, in principle, be monitored in real time via imaging of the fluorescent tag. However, despite the relative ease and simplicity of these methods they have obvious limitations; the elimination of tagged proteins is relatively slow and they are not readily reversible. For the rapid elimination of gene function, an auxin-inducible degradation system (AID) has recently been shown to work in *Drosophila* [58]. This system offers the ability to regulate the inducible and rapid degradation of specific target proteins by the recruitment of proteins with an auxin-inducible degron to the E3 ubiquitin ligase complex. A disadvantage of this system is the need to introduce a specific tag containing an auxin-inducible degron domain into the gene of interest, though this could be achieved for genes harbouring an appropriate MiMIC insertion.

3. CRISPR/Cas9

3.1. Overview

As the above discussion emphasises, the range of transposon based tools available to the *Drosophila* biologist is impressive, however, even here the landscape for genome engineering has changed dramatically with the introduction of the CRISPR-Cas9 system. Clustered, Regularly Interspaced, Short Palindromic Repeat (CRISPR)-based engineering has become the most popular and commonly used tool for genome manipulation over the last five years. Since its discovery, or rather understanding its potential, numerous studies have confirmed the wide applicability and adaptability of the system [59–61] and a number of different methods have been developed with different variants of the RNA-guided Cas9 endonuclease adopted for genome editing [62].

CRISPR-Cas systems are natural RNA-guided adaptive immune systems that provide protection mechanisms against viruses or other exogenous DNA entering bacterial and archaeal cells. The most studied and widely applied system for genome editing, Class 2 CRISPR-Cas from *Streptococcus pyogenes*, uses a guide RNA (gRNA) composed of a specific CRISPR RNA (crRNA) combined with a tracrRNA sequence to guide the Cas9 nuclease to a genomic target site where it introduces a double-strand break (DSB) that is repaired by endogenous repair pathways. The system requires only 20 nucleotides of homology to the target site in the gRNA, but the target must be adjacent to a three nucleotide (NGG) protospacer adjacent motif (PAM) in the genome (Figure 3a) [63,64]. Repair may be via the non-homologous end joining (NHEJ) pathway [65], leading to insertions and/or deletions (indels) at the target site. Alternatively, the activity of homology-directed repair (HDR) pathways can facilitate precise genome editing by introducing alternative sequences via a supplied donor DNA (Figure 3a) [66,67]. While the initial efficiency of specific targeting, especially when trying to introduce mutations or tags, was initially low and variable [68,69], various improved systems have been developed. The method originally relied on delivery of the nuclease and RNA components via microinjection of RNA or DNA, the development of transgenic lines expressing Cas9 under the control of germline promoters such as *nanos* (*nos*) or *vasa* as recipients for injection of plasmids expressing gRNAs from U6 promoters has considerably simplified the process and can generate desired events in the genome with higher efficiency than with the initial injection approach [70].

One of the major concerns in any genome engineering work, whether using CRISPR/Cas9 or any other system, is the possibility of off-target effects resulting from gRNA binding to imperfectly matched targets at other genomic locations [71]. While the extent of off-target effects has not been fully characterised, several studies that suggest that for relatively small genomes, such as that of *Drosophila*, off-targets may be less of an issue compared to organisms with larger genomes [72,73]. The problem may be, at least in part, overcome by using a mutant version of Cas9 developed by Cong and colleagues [74], Cas9D10A (Cas9n), that makes single strand nicks at the target locus [74–76]. When used in conjunction with two gRNAs flanking the desired target site and a donor repair template, repair is generally via the high-fidelity HDR pathway [69,77]. The nickase activity of Cas9n does not activate the NHEJ pathway and while Cas9n is certainly more specific than wild-type Cas9, DSBs are still detectable at target sites when only one gRNA is used and it is possible that Cas9n may cause indels at off-target sites due to either of the gRNAs. Encouragingly, the paired Cas9n system has been used to generate large deletions without observing unwanted translocations, and so forth (Figure 3b) [76,78], and a recent comparative study indicates that Cas9n can induce more HDR than NHEJ, at least in human HEK293 cells [79].

Concerns regarding the use of Cas9n have been raised due to the possibility of DNA rearrangements via single nicks [62]. These unwanted off-targets can result from the recognition of multiple genomic sites by gRNAs that that can subsequently be cleaved. In contrast to other nuclease systems for genome engineering, such as zinc finger nucleases or transcription activator-like effector nucleases (TALENs), that rely on dimerization for target cleavage, Cas9 acts as a monomer. In order to increase specificity, new Cas9 versions that only cleave when dimerised have been developed:

for example, a nuclease dead Cas9 (dCas9) fused to the non-specific endonuclease FokI has the gRNA directed specificity of the CRISPR system but relies on obligate FokI dimerization for cleavage (Figure 3c) [63,64,80,81]. A clear limitation of the paired dCas9-FolkI approach is the requirement for two appropriately spaced gRNA sites 15–25 bases apart [81]. This constrains the widest applicability of the method since these conditions cannot always be achieved. Alternatively, there have been efforts to develop engineered Cas9 enzymes (i.e., SpCas9-HF1) with increased fidelity to ensure more specific targeting [82], however, there are reports of lower efficiencies with such variants [83]. Irrespective of the engineering method used, effects of off-target events may be largely ameliorated by cleaning up stocks via standard back-crossing methods.

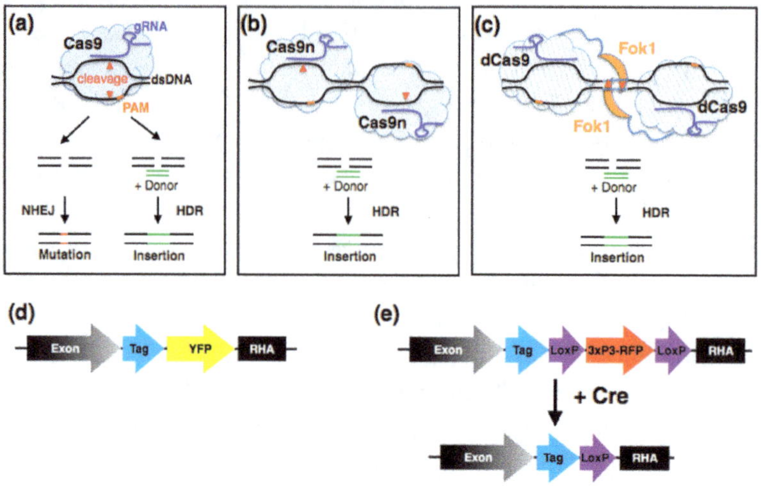

Figure 3. CRISPR Genome Engineering. (**a**) The wild-type Cas9 complex (light blue cloud) contains the Cas9 endonuclease and a guide RNA (gRNA) (blue) complementary to the target site adjacent to a PAM sequence (orange). The complex opens the DNA duplex and introduces a double strand break (red triangles). Repair by the non-homologous end joining (NHEJ) pathway may result in indel mutations whereas homology directed repair (HDR) in the presences of a donor template (green) generates insertions. (**b**) Using mutant Cas9n enzymes that make single strand cuts with two gRNAs (blue) direct the Cas9n complexes to make cuts (red triangles) separated by some distance. The gap may be repaired in the presence of a donor (green) to generate an insertion. (**c**) Dead Cas9 (dCas9) enzymes (which are unable to cleave DNA) are fused with FokI nuclease monomers (orange). When two gRNAs (blue) some distance apart are used, dCas9-FolkI monomers are brought into proximity allowing the FokI to dimerise and cleave in between. The resulting gap may be repaired in the presence of a donor (green) to generate an insertion. (**d**) Diagram of donor cassette for direct addition of a protein tag to the C-terminal of a coding exon. The donor DNA contains part of the exon sequence with a biochemistry tag and yellow fluorescent protein (YFP) in frame, followed by a right hand end homology arm (RHA) that can mediate the type of insertion event shown in (**a**). (**e**) Diagram of a donor sequence used to introduce biochemistry tags to the C-terminal of a coding exon along with a removable marker. The donor DNA contains part of the exon sequence with a biochemistry tag in frame. This is followed by an eye expressed red fluorescent protein (RFP) cassette flanked by *LoxP* sites, which are downstream of the splice donor site and thus within an intron, then a right hand homology arm. Once a verified insertion has been recovered, tracked by the acquisition of RFP, the RFP is removed by exposing to Cre recombinase and recovering flies who have lost RFP expression. The resulting lines have the tagged exon and a single intronic *LoxP* site.

The obvious restriction when using Cas9 for genome engineering is the requirement for a PAM sequence adjacent to the desired target site, which limits the number of Cas9 engineerable sites in a genome. To overcome this limitation several mutant versions of Cas9 with altered specificities have been engineered [84] and Cas9 homologs from other organisms that show different PAM specificities have been isolated. For example, a recently identified class 2 endonuclease, Cpf1 from *Francisella novicida*, has received some attention [85]. The Cpf1 PAM is TTN and is located 5′ to the target site (in contrast to the Cas9 PAM which is 3′ to the target), providing a new set of potential targets not accessible to Cas9 systems. Finally, other Class 2 CRISPR effectors, such as Cas13a (C2C2), have been shown to target RNA, however, few studies have explored the potential of this activity [62,86]. It should be remembered that the potential of CRISPR based systems has only been apparent for five years and the progress in developing variants systems with different properties has been tremendous and will surely continue to provide new enzymes and systems with improved specificity and fidelity.

3.2. Mutagenesis with CRISPR/Cas9 Systems

The most straight forward way to modify the genome using CRISPR/Cas9 based genome engineering is the introduction of short insertions/deletions (indels) by inducing NHEJ, which frequently leads to frameshift mutations and thus null alleles or truncations of the targeted gene product. All that is required is the Cas9 protein and a single gRNA targeting the gene of interest after the transcription start site [87]. Large deletions of 30 Mb have been generated in human cell lines and even chromosomal rearrangements such as translocations can be generated by providing two gRNAs, targeting either side of the region to be deleted or sites on two different chromosomes [87–89]. Insertions of up to 5.7 kb have been achieved by providing linear DNA during CRISPR/Cas9 induced NHEJ in Zebrafish [90], and up to 34 kb in human cell lines [91] using *piggyBac* or adenoviral vectors that were cleaved by Cas9, targeted via a vector specific gRNA. A drawback is that the position of the knock-in is not precisely defined, and thus a molecular screen for in-frame insertions must be performed. It remains to be determined to what extent NHEJ-driven rearrangement can be exploited in *Drosophila*.

The considerable versatility of CRISPR/Cas9 based genome editing lies in the utilization of homology directed repair (HDR) for the precise incorporation of a donor sequence in the form of ssDNA-oligonucleotides, PCR (Polymerase Chain Reaction) products or plasmids, that are carried between homology arms, which are used as templates during the double strand repair process rather than the homologous chromosome [69,92]. Depending on the target site(s), the gRNA(s), the homology arms used and the region between them, different outcomes are possible. An insertion is achieved when the donor carries a sequence between the homology arms that cover the regions directly adjacent to the double strand break, thus inserting the sequence (Figure 3a). In this way, for example, additional amino acids can be added to a coding exon, point mutations introduced, or new binding sites added to a regulatory region. The optimum length for homology arms appears to be approximately 1 kb [93]. If the homology arms are separated with respect to the target genomic DNA the donor DNA between the homology arms may replace the intervening genomic segment. Since HDR efficiency decreases with the distance of homology deviations from the double strand break, replacement or deletion of large regions can be obtained by providing two gRNAs, each next to a homology arm, thus actively excising the intervening region [93–95]. Again, the sequence between the homology arms is inserted. Note that if homologous recombination does not take place the region between the homology arms may still be deleted, which obviously generates a deletion and perhaps a null allele. While a null allele can also be produced when only one gRNA is used as indicated above, this is likely to generate a frame shift mutation, which may be less useful, especially in non-coding regions. If a specific point mutation at a locus is desirable, a dCas9-cytidine deaminase chimeric enzyme may be used rather than homology directed replacement. The deaminase catalyses the conversion of C to U residues thus generating C to T or G to A substitutions. This 'base editing' is generally more efficient than HDR-mediated point mutation and at the same time minimizes indel formation [96]. This has recently been complemented by conversion of A-T to G-C base pairs by using a transfer RNA specific adenosine deaminase fused to catalytically impaired CRISPR/Cas9 [97].

Initial screening for positive events can be done using standard PCR, high resolution melt analysis (HRM), or sometimes more easily with visual markers. These can include body or eye colour markers, or fluorescent proteins expressed, for example, in the eye via the 3XP3 promoter (Figure 3e). Such markers are usually flanked by *FRT* sites for subsequent removal to avoid interference with gene function or other unphysiological responses [98]. Whichever method is used, any mutations should be verified by sequencing. Experiments ranging from the generation of specific single base pair mutations to replacement of whole genes have been described [99–101]. The latter is especially relevant for evo-devo studies, for example, to define the degree of functional redundancy between paralogous or orthologous genes by swapping the coding sequences of one gene into the genomic regulatory context of another. An example of such an approach is a swap of *Sox3* with *Sox2* coding sequences in mouse, facilitated by the simple genomic architecture of this class of genes since they only contain a single exon [102]. We have recently used a similar approach with the fly orthologues of these genes (S. A. Koestler and S. Russell, unpublished). Such experiments become more complex for larger genes, which may consist of several exons separated by introns, especially if the donor or host genes have intronic regulatory sequences. In such cases addition of a cDNA to functionally replace a null allele at a donor locus may be considered, although it is probably prudent to first test whether the coding sequence of the host gene is sufficient to effect rescue. For example, using this approach, gene and domain swaps in *D. melanogaster* as well as between *D. melanogaster* and *Tribolium castaneum* have led to the dissection of the regulatory function and evolution of members of the *robo* gene family [103,104]. However, fully understanding redundancy requires the removal of all redundant members of a gene family [105]. While such efforts previously required crossing or recombination of single mutants, or removal of large chromosomal regions [106,107], CRISPR/Cas9 allows a much more precise and efficient approach. Encouragingly, as the requirement for ever more high-throughput analysis is increasing, up to four genes have already been targeted in parallel by multiplexing [108,109].

The major challenges currently encountered when applying CRISPR/Cas systems for generating mutations arise from the relatively low efficiency of the desired genome modifications and the lack of an obvious phenotype in heterozygous F1 progeny, hence the need for time-consuming molecular screening of positive variants mentioned above. Due to the laborious nature of such screens and the need to examine many progeny, several groups developed methods facilitating more efficient engineering and hence more rapid screening. Bullock and colleagues [70] developed a versatile toolbox for *Drosophila* genome engineering consisting of a set of evaluated transgenic Cas9 lines and gRNA-expression plasmids. Their studies indicated that injecting gRNA(s) into Cas9 transgenic flies increases the overall efficiency but a fully transgenic system where the gRNA expressing construct is also inserted in the genome results in much higher efficiency. In principle, this strategy requires the generation of a transgenic fly line for every gRNA of interest that subsequently needs to be crossed to a transgenic Cas9 line. Despite being more time consuming initially, the fully transgenic system means that all transgenic flies pass on mutant alleles to their progeny, with efficiencies approaching 100% observed with the majority of tested gRNAs. Consequently, there is a significant improvement in screening efficiency since the majority of progeny have an engineered chromosome. More recently, based on the fact that that multiple CRISPR events can occur in a single cell [74,110], the co-CRISPR or co-conversion method initially developed in *C. elegans* [111,112] has been also successfully applied in *Drosophila* [113]. The method relies on simultaneously targeting a gene of interest and an *ebony* marker gene: *nos-Cas9* embryos [70] are injected with a mix of gRNAs targeting both *ebony* and the gene of interest (with or without a donor template). It is expected that in any given cell where *ebony* is mutated it is likely that the Cas9 has also been active at the gene of interest, thus progeny showing loss of *ebony* are selected for molecular analysis of the target gene [113]. In *Drosophila*, the co-CRSIPR strategy was found to be more efficient for identifying NHEJ mutagenesis events rather than HDR knock-in events. A similar approach targeting *white* as an editing marker has also been reported [114], however, it should be noted that *white* is a much more common transgenesis marker than *ebony* and thus may preclude its use in some crossing schemes. These efficiency and marker gene approaches simplify the task of screening for desired engineering events by enriching for successfully edited progeny and provide an excellent framework for further development of strategies to rapidly isolate mutations.

The generation of stable stocks carrying defined gene modifications is most often achieved with a germline expressed Cas9, usually via the *vasa* or *nanos* promoters described above [70,90,115,116]. However, in some situations modification of the germline may be either lethal or undesirable and it is therefore necessary to generate mosaic animals. These can be made by simply injecting the gRNA and donor construct (if one is used) into embryos ubiquitously expressing Cas9 or alternatively by co-injecting a source of Cas9. For the detection of successfully modified cells in mosaics the use of a marker is strongly recommended. The marker can be a fluorescent protein under control of a ubiquitous promoter carried on the injected donor construct. Although using a marker greatly facilitates the screening process it is not always possible to apply and the other approaches described in the previous section need to be employed. An alternative is to utilise the Gal4/*UAS* system to drive Cas9/gRNA expression in a specific cell type or at a particular developmental stage. Conditional mutants can be established, but the time it takes until they take affect and the turnover rates of the involved agents have to be considered [69,117]. It must be emphasized that this approach has to be used with care due to complications arising from the possible introduction of multiple mutations and cellular heterogeneity, especially when NHEJ is the desired repair event or gRNAs are not highly specific. Additional control and flexibility in the design of experiments can be obtained by combining CRISPR/Cas9 with other genome engineering techniques, for example, by insertion of *FRT* or *attP* sites [69]. Zhang et al. took this a step further by combining CRISPR/Cas9 with RMCE [118] to facilitate the analysis of several different constructs at the same locus.

A given gene product often exerts its function in a variety of different cellular contexts, that is, a transcription factor used in different cells during development, and this is reflected in the control of its expression by a combination of multiple regulatory sequences [119,120]. To probe these non-coding, regulatory regions the same CRISPR/Cas9 approaches that are being applied to coding sequences are useful. These include deletion [121], insertion or swapping of regulatory elements to study functional redundancy, for example, of distributed (primary and shadow) enhancers [122–124]. Single base pair changes are suited to determine the relevance of specific positions in transcription factor binding motifs or to alter splice sites [125,126]. The readout, however, is distinct. Regulatory sequences and their combinatorial logic can be revealed by examination of expression patterns, transcription factor binding, their DNA binding dynamics, and chromatin state or three dimensional nuclear chromatin organization [88,127,128]. Knowledge of regulatory modules also allows generation of more refined tools by driving cell type and developmental stage specific expression of CRISPR/Cas9 components [123,129]. Databases such as the super enhancer archive (http://sea.edbc.org) [130] that integrate results from various types of experiments and CRISPR/Cas9 target sites can aid in the design of experiments. Integration of all these data with fluorescent tagging of gene products at their endogenous loci (see below) will facilitate measurement of local concentrations. Together, this will significantly contribute to a more comprehensive understanding of gene regulation, which will culminate in more precise mathematical models [120].

3.3. Protein Tagging with CRISPR/Cas9 Systems

As with transposon based approaches, the use of CRISPR/Cas9 genome engineering for protein tagging is beginning to provide new tools for developmental biology. In contrast to the methods described above, the CRISPR system facilities the precise introduction of protein tag sequences at endogenous loci [70]. Thus, tagged proteins are expressed under the control of endogenous regulatory sequences, more likely to be expressed at physiological levels and, if relevant information is available, may be engineered to minimise the impact on normal protein function. Introduced protein tags may be visible fluorescent proteins, for example, YFP, GFP, mCherry, and so forth [131] (Figure 3d) or short epitope tags (FLAG, STREPII, Myc, etc.) that, due to their small size, are less likely to perturb protein function. Tagged proteins may be visualised in vivo via fluorescence microscopy or immunohistochemistry and tags may also be used in biochemical studies, for example complex purification combined with mass spectrometry [48,132].

Due to the fact that the small size of epitope tags is likely to reduce perturbation since they can be more readily incorporated into a protein scaffold, versatile protein tagging approaches with split fluorescent proteins have been developed [133]. Such self-complementing split fluorescent proteins enable live cell imaging when they are used as epitope tags and have been shown to work in *Drosophila* cells [134]. The other advantage, aside from the small tag size, is the ease with which it is adaptable to genome engineering since successful tag knock-in via CRISPR-mediated HDR can be achieved with readily synthesised single stranded DNAs of ~200 nucleotides that include the tag and sufficient homology arms to mediate recombination. Alternatively, the tags can be introduced with removable markers that provide convenient reporters for successful recombination and subsequent genetic crossing. These can be readily removed from the genome via one of recombination systems (e.g., Cre-*Lox* or Flp-*FRT*) described above, leaving only the tag of interest and a short *LoxP* or *FRT* site [69,135] (Figure 3e). The development of 'scarless' techniques can further limit the introduction of unwanted additional sequences into the genome [98]. A method combining both CRISPR and *piggyBac* transposase utilises the pHD-ScarlessDsRed vector, available from the *Drosophila* Genomics Resource Center (http://flycrispr.molbio.wisc.edu/scarless). Another possibility is that *LoxP* or *FRT* sites could be replaced by a gRNA to efficiently mediate the marker removal via CRISPR, however, a limitation of this approach is that NHEJ events could lead to the recovery of INDELs.

The use of CRISPR/Cas9 for tagging proteins in situ is becoming widespread and opens up genome biology approaches to virtually any desired protein in the genome: for example, ChIP or RNA immunoprecipitation (RiP) methods for identifying the binding targets of DNA or RNA binding proteins respectively, rely on the availability of highly specific antibodies. Even when good antibodies are available, it is often difficult to make accurate comparisons between samples because different antibodies do not behave in the same way. The possibility of performing such experiments for multiple different proteins with a single antibody is clearly attractive and an obvious route to achieving this is by the introduction of in situ epitope tags. Multiply RNA binding proteins have been engineered with V5 and FLAG tags [136]. Similarly, methods for introducing FLAG (CETCh-seq) [137] or FLAG and fluorescence tags (cmChIP-Seq) [138] have also been developed recently [138]. The latter combines CRISPR with microhomology mediated end joining (MMEJ) to tag transcription factors and, while it has currently been shown to be successful in human cell lines, MMEJ is certainly active in *Drosophila* [139].

Koles et al. [140] developed a strategy to enable tissue specific tagging of endogenous loci in *Drosophila* (T-STEP) by combining CRISPR-Cas9 with a yeast recombinase (Rippase). In brief, a knock-in cassette comprising tandem Rippase-specific Recognition Sequences (RRS) in frame with the targeted protein brackets a strong lethality selection marker (Golic+ [141]) to ensure high efficiency targeting: an in frame GFP is 3′ to the RRS site. Tissue specific expression of the Rippase via the Gal4-*UAS* system allows removal of the T-STEP cassette and puts the GFP in frame with the protein. The authors used this approach for live imaging of endosomal components. In cultured cells, Kunzelmann et al. [142] recently described a simplified protocol for rapid genome engineering that indicates homology arms as short as 60 bp are sufficient for efficient HDR. Their work provides a number of improvements to the system in terms of gRNA design and expression, making routine tagging of proteins in cultured cells feasible. Interesting, these authors more recently showed that the presence of selection markers in vectors for engineering cell lines induced siRNA-mediated changes in gene expression and thus removal of the marker when using such approaches [143] is desirable.

While specific reports of in vivo tagging in *Drosophila* are currently limited, anecdotally it is clear that many laboratories are actively using the technology and it is obvious that CRISPR mediated strategies will be widely applicable for tagging protein isoforms. For example, we have recently used fluorescent or small epitope tags to specifically tag different isoforms of *shaggy*, *ventral nervous system defective* and *roadkill* amongst others, using both direct tagging and Cre-*Lox* directed marker removal, with high efficiency (D. Korona and S. Russell, unpublished).

4. Conclusions

Taken together, the methods we have described here are increasing the utility of the already well-established *Drosophila* model for exploring developmental processes. From the generation of specific mutations through to the ability to track proteins in real time in vivo, genome engineering is opening new possibilities for characterising the deployment of the genome: from the characterisation of gene regulatory networks through post-transcriptional gene control, to the cell biology of proteins and protein complexes. These developments will continue to keep the fly at the forefront as an experimentally tractable system for understanding metazoan biology.

Acknowledgments: Our work in this area is supported by grants for the UK Biotechnology and Biological Sciences Research Council to SR (BB/L002817/1 and BB/N007069/1).

Author Contributions: All authors contributed to the background research and writing.

Conflicts of Interest: The authors declare no conflict of interest.

References

1. Rubin, G.M.; Lewis, E.B. A brief history of *Drosophila*'s contributions to genome research. *Science* **2000**, *287*, 2216–2218. [CrossRef] [PubMed]
2. Arias, A.M. *Drosophila* melanogaster and the development of biology in the 20th century. *Methods Mol. Biol.* **2008**, *420*, 1–25. [CrossRef] [PubMed]
3. Adryan, B.; Russell, S. Genome mapping and genomics in *Drosophila*. In *Genome Mapping and Genomics in Laboratory Animals*; Denny, P., Kole, C., Eds.; Springer: Berilin, Germany, 2012; Volume 4, pp. 31–86, ISBN 978-3-642-31315-8.
4. Russell, S. From sequence to function: The impact of the genome sequence on *Drosophila* biology. *Brief. Funct. Genom.* **2012**, *11*, 333–335. [CrossRef] [PubMed]
5. Rong, Y.S.; Golic, K.G. Gene targeting by homologous recombination in *Drosophila*. *Science* **2000**, *288*, 2013–2018. [CrossRef] [PubMed]
6. Gong, W.J.; Golic, K.G. Ends-out, or replacement, gene targeting in *Drosophila*. *Proc. Natl. Acad. Sci. USA* **2003**, *100*, 2556–2561. [CrossRef] [PubMed]
7. Ryder, E.; Russell, S. Transposable elements as tools for genomics and genetics in *Drosophila*. *Brief. Funct. Genom. Proteom.* **2003**, *2*, 57–71. [CrossRef]
8. Venken, K.J.; Bellen, H.J. Chemical mutagens, transposons, and transgenes to interrogate gene function in *Drosophila* melanogaster. *Methods* **2014**, *68*, 15–28. [CrossRef] [PubMed]
9. Rubin, G.M.; Spradling, A.C. Genetic transformation of *Drosophila* with transposable element vectors. *Science* **1982**, *218*, 348–353. [CrossRef] [PubMed]
10. Brand, A.H.; Perrimon, N. Targeted gene expression as a means of altering cell fates and generating dominant phenotypes. *Development* **1993**, *118*, 401–415. [PubMed]
11. O'Kane, C.J.; Gehring, W.J. Detection in situ of genomic regulatory elements in *Drosophila*. *Proc. Natl. Acad. Sci. USA* **1987**, *84*, 9123–9127. [CrossRef] [PubMed]
12. Wilson, C.; Pearson, R.K.; Bellen, H.J.; O'Kane, C.J.; Grossniklaus, U.; Gehring, W.J. P-element-mediated enhancer detection: An efficient method for isolating and characterizing developmentally regulated genes in *Drosophila*. *Genes. Dev.* **1989**, *3*, 1301–1313. [CrossRef] [PubMed]
13. Rørth, P. A modular misexpression screen in *Drosophila* detecting tissue-specific phenotypes. *Proc. Natl. Acad. Sci. USA* **1996**, *93*, 12418–12422. [CrossRef] [PubMed]
14. Spradling, A.C.; Bellen, H.J.; Hoskins, R.A. *Drosophila* p elements preferentially transpose to replication origins. *Proc. Natl. Acad. Sci. USA* **2011**, *108*, 15948–15953. [CrossRef] [PubMed]
15. Loukeris, T.G.; Arcà, B.; Livadaras, I.; Dialektaki, G.; Savakis, C. Introduction of the transposable element minos into the germ line of *Drosophila* melanogaster. *Proc. Natl. Acad. Sci. USA* **1995**, *92*, 9485–9489. [CrossRef] [PubMed]
16. Lobo, N.; Li, X.; Fraser, M.J. Transposition of the piggybac element in embryos of *Drosophila* melanogaster, aedes aegypti and trichoplusia ni. *Mol. Gen. Genet.* **1999**, *261*, 803–810. [CrossRef] [PubMed]

17. Groth, A.C.; Fish, M.; Nusse, R.; Calos, M.P. Construction of transgenic *Drosophila* by using the site-specific integrase from phage phic31. *Genetics* **2004**, *166*, 1775–1782. [CrossRef] [PubMed]
18. Venken, K.J.; Sarrion-Perdigones, A.; Vandeventer, P.J.; Abel, N.S.; Christiansen, A.E.; Hoffman, K.L. Genome engineering: *Drosophila* melanogaster and beyond. *Wiley Interdiscip. Rev. Dev. Biol.* **2016**, *5*, 233–267. [CrossRef] [PubMed]
19. Golic, K.G.; Lindquist, S. The flp recombinase of yeast catalyzes site-specific recombination in the *Drosophila* genome. *Cell* **1989**, *59*, 499–509. [CrossRef]
20. Nakazawa, N.; Taniguchi, K.; Okumura, T.; Maeda, R.; Matsuno, K. A novel cre/loxp system for mosaic gene expression in the *Drosophila* embryo. *Dev. Dyn.* **2012**, *241*, 965–974. [CrossRef] [PubMed]
21. Del Valle Rodríguez, A.; Didiano, D.; Desplan, C. Power tools for gene expression and clonal analysis in *Drosophila*. *Nat. Methods* **2011**, *9*, 47–55. [CrossRef] [PubMed]
22. Potter, C.J.; Luo, L. Using the q system in *Drosophila* melanogaster. *Nat. Protoc.* **2011**, *6*, 1105–1120. [CrossRef] [PubMed]
23. Boulina, M.; Samarajeewa, H.; Baker, J.D.; Kim, M.D.; Chiba, A. Live imaging of multicolor-labeled cells in *Drosophila*. *Development* **2013**, *140*, 1605–1613. [CrossRef] [PubMed]
24. Griffin, R.; Binari, R.; Perrimon, N. Genetic odyssey to generate marked clones in *Drosophila* mosaics. *Proc. Natl. Acad. Sci. USA* **2014**, *111*, 4756–4763. [CrossRef] [PubMed]
25. Pantazis, P.; Supatto, W. Advances in whole-embryo imaging: A quantitative transition is underway. *Nat. Rev. Mol. Cell Biol.* **2014**, *15*, 327–339. [CrossRef] [PubMed]
26. Owald, D.; Lin, S.; Waddell, S. Light, heat, action: Neural control of fruit fly behaviour. *Phil. Trans. R. Soc. Lond. B Biol. Sci.* **2015**, *370*, 20140211. [CrossRef] [PubMed]
27. Viktorin, G. Using marcm to study *Drosophila* brain development. *Methods Mol. Biol.* **2014**, *1082*, 79–96. [CrossRef] [PubMed]
28. Parks, A.L.; Cook, K.R.; Belvin, M.; Dompe, N.A.; Fawcett, R.; Huppert, K.; Tan, L.R.; Winter, C.G.; Bogart, K.P.; Deal, J.E.; et al. Systematic generation of high-resolution deletion coverage of the *Drosophila* melanogaster genome. *Nat. Genet.* **2004**, *36*, 288–292. [CrossRef] [PubMed]
29. Ryder, E.; Ashburner, M.; Bautista-Llacer, R.; Drummond, J.; Webster, J.; Johnson, G.; Morley, T.; Chan, Y.S.; Blows, F.; Coulson, D.; et al. The drosdel deletion collection: A *Drosophila* genomewide chromosomal deficiency resource. *Genetics* **2007**, *177*, 615–629. [CrossRef] [PubMed]
30. Golic, K.G.; Golic, M.M. Engineering the *Drosophila* genome: Chromosome rearrangements by design. *Genetics* **1996**, *144*, 1693–1711. [PubMed]
31. Bieli, D.; Kanca, O.; Gohl, D.; Denes, A.; Schedl, P.; Affolter, M.; Müller, M. The *Drosophila* melanogaster mutants apblot and apxasta affect an essential apterous wing enhancer. *G3 Genes Genomes Genet.* **2015**, *5*, 1129–1143. [CrossRef] [PubMed]
32. Lee, H.; Cho, D.Y.; Whitworth, C.; Eisman, R.; Phelps, M.; Roote, J.; Kaufman, T.; Cook, K.; Russell, S.; Przytycka, T.; et al. Effects of gene dose, chromatin, and network topology on expression in *Drosophila* melanogaster. *PLoS Genet.* **2016**, *12*, e1006295. [CrossRef] [PubMed]
33. Meadows, L.A.; Chan, Y.S.; Roote, J.; Russell, S. Neighbourhood continuity is not required for correct testis gene expression in *Drosophila*. *PLoS Biol.* **2010**, *8*, e1000552. [CrossRef] [PubMed]
34. Cook, R.K.; Christensen, S.J.; Deal, J.A.; Coburn, R.A.; Deal, M.E.; Gresens, J.M.; Kaufman, T.C.; Cook, K.R. The generation of chromosomal deletions to provide extensive coverage and subdivision of the *Drosophila* melanogaster genome. *Genome Biol.* **2012**, *13*, R21. [CrossRef] [PubMed]
35. Bischof, J.; Maeda, R.K.; Hediger, M.; Karch, F.; Basler, K. An optimized transgenesis system for *Drosophila* using germ-line-specific phic31 integrases. *Proc. Natl. Acad. Sci. USA* **2007**, *104*, 3312–3317. [CrossRef] [PubMed]
36. Markstein, M.; Pitsouli, C.; Villalta, C.; Celniker, S.E.; Perrimon, N. Exploiting position effects and the gypsy retrovirus insulator to engineer precisely expressed transgenes. *Nat. Genet.* **2008**, *40*, 476–483. [CrossRef] [PubMed]
37. Knapp, J.M.; Chung, P.; Simpson, J.H. Generating customized transgene landing sites and multi-transgene arrays in *Drosophila* using phic31 integrase. *Genetics* **2015**, *199*, 919–934. [CrossRef] [PubMed]
38. Bateman, J.R.; Lee, A.M.; Wu, C.T. Site-specific transformation of *Drosophila* via phic31 integrase-mediated cassette exchange. *Genetics* **2006**, *173*, 769–777. [CrossRef] [PubMed]

39. Venken, K.J.; He, Y.; Hoskins, R.A.; Bellen, H.J. P[acman]: A bac transgenic platform for targeted insertion of large DNA fragments in *D. melanogaster*. *Science* **2006**, *314*, 1747–1751. [CrossRef] [PubMed]
40. Ejsmont, R.K.; Sarov, M.; Winkler, S.; Lipinski, K.A.; Tomancak, P. A toolkit for high-throughput, cross-species gene engineering in *Drosophila*. *Nat. Methods* **2009**, *6*, 435–437. [CrossRef] [PubMed]
41. Sarov, M.; Barz, C.; Jambor, H.; Hein, M.Y.; Schmied, C.; Suchold, D.; Stender, B.; Janosch, S.; Vikas, V.K.J.; Krishnan, R.T.; et al. A genome-wide resource for the analysis of protein localisation in *Drosophila*. *Elife* **2016**, *5*, e12068. [CrossRef] [PubMed]
42. Morin, X.; Daneman, R.; Zavortink, M.; Chia, W. A protein trap strategy to detect gfp-tagged proteins expressed from their endogenous loci in *Drosophila*. *Proc. Natl. Acad. Sci. USA* **2001**, *98*, 15050–15055. [CrossRef] [PubMed]
43. Aleksic, J.; Lazic, R.; Müller, I.; Russell, S.R.; Adryan, B. Biases in *Drosophila* melanogaster protein trap screens. *BMC Genom.* **2009**, *10*, 249. [CrossRef] [PubMed]
44. Buszczak, M.; Paterno, S.; Lighthouse, D.; Bachman, J.; Planck, J.; Owen, S.; Skora, A.D.; Nystul, T.G.; Ohlstein, B.; Allen, A.; et al. The carnegie protein trap library: A versatile tool for *Drosophila* developmental studies. *Genetics* **2007**, *175*, 1505–1531. [CrossRef] [PubMed]
45. Quiñones-Coello, A.T.; Petrella, L.N.; Ayers, K.; Melillo, A.; Mazzalupo, S.; Hudson, A.M.; Wang, S.; Castiblanco, C.; Buszczak, M.; Hoskins, R.A.; et al. Exploring strategies for protein trapping in *Drosophila*. *Genetics* **2007**, *175*, 1089–1104. [CrossRef] [PubMed]
46. Lowe, N.; Rees, J.S.; Roote, J.; Ryder, E.; Armean, I.M.; Johnson, G.; Drummond, E.; Spriggs, H.; Drummond, J.; Magbanua, J.P.; et al. Analysis of the expression patterns, subcellular localisations and interaction partners of *Drosophila* proteins using a pigp protein trap library. *Development* **2014**, *141*, 3994–4005. [CrossRef] [PubMed]
47. Lye, C.M.; Naylor, H.W.; Sanson, B. Subcellular localisations of the cpti collection of yfp-tagged proteins in *Drosophila* embryos. *Development* **2014**, *141*, 4006–4017. [CrossRef] [PubMed]
48. Rees, J.S.; Lowe, N.; Armean, I.M.; Roote, J.; Johnson, G.; Drummond, E.; Spriggs, H.; Ryder, E.; Russell, S.; St Johnston, D.; et al. In vivo analysis of proteomes and interactomes using parallel affinity capture (ipac) coupled to mass spectrometry. *Mol. Cell. Proteom.* **2011**, *10*. [CrossRef] [PubMed]
49. Nashchekin, D.; Fernandes, A.R.; St Johnston, D. Patronin/shot cortical foci assemble the noncentrosomal microtubule array that specifies the *Drosophila* anterior-posterior axis. *Dev. Cell* **2016**, *38*, 61–72. [CrossRef] [PubMed]
50. Byri, S.; Misra, T.; Syed, Z.A.; Bätz, T.; Shah, J.; Boril, L.; Glashauser, J.; Aegerter-Wilmsen, T.; Matzat, T.; Moussian, B.; et al. The triple-repeat protein anakonda controls epithelial tricellular junction formation in *Drosophila*. *Dev. Cell* **2015**, *33*, 535–548. [CrossRef] [PubMed]
51. Norman, M.; Vuilleumier, R.; Springhorn, A.; Gawlik, J.; Pyrowolakis, G. Pentagone internalises glypicans to fine-tune multiple signalling pathways. *Elife* **2016**, *5*. [CrossRef] [PubMed]
52. Choo, S.W.; White, R.; Russell, S. Genome-wide analysis of the binding of the hox protein ultrabithorax and the hox cofactor homothorax in *Drosophila*. *PLoS ONE* **2011**, *6*, e14778. [CrossRef] [PubMed]
53. Venken, K.J.; Schulze, K.L.; Haelterman, N.A.; Pan, H.; He, Y.; Evans-Holm, M.; Carlson, J.W.; Levis, R.W.; Spradling, A.C.; Hoskins, R.A.; et al. Mimic: A highly versatile transposon insertion resource for engineering *Drosophila* melanogaster genes. *Nat. Methods* **2011**, *8*, 737–743. [CrossRef] [PubMed]
54. Nagarkar-Jaiswal, S.; DeLuca, S.Z.; Lee, P.T.; Lin, W.W.; Pan, H.; Zuo, Z.; Lv, J.; Spradling, A.C.; Bellen, H.J. A genetic toolkit for tagging intronic mimic containing genes. *Elife* **2015**, *4*. [CrossRef] [PubMed]
55. Nagarkar-Jaiswal, S.; Lee, P.T.; Campbell, M.E.; Chen, K.; Anguiano-Zarate, S.; Gutierrez, M.C.; Busby, T.; Lin, W.W.; He, Y.; Schulze, K.L.; et al. A library of mimics allows tagging of genes and reversible, spatial and temporal knockdown of proteins in *Drosophila*. *Elife* **2015**, *4*. [CrossRef] [PubMed]
56. Caussinus, E.; Kanca, O.; Affolter, M. Fluorescent fusion protein knockout mediated by anti-gfp nanobody. *Nat. Struct. Mol. Biol.* **2011**, *19*, 117–121. [CrossRef] [PubMed]
57. Neumüller, R.A.; Wirtz-Peitz, F.; Lee, S.; Kwon, Y.; Buckner, M.; Hoskins, R.A.; Venken, K.J.; Bellen, H.J.; Mohr, S.E.; Perrimon, N. Stringent analysis of gene function and protein-protein interactions using fluorescently tagged genes. *Genetics* **2012**, *190*, 931–940. [CrossRef] [PubMed]
58. Trost, M.; Blattner, A.C.; Lehner, C.F. Regulated protein depletion by the auxin-inducible degradation system in *Drosophila* melanogaster. *Fly* **2016**, *10*, 35–46. [CrossRef] [PubMed]
59. Hsu, P.D.; Lander, E.S.; Zhang, F. Development and applications of crispr-cas9 for genome engineering. *Cell* **2014**, *157*, 1262–1278. [CrossRef] [PubMed]

60. Beumer, K.J.; Carroll, D. Targeted genome engineering techniques in *Drosophila*. *Methods* **2014**, *68*, 29–37. [CrossRef] [PubMed]
61. Yu, Z.; Chen, H.; Liu, J.; Zhang, H.; Yan, Y.; Zhu, N.; Guo, Y.; Yang, B.; Chang, Y.; Dai, F.; et al. Various applications of talen- and crispr/cas9-mediated homologous recombination to modify the *Drosophila* genome. *Biol. Open* **2014**, *3*, 271–280. [CrossRef] [PubMed]
62. Cebrian-Serrano, A.; Davies, B. Crispr-cas orthologues and variants: Optimizing the repertoire, specificity and delivery of genome engineering tools. *Mamm. Genome* **2017**. [CrossRef] [PubMed]
63. Gasiunas, G.; Barrangou, R.; Horvath, P.; Siksnys, V. Cas9-crrna ribonucleoprotein complex mediates specific DNA cleavage for adaptive immunity in bacteria. *Proc. Natl. Acad. Sci. USA* **2012**, *109*, E2579–E2586. [CrossRef] [PubMed]
64. Jinek, M.; Chylinski, K.; Fonfara, I.; Hauer, M.; Doudna, J.A.; Charpentier, E. A programmable dual-rna-guided DNA endonuclease in adaptive bacterial immunity. *Science* **2012**, *337*, 816–821. [CrossRef] [PubMed]
65. Bétermier, M.; Bertrand, P.; Lopez, B.S. Is non-homologous end-joining really an inherently error-prone process? *PLoS Genet.* **2014**, *10*, e1004086. [CrossRef] [PubMed]
66. Greene, E.C. DNA sequence alignment during homologous recombination. *J. Biol. Chem.* **2016**, *291*, 11572–11580. [CrossRef] [PubMed]
67. Greene, E.C. On the influence of protein-DNA register during homologous recombination. *Cell Cycle* **2016**, *15*, 172–175. [CrossRef] [PubMed]
68. Gratz, S.J.; Rubinstein, C.D.; Harrison, M.M.; Wildonger, J.; O'Connor-Giles, K.M. Crispr-cas9 genome editing in *Drosophila*. *Curr. Protoc. Mol. Biol.* **2015**, *111*. [CrossRef]
69. Gratz, S.J.; Ukken, F.P.; Rubinstein, C.D.; Thiede, G.; Donohue, L.K.; Cummings, A.M.; O'Connor-Giles, K.M. Highly specific and efficient crispr/cas9-catalyzed homology-directed repair in *Drosophila*. *Genetics* **2014**, *196*, 961–971. [CrossRef] [PubMed]
70. Port, F.; Chen, H.M.; Lee, T.; Bullock, S.L. Optimized crispr/cas tools for efficient germline and somatic genome engineering in *Drosophila*. *Proc. Natl. Acad. Sci. USA* **2014**, *111*, E2967–E2976. [CrossRef] [PubMed]
71. Schaefer, K.A.; Wu, W.H.; Colgan, D.F.; Tsang, S.H.; Bassuk, A.G.; Mahajan, V.B. Unexpected mutations after crispr-cas9 editing in vivo. *Nat. Methods* **2017**, *14*, 547–548. [CrossRef] [PubMed]
72. Lee, C.M.; Cradick, T.J.; Fine, E.J.; Bao, G. Nuclease target site selection for maximizing on-target activity and minimizing off-target effects in genome editing. *Mol. Ther.* **2016**, *24*, 475–487. [CrossRef] [PubMed]
73. Stella, S.; Montoya, G. The genome editing revolution: A crispr-cas tale off-target story. *Bioessays* **2016**, *38* (Suppl. 1), S4–S13. [CrossRef] [PubMed]
74. Cong, L.; Ran, F.A.; Cox, D.; Lin, S.; Barretto, R.; Habib, N.; Hsu, P.D.; Wu, X.; Jiang, W.; Marraffini, L.A.; et al. Multiplex genome engineering using crispr/cas systems. *Science* **2013**, *339*, 819–823. [CrossRef] [PubMed]
75. Mali, P.; Aach, J.; Stranges, P.B.; Esvelt, K.M.; Moosburner, M.; Kosuri, S.; Yang, L.; Church, G.M. Cas9 transcriptional activators for target specificity screening and paired nickases for cooperative genome engineering. *Nat. Biotechnol.* **2013**, *31*, 833–838. [CrossRef] [PubMed]
76. Ran, F.A.; Hsu, P.D.; Lin, C.Y.; Gootenberg, J.S.; Konermann, S.; Trevino, A.E.; Scott, D.A.; Inoue, A.; Matoba, S.; Zhang, Y.; et al. Double nicking by rna-guided crispr cas9 for enhanced genome editing specificity. *Cell* **2013**, *154*, 1380–1389. [CrossRef] [PubMed]
77. Overballe-Petersen, S.; Harms, K.; Orlando, L.A.; Mayar, J.V.; Rasmussen, S.; Dahl, T.W.; Rosing, M.T.; Poole, A.M.; Sicheritz-Ponten, T.; Brunak, S.; et al. Bacterial natural transformation by highly fragmented and damaged DNA. *Proc. Natl. Acad. Sci. USA* **2013**, *110*, 19860–19865. [CrossRef] [PubMed]
78. Cho, S.W.; Kim, S.; Kim, Y.; Kweon, J.; Kim, H.S.; Bae, S.; Kim, J.S. Analysis of off-target effects of crispr/cas-derived rna-guided endonucleases and nickases. *Genome Res.* **2014**, *24*, 132–141. [CrossRef] [PubMed]
79. Miyaoka, Y.; Berman, J.R.; Cooper, S.B.; Mayerl, S.J.; Chan, A.H.; Zhang, B.; Karlin-Neumann, G.A.; Conklin, B.R. Systematic quantification of hdr and nhej reveals effects of locus, nuclease, and cell type on genome-editing. *Sci. Rep.* **2016**, *6*, 23549. [CrossRef] [PubMed]
80. Doudna, J.A.; Charpentier, E. Genome editing. The new frontier of genome engineering with crispr-cas9. *Science* **2014**, *346*, 1258096. [CrossRef] [PubMed]
81. Guilinger, J.P.; Thompson, D.B.; Liu, D.R. Fusion of catalytically inactive cas9 to foki nuclease improves the specificity of genome modification. *Nat. Biotechnol.* **2014**, *32*, 577–582. [CrossRef] [PubMed]

82. Kleinstiver, B.P.; Pattanayak, V.; Prew, M.S.; Tsai, S.Q.; Nguyen, N.T.; Zheng, Z.; Joung, J.K. High-fidelity crispr-cas9 nucleases with no detectable genome-wide off-target effects. *Nature* **2016**, *529*, 490–495. [CrossRef] [PubMed]
83. Slaymaker, I.M.; Gao, L.; Zetsche, B.; Scott, D.A.; Yan, W.X.; Zhang, F. Rationally engineered cas9 nucleases with improved specificity. *Science* **2016**, *351*, 84–88. [CrossRef] [PubMed]
84. Kleinstiver, B.P.; Prew, M.S.; Tsai, S.Q.; Topkar, V.V.; Nguyen, N.T.; Zheng, Z.; Gonzales, A.P.; Li, Z.; Peterson, R.T.; Yeh, J.R.; et al. Engineered crispr-cas9 nucleases with altered pam specificities. *Nature* **2015**, *523*, 481–485. [CrossRef] [PubMed]
85. Zetsche, B.; Gootenberg, J.S.; Abudayyeh, O.O.; Slaymaker, I.M.; Makarova, K.S.; Essletzbichler, P.; Volz, S.E.; Joung, J.; van der Oost, J.; Regev, A.; et al. Cpf1 is a single rna-guided endonuclease of a class 2 crispr-cas system. *Cell* **2015**, *163*, 759–771. [CrossRef] [PubMed]
86. Shmakov, S.; Abudayyeh, O.O.; Makarova, K.S.; Wolf, Y.I.; Gootenberg, J.S.; Semenova, E.; Minakhin, L.; Joung, J.; Konermann, S.; Severinov, K.; et al. Discovery and functional characterization of diverse class 2 crispr-cas systems. *Mol. Cell.* **2015**, *60*, 385–397. [CrossRef] [PubMed]
87. Gratz, S.J.; Cummings, A.M.; Nguyen, J.N.; Hamm, D.C.; Donohue, L.K.; Harrison, M.M.; Wildonger, J.; O'Connor-Giles, K.M. Genome engineering of *Drosophila* with the crispr rna-guided cas9 nuclease. *Genetics* **2013**, *194*, 1029–1035. [CrossRef] [PubMed]
88. Hilton, I.B.; Gersbach, C.A. Enabling functional genomics with genome engineering. *Genome Res.* **2015**, *25*, 1442–1455. [CrossRef] [PubMed]
89. Essletzbichler, P.; Konopka, T.; Santoro, F.; Chen, D.; Gapp, B.V.; Kralovics, R.; Brummelkamp, T.R.; Nijman, S.M.; Burckstummer, T. Megabase-scale deletion using crispr/cas9 to generate a fully haploid human cell line. *Genome Res.* **2014**, *24*, 2059–2065. [CrossRef] [PubMed]
90. Auer, T.O.; Duroure, K.; De Cian, A.; Concordet, J.P.; Del Bene, F. Highly efficient crispr/cas9-mediated knock-in in zebrafish by homology-independent DNA repair. *Genome Res.* **2014**, *24*, 142–153. [CrossRef] [PubMed]
91. He, X.; Tan, C.; Wang, F.; Wang, Y.; Zhou, R.; Cui, D.; You, W.; Zhao, H.; Ren, J.; Feng, B. Knock-in of large reporter genes in human cells via crispr/cas9-induced homology-dependent and independent DNA repair. *Nucleic Acids Res.* **2016**, *44*, e85. [CrossRef] [PubMed]
92. Böttcher, R.; Hollmann, M.; Merk, K.; Nitschko, V.; Obermaier, C.; Philippou-Massier, J.; Wieland, I.; Gaul, U.; Förstemann, K. Efficient chromosomal gene modification with crispr/cas9 and pcr-based homologous recombination donors in cultured *Drosophila* cells. *Nucleic Acids Res.* **2014**, *42*, e89. [CrossRef] [PubMed]
93. Housden, B.E.; Lin, S.; Perrimon, N. Cas9-based genome editing in *Drosophila*. *Methods Enzymol.* **2014**, *546*, 415–439. [CrossRef] [PubMed]
94. Salsman, J.; Dellaire, G. Precision genome editing in the crispr era. *Biochem. Cell Biol.* **2017**, *95*, 187–201. [CrossRef] [PubMed]
95. Yang, L.; Guell, M.; Byrne, S.; Yang, J.L.; De Los Angeles, A.; Mali, P.; Aach, J.; Kim-Kiselak, C.; Briggs, A.W.; Rios, X.; et al. Optimization of scarless human stem cell genome editing. *Nucleic Acids Res.* **2013**, *41*, 9049–9061. [CrossRef] [PubMed]
96. Komor, A.C.; Badran, A.H.; Liu, D.R. Crispr-based technologies for the manipulation of eukaryotic genomes. *Cell* **2017**, *169*, 559. [CrossRef] [PubMed]
97. Gaudelli, N.M.; Komor, A.C.; Rees, H.A.; Packer, M.S.; Badran, A.H.; Bryson, D.I. Programmable base editing of A•T to G•C in genomic DNA without DNA cleavage. *Nature* **2017**. [CrossRef] [PubMed]
98. Lamb, A.M.; Walker, E.A.; Wittkopp, P.J. Tools and strategies for scarless allele replacement in *Drosophila* using crispr/cas9. *Fly* **2017**, *11*, 53–64. [CrossRef] [PubMed]
99. Byrne, S.M.; Ortiz, L.; Mali, P.; Aach, J.; Church, G.M. Multi-kilobase homozygous targeted gene replacement in human induced pluripotent stem cells. *Nucleic Acids Res.* **2015**, *43*, e21. [CrossRef] [PubMed]
100. Zheng, Q.; Cai, X.; Tan, M.H.; Schaffert, S.; Arnold, C.P.; Gong, X.; Chen, C.Z.; Huang, S. Precise gene deletion and replacement using the crispr/cas9 system in human cells. *Biotechniques* **2014**, *57*, 115–124. [CrossRef] [PubMed]
101. Zhao, Y.; Zhang, C.; Liu, W.; Gao, W.; Liu, C.; Song, G.; Li, W.X.; Mao, L.; Chen, B.; Xu, Y.; et al. An alternative strategy for targeted gene replacement in plants using a dual-sgrna/cas9 design. *Sci. Rep.* **2016**, *6*, 23890. [CrossRef] [PubMed]

102. Adikusuma, F.; Pederick, D.; McAninch, D.; Hughes, J.; Thomas, P. Functional equivalence of the sox2 and sox3 transcription factors in the developing mouse brain and testes. *Genetics* **2017**, *206*, 1495–1503. [CrossRef] [PubMed]
103. Evans, T.A. Crispr-based gene replacement reveals evolutionarily conserved axon guidance functions of *Drosophila* robo3 and tribolium robo2/3. *Evodevo* **2017**, *8*, 10. [CrossRef] [PubMed]
104. Spitzweck, B.; Brankatschk, M.; Dickson, B.J. Distinct protein domains and expression patterns confer divergent axon guidance functions for *Drosophila* robo receptors. *Cell* **2010**, *140*, 409–420. [CrossRef] [PubMed]
105. Ewen-Campen, B.; Mohr, S.E.; Hu, Y.; Perrimon, N. Accessing the phenotype gap: Enabling systematic investigation of paralog functional complexity with crispr. *Dev. Cell* **2017**, *43*, 6–9. [CrossRef] [PubMed]
106. Overton, P.M.; Meadows, L.A.; Urban, J.; Russell, S. Evidence for differential and redundant function of the sox genes dichaete and soxn during cns development in *Drosophila*. *Development* **2002**, *129*, 4219–4228. [PubMed]
107. Sánchez-Soriano, N.; Russell, S. Regulatory mutations of the *Drosophila* sox gene dichaete reveal new functions in embryonic brain and hindgut development. *Dev. Biol.* **2000**, *220*, 307–321. [CrossRef] [PubMed]
108. Ren, X.; Yang, Z.; Xu, J.; Sun, J.; Mao, D.; Hu, Y.; Yang, S.J.; Qiao, H.H.; Wang, X.; Hu, Q.; et al. Enhanced specificity and efficiency of the crispr/cas9 system with optimized sgrna parameters in *Drosophila*. *Cell Rep.* **2014**, *9*, 1151–1162. [CrossRef] [PubMed]
109. Port, F.; Bullock, S.L. Augmenting crispr applications in *Drosophila* with trna-flanked sgrnas. *Nat. Methods* **2016**, *13*, 852–854. [CrossRef] [PubMed]
110. Yang, H.; Wang, H.; Shivalila, C.S.; Cheng, A.W.; Shi, L.; Jaenisch, R. One-step generation of mice carrying reporter and conditional alleles by crispr/cas-mediated genome engineering. *Cell* **2013**, *154*, 1370–1379. [CrossRef] [PubMed]
111. Arribere, J.A.; Bell, R.T.; Fu, B.X.; Artiles, K.L.; Hartman, P.S.; Fire, A.Z. Efficient marker-free recovery of custom genetic modifications with crispr/cas9 in caenorhabditis elegans. *Genetics* **2014**, *198*, 837–846. [CrossRef] [PubMed]
112. Ward, J.D. Rapid and precise engineering of the caenorhabditis elegans genome with lethal mutation co-conversion and inactivation of nhej repair. *Genetics* **2015**, *199*, 363–377. [CrossRef] [PubMed]
113. Kane, N.S.; Vora, M.; Varre, K.J.; Padgett, R.W. Efficient screening of crispr/cas9-induced events in *Drosophila* using a co-crispr strategy. *G3 Genes Genomes Genet.* **2017**, *7*, 87–93. [CrossRef] [PubMed]
114. Ge, D.T.; Tipping, C.; Brodsky, M.H.; Zamore, P.D. Rapid screening for crispr-directed editing of the *Drosophila* genome using white coconversion. *G3 Genes Genomes Genet.* **2016**, *6*, 3197–3206. [CrossRef]
115. Kondo, S.; Ueda, R. Highly improved gene targeting by germline-specific cas9 expression in *Drosophila*. *Genetics* **2013**, *195*, 715–721. [CrossRef] [PubMed]
116. Ren, X.; Sun, J.; Housden, B.E.; Hu, Y.; Roesel, C.; Lin, S.; Liu, L.P.; Yang, Z.; Mao, D.; Sun, L.; et al. Optimized gene editing technology for *Drosophila* melanogaster using germ line-specific cas9. *Proc. Natl. Acad. Sci. USA* **2013**, *110*, 19012–19017. [CrossRef] [PubMed]
117. Xue, Z.; Wu, M.; Wen, K.; Ren, M.; Long, L.; Zhang, X.; Gao, G. Crispr/cas9 mediates efficient conditional mutagenesis in *Drosophila*. *G3 Genes Genomes Genet.* **2014**, *4*, 2167–2173. [CrossRef] [PubMed]
118. Zhang, X.; Koolhaas, W.H.; Schnorrer, F. A versatile two-step crispr- and rmce-based strategy for efficient genome engineering in *Drosophila*. *G3 Genes Genomes Genet.* **2014**, *4*, 2409–2418. [CrossRef] [PubMed]
119. Stern, D.L.; Frankel, N. The structure and evolution of cis-regulatory regions: The shavenbaby story. *Philos. Trans. R. Soc. Lond. Ser. B Biol. Sci.* **2013**, *368*, 20130028. [CrossRef] [PubMed]
120. Wilczynski, B.; Furlong, E.E. Challenges for modeling global gene regulatory networks during development: Insights from *Drosophila*. *Dev. Biol.* **2010**, *340*, 161–169. [CrossRef] [PubMed]
121. Mendoza-Garcia, P.; Hugosson, F.; Fallah, M.; Higgins, M.L.; Iwasaki, Y.; Pfeifer, K.; Wolfstetter, G.; Varshney, G.; Popichenko, D.; Gergen, J.P.; et al. The zic family homologue odd-paired regulates alk expression in *Drosophila*. *PLoS Genet.* **2017**, *13*, e1006617. [CrossRef] [PubMed]
122. Zandvakili, A.; Gebelein, B. Mechanisms of specificity for hox factor activity. *J. Dev. Biol.* **2016**, *4*, 16. [CrossRef] [PubMed]
123. Rogers, W.A.; Goyal, Y.; Yamaya, K.; Shvartsman, S.Y.; Levine, M.S. Uncoupling neurogenic gene networks in the *Drosophila* embryo. *Genes Dev.* **2017**, *31*, 634–638. [CrossRef] [PubMed]

124. Lagha, M.; Bothma, J.P.; Levine, M. Mechanisms of transcriptional precision in animal development. *Trends Genet.* **2012**, *28*, 409–416. [CrossRef] [PubMed]
125. Consortium, E.P. An integrated encyclopedia of DNA elements in the human genome. *Nature* **2012**, *489*, 57–74. [CrossRef]
126. Lee, Y.; Rio, D.C. Mechanisms and regulation of alternative pre-mrna splicing. *Annu. Rev. Biochem.* **2015**, *84*, 291–323. [CrossRef] [PubMed]
127. Beagan, J.A.; Phillips-Cremins, J.E. Crispr/cas9 genome editing throws descriptive 3-D genome folding studies for a loop. *Wiley Interdiscip. Rev. Syst. Biol. Med.* **2016**, *8*, 286–299. [CrossRef] [PubMed]
128. Mir, M.; Reimer, A.; Haines, J.E.; Li, X.Y.; Stadler, M.; Garcia, H.; Eisen, M.B.; Darzacq, X. Dense bicoid hubs accentuate binding along the morphogen gradient. *Genes Dev.* **2017**, *31*, 1784–1794. [CrossRef] [PubMed]
129. Crews, S. Creating cell type-specific mutants by enhancer mutagenesis. *Genes Dev.* **2017**, *31*, 629–631. [CrossRef] [PubMed]
130. Wei, Y.; Zhang, S.; Shang, S.; Zhang, B.; Li, S.; Wang, X.; Wang, F.; Su, J.; Wu, Q.; Liu, H.; et al. Sea: A super-enhancer archive. *Nucleic Acids Res.* **2016**, *44*, D172–D179. [CrossRef] [PubMed]
131. Thorn, K. Genetically encoded fluorescent tags. *Mol. Biol. Cell* **2017**, *28*, 848–857. [CrossRef] [PubMed]
132. Rees, J.S.; Lilley, K.S.; Jackson, A.P. Silac-ipac: A quantitative method for distinguishing genuine from non-specific components of protein complexes by parallel affinity capture. *J. Proteom.* **2015**, *115*, 143–156. [CrossRef] [PubMed]
133. Cabantous, S.; Terwilliger, T.C.; Waldo, G.S. Protein tagging and detection with engineered self-assembling fragments of green fluorescent protein. *Nat. Biotechnol.* **2005**, *23*, 102–107. [CrossRef] [PubMed]
134. Kamiyama, D.; Sekine, S.; Barsi-Rhyne, B.; Hu, J.; Chen, B.; Gilbert, L.A.; Ishikawa, H.; Leonetti, M.D.; Marshall, W.F.; Weissman, J.S.; et al. Versatile protein tagging in cells with split fluorescent protein. *Nat. Commun.* **2016**, *7*, 11046. [CrossRef] [PubMed]
135. Li, Q.; Barish, S.; Okuwa, S.; Volkan, P.C. Examination of endogenous rotund expression and function in developing *Drosophila* olfactory system using crispr-cas9-mediated protein tagging. *G3 Genes Genomes Genet.* **2015**, *5*, 2809–2816. [CrossRef] [PubMed]
136. Van Nostrand, E.L.; Gelboin-Burkhart, C.; Wang, R.; Pratt, G.A.; Blue, S.M.; Yeo, G.W. Crispr/cas9-mediated integration enables tag-eclip of endogenously tagged rna binding proteins. *Methods* **2017**, *118–119*, 50–59. [CrossRef] [PubMed]
137. Savic, D.; Partridge, E.C.; Newberry, K.M.; Smith, S.B.; Meadows, S.K.; Roberts, B.S.; Mackiewicz, M.; Mendenhall, E.M.; Myers, R.M. Cetch-seq: Crispr epitope tagging chip-seq of DNA-binding proteins. *Genome Res.* **2015**, *25*, 1581–1589. [CrossRef] [PubMed]
138. Xiong, X.; Zhang, Y.; Yan, J.; Jain, S.; Chee, S.; Ren, B.; Zhao, H. A scalable epitope tagging approach for high throughput chip-seq analysis. *ACS Synth. Biol.* **2017**, *6*, 1034–1042. [CrossRef] [PubMed]
139. Yu, A.M.; McVey, M. Synthesis-dependent microhomology-mediated end joining accounts for multiple types of repair junctions. *Nucleic Acids Res.* **2010**, *38*, 5706–5717. [CrossRef] [PubMed]
140. Koles, K.; Yeh, A.R.; Rodal, A.A. Tissue-specific tagging of endogenous loci in *Drosophila* melanogaster. *Biol. Open* **2015**, *5*, 83–89. [CrossRef] [PubMed]
141. Chen, H.M.; Huang, Y.; Pfeiffer, B.D.; Yao, X.; Lee, T. An enhanced gene targeting toolkit for *Drosophila*: Golic+. *Genetics* **2015**, *199*, 683–694. [CrossRef] [PubMed]
142. Kunzelmann, S.; Böttcher, R.; Schmidts, I.; Förstemann, K. A comprehensive toolbox for genome editing in cultured *Drosophila* melanogaster cells. *G3 Genes Genomes Genet.* **2016**, *6*, 1777–1785. [CrossRef] [PubMed]
143. Kunzelmann, S.; Förstemann, K. Reversible perturbations of gene regulation after genome editing in *Drosophila* cells. *PLoS ONE* **2017**, *12*, e0180135. [CrossRef] [PubMed]

3

The Roles of the Wnt-Antagonists Axin and Lrp4 during Embryogenesis of the Red Flour Beetle *Tribolium castaneum*

Romy Prühs [1], Anke Beermann [2] and Reinhard Schröder [1,*]

[1] Department of Genetics, Biological Sciences, University of Rostock, Albert-Einstein-Str. 3, D 18059 Rostock, Germany; romy.pruehs@gmx.de

[2] Animal Genetics, University of Tübingen, Auf der Morgenstelle 15, D 72076 Tübingen, Germany; anke.beermann@t-online.de

* Correspondence: reinhard.schroeder@uni-rostock.de

Abstract: In both vertebrates and invertebrates, the Wnt-signaling pathway is essential for numerous processes in embryogenesis and during adult life. Wnt activity is fine-tuned at various levels by the interplay of a number of Wnt-agonists (Wnt ligands, Frizzled-receptors, Lrp5/6 coreceptors) and Wnt-antagonists (among them Axin, Secreted frizzled and Lrp4) to define anterior–posterior polarity of the early embryo and specify cell fate in organogenesis. So far, the functional analysis of Wnt-pathway components in insects has concentrated on the roles of Wnt-agonists and on the Wnt-antagonist Axin. We depict here additional features of the Wnt-antagonist Axin in the flour beetle *Tribolium castaneum*. We show that *Tc-axin* is dynamically expressed throughout embryogenesis and confirm its essential role in head development. In addition, we describe an as yet undetected, more extreme *Tc-axin* RNAi-phenotype, the ectopic formation of posterior abdominal segments in reverse polarity and a second hindgut at the anterior. For the first time, we describe here that an *lrp4* ortholog is involved in axis formation in an insect. The *Tribolium* Lrp4 ortholog is ubiquitously expressed throughout embryogenesis. Its downregulation via maternal RNAi results in the reduction of head structures but not in axis polarity reversal. Furthermore, segmentation is impaired and larvae develop with a severe gap-phenotype. We conclude that, as in vertebrates, *Tc-lrp4* functions as a Wnt-inhibitor in *Tribolium* during various stages of embryogenesis. We discuss the role of both components as negative modulators of Wnt signaling in respect to axis formation and segmentation in *Tribolium*.

Keywords: *Tribolium*; Wnt signaling; Wnt-antagonist; Axin; LRP4; AP-polarity; double abdomen

---

1. Introduction

During embryogenesis and adult life, the precise regulation of the Wnt-signaling pathway is essential for a wide variety of processes like cell-fate specification, axis formation, tissue and organ development. In the canonical or cell-fate pathway, signaling is triggered when a Wnt-ligand binds to a Frizzled (Fz) receptor and a coreceptor, the low-density lipoprotein receptor-related protein (Lrp) 5, 6 or Arrow. This ternary complex represents the "ON"-status and leads to the disintegration of the β-catenin destruction complex by recruiting one of its components, Axin, to the cell membrane. As a consequence, β-catenin becomes stabilised in the cytoplasm and subsequently undergoes nuclear localization to regulate Wnt-target genes. In the absence of a Wnt-ligand ("OFF"-status), β-catenin has a very short half-life in the cytoplasm as it becomes marked for destruction by the proteasome by components of the destruction complex APC, Axin, Gsk3β and CK1α [1,2]. The active Wnt pathway regulates the increase in cell proliferation and motility [3], and unregulated elevated Wnt activity can eventually result in various diseases, including the formation of tumours. Thus, the negative modulation of Wnt activity is equally important for tissue differentiation and homeostasis. In vertebrates, an array of Wnt-inhibitor molecules is known that antagonistically interfere with this signaling pathway at various levels, like Wnt-secretion, intracellular Wnt-transport, Wnt-modification and Wnt-reception. At the receptor level, Wnt-induced Fz-Lrp 5/6 dimerization is antagonized by proteins such as WISE, SOST or DKK [4]. Another way to downregulate Wnt activity is sequestering Wnts by Wnt-binding proteins that are unable to transmit the signal to the nucleus and, in this way, limiting the number of Wnt-ligands available for the Fz-receptor. To this class of "dominant-negative" acting receptors belong the secreted frizzled-related proteins (sFRPs) and Lrp4 [5]. Decreasing the function of Wnt-inhibitors either experimentally or by mutation results in the upregulation of Wnt activity, a situation seen in various forms of cancer [4]. One important step in early embryogenesis it the establishment of anterior–posterior polarity of the body axis and of the nervous system [2,6,7]. In both normal development and during regeneration, high levels of Wnt activity are required for the specification of posterior fate, while the absence of Wnt signaling enables the development of anterior structures in all extant animals [8–11]. At the posterior, a Wnt-signaling centre is established through expressed Wnt-ligands that activates targets like Brachyury (Byn) expression, which itself activates Wnt signaling. In this way, a self-perpetuating Wnt–Byn circuit is established that maintains posterior fate [2]. Consequently, activating Wnt signaling ectopically at the anterior pole results in the development of ectopic tail structures at the expense of anterior ones. To allow for the specification of anterior fate, Wnt activity has to be efficiently downregulated in regions where head—or more generally—anterior development should take place. This can be ensured by anteriorly expressed Wnt-antagonists. Most of such antagonists known from vertebrates are, however, absent from the genome of insects. In the short germband insect *Tribolium*, Wnt-ligands are expressed at the posterior of the early blastoderm embryo [12] and required for axial elongation [13–15]. Breaking the symmetry of Wnt activity along the main body axis in *Tribolium* is achieved by the anterior asymmetric expression of Axin, a crucial component of the destruction complex. Here *Tc-axin* is required for keeping the anterior egg free of Wnt activity and in this way, allows for head development. Consequently, downregulation of *Tc-axin* resulted in severely affected embryos consisting of posterior abdominal segments only [16]. The *Drosophila* Lrp4 ortholog is expressed in the adult brain and functions independently of a Wnt-ligand in synaptogenesis [17]. Lrp4 in vertebrates can interact with Wnts [18] and other ligands and has been characterized as a Wnt-modulator in various processes such as the development of skin-appendages [19], as well as bone [20] and tooth formation [21].

Here, we extend a previous study on the Wnt-antagonist *axin* in *Tribolium* [16] and show that, also later during embryogenesis, distinctly expressed *Tc-axin* defines regions where Wnt activity is modulated. Furthermore, we show here that the downregulation of *Tc-axin* results in ectopic formation of a second hindgut and the development of an ectopic tail with reversed polarity at the anterior pole, resembling the *bicoid* phenotype in *Drosophila* [22]. We also have characterised another antagonist of Wnt activity, the *Tribolium* Lrp4-ortholog. In contrast to *Tc-axin*, *Tc-lrp4* is ubiquitously expressed in the egg and throughout later embryogenesis. RNAi knockdown of *Tc-lrp4* resulted in anterior truncations, as observed with *Tc-axin* RNAi. Interestingly, strong segmentation defects seen are likely caused by the upregulation of segment polarity genes.

Together, we demonstrate here, that the concerted interaction of Wnt-agonists, and Wnt-antagonists fine-tune early pattern formation in a short germband insect.

2. Materials and Methods

Animal husbandry. Beetles were reared on a regular changing diet of whole wheat and instant flour in a 30 °C incubator.

Molecular biology. DNA fragments corresponding to the respective genes were amplified from cDNA by PCR, and cloned according to standard procedures. Each clone was verified by sequencing (LGC-Genomics). Injections of dsRNA generated from non-overlapping constructs were performed to avoid off-target effects. All tested dsRNA showed the same range of phenotypes for the respective gene. Injections were usually performed into the wild-type strain (San Bernadino) as described [23].

dsRNA generation used for parental RNAi. The position of the fragments relative to the exon–intron structure of *Tc-axin* is indicated in Figure S1A. dsRNA for *Tc-axin* (TC006314): *axin*-iB1 (position coding sequence 184–695) and *axin*-iB2 (position cds 990–1213) were obtained from a company (Eupheria) [16]. One *Tc-axin* cDNA-clone (position cds 523–1344) partially overlapping with both *Tc-axin*-iB fragments served as a template for the generation of dsRNA. We injected various concentrations of *Tc-axin*-dsRNA (10–1000 ng/µL) and got the full range of phenotypes between 20 ng/µL and 50 ng/µL. Results of *axin* RNAi were also reproduced in another beetle strain EFA-nGFP [24] to exclude strain-specific background effects.

The *Tc-lrp4* ortholog (TC007146) is present on the minus-strand on ChLG4. The position of cloned cDNA fragments relative to the gene structure is indicated in Figure S2A. The following primers were used (position number according to the coding sequence): *Tc-lrp4* #1: fw: GGACGCGCTACCCCTTTGA (473–491); rev: CGCCGTCGCATTTGAACTTTA (1223–1243), fragment length 771bp; *Tc-lrp4* #2: fw: ACCGCGCTGATAAAACCAAT (4448–4467); rev: TCGCGTCAATCCAATAGAGG (5205–5224)), fragment length 777bp; *Tc-lrp4* #3 (iBeetle fragment iB_01131): fw: GTGTAGAAGGCGAACCCAAA (5564–5583); rev: CAGTTGGAGCTGCAAGGAGT 6019–6038), fragment length 475bp. Concentrations of *Tc-lrp4*-dsRNA varied from 340 ng/µL to 3 µg/µL. Details are listed in Supplementary Materials (Tables S1–S3). The same low amount of wild-type cuticles and all phenotypic classes have been observed with all concentrations of dsRNA tested. Eggs and embryos used for marker-gene analyses came from females injected with dsRNA with a concentration of 800 ng/µL.

Phenotypic analysis. The larval cuticle phenotype was inspected making use of their autofluorescence under UV light to visualize all cuticularised structures including the gut. In situ hybridizations and antibody staining on wild-type and RNAi embryos were performed as described previously [25]. Each probe/antibody was carefully titrated to ensure an optimal signal-to-background ratio.

Analysis of protein-structure was performed using SMART (http://smart.embl-heidelberg.de) or motif-search (www.genome.jp/tools/motif/).

3. Results

3.1. The Wnt-Inhibitor Tc-axin Is Asymmetrically Expressed in the Egg and Dynamically with Distinct Patterns during Further Embryogenesis

As described previously [16], the initial expression of the Wnt-inhibitor *Tc-axin* (*Tc-ax*) was observed asymmetrically in a small anterior crescent of a freshly laid egg (Figure 1A), and more broadly covers the anterior third of the egg throughout the early cleavage stages. At the beginning of gastrulation, *Tc-axin* was lost from the anterior region and expression was seen in the posterior growth zone of the young germ rudiment (Figure 1B). During axis elongation, *Tc-axin* remained strongly expressed in a longitudinal domain within the growth zone and weakly along the midline in the non-segmented region anterior to the growth zone, the presegmented region (Figure 1C). Within the thorax and the head, weak additional *Tc-axin* expression domains in the limb fields were visible. During segmentation of the germband, *Tc-axin* expression sites became stronger and more pronounced in the head lobes and in the distal parts of the outgrowing appendages of the gnathum and the thorax, while in the abdominal segments new dot-shaped expression domains appeared. At this stage, *Tc-axin* expression at the posterior is confined to a subterminal stripe (Figure 1D,E). After germband retraction, *Tc-axin* was seen ventrally in the thoracic appendages as well as in the anlagen of the hindgut (Figure 1F,G). The expression domains within the distal head lobes and the gnathal appendages remained while in each segment one inner and one outer pair of *Tc-axin* positive groups of cells were visible (Figure 1G).

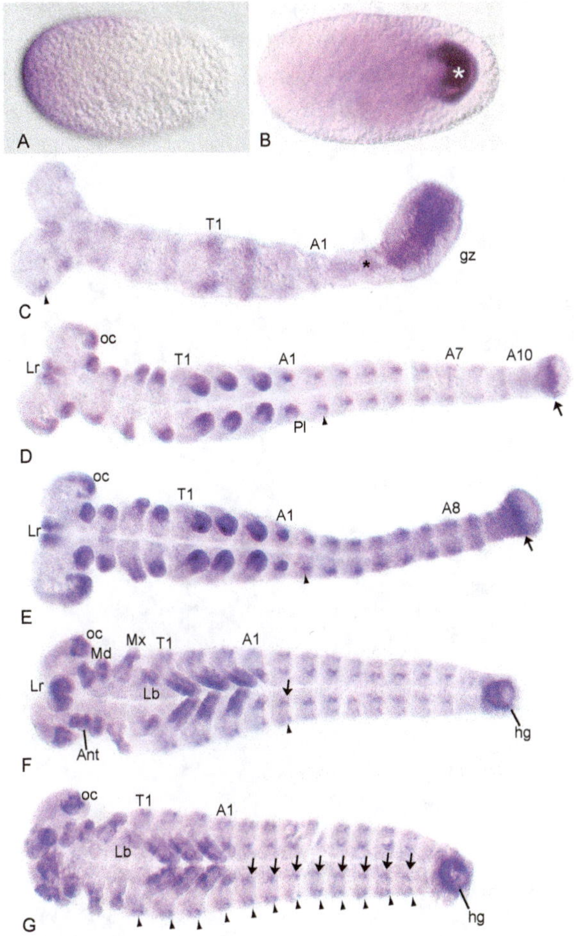

Figure 1. *Tc-axin* shows a distinct, dynamic expression pattern throughout embryonic development. *Tc-axin*-mRNA localisation by in-situ hybridisation in wild-type embryos. (**A**) Early cleavage stage embryo with asymmetric anterior expression of *Tc-axin* mRNA (anterior points to the right). (**B**) Gastrulating embryo (ventral view) with strong upregulation of *Tc-axin* expression in the posterior invaginating cells (star). (**C**) Embryo undergoing germband extension. A longitudinal *Tc-axin* domain within the growth zone shows slight upregulation at its border. Weak expression is seen along the midline within the presegmented region (star). In addition, segmental stripes within thoracic segment 1 and 2 and spotted domains (arrowhead) within the head are visible. (**D**,**E**) Fully segmented embryos with expression of *Tc-axin* in the distal parts of the head-, gnathal- and thoracic appendages and in the pleuropodia, the appendages of abdominal segment 1. Within each abdominal segment a pair of *axin*-expressing cells become apparent (arrowhead). Arrow marks cells likely belonging to the hindgut anlage. (**F**,**G**) In the retracting germband, *Tc-axin* is expressed ventrally in the thoracic legs while in the other appendages no regionalized expression is seen. Expression strength diminished in the maxillae (Mx) and the labral appendages (Lr). During formation of the hindgut (hg), *Tc-axin* was seen around the future proctodeal opening. Within each abdominal segment as well as in the thorax, a further *axin* expression domain marks a group of lateral cells (arrowhead) in addition to *axin*-positive cells near the midline (arrow). In the fully segmented embryo, the ocular region of the head (oc) permanently expresses *Tc-axin* (**D**–**G**). A: abdominal segment; gz: growth-zone; hg: hindgut; Lb: labium; Lr: labrum Md: mandible; Mx: maxilla; oc: ocular segment; Pl: pleuropodium.

3.2. Reduction of Tc-axin Function Results in Anterior Truncated, Mild Bicaudal and Symmetric Double Abdomen Embryos

By knocking down *Tc-axin* through RNA interference, we obtained embryos and larval cuticles missing anterior pattern elements and classified them according to their phenotypic strength in two different classes. When compared to the wild-type cuticle (Figure 2A), class I *Tc-axin*^{RNAi} embryos and larvae were identified by the loss of anterior structures to various degrees ranging from reduced head parts only, the loss of the head and some thoracic segments to the loss of all anterior segments including head, thorax and the anterior part of the abdomen (Figure S3); see also [16].

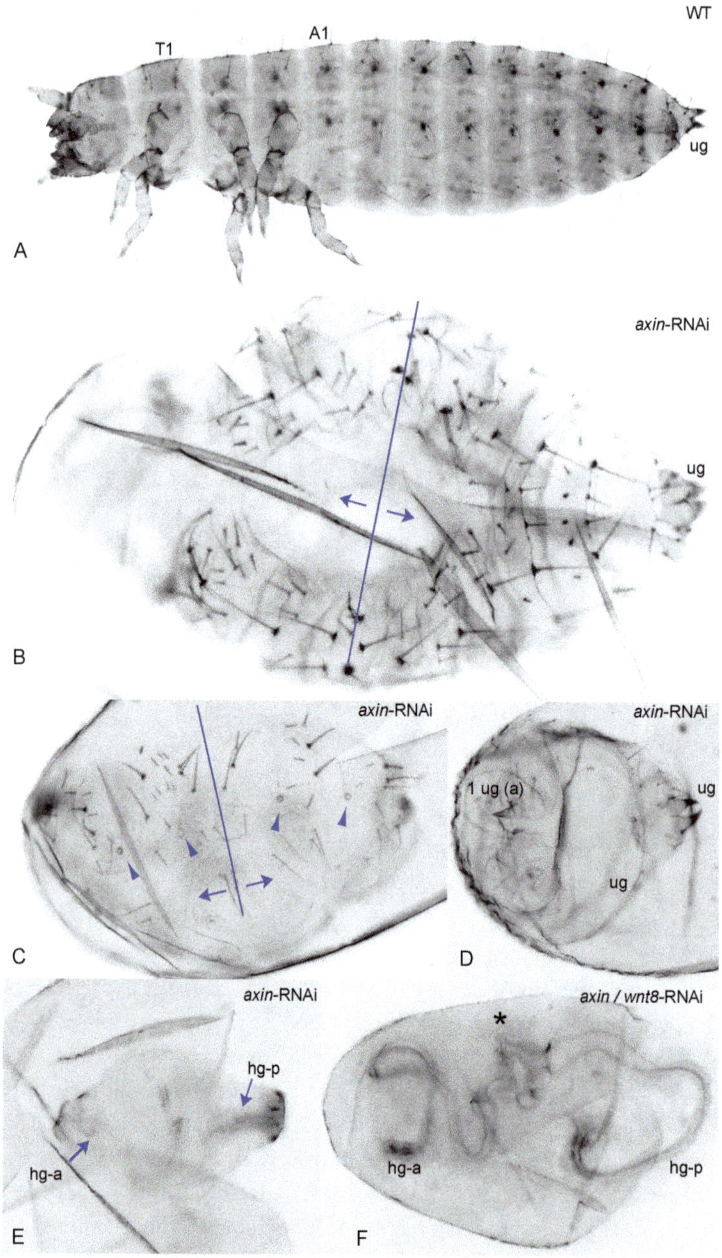

Figure 2. Polarity reversal in *Tc-axin*^{RNAi} and *Tc-axin/wnt8*^{RNAi} cuticles. (**A**) Wild-type cuticle (ug: urogomphi) (**B–E**) *Tc-axin*^{RNAi} polarity reversal phenotype. (**B**,**C**) Arrows indicate the polarity of the segments as judged by the orientation of the larger bristles that always point to the posterior pole. Plane of polarity reversal is indicated (line). (**B**) Posterior abdomen consists of five segments; at the anterior, at least three abdominal segments of reversed polarity can be identified. (**C**) Mirror-image double abdomen with two abdominal segments at each half. Arrowheads point to the tracheal openings. (**D**) One additional urogomphus (ug) has formed at the anterior. (**E**) Symmetric *axin*-RNAi cuticle of the double-abdomen phenotype with two hindguts on opposing sites of the unsegmented cuticle. (**F**) *axin-wnt8* double-RNAi: cuticle remnant (star) with two hindguts (hg-a, hg-p).

Embryos and cuticles belonging to the second class showed a different phenotype: instead of anterior structures, posterior abdominal segments with opposing polarity formed anterior to abdominal segments with original orientation (Figure 2B). Such polarity reversal along the AP-axis was classified as a mild double-abdomen (DA) phenotype. A weak, asymmetric DA phenotype is marked by the polarity-reversal of abdominal bristles belonging to one or two abdominal segments at the anterior of an otherwise normally formed abdomen with fewer or the normal number of segments (Figure 2B). A more distinct phenotype was represented by cuticles with complete mirror symmetry that displayed two partial posterior abdomens opposing each other (Figure 2C–F). In such cuticles, each half consisted of 2–3 abdominal segments where the ectopically formed, duplicated abdomen mostly also included internal structures like the hindgut (Figure 2E,F). The most extreme double abdomen phenotype displayed only remnants of a posterior cuticle, including two hindgut tubes that formed at opposite sites, and was seen not only in *Tc-axin*^{RNAi} but also in *axin-Wnt8* double RNAi experiments (Figure 2F).

Since 46% N = 185) of the analysed eggs were empty (Table 1) and did not develop to an analysable cuticle, we asked whether the formation of the ectopically formed abdomen was the result of early patterning events or is caused only secondarily later in development during germband elongation or cuticle formation. Therefore, we analysed the expression of posteriorly expressed genes as well as segmental marker genes in *Tc-axin*^{RNAi} embryos. First, we visualized the read-out of the Wnt-signaling pathway in *Tc-axin*^{RNAi} embryos by looking for the expression of *Tc-caudal*, *Tc-brachyenteron*, *Tc-wnt8* and *Tc-AbdB* as marker genes.

Posteriorization of early *Tc-axin*^{RNAi} embryos was documented by the expansion of the Wnt-target gene *Tc-caudal* already in the blastoderm embryo (compare Figure 3A,D). In contrast to the wild-type, where *Tc-caudal* is expressed in the posterior half between 44% and 58% egg-length (Figure 3A, mean: 51.5%, posterior pole = 100% egg length; N = 23) [26], *Tc-caudal* expression in *Tc-axin*^{RNAi} blastoderm embryos was expanded to the anterior to 63% egg-length (Figure 3D, range: 57–73%; N = 24) and covered nearly the complete prospective embryonic region of the blastoderm embryo. However, in wild-type embryos of the germband stage, *Tc-caudal* expression was restricted to the posterior growth zone (Figure 3B), but was distributed within nearly the complete embryo after *Tc-axin*-RNAi (Figure 3E). Later, in the fully segmented wild-type embryo, *Tc-caudal* expression was restricted to a small band in the posterior gut region (Figure 3C), but, after *Tc-axin*-RNAi, was also detected ectopically at the anterior indicating the formation of tissue with hindgut properties (Figure 3F). Indeed, the exclusive marker for the hindgut precursor *Tc-brachyenteron* (Figure 3G,H) [27] was also observed ectopically at the anterior in *Tc-axin*^{RNAi} embryos, either in a single domain (N = 8/20) or in two spots (12/20) at the newly formed posterior end (Figure 3I–L), otherwise never observed in wild-type embryo. In the *Tc-axin*^{RNAi} embryo of the early germband stage, *Tc-byn* expression itself was highly upregulated and encompassed the complete posterior region (Figure 3I). In the single *Tc-axin*-RNAi experiment, the strongest double abdomen phenotype (two opposing hindguts, Figure 2F) in the cuticle preparations showing polarity reversal was obtained with the frequency of 47% (N = 9, Figure S4). That Wnt activity, indeed, was elevated in *Tc-axin*-RNAi embryos and was visualised by the ubiquitous expression in the germband and with the ectopic upregulation at the anterior of the embryo (Figure 3M). By analysing *Tc-Abdominal B,* we could also show that posterior expression domain was slightly expanded and an ectopic domain formed at the anterior in weaker affected embryos (Figure 3O). Embryos with a stronger RNAi-phenotype showed nearly ubiquitous *Tc-AbdB* expression (Figure 3P). We then asked whether the frequency of the strongest double abdomen phenotype could be enhanced, and carried out *Tc-axin* RNAi in combination with knocking down *Tc-wnt8* and *Tc-arrow*, respectively. In the wild-type blastoderm embryo, posteriorly expressed *Tc-wnt8* [12] showed a limited contribution to the axis elongation process, since only a minor fraction of *Tc-wnt8*^{RNAi} embryos (6%, N = 8) were posteriorly truncated [13]. The frequency of embryos with an axis-elongation phenotype was only enhanced in the double RNAi with *Tc-wntless* dsRNA [13]. On the contrary, the Wnt-coreceptor *Tc-arrow* is an obligatory component for Wnt-dependent developmental processes, including abdominal elongation [14]. For the double knock-down, the same range of phenotypes as in the single *Tc-axin* RNAi experiment has been obtained in both double RNAi combinations. However, the frequency of the DA-phenotype was found to be remarkably higher, 75.5% (N = 62) in *Tc-axin*/*Tc-wnt8*^{RNAi} and 62% (N = 21) in *Tc-axin*/*Tc-arrow* ^{RNAi} embryos, in both of the respective double-RNAi experiments (see for an example Figure 2C; Table 1). Moreover, the expression of marker genes analysed in the combined RNAi experimental embryos showed the same characteristic changes as observed in *Tc-axin* single RNAi embryos: the enhancement of *Tc-caudal*, *Tc-wnt8* and ectopic expression of the hindgut marker *Tc-byn*.

Table 1. Quantitative analysis of *Tc-axin*-RNAi.

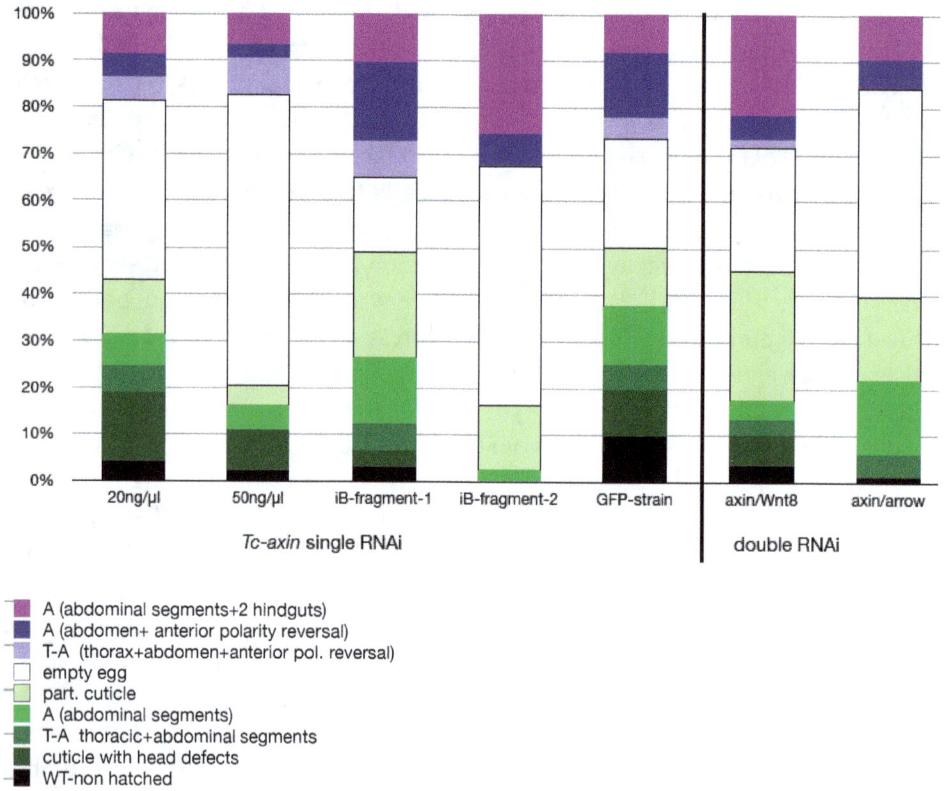

Numbers (N =)	*axin*^{RNAi}_20	*axin*^{RNAi}_50	iB-04108-1	iB-04108-2	*axin*^{RNAi} GFP-Strain	*axin/Wnt8* Double RNAi	*axin/Arrow* Double RNAi
WT	93	91	44	-	167	126	25
K-T-A	15	8	4	-	7	19	1
T-A (−)	6	-	6	-	4	9	10
A (−)	7	5	15	1	9	12	34
part. cuticle	12	4	24	6	9	80	39
T-A (+)	5	7	8	-	3	5	-
A (+)	5	3	18	3	10	15	12
A (+2 hg)	9	6	11	11	6	62	21
cuticles with polarity reversal	19	16	37	14	19	82	33
affected cuticles	59	33	86	21	48	202	117
all cuticles	152	124	130	21	215	328	142
empty eggs	39	58	17	22	17	76	97

Green indicates cuticles missing anterior structures; pink highlights the cuticles showing polarity reversal; (−) cuticles without polarity reversal; (+) cuticles displaying polarity reversal.

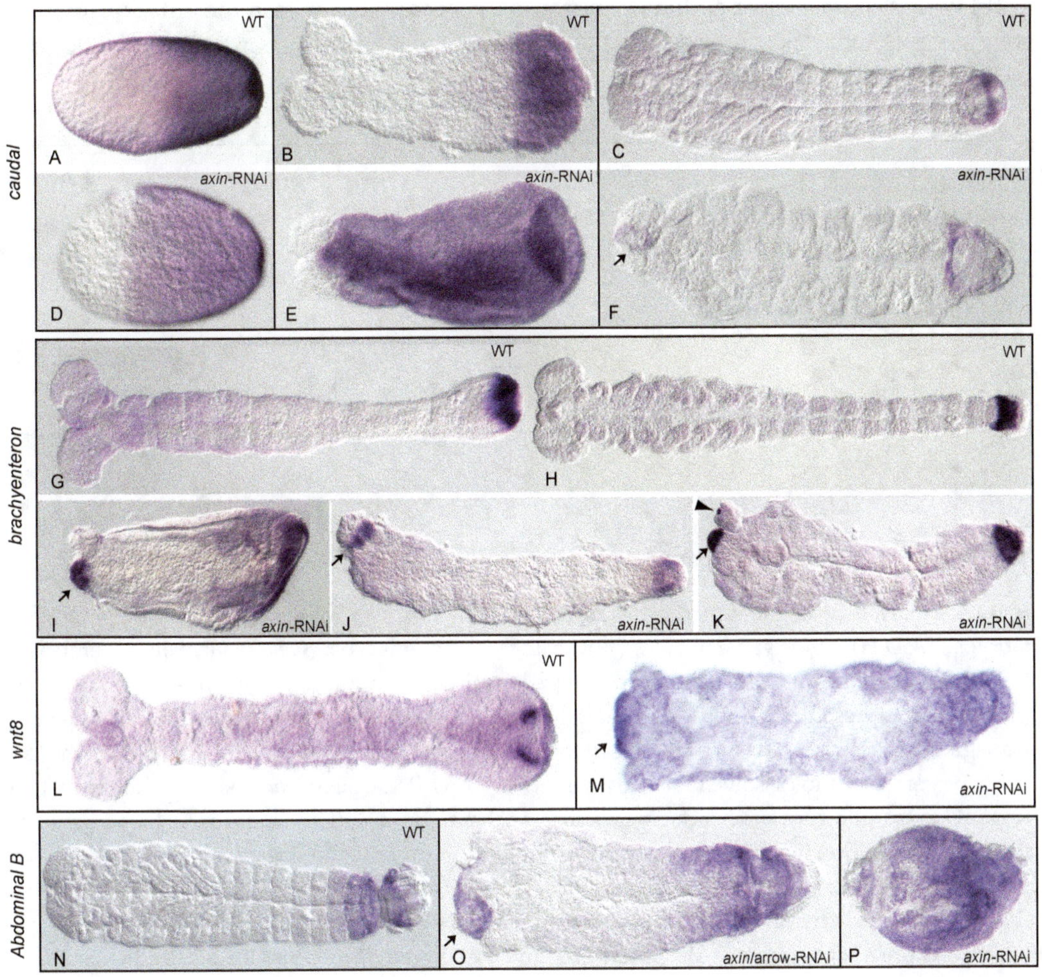

Figure 3. Analysis of posterior marker gene expression in *Tc-axin*^{RNAi} embryos. (**A–C**) *Tc-caudal* expression in successively older wild-type embryos (**D**) *Tc-axin*^{RNAi} blastoderm embryo: anterior expansion of *Tc-caudal* expression covering the embryonic region. (**E**) *Tc caudal* is highly overexpressed in a *Tc-axin*^{RNAi} embryo leaving free only the most anterior tissue. (**F**) *Tc-axin*^{RNAi} embryo after completion of segmentation: ectopic *Tc-cad* expression at the anterior (arrow). (**G,H**) In the wild-type, *Tc-brachyenteron* (*Tc-byn*) is exclusively expressed in the precursor of the hindgut at the posterior. (**I–K**) *Tc-axin*^{RNAi} embryos showing ectopic *Tc-byn* expression at the anterior (arrow). (**I**) *Tc-byn* expression is strongly expanded at the posterior of a young germ anlage. (**K**) Two ectopic *Tc-byn* domains are visible at the anterior (arrow, arrow-head); (**L**) *Tc-wnt8* is detectable in two spots in the growth zone in the wild-type but diffuse throughout a *Tc-axin* RNAi-embryo (**M**) and upregulated at the anterior (arrow). (**N**) *Tc-Abdominal B* is expressed in the posterior segments and the hindgut anlagen in the wild-type. (**O**) *Tc-axin-arrow*-double RNAi: *AbdB* expression is seen expanded in the posterior abdomen and ectopic at the anterior (arrow). (**P**) *Tc-axin*-RNAi upregulated throughout the embryo.

To reveal segment formation in embryos with elevated Wnt activity at the anterior, we analysed the expression patterns of the segmentation gene products Even-skipped and Engrailed in *Tc-axin*^{RNAi} embryos. The expression of the pair-rule gene product Tc-Eve in the wild-type embryo developed its segmental pattern from primary, broad, double segmental stripes within the growth zone of the embryo that then split into secondary, segmental stripes (Figure 4A) [28,29].

Remarkably, in *Tc-axin*^{RNAi} embryos, Tc-Eve was expressed in young germband embryos along the AP axis in dense blocks in the mesodermal region, ectopically at the anterior and in the growth zone in solid stripes that apparently did not dissolve into secondary, segmental stripes (Figure 4C). The wild-type expression pattern of Tc-En was seen in *Tc-axin*^{RNAi} embryos as segmental stripes

(Figure 4F) within the posterior abdomen, but was uncoordinated within the anterior region of the embryo (Figure 4F, star). Stronger affected short embryos display uncoordinated patches of Tc-En expression (Figure 4G).

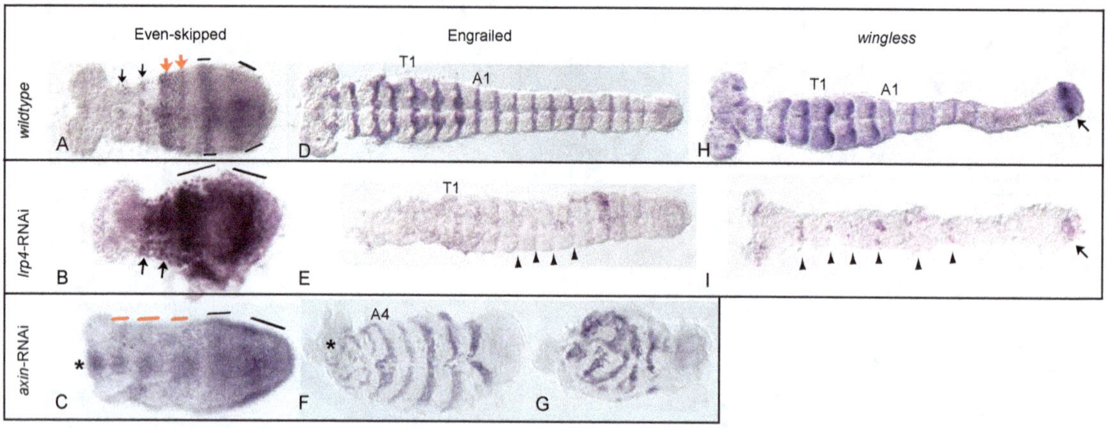

Figure 4. Analysis of segmentation genes in *Tc-lrp4* and *Tc-axin*^{RNAi}-embryos. Preparation of fixed embryos. (**A**,**D**,**H**): wild-type embryos; (**B**,**E**,**I**): *Tc-lrp4*^{RNAi}-embryos; (**C**,**F**,**G**): *Tc-axin*^{RNAi}-embryos. (**A**) Expression of the pair-rule gene Even-skipped (antibody staining). In the wild-type, two broad domains that represent the double-segmented phase are seen in the growth-zone (bars). The split into segmented stripes starts just anterior to the growth-zone (red arrows). The first two segmental eve stripes are about to dissolve (black arrows). (**B**) *Tc-lrp4*-RNAi: Eve expression is stronger when compared to the wild-type and hence, single stripes are hardly visible. (**C**) *Tc-axin*-RNAi: broad domains (black bars) in the growth-zone appear but do not dissolve into segmental stripes; instead, block-shaped domains in the region where the mesoderm forms are visible (red bars). Ectopically, Eve-positive cells are visible in a medial, anterior position (red star). (**D**): Fully segmented wild-type embryo. Engrailed expression demarcates the posterior segmental border. (**E**): *Tc-lrp4*-RNAi: headless embryo lacking clearly identifiable En-stripes but with clearly visible segmental grooves (see arrowhead for examples). (**F**,**G**) *Tc-axin*-RNAi: scrambled En stripes at the anterior part of the embryos (red star in (**F**)). No polarity information is recognizable. (**H**) *Tc-wingless* (*wg*)/*Wnt1* expression (in-situ-hybridization) in the wild-type undergoing axis elongation. (**I**): *Tc-lrp4*-RNAi: remnants of *wg* positive cells are visible in the segmental pattern (arrowheads). Posterior *wg* domain (arrow) seems weaker than in the wild-type but is clearly visible.

3.3. Structure, Expression and Functional Analysis of Tc-lrp4

The gene TC007146 was identified as the single Lrp4 ortholog in the *Tribolium* genome [14] that codes for a 2042 amino acid long, single-pass transmembrane protein. It is structured as a typical LDL (low density lipoprotein) receptor [30] with 8 LDL-A repeats, followed by 1 EGF repeat, four (1 EGF-5 YMTD) repeats and the transmembrane domain (Figure S2B). Crucially involved in binding WNT proteins are the first two of the YWTD and the EGF domains [31–33]. Intracellularly, Lrp4 lacks the highly conserved PPT(S/T)Px(S/T) motifs required for Axin-binding present in the Lrp5/6/Arrow class of WNT-coreceptors [14,34] (Figure S2C).

Throughout embryogenesis, and in contrast to *Tc-axin*, *Tc-lrp4* is ubiquitously expressed (Figure S5). To analyse its function during embryogenesis, we performed *Tc-lrp4* knock-down experiments applying maternal RNAi. We observed various segmentation defects along the AP axis but not polarity reversal as seen in *Tc-axin*-RNAi experiments (see above). The variety of *Tc-lrp4*-RNAi cuticle-phenotypes was grouped into three different categories (Figure 5). Class Ia was characterised by missing a single or more gnathal segments up to the loss of the complete head (Figure 5A) and the thorax resulting in an abdomen-only cuticle (Class Ib) (Figure 5C). Class II phenotypes displayed constrictions within the thorax or the abdomen of the larval cuticle. This was accompanied by the loss of a single or a number of adjacent body segments resulting in a gap within the anterior-posterior pattern and, therefore, classified as "gap"-phenotype (Figure 5B,D). The range of gap-phenotypes are displayed in Figure S6. Class III comprised cuticles that consisted of separate pieces with ball-shaped cuticles of unknown segment identity or cuticle remnants with hindgut structures (Figure 5E,F).

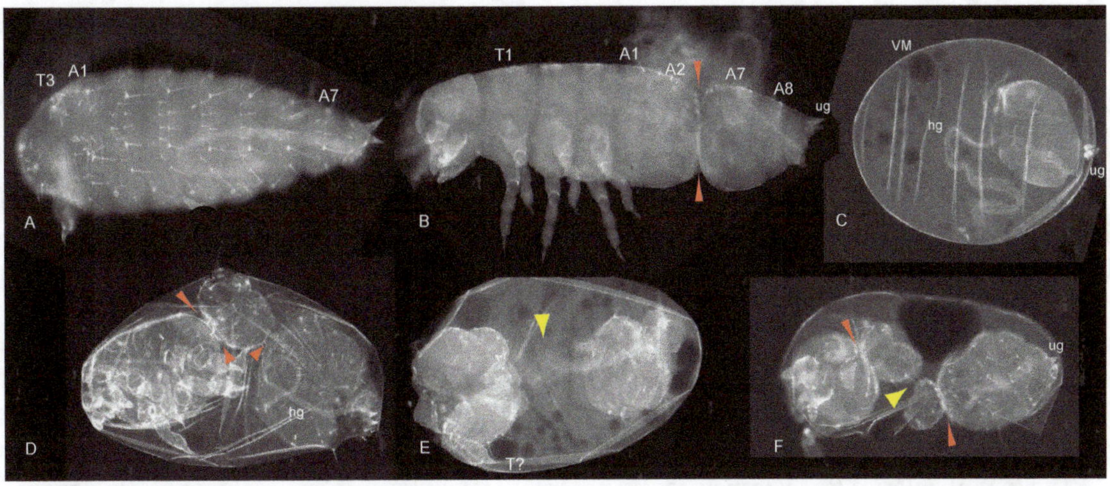

Figure 5. *Tc-lrp4* is essential for patterning the anterior-posterior axis and for segmentation. (**A–F**: Cuticle preparations). (**A**) Headless *Tc-lrp4*^{RNAi} larval cuticle. (**B**) *Tc-lrp*^{RNAi} phenotype with a constricted abdomen (arrow heads) missing abdominal segments A3–A6 ("gap-phenotype"). (**C**) Remnant of the posterior cuticle including the urogomphi and the hindgut within the vitelline membrane. (**D**,**E**) Gap-phenotypes are characterized by constrictions and loss of body parts (shown by arrowheads). (**D**) Constrictions between thorax and abdomen and within the abdomen (arrowheads). (**E**) Larval cuticle consists of two pieces (yellow arrowhead indicates absence of connection). The anterior part developed with one pair of malformed legs of unknown identity (T?). (**F**) Cuticle consists of two pieces, each in addition shows constrictions (red arrowheads). Separation is indicated by the yellow arrowhead. A1: abdominal segment 1, HD: hindgut, Lb: labial segment; Mx: maxillar segment, T1: thoracic segment 1. ug: urogomphi. VM: vitelline membrane.

3.4. Pair-Rule and Segment-Polarity Gene Expression in Tc-lrp4 RNAi Embryos Reveal Instability of the Segment Boundary

In RNAi embryos with impaired Lrp4 function, misexpression of *Tc-caudal* was not as drastic when compared to *Tc-axin* RNAi. Nevertheless, the *Tc-caudal* domain was seen enlarged towards the anterior, covering at most 50% in the growing germband (Figure S7). On the other hand, Tc-Eve expression was seen to be strongly enhanced, and also inter-stripe cells showed to various degrees Eve protein expression so that the stripes appeared to be fused to form a single, large posterior expression domain (Figure 4B). Surprisingly, in *Tc-lrp4* RNAi embryos, the expression of the segment polarity genes *wingless/Wnt1* and *engrailed*, however, are strongly reduced, irregular or absent (Figure 4E,I).

4. Discussion

In this work, we analysed in the beetle *Tribolium* the function of Wnt-antagonists that act at the level of the destruction complex and at Wnt-reception. Knocking down the asymmetrically distributed Wnt-inhibitor *Tc-axin* resulted in the polarity reversal phenotype, not observed before. For the first time in an insect, we show that the Wnt-inhibitor Lrp4 acts globally along the anterior–posterior axis.

The complex expression pattern of *Tc-axin* reveals sites for the modulation of Wnt activity.

During early embryogenesis, the Wnt-antagonist *Tc-axin* is asymmetrically expressed at the anterior pole of the pre-blastoderm embryo [16]. In addition to this early phase of expression, we show here that *Tc-axin* was also expressed during later embryonic stages in various tissues such as the growth zone of the elongating embryo, in developing appendages and in the abdominal segments. The dynamic expression of *Tc-axin* marks regions and tissues in the embryo where the level of Wnt activity is tightly balanced by the Wnt-destruction complex during embryogenesis. In the pre-blastoderm embryo, *Tc-axin* establishes asymmetry along the AP axis by lowering or eliminating Wnt activity off the anterior pole of the egg. The clearance of Wnt activity by *axin* represents an evolutionary, ancient principle for enabling the development of anterior structures in vertebrates and invertebrates including *Tribolium* [6,16,35]. By regulating the stability of β-catenin, Axin is only indirectly and not instructively required for anterior patterning. One such factor required for anterior development in *Tribolium* and in other invertebrate species outside *Drosophila* [36,37] is exemplified by the *orthodenticle* gene [38–40].

Later during embryogenesis, *Tc-axin* is expressed in the growth zone of the elongating embryo in a longitudinal stripe. None of the Wnt ligands is expressed in this region (Figure 1C) flanked by *Tc-fgf8* expression [41]. We speculate that, in this way, this part of the growth zone is kept clear of Wnt activity. Interestingly, the expression sites of *Tc-axin* coincide in part with many regions that also express Wnt ligands [12] and hints at the requirement to tightly control the extent of Wnt signaling in tissues like the growth zone and the elongating appendages.

Embryos with elevated but flattened Wnt activity posteriorise and also develop the double abdomen/double hindgut phenotype.

The loss of anterior larval structures by parental *Tc-axin* RNA interference has been described previously [16] and is shown here (Figure S3B–D). This posteriorisation phenotype is explained by ectopic elevation of Wnt activity in the anterior egg to a level normally found in a more posterior position (Figure 6A, curves II–III) and interferes with anterior pattering. That posterior fate, indeed, has expanded anteriorly, visualised by the ectopic expression of the Wnt target *Tc-caudal* [16] (Figure 3D–F), of *Tc-wnt8* (Figure 3M), and of *Tc-Abdominal B* (Figure 3P).

In addition, we show here that a substantial proportion of *Tc-axin*^{RNAi} larvae displayed polarity reversal of anterior abdominal segments, the double-abdomen phenotype (Figure 2B–F, Table 1). In embryos with an asymmetric double-abdomen (Figure 2B) the downregulation of *Tc-axin* likely was most effective at the anterior pole, but did not affect the architecture (steepness) of the Wnt-activity gradient in the posterior egg. In such embryos, no recognisable posterior-most structures such as a hindgut or ectopic urogomphi have developed, indicating that Wnt activity at the anterior was too low (Figure 6B, curve V). In mirror-symmetric embryos and larval cuticles, where both abdomens consisted of only 2–3 abdominal segments, the slope of Wnt activity was not steep enough to provide sufficient positional information for abdominal segments of more anterior identity (Figure 2C). In even more extreme cases of the double posterior phenotype, high levels of Wnt activity are even more flattened along the anterior–posterior axis. This "Wnt-flat^{HIGH}" situation results in the formation of two posterior centres at both poles that each develops with a hindgut recognised in the cuticle (Figure 2E,F) and the expression of the hindgut-specific marker *Tc-brachyenteron* (Figure 3I–K). The lack of steepness of the Wnt-gradient, and thus the lack of sufficient positional information, only enables the formation of a small cuticle remnant (Figures 2F and 6B, curve VI).

The rationale of the double RNAi experiments knocking down the Wnt-antagonist *Tc-axin* and at the same time a Wnt-agonist like *Tc-wnt8* or *Tc-arrow*, was to reduce overall Wnt activity and thereby support the flattening of Wnt activity. The fact that slightly more embryos with duplicated posterior structures developed showed that this situation, indeed, destabilises the embryonic axis. The formation of posterior structures and segments could be under the control of the terminal system [42].

Strongly affected *Tc-axin* single-RNAi- as well as in *Tc-axin/Tc-wnt8* or in *Tc-axin/Tc-arrow* double-RNAi embryos develop with few if any abdominal segments, indicating that high Wnt levels interfere with the segmentation process. We found that the primary pair-rule stripes of Tc-Eve do not resolve into secondary, segmental stripes [28,29] but remain as solid blocks within the embryo along the AP axis (Figure 4C). Obviously, in embryos with **up**regulated Wnt activity, the segmentation process is not primarily affected and repeatedly generates primary Eve-domains. These, however, do not resolve into secondary ones but remain stable and become ectopically placed in the anterior embryo. We have shown, that in *Tc-frizzled*1/2^{RNAi}-embryos where Wnt activity is **down**regulated, the transition of pair-rule gene expression from primary to the secondary pair-rule phase is also impaired [14]. Therefore, we conclude, that the level of Wnt activity has to be tightly balanced by agonists and antagonists within the anterior part of the growth zone, and likely in other organs such as the appendages, to ensure normal development.

Mirror-type duplications of posterior pattern elements in insects by experimental perturbation were explained by the double gradient hypothesis [43,44], but were at that time not understood at the molecular level. We propose that an inhibitor such as *axin* could fulfil the requirement of one component of this system also in other organisms.

Lrp4 as a Wnt-inhibitor in *Tribolium*?

In the mouse, LRP4 acts at the neuromuscular junction as a receptor for the motoneuron-derived protein Agrin that is absent from the *Drosophila* genome [45]. An Lrp4 ortholog in *Drosophila* has been identified as a regulator of synaptic physiology in the central nervous system in a Wnt-independent fashion [17]. The reduction of Lrp4 function in *Drosophila* results in the loss of certain synapses but has no implications for the segmentation process [17].

Structurally, Tc-Lrp4 could function as a Wnt-receptor that is unable to transmit its signal to the nucleus due to the lack of an Axin-binding domain. In this way, Tc-Lrp4 would function in a dominant-negative fashion to negatively modulate Wnt signaling. In fact, Lrp4 can compete with the Wnt-coreceptors Lrp5/6 for Wnt ligands, as this has been described in the mouse system [46]. The function of vertebrate Lrp4s as a negative modulator of Wnt activity has been also shown for other processes [19–21,47]. Also in vertebrates, Lrp4 functions as a receptor for Wnt-inhibitors like WISE, DKK or Sclerotin. It is discussed that Lrp4 bind e.g., WISE, presents it either to the LRP5/6 coreceptor or the Frizzled-receptor, and in these ways competes with Wnt-loaded Lrp5/6 and antagonizes Wnt activity [21]. Since the respective orthologues of WISE, DKK and Sclerotin seem to be missing from the *Tribolium* genome, *Tc*-Lrp4 could function as a dominant-negative WNT co-receptor in *Tribolium*. In this scenario, which has also been discussed for the vertebrate system [21], Lrp4 directly competes with Lrp5/6 for Wnt-ligands and/or Frizzled receptors. Once the Wnt/Lrp4 dimer has formed, less Wnt signal is available and hence, Wnt activity is attenuated. In the experimental situation of downregulating a Wnt-inhibitor, in this case Lrp4, Wnt signaling is expected to become upregulated. That this, indeed, was the case was seen at the anterior through the upregulation of *brachyenteron* and the loss of head- and thoracic structures. At the posterior, the upregulation of Wnt activity results in high levels of *caudal* expression and, as a consequence, elevated *even-skipped* levels were observed in *Tc-axin* (see above) and also, but not as pronounced, in *Tc-lrp4*-RNAi embryos. *caudal/Cdx genes* have been characterized as direct targets of the Wnt pathway [48–50] and can be seen as a read-out of Wnt activity. In the *Tribolium* wild-type embryo, the pair-rule circuit that involves *eve, runt* and *paired* leads to the activation of *Tc-engrailed* and represses *Tc-wingless,* resulting in an alternating Wg–En expression pattern [51]. Although segments were morphologically visible in *Tc-lrp4*-RNAi embryos, segmental borders are obviously not maintained during development until the cuticle has formed, due to the lack or strong reduction of Wg and En in *Tc-lrp4*-RNAi embryos (Figure 4E,I). Without an established connection between two adjacent segments, neighboured segments might collapse and eventually fuse ("gap-phenotype"), and the cuticle displays severe constrictions when only part of a segment border forms or the larval cuticle falls apart and consists of separate pieces that we observed within the vitelline membrane.

Since the exact binding partner of *Tc*-Lrp4 has not been identified, other ligands but also other downstream events could be elicited by the activated Lrp4 receptor in *Tribolium* as in *Drosophila* Lrp4 signals through a synapse-specific kinase in a Wnt-independent manner. Interestingly, an Agrin-ortholog has been identified in *Tribolium* [52] that, indeed, could function as a ligand for *Tc*-Lrp4. Taking into account that such a scenario is given for the neuromuscular junction in the mouse, a contribution of Agrin in segmentation in *Tribolium* seems unlikely.

No polarity reversal was observed after *Tc-lrp4*-RNAi. In contrast to *Tc-axin*, *Tc-lrp4* was shown to be ubiquitously expressed throughout the egg and the embryo. In the *lrp4*-RNAi embryo, mRNA became evenly reduced along the complete AP axis to a lower level, so that Wnt activity becomes elevated without altering the steepness of its activity gradient and, importantly, without accumulating at the anterior pole to a level that could result in polarity reversal (Figure 5A; Figure 6A, curve II). The elevation of Wnt activity at the anterior pole above a certain threshold, however, interferes with head and thorax development without altering the overall polarity of the embryo. When the Wnt-activity level rose more in *Tc-lrp4* RNAi embryos, only posterior abdominal segments can form but still exhibited normal polarity (Figure 6A, curve III). A different situation is seen in *Tc-axin*-RNAi embryos where asymmetric expression of *Tc-axin* is crucial for downregulating Wnt activity at the anterior pole of the early wild-type embryo. In the *Tc-axin*-RNAi experiment, anteriorly expressed maternal *axin* is downregulated and results in an increase of Wnt activity at the anterior (Figure 6B, curve IV). Here, polarity information on anterior-forming abdominal segments could, additionally, come from the terminal system [42], resulting in hindgut development at the anterior pole. In the extreme situation, the height and the lack of steepness of the Wnt activity level (Figure 6B, curve V, VI) allows for the formation of only few abdominal segments of posterior-most identity, with reversed polarity also seen in the *Drosophila bicaudal* mutant [53,54] or in *panish*-RNAi *Chironomous* larvae [55] and not for two opposing abdomens with the full set of eight abdominal segments.

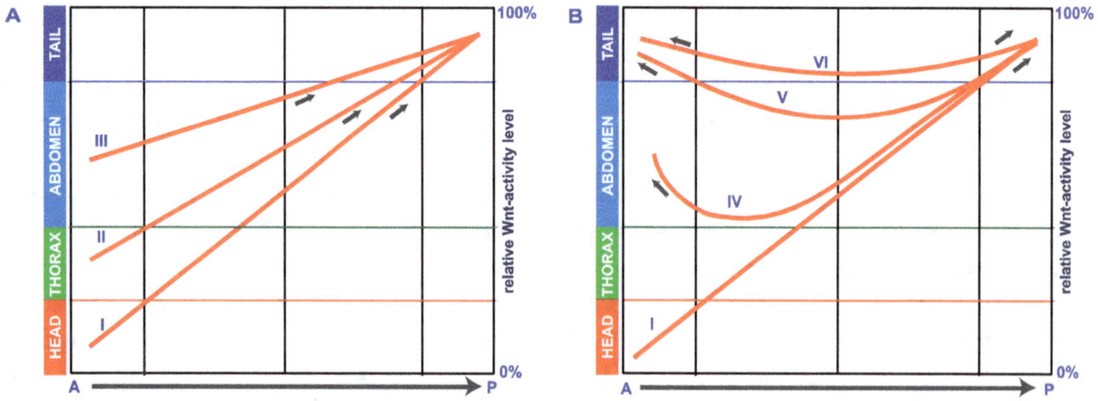

Figure 6. Schematic illustration of Wnt activity along the AP-axis in wild-type- and RNAi-embryos. (**A**) Wnt activity along the anterior-posterior axis is graded from low to high in the wild-type embryo (curve I) resulting in the formation of all pattern elements head, thorax and abdomen. Upon *Tc-axin*-RNAi (weak effect) or LRP4-RNAi, Wnt activity at the anterior becomes de-repressed and thus is elevated in the anterior egg to a level that is not compatible with head- (curve II) nor head/thorax development (curve III). AP-polarity, however, remains normal due to the graded distribution of Wnt activity. (**B**) Embryos that not only miss anterior structures but in addition show development of posterior fate at the anterior pole (curves IV–VI) observed in RNAi experiments knocking down *Tc-axin* alone or in combination with either *Tc-wnt8* or *Tc-arrow*. Such embryos developed polarity reversal of only few abdominal bristles or abdominal segments with reverse polarity at the anterior (curve IV) or a mirror image double abdomen phenotype (curve V). Two opposing hindguts connected by a small cuticle were seen predominantly in the combinatorial RNAi experiments (curve VI), the Wnt-flat^{HIGH} situation.

5. Conclusions

The distinct expression sites of the Wnt-antagonist *Tc-axin* seen in the developing embryo reveal tissues and cells where Wnt signaling possibly needs to become fine-tuned to ensure proper development. Apparently and unexpectedly, negative modulation of Wnt activity likely also occurs close to or coinciding with Wnt-agonist expression sites, such as in a central stripe within the growth zone or in the ventral appendage. Experimentally, we found that downregulation via maternal RNAi,

Tc-axin not only led to the repression of anterior development but, in addition, to the formation of an ectopic tail at the anterior pole with reversed polarity and a hindgut, resembling the *Drosophila bicoid* or *bicaudal* mutant phenotype. For the first time, we could show that an insect Lrp4 orthologue functions along the anterior–posterior axis, coordinating axis formation and segmentation possibly by antagonising Wnt signaling. This work broadens our view on the modulation of Wnt activity by the negative Wnt-regulators Axin and LRP4 to ensure proper embryonic development.

Supplementary Materials: The Supplementary Materials are available online at www.mdpi.com/2221-3759/5/4/10/s1.

Acknowledgments: The authors wish to thank H. Brandt and K. Schwandt for technical assistance and the German Research Foundation for continuous support through a grant to RS (SCHR 435 2-1).

Author Contributions: R.P., R.S. and A.B. conceived and designed the experiments; R.P. and A.B. performed the experiments; R.P., A.B. and R.S. analyzed the data; R.S. wrote the paper.

Conflicts of Interest: The authors declare no conflict of interest. The founding sponsors had no role in the design of the study; in the collection, analyses, or interpretation of data; in the writing of the manuscript, and in the decision to publish the results.

References

1. Nusse, R.; Clevers, H. Wnt/beta-Catenin signaling, disease, and emerging therapeutic modalities. *Cell* **2017**, *169*, 985–999. [CrossRef] [PubMed]
2. Loh, K.M.; van Amerongen, R.; Nusse, R. Generating Cellular Diversity and Spatial Form: Wnt Signaling and the Evolution of Multicellular Animals. *Dev. Cell* **2016**, *38*, 643–655. [CrossRef] [PubMed]
3. Moon, R.T.; Kohn, A.D.; De Ferrari, G.V.; Kaykas, A. WNT and beta-catenin signalling: Diseases and therapies. *Nat. Rev. Genet.* **2004**, *5*, 691–701. [CrossRef] [PubMed]
4. Clevers, H.; Nusse, R. Wnt/beta-catenin signaling and disease. *Cell* **2012**, *149*, 1192–1205. [CrossRef] [PubMed]
5. Lara-Castillo, N.; Johnson, M.L. LRP receptor family member associated bone disease. *Rev. Endocr. Metab. Disord.* **2015**, *16*, 141–148. [CrossRef] [PubMed]
6. Green, D.; Whitener, A.E.; Mohanty, S.; Lekven, A.C. Vertebrate nervous system posteriorization: Grading the function of Wnt signaling. *Dev. Dyn.* **2015**, *244*, 507–512. [CrossRef] [PubMed]
7. Niehrs, C. Regionally specific induction by the Spemann-Mangold organizer. *Nat. Rev. Genet.* **2004**, *5*, 425–434. [CrossRef] [PubMed]
8. Hikasa, H.; Sokol, S.Y. Wnt signaling in vertebrate axis specification. *Cold Spring Harb. Perspect. Biol.* **2013**, *5*. [CrossRef] [PubMed]
9. Umesono, Y.; Tasaki, J.; Nishimura, Y.; Hrouda, M.; Kawaguchi, E.; Yazawa, S.; Nishimura, O.; Hosoda, K.; Inoue, T.; Agata, K. The molecular logic for planarian regeneration along the anterior-posterior axis. *Nature* **2013**, *500*, 73–76. [CrossRef] [PubMed]
10. Glinka, A.; Wu, W.; Onichtchouk, D.; Blumenstock, C.; Niehrs, C. Head induction by simultaneous repression of Bmp and Wnt signalling in *Xenopus*. *Nature* **1997**, *389*, 517–519. [PubMed]
11. Niehrs, C. On growth and form: A Cartesian coordinate system of Wnt and BMP signaling specifies bilaterian body axes. *Development* **2010**, *137*, 845–857. [CrossRef] [PubMed]
12. Bolognesi, R.; Beermann, A.; Farzana, L.; Wittkopp, N.; Lutz, R.; Balavoine, G.; Brown, S.J.; Schröder, R. *Tribolium* Wnts: Evidence for a larger repertoire in insects with overlapping expression patterns that suggest multiple redundant functions in embryogenesis. *Dev. Genes Evol.* **2008**, *218*, 193–202. [CrossRef] [PubMed]
13. Bolognesi, R.; Farzana, L.; Fischer, T.D.; Brown, S.J. Multiple Wnt genes are required for segmentation in the short-germ embryo of *Tribolium castaneum*. *Curr. Biol.* **2008**, *18*, 1624–1629. [CrossRef] [PubMed]
14. Beermann, A.; Prühs, R.; Lutz, R.; Schröder, R. A context-dependent combination of Wnt receptors controls axis elongation and leg development in a short germ insect. *Development* **2011**, *138*, 2793–2805. [CrossRef] [PubMed]
15. Bolognesi, R.; Fischer, T.D.; Brown, S.J. Loss of *Tc-arrow* and canonical Wnt signaling alters posterior morphology and pair-rule gene expression in the short-germ insect, *Tribolium castaneum*. *Dev. Genes Evol.* **2009**, *219*, 369–375. [CrossRef] [PubMed]

16. Fu, J.; Posnien, N.; Bolognesi, R.; Fischer, T.D.; Rayl, P.; Oberhofer, G.; Kitzmann, P.; Brown, S.J.; Bucher, G. Asymmetrically expressed *axin* required for anterior development in *Tribolium*. *Proc. Natl. Acad. Sci. USA* **2012**, *109*, 7782–7786. [CrossRef] [PubMed]
17. Mosca, T.J.; Luginbuhl, D.J.; Wang, I.E.; Luo, L.Q. Presynaptic LRP4 promotes synapse number and function of excitatory CNS neurons. *eLife* **2017**, *6*. [CrossRef] [PubMed]
18. Zhang, B.; Liang, C.; Bates, R.; Yin, Y.M.; Xiong, W.C.; Mei, L. Wnt proteins regulate acetylcholine receptor clustering in muscle cells. *Mol. Brain* **2012**, *5*, 7. [CrossRef] [PubMed]
19. Ahn, Y.; Sims, C.; Logue, J.M.; Weatherbee, S.D.; Krumlauf, R. Lrp4 and Wise interplay controls the formation and patterning of mammary and other skin appendage placodes by modulating Wnt signaling. *Development* **2013**, *140*, 583–593. [CrossRef] [PubMed]
20. Choi, H.Y.; Dieckmann, M.; Herz, J.; Niemeier, A. Lrp4, a novel receptor for Dickkopf 1 and sclerostin, is expressed by osteoblasts and regulates bone growth and turnover in vivo. *PLoS ONE* **2009**, *4*, e7930. [CrossRef] [PubMed]
21. Ahn, Y.; Sims, C.; Murray, M.J.; Kuhlmann, P.K.; Fuentes-Antras, J.; Weatherbee, S.D.; Krumlauf, R. Multiple modes of Lrp4 function in modulation of Wnt/beta-catenin signaling during tooth development. *Development* **2017**, *144*, 2824–2836. [CrossRef] [PubMed]
22. Driever, W.; Nüsslein-Volhard, C. The Bicoid protein determines position in the *Drosophila* embryo in a concentration-dependent manner. *Cell* **1988**, *54*, 95–104. [CrossRef]
23. Bucher, G.; Scholten, J.; Klingler, M. Parental RNAi in *Tribolium* (Coleoptera). *Curr. Biol.* **2002**, *12*, R85–R86. [CrossRef]
24. Sarrazin, A.F.; Peel, A.D.; Averof, M. A segmentation clock with two-segment periodicity in insects. *Science* **2012**, *336*, 338–341. [CrossRef] [PubMed]
25. Sharma, R.; Beermann, A.; Schröder, R. The dynamic expression of extraembryonic marker genes in the beetle *Tribolium castaneum* reveals the complexity of serosa and amnion formation in a short germ insect. *Gene Expr. Patterns* **2013**, *13*, 362–371. [CrossRef] [PubMed]
26. Schulz, C.; Schröder, R.; Hausdorf, B.; Wolff, C.; Tautz, D. A *caudal* homologue in the short germ band beetle *Tribolium* shows similarities to both, the *Drosophila* and the vertebrate *caudal* expression patterns. *Dev. Genes Evol.* **1998**, *208*, 283–289. [CrossRef] [PubMed]
27. Berns, N.; Kusch, T.; Schröder, R.; Reuter, R. Expression, function and regulation of Brachyenteron in the short germband insect *Tribolium castaneum*. *Dev. Genes Evol.* **2008**, *218*, 169–179. [CrossRef] [PubMed]
28. Patel, N.H.; Condron, B.G.; Zinn, K. Pair-rule expression patterns of *even-skipped* are found in both short- and long-germ beetles. *Nature* **1994**, *367*, 429–434. [CrossRef] [PubMed]
29. Choe, C.P.; Miller, S.C.; Brown, S.J. A pair-rule gene circuit defines segments sequentially in the short-germ insect *Tribolium castaneum*. *Proc. Natl. Acad. Sci. USA* **2006**, *103*, 6560–6564. [CrossRef] [PubMed]
30. Springer, T.A. An extracellular beta-propeller module predicted in lipoprotein and scavenger receptors, tyrosine kinases, epidermal growth factor precursor, and extracellular matrix components. *J. Mol. Biol.* **1998**, *283*, 837–862. [CrossRef] [PubMed]
31. Itasaki, N.; Jones, C.M.; Mercurio, S.; Rowe, A.; Domingos, P.M.; Smith, J.C.; Krumlauf, R. Wise, a context-dependent activator and inhibitor of Wnt signalling. *Development* **2003**, *130*, 4295–4305. [CrossRef] [PubMed]
32. Li, X.; Zhang, Y.; Kang, H.; Liu, W.; Liu, P.; Zhang, J.; Harris, S.E.; Wu, D. Sclerostin binds to LRP5/6 and antagonizes canonical Wnt signaling. *J. Biol. Chem.* **2005**, *280*, 19883–19887. [CrossRef] [PubMed]
33. Mao, J.; Wang, J.; Liu, B.; Pan, W.; Farr, G.H., 3rd; Flynn, C.; Yuan, H.; Takada, S.; Kimelman, D.; Li, L.; et al. Low-density lipoprotein receptor-related protein-5 binds to Axin and regulates the canonical Wnt signaling pathway. *Mol. Cell* **2001**, *7*, 801–809. [CrossRef]
34. Tamai, K.; Zeng, X.; Liu, C.; Zhang, X.; Harada, Y.; Chang, Z.; He, X. A mechanism for Wnt coreceptor activation. *Mol. Cell* **2004**, *13*, 149–156. [CrossRef]
35. Petersen, C.P.; Reddien, P.W. Wnt signaling and the polarity of the primary body axis. *Cell* **2009**, *139*, 1056–1068. [CrossRef] [PubMed]
36. Nakao, H. Anterior and posterior centers jointly regulate *Bombyx* embryo body segmentation. *Dev. Biol.* **2012**, *371*, 293–301. [CrossRef] [PubMed]
37. Lynch, J.A.; Brent, A.E.; Leaf, D.S.; Pultz, M.A.; Desplan, C. Localized maternal *orthodenticle* patterns anterior and posterior in the long germ wasp *Nasonia*. *Nature* **2006**, *439*, 728–732. [CrossRef] [PubMed]

38. Schröder, R. The genes *orthodenticle* and *hunchback* substitute for *bicoid* in the beetle *Tribolium*. *Nature* **2003**, *422*, 621–625. [CrossRef] [PubMed]
39. Kotkamp, K.; Klingler, M.; Schoppmeier, M. Apparent role of *Tribolium orthodenticle* in anteroposterior blastoderm patterning largely reflects novel functions in dorsoventral axis formation and cell survival. *Development* **2010**, *137*, 1853–1862. [CrossRef] [PubMed]
40. Schinko, J.B.; Kreuzer, N.; Offen, N.; Posnien, N.; Wimmer, E.A.; Bucher, G. Divergent functions of *orthodenticle*, *empty spiracles* and buttonhead in early head patterning of the beetle *Tribolium castaneum* (Coleoptera). *Dev. Biol.* **2008**, *317*, 600–613. [CrossRef] [PubMed]
41. Beermann, A.; Schröder, R. Sites of Fgf signalling and perception during embryogenesis of the beetle *Tribolium* castaneum. *Dev. Genes Evol.* **2008**, *218*, 153–167. [CrossRef] [PubMed]
42. Schoppmeier, M.; Schröder, R. Maternal Torso signaling controls body axis elongation in a short germ insect. *Curr. Biol.* **2005**, *15*, 2131–2136. [CrossRef] [PubMed]
43. Sander, K. Specification of the basic body pattern in insect embryogenesis. *Adv. Insect Physiol.* **1976**, *12*, 125–238.
44. Lawrence, P.A. Background to *bicoid*. *Cell* **1988**, *54*, 1–2. [CrossRef]
45. Zong, Y.; Zhang, B.; Gu, S.; Lee, K.; Zhou, J.; Yao, G.; Figueiredo, D.; Perry, K.; Mei, L.; Jin, R. Structural basis of agrin-LRP4-MuSK signaling. *Genes Dev.* **2012**, *26*, 247–258. [CrossRef] [PubMed]
46. Johnson, E.B.; Hammer, R.E.; Herz, J. Abnormal development of the apical ectodermal ridge and polysyndactyly in Megf7-deficient mice. *Hum. Mol. Genet.* **2005**, *14*, 3523–3538. [CrossRef] [PubMed]
47. Zhang, B.; Luo, S.; Wang, Q.; Suzuki, T.; Xiong, W.C.; Mei, L. LRP4 serves as a coreceptor of agrin. *Neuron* **2008**, *60*, 285–297. [CrossRef] [PubMed]
48. Lickert, H.; Domon, C.; Huls, G.; Wehrle, C.; Duluc, I.; Clevers, H.; Meyer, B.I.; Freund, J.N.; Kemler, R. Wnt/(beta)-catenin signaling regulates the expression of the homeobox gene Cdx1 in embryonic intestine. *Development* **2000**, *127*, 3805–3813. [PubMed]
49. Pilon, N.; Oh, K.; Sylvestre, J.R.; Bouchard, N.; Savory, J.; Lohnes, D. Cdx4 is a direct target of the canonical Wnt pathway. *Dev. Biol.* **2006**, *289*, 55–63. [CrossRef] [PubMed]
50. Shinmyo, Y.; Mito, T.; Matsushita, T.; Sarashina, I.; Miyawaki, K.; Ohuchi, H.; Noji, S. *Caudal* is required for gnathal and thoracic patterning and for posterior elongation in the intermediate-germband cricket Gryllus bimaculatus. *Mech. Dev.* **2005**, *122*, 231–239. [CrossRef] [PubMed]
51. Choe, C.P.; Brown, S.J. Genetic regulation of *engrailed* and *wingless* in *Tribolium* segmentation and the evolution of pair-rule segmentation. *Dev. Biol.* **2009**, *325*, 482–491. [CrossRef] [PubMed]
52. Banyai, L.; Sonderegger, P.; Patthy, L. Agrin binds BMP2, BMP4 and TGFbeta1. *PLoS ONE* **2010**, *5*, e10758. [CrossRef] [PubMed]
53. Bull, A. *Bicaudal*, a genetic factor which affects the polarity of the embryo in *Drosophila melanogaster*. *J. Exp. Zool.* **1966**, *161*, 221–241. [CrossRef]
54. Mohler, J.; Wieschaus, E.F. Dominant maternal-effect mutations of *Drosophila melanogaster* causing the production of double-abdomen embryos. *Genetics* **1986**, *112*, 803–822. [PubMed]
55. Klomp, J.; Athy, D.; Kwan, C.W.; Bloch, N.I.; Sandmann, T.; Lemke, S.; Schmidt-Ott, U. Embryo development. A cysteine-clamp gene drives embryo polarity in the midge *Chironomus*. *Science* **2015**, *348*, 1040–1042. [CrossRef] [PubMed]

4

Unanswered Questions Regarding Sex and BMP/TGF-β Signaling

Tapan A. Shah and Melissa B. Rogers *

Rutgers—New Jersey Medical School, Microbiology, Biochemistry, & Molecular Genetics, Newark, NJ 07103, USA; shaht3@njms.rutgers.edu
* Correspondence: rogersmb@njms.rutgers.edu

Academic Editor: Samuel J. Pleasure

Abstract: Crosstalk between the BMP and TGF-β signaling pathways regulates many complex developmental processes from the earliest stages of embryogenesis throughout adult life. In many situations, the two signaling pathways act reciprocally. For example, TGF-β signaling is generally pro-fibrotic, whereas BMP signaling is anti-fibrotic and pro-calcific. Sex-specific differences occur in many diseases including cardiovascular pathologies. Differing ratios of fibrosis and calcification in stenotic valves suggests that BMP/TGF-β signaling may vary in men and women. In this review, we focus on the current understanding of the interplay between sex and BMP/TGF-β signaling and pose several unanswered questions.

Keywords: BMP; TGF-β; signaling; sex; chromosomes; XIST; genomic imprinting; hormones; fibrosis

---

1. Introduction

The distinct developmental mechanisms that bring about the dramatic differences in male and female characteristics are well studied. However, understanding of the impact of sex-associated signaling on the bone morphogenetic protein (BMP), transforming growth factor (TGF)-β and other pathways in developing animals and during adult life is incomplete. Cardiac valvulogenesis is just one of the many complex developmental processes where both BMP and TGF-β signals, along with WNT, fibroblast growth factor (FGF), NOTCH, vascular endothelial growth factor (VEGF) and epidermal growth factor (EGF) signals, orchestrate differentiation and morphology [1–3]. As in many adult diseases, processes that control normal embryonic processes also are involved in valve pathologies. Recent reports described dissimilarities in the ratios of fibrotic to calcified tissue in equally stenotic valves from men versus women [4–7]. Because TGF-β signaling is generally pro-fibrotic and BMP signaling is generally anti-fibrotic, we wondered if sex-associated changes in this balance contributed to the skewed sex distribution of valvular heart disease [4,7–11]. Here, we review the current state of understanding regarding the impact of sex on BMP and TGF-β signaling and identify several unanswered questions.

2. Sex Chromosomes

Most biological discussions of sex start with the X- and Y-chromosomes. Basic mechanisms that may influence BMP and TGF-β signaling would include X- or Y-linked inheritance of variant alleles and differences in X-chromosome inactivation. Of 64 mammalian BMP/TGF-β ligands, receptors, canonical signal mediators and extracellular and intracellular antagonists, only *Bmp15* (GDF9B) maps to the X chromosome in humans, rodents and sheep (Supplemental Table S1 and [12–15]). BMP15, an oocyte-derived growth and differentiation factor, is essential for folliculogenesis, granulosa cell function and many aspects of female fertility [16]. Interestingly, although clearly essential, mutations of the *Bmp15* gene can have the opposite effects on fecundity in different species [17–19]. Such differences in the impact of BMP15 underscore the challenges of translating how signaling pathways crosstalk in model organisms to agriculturally important species and humans.

Of greater overall impact to female biology, both the BMP and TGF-β signaling pathways regulate a key factor in X-chromosome inactivation. This critical process assures similar ratios of X to autosome gene expression between XY males and XX females. In female cells, one of the two X-chromosomes is transcriptionally inactivated by a mechanism involving the noncoding RNA XIST. Six members of the BMP/TGF-β signaling pathway (ACVR1B, BMPR2, SMAD2, SNIP1 and ZFYVE9) were identified in a screen for X-chromosome inactivation modulators [20]. Chromatin immunoprecipitation experiments determined that BMP signaling directly induced the expression of XIST [20]. In contrast, TGF-β1 downregulated XIST in vitro and in vivo. Significantly, TSIX, a negative regulator of XIST, was upregulated upon TGF-β stimulation [21]. Antagonism between the BMP and TGF-β pathways would profoundly influence X-chromosome dosage compensation and thus female biology on a cellular level [20].

Unanswered questions: Human X-chromosome inactivation and reactivation have profound consequences on cellular reprogramming and disease [22]. Likewise, BMP/TGF-β signaling strongly impacts pluripotency and differentiation. How does the upstream modulation of XIST expression by BMP/TGF-β signaling impact the downstream processes affected by these pathways? Does the ratio of BMP/TGF-β signaling affect the 15% of X-linked genes that escape from X-inactivation in human females, many in a tissue-specific pattern [23]? What is the relationship between pathological conditions that alter X-chromosome inactivation and BMP/TGF-β signaling?

3. Genetic Imprinting

Genetic Imprinting is another influential cellular process controlled by sex, in this case, that of the parents. Imprinted genes are expressed on either the maternal or paternal allele, but not both. This monoallelic expression causes a variant or mutated allele to produce a different phenotype based on the parental origin of that imprinted gene. Two hundred and fifty human and 150 mouse imprinted genes were surveyed (Supplemental Table S1 and [24,25]). The ligand BMP8B is predicted to have a paternal genetic imprint in humans. The extracellular antagonist Decorin is maternally imprinted in mouse, but not humans [25]. Decorin binds and sequesters the TGF-β ligand, thereby blocking the ligand-receptor interaction and the downstream signaling transduction [26]. Furthermore, at least two imprinted long noncoding (lnc) RNAs (H19 and MEG3) and their miRNA derivatives (e.g., miR-675-3p and -5p) have been shown to regulate key BMP/TGF-β ligands and signaling intermediaries, including BMP4, SMAD1 and SMAD5 [27,28]. These two lncRNAs are imprinted in both humans and mice and have been shown to influence mesenchymal stem cell lineage decisions such as myogenesis, adipogenesis and osteogenesis that BMP/TGF-β ligands also direct [27,28].

Unanswered questions: The epigenetic regulation of imprinted genes controls fetal and postnatal growth, with lifelong metabolic consequences such as obesity that impact health [29]. BMP and TGF-β signaling govern the differentiation of cells into myoblasts, adipocytes, chondrocytes or osteoblasts. How do parentally-imprinted regulators impact critical differentiation choices controlled by BMP/TGF-β signaling?

4. Hormones

Beyond the cell-intrinsic impact of the sex chromosomes and imprinting, hormones such as estrogen and androgens are essential drivers of female and male characteristics and function. Furthermore, the natural developmental variation in the hormonal milieu, for example during puberty, pregnancy, lactation and menopause, is substantial. In contrast to X-chromosome inactivation and imprinting, which alter the intrinsic nature of each cell, hormones are extrinsic factors that coordinate cell behaviors on a physiological scale. Unsurprisingly, sex hormones directly regulate many members of the BMP/TGF-β signaling pathways (Table 1). For example, estrogen directly induces *Bmp2* and *Bmp6* transcription [30,31]. In contrast, estrogen inhibits TGF-β signaling by stimulating SMAD2/3 protein degradation [32]. Testosterone was shown to significantly alter the expression of 20 members of the BMP/TGF-β pathway in skeletal muscle progenitors (satellite cells [33]). BMPs and TGF-βs have a long history in reproductive endocrinology. For example, the anti-Müllerian hormone (AMH), also known as the Müllerian-inhibiting substance (MIS), is a member of the superfamily that was discovered in 1947 [34]. An exceedingly complex network of steroid hormone interactions with BMP/TGF-β pathways has been described in reproductive organs. The many highly conserved roles in germ cell-specific and reproductive biology across all animals have been extensively reviewed recently [35–38]. Estrogen, progesterone and androgens interact with nearly all the members of the TGF-β superfamily (TGF-βs, BMPs, activin, inhibins, anti-Müllerian hormone, growth differentiation factors (GDFs), LEFTY and NODAL) in females [16]. Although possibly more extensively studied in the context of female reproduction, a complex network of BMP/TGF-β signaling is also essential in male reproductive biology [39,40].

Unanswered questions: A daunting web of BMP/TGF-β additive, synergistic and antagonistic actions among members of the ligand superfamily and signal mediators occurs in reproductive tissues. Each interaction within this panoply "may" occur in other tissues. The challenge is to identify which interactions also occur in other tissues.

Table 1. BMP/TGF-β signaling pathway members with molecular evidence of direct regulation by sex-related steroids.

Protein	Steroid	Receptor	Effect (↑ or ↓)	Evidence	Cell or Tissue Type	Reference
Ligands						
AMH (MIS)	Estrogen	Estrogen receptor α	↑	Luciferase reporter assay	KK1 cells	[24]
AMH (MIS)	Estrogen	Estrogen receptor β	↓	Luciferase reporter assay	KK1 cells	[24]
BMP2	Estrogen	Estrogen receptor α	↑	Luciferase reporter assay, ovariectomy	C3H10T1/2 cells, bone marrow mesenchymal stem cells	[30,41]
BMP6	Estrogen	Estrogen receptor α	↑	Luciferase reporter assay	MCF-7, T47-D cells, and HepG2 cells	[31]
INHβA (ACTA)	Estrogen	-	↓	Luciferase reporter assay	GRMO2 granulosa cells	[42]
INHβB (ACTB)	Estrogen	-	↓	Luciferase reporter assay	GRMO2 granulosa cells	[42]
TGF-β1	Dihydrotestosterone, R1881 synthetic androgen	Androgen receptor	↑ (PC3mm2 cells), ↓ (primary osteoblasts and LNCaP cells)	Luciferase reporter assay, Chromatin Immunoprecipitation	PC3mm2 cells, LNCaP cells, primary osteoblasts	[43–45]
TGF-β3	Estrogen	Estrogen receptor	↑	Chloramphenicol acetyl transferase (CAT) reporter assay	Human MG63 osteosarcoma cells	[46]
Extracellular Inhibitors						
Decorin	Progesterone, dienogest synthetic progestin	Progesterone receptor	↑	Chromatin immunoprecipitation	EMOsis cc/TERT and CRL-4003 cells.	[47]
Receptors						
TGFβR1 (ALK5)	Estrogen	Estrogen receptor α	↑	Luciferase reporter assay	osteoblasts	[48]
Intracellular Signal Transducers						
SMAD3	Dihydrotestosterone	Androgen receptor	↓	Luciferase reporter assay	prostate cancer cell lines	[49]
Intracellular Inhibitors						
SMURF1	Mibolerone synthetic androgen	Androgen receptor	↑	Chromatin immunoprecipitation	LNCaP cells	[50]

5. Crosstalk and Balance

In just the context of cardiovascular biology, sex profoundly influences heart and vascular health, with estrogen generally playing a protective role [51,52]. That a "one size fits all" approach to cardiovascular treatment cannot work for men and women is now widely recognized [52]. Several cardiovascular diseases, including aortic valve stenosis, exhibit sex-specific differential levels of calcified and fibrotic tissues. The reduced blood flow associated with aortic valve stenosis is life-threatening. Unfortunately, treatment is limited to surgical replacement. Both calcification and fibrosis impair valve leaflet mobility. Studies have shown that valves from men typically have greater aortic valve calcification, whereas women have a greater fibrosis score despite equal levels of stenosis and loss of function [5–7]. We postulate that the balance of BMP/TGF-β signaling may differ in the valves from males and females. Many members of the TGF-β superfamily, particularly the founding members TGF-β1 and -β2, but also activin A, myostatin and BMP9, promote fibrosis in various tissues [4,53]. Others, for example TGF-β3, BMP2 and BMP7, oppose fibrosis in multiple organs [53–63]. Different ligands often promote alternative lineage choices. For example, TGF-β1 inhibits calcific nodule formation in aortic valves in vivo by inducing SOX9, a pro-chondrogenic, anti-osteogenic transcription factor [64]. On the other hand, BMP2 and its downstream effectors, e.g., phosphorylated SMAD1/5/8(9), are potent pro-osteogenic signals strongly implicated in pathological calcification [65–70]. In a few models of organ fibrosis, the antagonistic nature of TGF-β and BMP signaling has been directly observed in the same tissues. In these in vitro and in vivo studies, TGF-β signaling promoted extracellular matrix synthesis and epithelial-mesenchymal transition (EMT), whereas BMP2 negatively regulated these pro-fibrotic processes [54,71,72]. Understanding this interplay between BMP and TGF-β signaling will potentially reveal potential strategies to control fibrotic, calcific and other pathologies [53,73].

Unanswered questions: Clear sex-specific differences in the relative levels of calcification and fibrosis in aortic valves occur. Although BMP/TGF-β signaling strongly influences these processes, few studies have addressed the regulation of these pathways in each sex. What are the relative levels of each BMP/TGF-β ligand in healthy and diseased valves in men and woman? How do sex and hormonal status influence the relative activities of signaling mediators and extracellular and intracellular antagonists of signaling? Most importantly, what therapeutic strategies may modulate the balance of BMP/TGF-β signaling optimally for each sex?

6. Concluding Remarks

Many factors lead to sex-specific differences in disease incidences and manifestations and in therapeutic efficacy [74]. These include cell-intrinsic genetic and cell-extrinsic physiological dissimilarities, as well as environmental circumstances such as healthcare inequities [75]. Although increased attention is now paid to social and organismal contrasts between males and females, far less is known regarding the impact of sex on biochemical signaling mechanisms. Despite extensive differences in many diseases, preclinical studies often ignore sex as an important biological variable. Studies often use only male animals or fail to report sex at all. Full understanding of disease processes will only be possible when the effect of sex on signal crosstalk is elucidated. The potential reward will be therapeutic methods to fine-tune the balance of networks involving BMPs, TGF-βs and other signals in both men and women.

Supplementary Materials: The Supplementary Materials are available online at http://www.mdpi.com/2221-3759/6/2/14/s1.

Author Contributions: T.A.S. and M.B.R. wrote the manuscript.

Funding: This work was funded by the National Heart, Lung and Blood Institute R01HL134947 and the National Institutes of Aging R56AG050762 awards to M.B.R.

Conflicts of Interest: The authors declare no conflict of interest.

References

1. Wu, B.; Wang, Y.; Xiao, F.; Butcher, J.T.; Yutzey, K.E.; Zhou, B. Developmental Mechanisms of Aortic Valve Malformation and Disease. *Annu. Rev. Physiol.* **2017**, *79*, 21–41. [CrossRef] [PubMed]
2. Kruithof, B.P.; Duim, S.N.; Moerkamp, A.T.; Goumans, M.J. TGFbeta and BMP signaling in cardiac cushion formation: Lessons from mice and chicken. *Differentiation* **2012**, *84*, 89–102. [CrossRef] [PubMed]
3. Dutta, P.; Lincoln, J. Calcific Aortic Valve Disease: A Developmental Biology Perspective. *Curr. Cardiol. Rep.* **2018**, *20*, 21. [CrossRef] [PubMed]
4. Sritharen, Y.; Enriquez-Sarano, M.; Schaff, H.V.; Casaclang-Verzosa, G.; Miller, J.D. Pathophysiology of Aortic Valve Stenosis: Is It Both Fibrocalcific and Sex Specific? *Physiology* **2017**, *32*, 182–196. [CrossRef] [PubMed]
5. Aggarwal, S.R.; Clavel, M.A.; Messika-Zeitoun, D.; Cueff, C.; Malouf, J.; Araoz, P.A.; Mankad, R.; Michelena, H.; Vahanian, A.; Enriquez-Sarano, M. Sex differences in aortic valve calcification measured by multidetector computed tomography in aortic stenosis. *Circ. Cardiovasc. Imaging* **2013**, *6*, 40–47. [CrossRef] [PubMed]
6. Thaden, J.J.; Nkomo, V.T.; Suri, R.M.; Maleszewski, J.J.; Soderberg, D.J.; Clavel, M.A.; Pislaru, S.V.; Malouf, J.F.; Foley, T.A.; Oh, J.K.; et al. Sex-related differences in calcific aortic stenosis: Correlating clinical and echocardiographic characteristics and computed tomography aortic valve calcium score to excised aortic valve weight. *Eur. Heart J.* **2016**, *37*, 693–699. [CrossRef] [PubMed]
7. Simard, L.; Cote, N.; Dagenais, F.; Mathieu, P.; Couture, C.; Trahan, S.; Bosse, Y.; Mohammadi, S.; Page, S.; Joubert, P.; et al. Sex-Related Discordance Between Aortic Valve Calcification and Hemodynamic Severity of Aortic Stenosis: Is Valvular Fibrosis the Explanation? *Circ. Res.* **2017**, *120*, 681–691. [CrossRef] [PubMed]
8. Andell, P.; Li, X.; Martinsson, A.; Andersson, C.; Stagmo, M.; Zoller, B.; Sundquist, K.; Smith, J.G. Epidemiology of valvular heart disease in a Swedish nationwide hospital-based register study. *Heart* **2017**, *103*, 1696–1703. [CrossRef] [PubMed]
9. Kong, W.K.; Regeer, M.V.; Ng, A.C.; McCormack, L.; Poh, K.K.; Yeo, T.C.; Shanks, M.; Parent, S.; Enache, R.; Popescu, B.A.; et al. Sex Differences in Phenotypes of Bicuspid Aortic Valve and Aortopathy: Insights From a Large Multicenter, International Registry. *Circ. Cardiovasc. Imaging* **2017**, *10*, e005155. [CrossRef] [PubMed]
10. Michelena, H.I.; Mankad, S.V. Sex Differences in Bicuspid Aortic Valve Adults: Who Deserves Our Attention, Men or Women? *Circ. Cardiovasc. Imaging* **2017**, *10*, e006123. [CrossRef] [PubMed]
11. Porras, A.M.; McCoy, C.M.; Masters, K.S. Calcific Aortic Valve Disease: A Battle of the Sexes. *Circ. Res.* **2017**, *120*, 604–606. [CrossRef] [PubMed]
12. Salazar, V.S.; Gamer, L.W.; Rosen, V. BMP signalling in skeletal development, disease and repair. *Nat. Rev. Endocrinol.* **2016**, *12*, 203–221. [CrossRef] [PubMed]
13. Kanehisa, M.; Furumichi, M.; Tanabe, M.; Sato, Y.; Morishima, K. KEGG: New perspectives on genomes, pathways, diseases and drugs. *Nucleic Acids Res.* **2017**, *45*, D353–D361. [CrossRef] [PubMed]
14. Qiagen Ingenuity Pathway Analysis. Available online: https://www.qiagenbioinformatics.com/products/ingenuity-pathway-analysis/ (accessed on 7 May 2018).
15. Sanchez-Duffhues, G.; Hiepen, C.; Knaus, P.; Ten Dijke, P. Bone morphogenetic protein signaling in bone homeostasis. *Bone* **2015**, *80*, 43–59. [CrossRef] [PubMed]
16. Ni, N.; Li, Q. TGFbeta superfamily signaling and uterine decidualization. *Reprod. Biol. Endocrinol.* **2017**, *15*, 84. [CrossRef] [PubMed]
17. Monsivais, D.; Matzuk, M.M.; Pangas, S.A. The TGF-beta Family in the Reproductive Tract. *Cold Spring Harb. Perspect. Biol.* **2017**, *9*, a022251. [CrossRef] [PubMed]
18. Galloway, S.M.; McNatty, K.P.; Cambridge, L.M.; Laitinen, M.P.; Juengel, J.L.; Jokiranta, T.S.; McLaren, R.J.; Luiro, K.; Dodds, K.G.; Montgomery, G.W.; et al. Mutations in an oocyte-derived growth factor gene (BMP15) cause increased ovulation rate and infertility in a dosage-sensitive manner. *Nat. Genet.* **2000**, *25*, 279–283. [CrossRef] [PubMed]
19. Yan, C.; Wang, P.; DeMayo, J.; DeMayo, F.J.; Elvin, J.A.; Carino, C.; Prasad, S.V.; Skinner, S.S.; Dunbar, B.S.; Dube, J.L.; et al. Synergistic roles of bone morphogenetic protein 15 and growth differentiation factor 9 in ovarian function. *Mol. Endocrinol.* **2001**, *15*, 854–866. [CrossRef] [PubMed]

20. Sripathy, S.; Leko, V.; Adrianse, R.L.; Loe, T.; Foss, E.J.; Dalrymple, E.; Lao, U.; Gatbonton-Schwager, T.; Carter, K.T.; Payer, B.; et al. Screen for reactivation of MeCP2 on the inactive X chromosome identifies the BMP/TGF-beta superfamily as a regulator of XIST expression. *Proc. Natl. Acad. Sci. USA* **2017**, *114*, 1619–1624. [CrossRef] [PubMed]
21. Zhongzhi, W.; Masatoshi, J.; Kayo, N.; Miho, H.; Hideo, K.; Wakana, N.; Kuniko, I.; Taiji, N.; Noritoshi, H.; Satoshi, F.; et al. Long non-coding RNA TSIX is upregulated in scleroderma dermal fibroblasts and controls collagen mRNA stabilization. *Exp. Dermatol.* **2016**, *25*, 131–136.
22. Cantone, I.; Fisher, A.G. Human X chromosome inactivation and reactivation: Implications for cell reprogramming and disease. *Phil. Trans. R. Soc. B* **2017**, *372*, 20160358. [CrossRef] [PubMed]
23. Disteche, C.M.; Berletch, J.B. X-chromosome inactivation and escape. *J. Genet.* **2015**, *94*, 591–599. [CrossRef] [PubMed]
24. Grynberg, M.; Pierre, A.; Rey, R.; Leclerc, A.; Arouche, N.; Hesters, L.; Catteau-Jonard, S.; Frydman, R.; Picard, J.Y.; Fanchin, R.; et al. Differential regulation of ovarian anti-mullerian hormone (AMH) by estradiol through alpha- and beta-estrogen receptors. *J. Clin. Endocrinol. Metab.* **2012**, *97*, E1649–E1657. [CrossRef] [PubMed]
25. Monk, D.; Arnaud, P.; Apostolidou, S.; Hills, F.A.; Kelsey, G.; Stanier, P.; Feil, R.; Moore, G.E. Limited evolutionary conservation of imprinting in the human placenta. *Proc. Natl. Acad. Sci. USA* **2006**, *103*, 6623–6628. [CrossRef] [PubMed]
26. Hildebrand, A.; Romaris, M.; Rasmussen, L.M.; Heinegard, D.; Twardzik, D.R.; Border, W.A.; Ruoslahti, E. Interaction of the small interstitial proteoglycans biglycan, decorin and fibromodulin with transforming growth factor beta. *Biochem. J.* **1994**, *302 Pt 2*, 527–534. [CrossRef] [PubMed]
27. Dey, B.K.; Pfeifer, K.; Dutta, A. The H19 long noncoding RNA gives rise to microRNAs miR-675-3p and miR-675-5p to promote skeletal muscle differentiation and regeneration. *Genes Dev.* **2014**, *28*, 491–501. [CrossRef] [PubMed]
28. Peng, S.; Cao, L.; He, S.; Zhong, Y.; Ma, H.; Zhang, Y.; Shuai, C. An Overview of Long Noncoding RNAs Involved in Bone Regeneration from Mesenchymal Stem Cells. *Stem Cells Int.* **2018**, *2018*, 8273648. [CrossRef] [PubMed]
29. Cassidy, F.C.; Charalambous, M. Genomic imprinting, growth and maternal-fetal interactions. *J. Exp. Biol.* **2018**, *221*. [CrossRef] [PubMed]
30. Zhou, S.; Turgeman, G.; Harris, S.E.; Leitman, D.C.; Komm, B.S.; Bodine, P.V.; Gazit, D. Estrogens activate bone morphogenetic protein-2 gene transcription in mouse mesenchymal stem cells. *Mol. Endocrinol.* **2003**, *17*, 56–66. [CrossRef] [PubMed]
31. Ong, D.B.; Colley, S.M.; Norman, M.R.; Kitazawa, S.; Tobias, J.H. Transcriptional regulation of a BMP-6 promoter by estrogen receptor alpha. *J. Bone Miner. Res.* **2004**, *19*, 447–454. [CrossRef] [PubMed]
32. Ito, I.; Hanyu, A.; Wayama, M.; Goto, N.; Katsuno, Y.; Kawasaki, S.; Nakajima, Y.; Kajiro, M.; Komatsu, Y.; Fujimura, A.; et al. Estrogen inhibits transforming growth factor beta signaling by promoting Smad2/3 degradation. *J. Biol. Chem.* **2010**, *285*, 14747–14755. [CrossRef] [PubMed]
33. Braga, M.; Bhasin, S.; Jasuja, R.; Pervin, S.; Singh, R. Testosterone inhibits transforming growth factor-beta signaling during myogenic differentiation and proliferation of mouse satellite cells: Potential role of follistatin in mediating testosterone action. *Mol. Cell. Endocrinol.* **2012**, *350*, 39–52. [CrossRef] [PubMed]
34. Teixeira, J.; Maheswaran, S.; Donahoe, P.K. Mullerian inhibiting substance: An instructive developmental hormone with diagnostic and possible therapeutic applications. *Endocr. Rev.* **2001**, *22*, 657–674. [PubMed]
35. Lochab, A.K.; Extavour, C.G. Bone Morphogenetic Protein (BMP) signaling in animal reproductive system development and function. *Dev. Biol.* **2017**, *427*, 258–269. [CrossRef] [PubMed]
36. Wijayarathna, R.; de Kretser, D.M. Activins in reproductive biology and beyond. *Hum. Reprod. Update* **2016**, *22*, 342–357. [CrossRef] [PubMed]
37. Young, J.C.; Wakitani, S.; Loveland, K.L. TGF-beta superfamily signaling in testis formation and early male germline development. *Semin. Cell Dev. Biol.* **2015**, *45*, 94–103. [CrossRef] [PubMed]
38. Roly, Z.Y.; Backhouse, B.; Cutting, A.; Tan, T.Y.; Sinclair, A.H.; Ayers, K.L.; Major, A.T.; Smith, C.A. The cell biology and molecular genetics of Mullerian duct development. *Wiley Interdiscip. Rev. Dev. Biol.* **2018**, *7*, e310. [CrossRef] [PubMed]
39. Murashima, A.; Kishigami, S.; Thomson, A.; Yamada, G. Androgens and mammalian male reproductive tract development. *Biochim. Biophys. Acta* **2015**, *1849*, 163–170. [CrossRef] [PubMed]

40. Ciller, I.M.; Palanisamy, S.K.; Ciller, U.A.; McFarlane, J.R. Postnatal expression of bone morphogenetic proteins and their receptors in the mouse testis. *Physiol. Res.* **2016**, *65*, 673–682. [PubMed]
41. Zhou, S.; Zilberman, Y.; Wassermann, K.; Bain, S.D.; Sadovsky, Y.; Gazit, D. Estrogen modulates estrogen receptor alpha and beta expression, osteogenic activity, and apoptosis in mesenchymal stem cells (MSCs) of osteoporotic mice. *J. Cell. Biochem.* **2001**, *81*, 144–155. [CrossRef]
42. Kipp, J.L.; Kilen, S.M.; Bristol-Gould, S.; Woodruff, T.K.; Mayo, K.E. Neonatal exposure to estrogens suppresses activin expression and signaling in the mouse ovary. *Endocrinology* **2007**, *148*, 1968–1976. [CrossRef] [PubMed]
43. Qi, W.; Gao, S.; Wang, Z. Transcriptional regulation of the TGF-beta1 promoter by androgen receptor. *Biochem. J.* **2008**, *416*, 453–462. [CrossRef] [PubMed]
44. McCarthy, T.L.; Centrella, M. Androgen receptor activation integrates complex transcriptional effects in osteoblasts, involving the growth factors TGF-beta and IGF-I, and transcription factor C/EBPdelta. *Gene* **2015**, *573*, 129–140. [CrossRef] [PubMed]
45. Chipuk, J.E.; Cornelius, S.C.; Pultz, N.J.; Jorgensen, J.S.; Bonham, M.J.; Kim, S.J.; Danielpour, D. The androgen receptor represses transforming growth factor-beta signaling through interaction with Smad3. *J. Biol. Chem.* **2002**, *277*, 1240–1248. [CrossRef] [PubMed]
46. Yang, N.N.; Bryant, H.U.; Hardikar, S.; Sato, M.; Galvin, R.J.; Glasebrook, A.L.; Termine, J.D. Estrogen and raloxifene stimulate transforming growth factor-beta 3 gene expression in rat bone: A potential mechanism for estrogen- or raloxifene-mediated bone maintenance. *Endocrinology* **1996**, *137*, 2075–2084. [CrossRef] [PubMed]
47. Ono, Y.J.; Terai, Y.; Tanabe, A.; Hayashi, A.; Hayashi, M.; Yamashita, Y.; Kyo, S.; Ohmichi, M. Decorin induced by progesterone plays a crucial role in suppressing endometriosis. *J. Endocrinol.* **2014**, *223*, 203–216. [CrossRef] [PubMed]
48. McCarthy, T.L.; Chang, W.Z.; Liu, Y.; Centrella, M. Runx2 integrates estrogen activity in osteoblasts. *J. Biol. Chem.* **2003**, *278*, 43121–43129. [CrossRef] [PubMed]
49. Song, K.; Wang, H.; Krebs, T.L.; Wang, B.; Kelley, T.J.; Danielpour, D. DHT selectively reverses Smad3-mediated/TGF-beta-induced responses through transcriptional down-regulation of Smad3 in prostate epithelial cells. *Mol. Endocrinol.* **2010**, *24*, 2019–2029. [CrossRef] [PubMed]
50. Gang, X.; Wang, G.; Huang, H. Androgens regulate SMAD ubiquitination regulatory factor-1 expression and prostate cancer cell invasion. *Prostate* **2015**, *75*, 561–572. [CrossRef] [PubMed]
51. Boese, A.C.; Kim, S.C.; Yin, K.J.; Lee, J.P.; Hamblin, M.H. Sex differences in vascular physiology and pathophysiology: Estrogen and androgen signaling in health and disease. *Am. J. Physiol. Heart Circ. Physiol.* **2017**, *313*, H524–H545. [CrossRef] [PubMed]
52. den Ruijter, H.M.; Haitjema, S.; Asselbergs, F.W.; Pasterkamp, G. Sex matters to the heart: A special issue dedicated to the impact of sex related differences of cardiovascular diseases. *Atherosclerosis* **2015**, *241*, 205–207. [CrossRef] [PubMed]
53. Walton, K.L.; Johnson, K.E.; Harrison, C.A. Targeting TGF-beta Mediated SMAD Signaling for the Prevention of Fibrosis. *Front. Pharmacol.* **2017**, *8*, 461. [CrossRef] [PubMed]
54. Wang, S.; Sun, A.; Li, L.; Zhao, G.; Jia, J.; Wang, K.; Ge, J.; Zou, Y. Up-regulation of BMP-2 antagonizes TGF-beta1/ROCK-enhanced cardiac fibrotic signalling through activation of Smurf1/Smad6 complex. *J. Cell. Mol. Med.* **2012**, *16*, 2301–2310. [CrossRef] [PubMed]
55. Izumi, M.; Masaki, M.; Hiramoto, Y.; Sugiyama, S.; Kuroda, T.; Terai, K.; Hori, M.; Kawase, I.; Hirota, H. Cross-talk between bone morphogenetic protein 2 and leukemia inhibitory factor through ERK 1/2 and Smad1 in protection against doxorubicin-induced injury of cardiomyocytes. *J. Mol. Cell. Cardiol.* **2006**, *40*, 224–233. [CrossRef] [PubMed]
56. Ebelt, H.; Hillebrand, I.; Arlt, S.; Zhang, Y.; Kostin, S.; Neuhaus, H.; Muller-Werdan, U.; Schwarz, E.; Werdan, K.; Braun, T. Treatment with bone morphogenetic protein 2 limits infarct size after myocardial infarction in mice. *Shock* **2013**, *39*, 353–360. [CrossRef] [PubMed]
57. Lepparanta, O.; Tikkanen, J.M.; Bespalov, M.M.; Koli, K.; Myllarniemi, M. Bone morphogenetic protein-inducer tilorone identified by high-throughput screening is antifibrotic in vivo. *Am. J. Respir. Cell Mol. Biol.* **2013**, *48*, 448–455. [CrossRef] [PubMed]
58. Koli, K.; Myllarniemi, M.; Vuorinen, K.; Salmenkivi, K.; Ryynanen, M.J.; Kinnula, V.L.; Keski-Oja, J. Bone morphogenetic protein-4 inhibitor gremlin is overexpressed in idiopathic pulmonary fibrosis. *Am. J. Pathol.* **2006**, *169*, 61–71. [CrossRef] [PubMed]

59. Myllarniemi, M.; Lindholm, P.; Ryynanen, M.J.; Kliment, C.R.; Salmenkivi, K.; Keski-Oja, J.; Kinnula, V.L.; Oury, T.D.; Koli, K. Gremlin-mediated decrease in bone morphogenetic protein signaling promotes pulmonary fibrosis. *Am. J. Respir. Crit. Care Med.* **2008**, *177*, 321–329. [CrossRef] [PubMed]
60. De Langhe, E.; Cailotto, F.; De Vooght, V.; Aznar-Lopez, C.; Vanoirbeek, J.A.; Luyten, F.P.; Lories, R.J. Enhanced endogenous bone morphogenetic protein signaling protects against bleomycin induced pulmonary fibrosis. *Respir. Res.* **2015**, *16*, 38. [CrossRef] [PubMed]
61. Gao, X.; Cao, Y.; Staloch, D.A.; Gonzales, M.A.; Aronson, J.F.; Chao, C.; Hellmich, M.R.; Ko, T.C. Bone morphogenetic protein signaling protects against cerulein-induced pancreatic fibrosis. *PLoS ONE* **2014**, *9*, e89114. [CrossRef] [PubMed]
62. Myllarniemi, M.; Vuorinen, K.; Pulkkinen, V.; Kankaanranta, H.; Aine, T.; Salmenkivi, K.; Keski-Oja, J.; Koli, K.; Kinnula, V. Gremlin localization and expression levels partially differentiate idiopathic interstitial pneumonia severity and subtype. *J. Pathol.* **2008**, *214*, 456–463. [CrossRef] [PubMed]
63. Farkas, L.; Farkas, D.; Gauldie, J.; Warburton, D.; Shi, W.; Kolb, M. Transient overexpression of Gremlin results in epithelial activation and reversible fibrosis in rat lungs. *Am. J. Respir. Cell. Mol. Biol.* **2011**, *44*, 870–878. [CrossRef] [PubMed]
64. Huk, D.J.; Austin, B.F.; Horne, T.E.; Hinton, R.B.; Ray, W.C.; Heistad, D.D.; Lincoln, J. Valve Endothelial Cell-Derived Tgfbeta1 Signaling Promotes Nuclear Localization of Sox9 in Interstitial Cells Associated With Attenuated Calcification. *Arterioscler. Thromb. Vasc. Biol.* **2016**, *36*, 328–338. [CrossRef] [PubMed]
65. Bostrom, K.; Watson, K.E.; Horn, S.; Wortham, C.; Herman, I.M.; Demer, L.L. Bone morphogenetic protein expression in human atherosclerotic lesions. *J. Clin. Investig.* **1993**, *91*, 1800–1809. [CrossRef] [PubMed]
66. Kaden, J.J.; Bickelhaupt, S.; Grobholz, R.; Vahl, C.F.; Hagl, S.; Brueckmann, M.; Haase, K.K.; Dempfle, C.E.; Borggrefe, M. Expression of bone sialoprotein and bone morphogenetic protein-2 in calcific aortic stenosis. *J. Heart Valve Dis.* **2004**, *13*, 560–566. [PubMed]
67. Yang, X.; Meng, X.; Su, X.; Mauchley, D.C.; Ao, L.; Cleveland, J.C., Jr.; Fullerton, D.A. Bone morphogenic protein 2 induces Runx2 and osteopontin expression in human aortic valve interstitial cells: Role of Smad1 and extracellular signal-regulated kinase 1/2. *J. Thorac. Cardiovasc. Surg.* **2009**, *138*, 1008–1015. [CrossRef] [PubMed]
68. Balachandran, K.; Sucosky, P.; Jo, H.; Yoganathan, A.P. Elevated cyclic stretch induces aortic valve calcification in a bone morphogenic protein-dependent manner. *Am. J. Pathol.* **2010**, *177*, 49–57. [CrossRef] [PubMed]
69. Miller, J.D.; Weiss, R.M.; Serrano, K.M.; Castaneda, L.E.; Brooks, R.M.; Zimmerman, K.; Heistad, D.D. Evidence for active regulation of pro-osteogenic signaling in advanced aortic valve disease. *Arterioscler. Thromb. Vasc. Biol.* **2010**, *30*, 2482–2486. [CrossRef] [PubMed]
70. Seya, K.; Yu, Z.; Kanemaru, K.; Daitoku, K.; Akemoto, Y.; Shibuya, H.; Fukuda, I.; Okumura, K.; Motomura, S.; Furukawa, K. Contribution of bone morphogenetic protein-2 to aortic valve calcification in aged rat. *J. Pharmacol. Sci.* **2011**, *115*, 8–14. [CrossRef] [PubMed]
71. Flechsig, P.; Dadrich, M.; Bickelhaupt, S.; Jenne, J.; Hauser, K.; Timke, C.; Peschke, P.; Hahn, E.W.; Grone, H.J.; Yingling, J.; et al. LY2109761 attenuates radiation-induced pulmonary murine fibrosis via reversal of TGF-beta and BMP-associated proinflammatory and proangiogenic signals. *Clin. Cancer Res.* **2012**, *18*, 3616–3627. [CrossRef] [PubMed]
72. Yang, Y.L.; Liu, Y.S.; Chuang, L.Y.; Guh, J.Y.; Lee, T.C.; Liao, T.N.; Hung, M.Y.; Chiang, T.A. Bone morphogenetic protein-2 antagonizes renal interstitial fibrosis by promoting catabolism of type I transforming growth factor-beta receptors. *Endocrinology* **2009**, *150*, 727–740. [CrossRef] [PubMed]
73. Akhurst, R.J.; Hata, A. Targeting the TGFbeta signalling pathway in disease. *Nat. Rev. Drug Discov.* **2012**, *11*, 790–811. [CrossRef] [PubMed]
74. Institute of Medicine. *Exploring the Biological Contributions to Human Health: Does Sex Matter?* The National Academies Press: Washington, DC, USA, 2001; p. 288.
75. Shaw, L.J.; Pepine, C.J.; Xie, J.; Mehta, P.K.; Morris, A.A.; Dickert, N.W.; Ferdinand, K.C.; Gulati, M.; Reynolds, H.; Hayes, S.N.; et al. Quality and Equitable Health Care Gaps for Women: Attributions to Sex Differences in Cardiovascular Medicine. *J. Am. Coll. Cardiol.* **2017**, *70*, 373–388. [CrossRef] [PubMed]

5

Specification and Patterning of *Drosophila* Appendages

Mireya Ruiz-Losada, David Blom-Dahl, Sergio Córdoba and Carlos Estella *

Departamento de Biología Molecular and Centro de Biología Molecular Severo Ochoa, Universidad Autónoma de Madrid (UAM/CSIC), Nicolás Cabrera 1, 28049 Madrid, Spain; mireya.ruiz@cbm.csic.es (M.R.-L.); david.blom-dahl@cbm.csic.es (D.B.-D.); scordoba@cbm.csic.es (S.C.)

* Correspondence: cestella@cbm.csic.es

Abstract: Appendages are external projections of the body that serve the animal for locomotion, feeding, or environment exploration. The appendages of the fruit fly *Drosophila melanogaster* are derived from the imaginal discs, epithelial sac-like structures specified in the embryo that grow and pattern during larva development. In the last decades, genetic and developmental studies in the fruit fly have provided extensive knowledge regarding the mechanisms that direct the formation of the appendages. Importantly, many of the signaling pathways and patterning genes identified and characterized in *Drosophila* have similar functions during vertebrate appendage development. In this review, we will summarize the genetic and molecular mechanisms that lead to the specification of appendage primordia in the embryo and their posterior patterning during imaginal disc development. The identification of the regulatory logic underlying appendage specification in *Drosophila* suggests that the evolutionary origin of the insect wing is, in part, related to the development of ventral appendages.

Keywords: *Drosophila*; appendages; imaginal discs; transcription factors; signaling pathways; proximodistal axis; wing; leg; antenna; appendage primordia

1. Introduction

Arthropods are the most successful group of animals and represent approximately three-quarters of the animal species currently living on Earth [1]. A key factor contributing to their success is their ability to exploit multiple ecological niches, most likely due to body plan modifications and innovations generating vast morphological diversity. Arthropod appendages show great variation in number, shape, and function, allowing multiple purposes from locomotion to feeding or environment sensing. However, besides their morphological disparity, it is likely that all appendages share common generative rules. Appendages have a proximodistal (P-D) axis that forms *de novo* orthogonal to the main body axes, the anteroposterior (A-P) and the dorsoventral (D-V). Molecular and genetic studies in the fruit fly, *Drosophila melanogaster,* have provided the basis of appendage development and served for comparative analysis with other arthropods [2]. Interestingly, many of the signaling pathways and patterning genes identified and characterized in *Drosophila* have comparable functions during vertebrate appendage development [3,4]. Thus, vertebrate and invertebrate appendages are built using similar underlying genetic programs, even though they are not homologous structures in the classic sense [5,6]. These similarities are often referred to as 'deep homology' [7].

Primitive insects develop appendages directly from embryonic limb buds that grow as external projections, while more derived insect species with complete metamorphosis, such as *Drosophila,* develop their appendages from imaginal precursors. The external adult body of *Drosophila* is formed by two different sets of cells, the imaginal discs and the histoblasts. The imaginal discs are specialized epithelial sheets specified in the embryo that grow and become patterned inside the larva. During the pupal stage, imaginal discs evert and differentiate to form the adult structures. The histoblasts are the precursors of the fly abdomen that, in contrast to the imaginal discs, only proliferate during pupal development. There are 19 imaginal discs in the larvae, 9 appearing in pairs, and the genitalia. The wing and haltere discs form the corresponding appendages and also the dorsal thorax. In a similar manner, the leg discs develop the appendage proper and the ventral pleura of the thorax. While thoracic imaginal precursors (wing, haltere, and legs) originate from a single embryonic segment [8,9], the genital disc primordia is a sexually dimorphic compound primordia derived from three abdominal segments (A8, A9, and A10) (reviewed in [10]). In a similar fashion to the genital primordia, the cells from different cephalic segmental identities coalesce together to form the eye-antennal disc [11]. The eye-antennal disc is also a compound structure that gives rise to the olfactory (antenna and maxillary palps) and visual (compound eyes and ocelli) organs plus the head epidermis [12]. We can group the appendages in dorsal or ventral depending on their relative positions within the body and their homology. Therefore, ventral appendages include the legs, antenna, and genitalia, while the wings and halteres are dorsal. In this review, we will focus on the patterning of the thoracic appendages and the antenna.

At the molecular level, a small number of signaling pathways are reiteratively used throughout development to specify and pattern the appendages in *Drosophila* and in vertebrates [3,4]. The function of signaling pathways such as hedgehog (Hh), decapentaplegic (Dpp), Wingless (Wg), Epidermal Growth Factor Receptor (EGFR) and Notch, and the transcription factors Distal-less (Dll), Homothorax (Hth) and members of the Sp family (Btd/Sp1) is fundamental in the formation of appendages. These signaling pathways and transcription factors are linked in regulatory networks, and cooperate to subdivide the developing appendages in different domains of gene expression. The identification and dissection of the *cis*-regulatory modules (CRMs) that controls the expression of genes required for wing, leg, or antenna development has been fundamental to decipher the regulatory networks that direct appendage formation [13–18]. Below, we discuss the regulatory logic behind appendage specification and patterning, and the implications in the evolutionary origin of the insect wing.

2. Allocation of the Thoracic Appendage Primordia

The development of imaginal discs is a progressive process initiated during embryogenesis with the specification of the imaginal primordia. Each primordium is formed by a characteristic number of cells located in the embryonic ectoderm in precise A-P and D-V location (Figure 1) [8,19]. All imaginal primordia are easily recognized by the restricted expression of the genes *escargot* (*esg*) and *headcase* (*hdc*), which maintain the imaginal state through the repression of endoreplication [20,21]. The segmental identity and position along the A-P axis of the different imaginal primordia is provided by the activity of Hox proteins [8,13,22–24]. The Hox genes encode homeodomain containing transcription factors that confer A-P identity along the body of the fly, defining unique developmental programs to each segment (reviewed in [25]). Therefore, the thoracic appendage primordia (legs, wing, and haltere) are restricted to their corresponding thoracic segments through the repression exerted by the abdominal Hox proteins (see below) [8,13,22,23].

The leg and wing/haltere presumptive imaginal discs are recognized in the thoracic segments of a late embryo as a genetically and morphologically distinct group of cells that form the ventral (VP) and dorsal primordia (DP), respectively (Figure 1A). The DP activates the expression of the *snail* (*sna*) and *vg* (*vestigial*) genes, while the VP expresses *Distal-less* (*Dll*) and the Sp family members *Sp1* and *buttonhead* (*btd*) [26–30]. The first molecular sign of appendage specification is the activation of the homeobox-containing gene *Dll* [5,8,19,31]. *Dll* is expressed in the outgrowths of many animals and is essential for appendage formation [5,32]. In *Drosophila*, *Dll* expression is initiated in a group of about 30 cells in each thoracic hemisegment at stage 10 of embryogenesis [19]. Multiple *Dll* CRMs have been identified that partially reproduce the spatial and temporal expression of *Dll*. The characterization of these *Dll* CRMs reveals not only the molecular logic of *Dll* regulation but also allows the description of the developmental fate of *Dll* expressing cells [14–16,23]. At stage 10–11, an early *Dll* CRM, named *Dll*-304, is activated in a pattern similar to *Dll* and its activity decays some hours later [8,23]. These cells are defined as thoracic primordia (TP) because their progeny will contribute to both the VP (leg) and DP (wing/haltere) (see below). At this stage, *Dll* expression is positively regulated by Wg, that is expressed in D-V stripes in the anterior compartment of each thoracic segment [19]. The precise localization of *Dll* expression along the D-V axis in the thoracic embryo epidermis is also regulated by repression mediated by Decapentaplegic (Dpp) and the Epidermal Growth Factor Receptor (EGFR) pathways in dorsal and ventral cells, respectively [33] (Figure 1B). The regulation exerted by the Wg, Dpp, and EGFR signaling pathways is present in the thoracic and abdominal segments, although *Dll* expression is restricted to the thorax by repression mediated by the abdominal Hox proteins [23,34,35]. The Hox proteins Ultrabithorax (Ubx), Abdominal-A (Abd-A), and Abdominal B (Abd-B) directly bind to the early *Dll*-304 CRM in a compartment specific manner [35]. Although *Dll* could be activated in the absence of Antenapedia (Antp), the thoracic Hox protein, it has been recently shown that Antp plays a positive role in enhancing *Dll* expression levels [36].

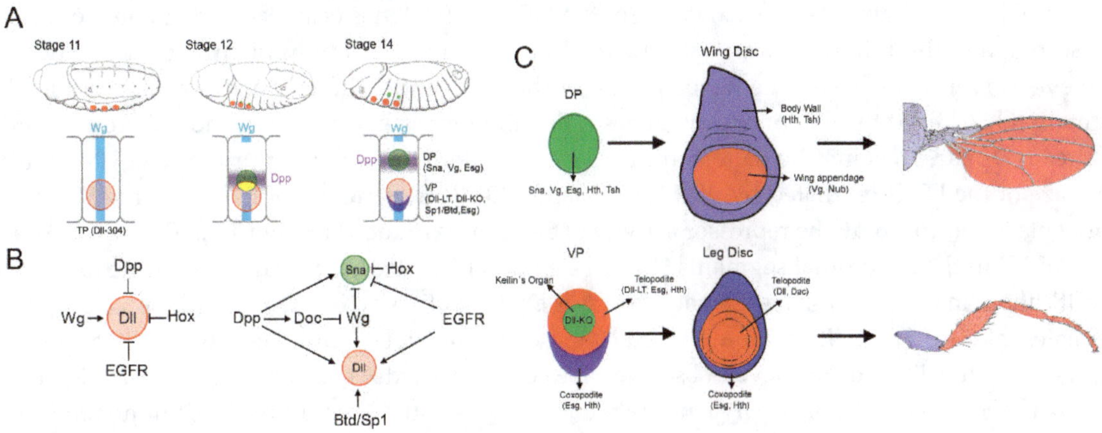

Figure 1. Specification of the thoracic appendages. (**A**) Cartoon showing three representative stages of *Drosophila* embryogenesis and the sequential appearance of the thoracic primordia (TP, red circle), the ventral primordia (VP, red circle) and the dorsal primordia (DP, green circle). Below is presented a schematic representation of the second thoracic segment where the expression of TP-, DP-, and VP-specific genes is related to the Wg and Dpp signaling pathways. Note that the DP originates from two populations of cells: one within the TP (drawn in yellow) and another one dorsal to the TP; (**B**) Genetic inputs that control appendage specification when *Dll* is first activated (stage 10–11) and during activation of the DP gene *sna* (stage 12 onwards). (**C**) Schematic representation of the DP and VP and the imaginal discs and adult structures that they will give rise to (wing and leg, respectively). The subdivision of the wing and leg imaginal discs in the body wall and appendage territories is also shown.

2.1. *Specification of the Ventral and Dorsal Primordia*

Classic lineage experiments using X-ray somatic recombination and gynandromorphs already suggested that the dorsal and ventral primordia originated at the blastoderm stage in close proximity and presumably derive from an overlapping population of cells [37–39]. In accordance, more recent lineage-tracing experiments revealed that the progeny of *Dll* or *Dll*-304 expressing cells (TP) not only contribute to the entire leg imaginal disc, including the ventral body wall and the appendage, but also to the wing disc [13,16]. The initial group of *Dll* expressing cells (TP) is subdivided as embryogenesis progresses in cells that will contribute to the DP (wing and haltere) and the VP (legs). This process requires a correct balance between the D-V positional cues provided by the Dpp, EGFR, and Wg pathways [8,13,33,40,41] (Figure 1A,B). *dpp* expression at the dorsal side of the TP field generates a concentration gradient that promotes DP fate at high levels and specifies the VP fate at intermediate levels [33]. Simultaneously, the ventral activation of the EGFR pathway antagonizes Dpp signaling and restricts the activation of the wing promoting genes to the dorsal domain of the TP [40]. Dpp also activates the *Dorsocross* (*Doc*) genes in a broad lateral stripe that partially overlaps with the dorsal domain of the TP [42]. This leads to the repression of *wg* expression in the lateral ectoderm [13,43,44] (Figure 1A,B). In this manner, dorsal TP cells are exposed to high levels of both Doc and Dpp signaling and low Wg and EGFR. These cells will contribute to form the DP, lose *Dll* expression and activate the DP developmental program. In contrast, ventral TP cells retain high levels of Wg and EGFR, maintain *Dll* expression, and are fated to form the VP. Contrary to the opposite roles that the Dpp and EGFR pathways have on DP formation, both pathways positively direct VP fates [8,13,40,41]. At this stage, the role of the Wg pathway is restricted to VP, as it is necessary for *Dll* expression but not for wing formation [41].

The repression of Wg and the presence of high Dpp levels are a prerequisite for the activation of the DP promoting genes *sna*, *esg*, and *vg* in cells of the lateral ectoderm and dorsal to *Dll* expressing cells (Figure 1A) [13,33,40,41]. The activation of *sna* and *vg* is only observed in Dll-negative cells after the separation of the dorsal and ventral primordia. In contrast to *Dll*, which is expressed in the three thoracic segments, the activation of the DP specific genes (*sna*, *esg*, and *vg*) only occurs in the second and third thoracic segments (T2 and T3). *sna* and *esg* act redundantly to specify DP fate while *vg* is required later for the development of the wing and haltere appendages but not for DP formation [27,28]. At the end of embryogenesis, the different size of the wing and haltere primordia is already apparent, being the haltere approx. 30% smaller than the wing primordia. The number and size of the DP is regulated by the Hox proteins [13,22]. Restriction of the DP to the T2 and T3 segments is controlled by the repressor activity of Sex combs reduced (Scr) in T1 and by Ubx, Abd-A, and Abd-B in the abdominal segments [13,22]. Remarkably, Ubx has two different functions in DP specification: in the abdominal segments represses DP formation, while in the third thoracic segment regulates the size of the haltere. How Ubx exerts these two functions is unknown, but could be possibly explained by the different Ubx levels observed between the T3 and A1 segments [13,22,24]. High levels of Ubx in the A1 could repress DP genes while lower levels in T3 would reduce, but not eliminate, their activation. Another possibility is that the different behavior of Ubx would be explained by the use of diverse Hox cofactors in each of these segments.

2.2. Dual Developmental Origin of the Drosophila Wing and Its Evolutionary Implications

The origin of the insect wing is one of the most intriguing evolutionary events in biology, and is still under debate. Two main theories have been proposed to explain the origin of this morphological innovation (reviewed in [45]). The paranotal or tergal hypothesis proposes that wings are created *de novo* as an extension of the dorsal thorax or tergum [46]. The alternative theory, the gill-exite or pleural hypothesis evolutionarily connects the wings to a preexisting structure, the gill, present in the proximal base of an ancestral leg [47–49]. Studies in *Drosophila* and other insects have contributed significantly to decipher the origin and specification of insect appendages [8,44,50–53]. Recently, thanks to the identification and characterization of the *sna* dorsal primordia CRM (*sna*-DP), a clear picture of the origin and the genetic inputs that control DP specification was obtained [13]. *sna*-DP is first activated in a few cells dorsal to *Dll* expressing cells in T2 and T3 that later label the entire DP. A cell lineage analysis of *sna*-DP cells demonstrated that this CRM labels the entire wing disc cells including the progenitors of the thorax and appendage. Importantly, cells that have expressed—but no longer actively transcribe—*Dll* populate half of the DP [8,13,41]. These cells can be located at any part of the wing disc, although they are found preferentially in the ventral domain of the disc [13,16]. This bias is probably due to the relative location of the DP in relation to the VP in the embryo rather than to a preexisting genetic determination of these cells. In addition, genetic ablation of the TP domain reduces, but does not eliminate the *sna*-DP cells, suggesting the existence of DP cells that form independently of the TP [13]. Moreover, DP fates could be specified in the absence of *Dll* or the ventral appendage selector genes *buttonhead* (*btd*) and *Sp1* [8,13,41,54]. All these experiments and the lineage tracing analysis indicate that the DP is derived from two populations of cells: one that derives from early *Dll* expressing cells (TP), and another that arises close to, but is independent of *Dll* expressing cells (Figure 1A). Importantly, as discussed above, both populations of cells that conform the DP require the same molecular signals for their specification: high levels of Dpp and low Wg [13] (Figure 1B). This dual developmental origin of the wing described for *Drosophila* is in accordance with recent evo-devo studies that support the unification of the paranotal and pleural hypotheses [44,50,51,53,55].

2.3. Proximo-Distal Subdivision of the Ventral Primordia

In contrast to the DP, where distinction between trunk and appendage occurs later in imaginal development (see below), the VP is organized in gene expression domains that subdivide the primordia in different territories (Figure 1C) [16,56–58]. The development of the VP requires the function of *Dll* and of the two paralogous genes *btd* and *Sp1*. Btd and Sp1 are members of the highly conserved Sp family of transcription factors required for appendage formation in vertebrates and invertebrates [30,59–63]. *btd/Sp1* are expressed in the embryonic progenitors of the leg imaginal discs starting around stage 10-11 [29,30,54,64]. As it is the case with *Dll*, *btd* is activated by Wg and repressed dorsally and in the abdominal segments by Dpp and Ubx, respectively [30]. At stage 11, *btd*, *Sp1* and *Dll* (*Dll*-304) are activated in parallel in the TP and are genetically independent of each other. Importantly, as mentioned before, these TP cells could contribute to the DP and to the entire VP. Some hours later, at stage 14, VP and DP are already separated and the activity of the *Dll*-304 CRM decays. At this time point, *Dll* expression is controlled by two *Dll* CRMs with mutually exclusive patterns at the VP, the leg trigger (*Dll*-LT) and the Keilin organ (*Dll*-KO) elements (Figure 1C) [16]. The progressive refinement of the appendage primordium developmental potential is reflected by the *cis*-architecture of the *Dll* gene [14,16,65]. At the end of embryogenesis, the VP is genetically subdivided in domains with different cell fates that correspond to the P-D subdivision of the arthropod leg proposed by Robert E. Snodgrass (see below) [66,67] (Figure 1C). VP cells that express *esg* but do not activate *Dll* will form the coxopodite, the most proximal and unsegmented domain of the appendage that forms as an outgrowth of the bodywall, including the coxa. VP cells co-expressing *esg* and *Dll* (through the *Dll*-LT CRM) will form the telopodite or the leg proper, which includes all the distal leg segments that are articulated. Additionally, cells at the center of the VP that are *esg* negative and Dll positive (*Dll*-KO) are fated to form a larval mechanosensory structure that shares a common lineage with the leg disc called the Keilin's organ (KO) [16,67]. While *Dll*-KO cells do not contribute to the leg imaginal disc, the progeny of *Dll*-LT activating cells form the entire telopodite. Consistently, in *Dll* mutants the telopodite is lost while the coxopodite is present [8,16,68,69]. *Dll*-LT is positively regulated by Wg and Dpp and repressed in the center of the VP by members of the *achaete–scute complex* (ASC) [14,16,67]. At the same time, Dll and ASC positively regulate the *Dll*-KO CRM restricting its activity to the VP [16]. At this stage *btd* and *Sp1* act upstream of *Dll* and only the elimination of both genes suppresses *Dll* expression and the activity of the *Dll*-LT and *Dll*-KO enhancers [14,16,54]. Initially, some studies proposed that the two related homeobox genes Homothorax (Hth) and nuclear Extradenticle (Exd) were markers, along with *esg*, of coxopodite fate [33,40,41,70,71]. However, a reevaluation of the expression of *hth*, *esg* and *Dll* and their specific cell fate helped define a high-resolution map of the VP fates formed by three molecularly different domains with distinct developmental potential [16,67].

3. Specification of the Appendage Domain and Patterning of the Leg, Antenna and Wing Imaginal Discs

After the specification stage in the embryonic ectoderm, each imaginal primordium follows a developmental program in which the same signaling pathways and an increasing number of transcription factors cooperate to pattern the growing epithelium. In this process the size of each disc increases by cell proliferation, and the epithelium is progressively divided into domains of gene expression that confer each territory with particular cell identities. A common aspect of all thoracic discs is their division in A and P compartments conferred by the expression of *engrailed* (*en*) in posterior cells. Compartment boundaries are the source of signaling molecules that organize the pattern and growth of the imaginal disc [72]. In this section we will describe the molecular mechanisms that distinguish between trunk and appendage and the subsequent patterning of appendages culminating in the assignation of cellular fates that will specify the future domains of the adult organs.

3.1. *Patterning of the Leg Imaginal Disc*

The third instar leg imaginal disc is roughly circular in shape with the distal-most region of the leg located in the center of the disc, and the most proximal leg segments and the pleura arising from the periphery of the disc. *Antp* is the selector Hox gene for leg identity that is expressed in the leg imaginal disc but not in the antenna disc. As expected for its selector function, *Antp* mutant legs are transformed into antenna and *Antp* ectopic expression in the antenna converts it into leg [73]. The leg P-D axis is formed orthogonally to the previously established A-P and D-V axis, and its formation is intimately connected with the A-P compartment subdivision. Thus, posterior cells express *hedgehog* (*hh*) which acts as a short-range signal that activates the expression of *dpp* and *wg* in the anterior dorsal and ventral halves of the disc, respectively [74]. Dpp and Wg are both necessary to establish and pattern the D-V and the P-D axes. The D-V axis is organized by the mutually antagonistic repression of Wg and Dpp and interactions between its downstream genes (Figure 2A) [75–80]. Briefly, Wg specifies ventral fates through the activation of *H15* and *midline* (*mid*) and the repression of dorsal genes such as *Doc* and *optomotor blind* (*omb*) [43,81]. Therefore, in *wg* mutants all the ventral structures are lost and replaced by a mirror duplication of dorsal ones and the reverse phenotypes are observed for *dpp* mutants [75,76,78,80,82].

The P-D axis is initiated at the center of the disc where high levels of Wg and Dpp activate downstream genes such as *Dll* [14,83–86]. Dll in turn activates *dachshund* (*dac*) in the medial domain of the leg disc, whereas high levels of Wg and Dpp represses its expression in the distal tip [85,87]. Once activated, *Dll* and *dac* expression is maintained in part by autoregulatory mechanisms [14,87]. Simultaneously, the expression of *hth* is restricted to the periphery of the disc by a Wg and Dpp dependent repression mechanism [88,89]. This mechanism could be mediated by the zinc finger proteins encoded by the *elbow* (*el*) and *no ocelli* (*noc*) genes that act in the leg and the wing discs to repress *hth* expression, and therefore body wall fates [90]. Consequently, as a result of the A-P compartment subdivision and the restricted expression of Hh, Dpp, and Wg the leg is divided along the P-D axes in proximal, medial, and distal domains defined by the expression of *hth*, *dac*, and *Dll*, respectively (Figure 2A). At this stage *Antp* prevents *hth* and *Dll* coexpression in the leg, suppressing antennal identity [91–94]. Importantly, the ancestral subdivision of the leg in the coxopodite and telopodite is reflected at the molecular level by the differential response of the leg imaginal disc cells to the Wg and Dpp signaling pathways. Therefore, cells that receive both Wg and Dpp activate the *Dll* and *dac* genes and specify the telopodite. In contrast, cells that are unable to respond to these pathways are specified as coxopodite and activate *hth* [70,71] (Figure 1C).

The distal domain of the leg is further patterned by a P-D gradient of EGFR signaling, mediated by the EGFR ligand Vein (Vn), that regulates the expression in a distal to proximal manner of *aristaless* (*al*), *Bar* (*B*), and *rotund* (*rn*) among others [95,96] (Figure 2A,B). A complex cross-regulation between these transcription factors ensure the segmental subdivision of the distal domain in five tarsi (reviewed in [97,98]). Besides providing unique segmental identity along the P-D axis of the leg, this regulatory network is responsible for the segmentally repeated expression in concentric rings of the Notch ligands *Delta* (*Dl*) and *Serrate* (*Ser*) [99,100]. Notch activation at the distal-end of each presumptive segment is evident at the end of larval development and directs the formation of the joints and the non-autonomous growth of the leg [101–103] (Figure 2A,B). The ventral appendage selector-like genes *btd* and *Sp1* are co-opted later during imaginal development to control the growth and pattern of the leg, in part through the regulation of *Ser* expression and therefore Notch signaling [100]. Two types of joints with different morphologies and evolutionary origin could be found in the adult leg. The proximal or true joints are asymmetrical and have associated musculature, whereas the distal joints are symmetrical and not attached to muscles [66,104,105]. Notch controls joint morphogenesis through the activation of subsidiary transcription factors required for distal joints such as *dysfusion* (*dysf*), proximal joints such as the *odd-skipped* gen family of for all joints like *dAp-2* ([103,106,107] and reviewed in [108]).

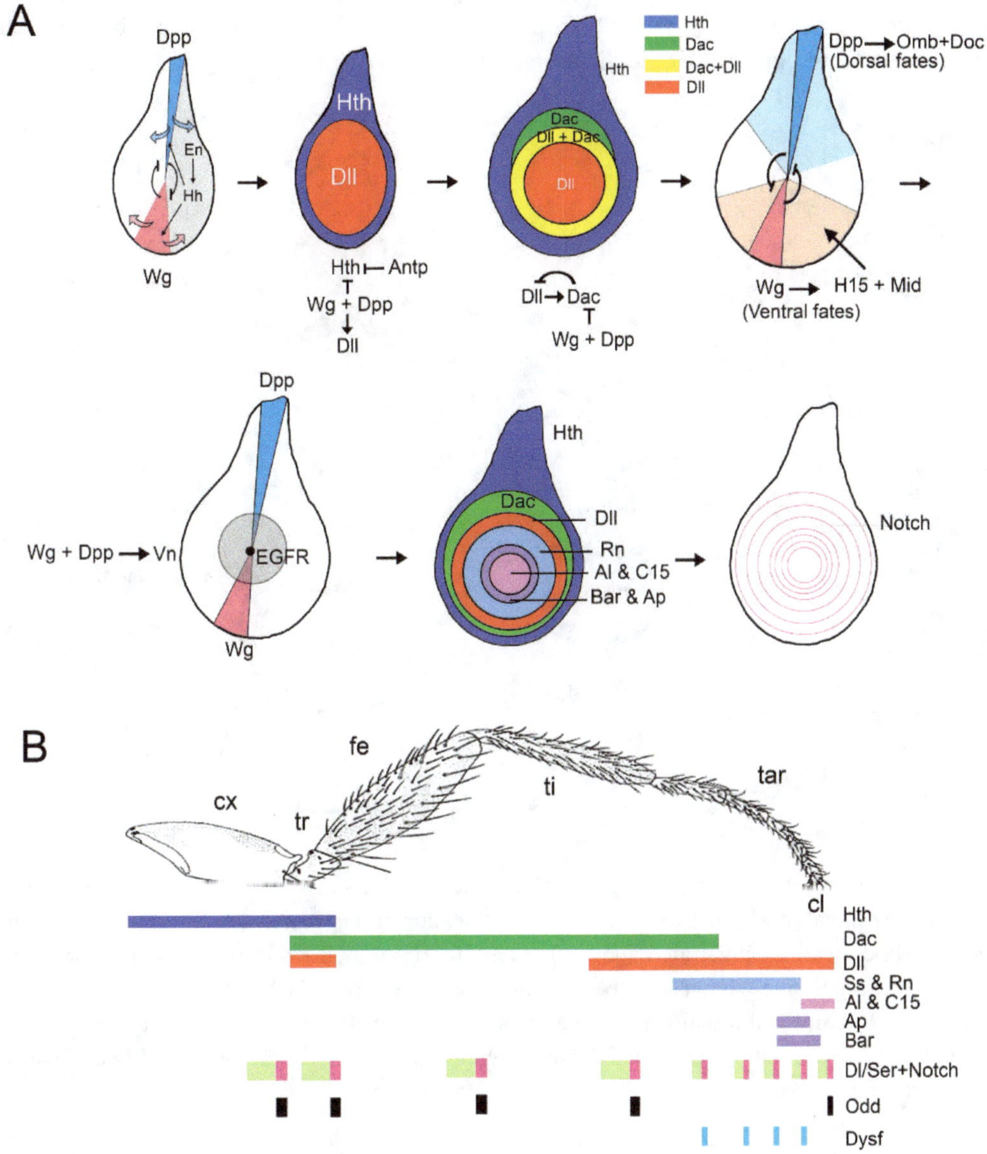

Figure 2. Patterning of the *Drosophila* leg. (**A**) Overview of the proximo-distal (P-D) and dorso-ventral (D-V) axes formation during leg development. The restricted expression of *dpp* and *wg* in the dorsal and ventral halves of the leg disc is required to initiate and to pattern the P-D and D-V axes. The first two leg imaginal discs from the left correspond to early second instar discs, while the rest depict third instar leg discs. Some of the genetic interactions that lead to the leg patterning are indicated. (**B**) The 'genetic code' generated by the expression of several transcription factors specify the future segments of the adult leg along the P-D axis. The expression of the Notch ligands *Dl* and *Ser*, and the activation of the Notch pathway and its target genes is necessary for the formation of the joints that separate the adjacent segments of the leg.

3.2. Patterning of the Wing Imaginal Disc

The DP contains the progenitors of both the wing and the body wall structures, however in contrast to early P-D specification of the VP, the trunk/appendage distinction is only evident later during larval development (Figures 1C and 3). The mature wing disc is formed by two distinct domains, the body wall and the appendage, which in turn is subdivided in the proximal hinge and the wing blade. The specification of the body wall and the wing appendage is made by the mutually repressive activities of two signaling pathways: the Vn/EGFR acting in the proximal domain of the disc that specifies body wall fates, and Wg at the distal region that will form the wing field (Figure 3) [109–112].

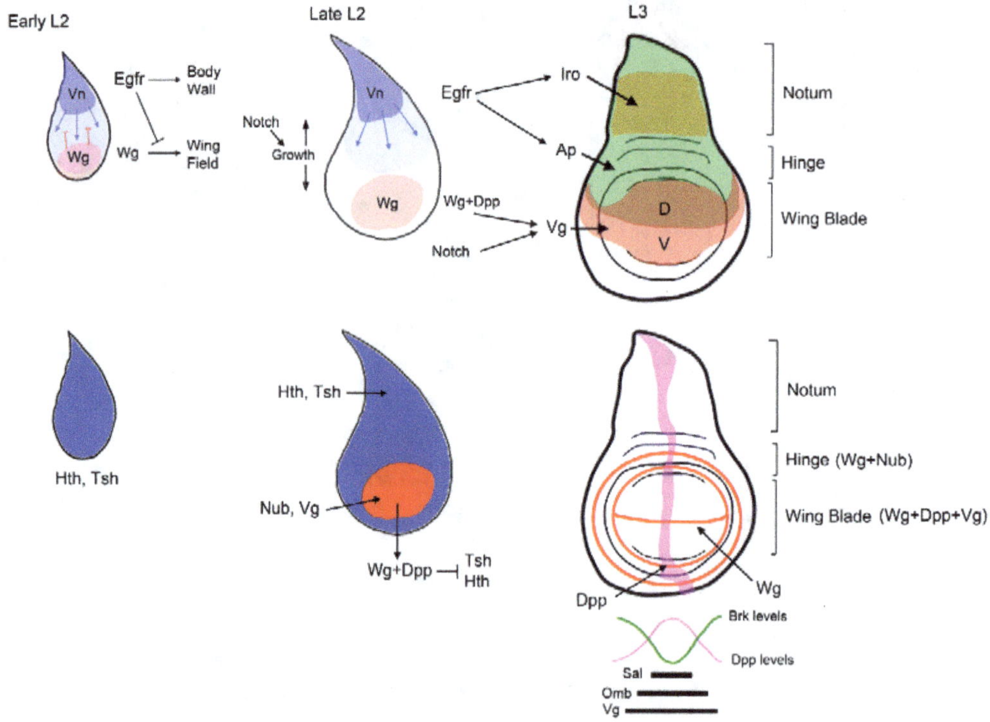

Figure 3. Patterning of the *Drosophila* wing. Three representative stages of wing disc development are depicted. The signals and genes that direct the specification and patterning of the appendage (wing) versus the body wall are illustrated above. The activity of the EGFR pathway promotes body wall identity while repressing wing fates. Notch induces the growth of the wing disc, which separates the source of the Wg and EGFR pathways and allows the specification of the wing field by Wg and the activation of wing promoting genes. Below are shown the expressions of the Dpp and Wg pathways in a third instar disc and the relative position of the Dpp target genes *sal*, *omb*, and *brk*.

Consequently, in the absence of Wg the wing field is lost and a duplication of proximal structures is observed [56,58]. In the early primordium, the activity of the EGFR pathway represses the ability to induce wing fates by Wg and promotes notum fates [110,111]. As the disc grows in size induced by the activity of Notch, the signaling sources of the EGFR and Wg pathways become separated allowing the specification of the wing field by Wg and the activation of wing promoting genes such as *nubbin* (*nub*) and *vg* [58,113]. The activation of *nub* and *vg* is accompanied by the repression of the body wall genes *teashirt* (*tsh*) and *hth* (Figure 3) [114,115]. The different P-D expression of *hth/tsh* (notum and hinge) and *vg* (wing blade) is analogous to the leg disc, where *hth/tsh* are also restricted to the proximal domain to promote body wall structures (Figure 1C). It has been proposed that as in the leg, Wg and Dpp in the wing disc are required, through the activation of *el* and *noc*, to repress *hth* and *tsh* expression from the wing territory [90,114,116]. However other mechanisms have also been proposed [115]. The EGFR pathway is also required to activate the dorsal selector gene *apterous* (*ap*) that subdivide the thorax and the wing field in dorsal (D) and ventral (V) compartments [17,111,117,118]. This D-V subdivision is absolutely necessary for *vg* expression and wing outgrowth [119]. Ap regulates the expression of the Notch ligands Dl and Ser that in turn activate the Notch pathway at the D-V boundary. Wg and Notch cooperate to induce the expression of *vg* in the cells that will become the wing blade [120–122]. The mutual antagonism between *hth* and *vg* subdivide the wing field in the pouch and the hinge [116]. The hinge region of the wing field is specified by the combined action of Wg, Hth, and Tsh [92,114,121]. In the proximal region of the disc, the EGFR pathway induces the expression of the notum specifying genes of the *Iroquois-complex* (*Iro-C*) that are repressed in the wing field by Dpp secreted in the A-P boundary (Figure 3) [123–126]. Once specified, the notum is further subdivided in medial and lateral domains by the activity of the transcription factor Pannier (Pnr) that promotes medial *vs* lateral fates [127].

As in the leg disc, *en* expression subdivides the wing disc in two populations of cells, the A and P compartments. The A-P boundary acts a source of positional information where Hh secreted from the posterior compartment activates the expression of *dpp* in A cells. Dpp and Hh are both required to pattern the wing blade along the A-P axis (Figure 3). Dpp diffusion from the A-P boundary restricts the expression of *brinker* (*brk*) to the lateral domain of the wing [128–130]. Dpp and Brk in turn, regulates the expression of *omb* and the *spalt* genes (*sal*) in nested domains [130]. The Hh and Dpp signals act coordinately with the EGFR signaling pathway to position and maintain vein territories in the developing wing (reviewed by [131,132]).

3.3. Patterning of the Eye-Antennal Disc

The eye-antennal disc (EAD) is formed by different populations of cells located in the cephalic segments that coalesce together [11]. The EAD is a compound disc that gives rise to the eye, the ocelli, the palpus and the antenna plus the head cuticle that surrounds them. Homeotic transformations from antenna to leg have been described for several mutations, including *hth*, *spineless* (*ss*), and *Antp* that converts every antenna part to its corresponding leg segment [133,134] (Figure 4B). Therefore, the antenna and leg appendages have been considered serial homologous structures that share a similar developmental program. Thus, variations in this developmental program are responsible for the morphological differences between these two appendages.

Early on larval development (L1) the EAD shows no sign of regional specification and presents uniform expression of several transcription factors including *hth* and *tsh* and the eye determinants *eyeless* (*ey*) and *sine oculis* (*so*) (Figure 4A) [135–137]. The two main developmental fields within the disc, the antenna and the eye, are segregated during larval stages (L2) at the time when specific eye and antenna determinants are restricted to the corresponding EAD domain [136–140]. The transcription factors Ey, Eyes absent (Eya) and So are selectors for eye development while Hth, Dll, Ss, and Cut have been proposed to select for antennal fates. The antagonistic interactions between the Notch and the EGFR pathways regulate the eye and antennal fate decision [135] (Figure 4A). Accordingly, the downregulation of the Notch pathway or the activation of the EGFR in the eye field transforms it into an antenna [135]. However, the role of Notch as an eye identity inducer has been debated [136]. In this view, Notch influences eye field specification through its control on cell proliferation, thus modulating the activity of the Dpp and Wg signaling pathways located at opposing sides of the early EA disc [136,141]. Wg is a negative regulator of eye development while Dpp induces the expression of the eye determinant *eya* ([136,142] and reviewed in [143]). Once the two morphogenetic fields have been established, the restricted expression of antenna and eye promoting genes is observed (Figure 4A). For example, Cut and Hth represses the eye specification genes *ey* and *so* while the latter represses *hth* and *cut* [137]. The patterning logic that controls the P-D axis formation in the antenna is similar to what can be found in the leg [84]. Hh secretion from posterior cells activates the expression of *dpp* and *wg* in dorsal and ventral anterior cells, respectively that in turn directs P-D axis formation through the regulation of *Dll*, *Dac*, and *hth*. As in the leg, Dll specifies distal fates while Hth promotes proximal identities. However, in contrast to the leg where the domains of Dll+Dac and Hth are mostly exclusive, these genes are co-expressed in a large territory in the antenna (Figure 4A) [91,144,145]. The presence in the leg of the selector gene Antp is a key factor for *hth* repression while its absence in the antenna allows *Dll* and *hth* coexpression [91,94,146]. Antennal identity depends on the coordinated action of Hth and Dll, which together activate the expression of antennal determinants such as *ss or spalt* (*sal*) [145,147]. Ss in turns control the expression of *distal antenna* (*dan*) and *distal antenna related* (*danr*) genes, responsible for the specification of distal antenna identity [148,149].

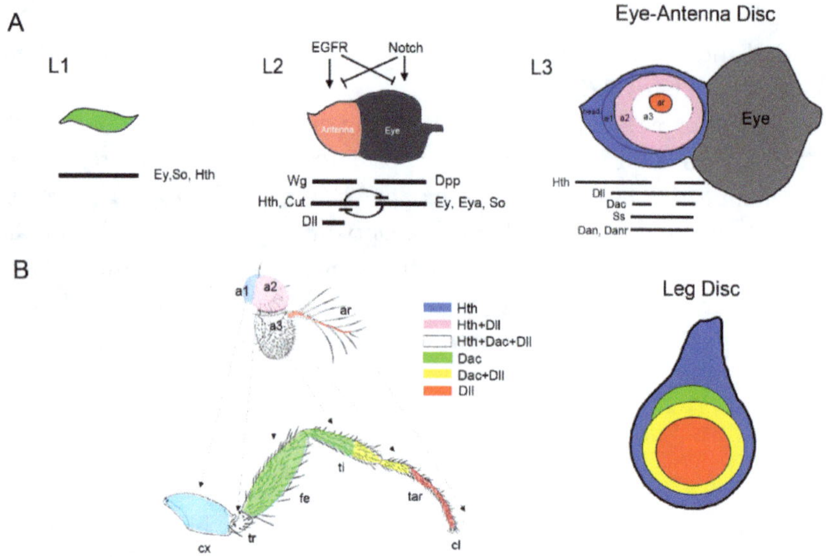

Figure 4. Specification and patterning of the *Drosophila* antenna. (**A**) Model of progressive specification of the eye and antenna territories and P-D subdivision of the antenna disc. Schematic representation of three representative stages of EAD development is shown. In each disc the expression and interactions of genes and signaling pathways required for eye and antenna field specification are schematized. (**B**) The antenna and leg appendages are homologous structures. Arrows indicate the correspondence between the antenna and leg domains (Postlethwait and Schneiderman, 1971). The expression of the P-D genes *Dll*, *dac*, and *hth* and their overlapping domains are represented by a color code. Compare the relative expression of *Dll*, *dac*, and *hth* in the antenna and the leg imaginal discs. Note that *hth* and *Dll* coexpress in a large domain in the antenna while in the leg these genes are expressed in almost exclusive domains.

4. Perspectives

In this review, we have discussed how the study of *Drosophila melanogaster* has provided a detailed picture of the molecular mechanisms that lead to the specification and patterning of the adult appendages. Importantly, many of the signaling pathways and transcription factors identified in the fly have similar functions during vertebrate limb development. However, many important questions remain unsolved. First, the target genes and the regulatory networks controlled by the transcription factors that subdivide the appendages along the different axes are still largely unknown. This includes the analysis of the molecular interactions between patterning genes and signaling pathways. Second, further investigation would be required to unravel how the patterning information is translated into cellular behaviors (cell growth, division, and cytoskeleton organization, among others), which are ultimately responsible for the characteristic shape and size of the different appendages of the fly. The use of whole genome techniques such as ChIP (Chromatin Immunoprecipitation) or ATAC (assay for transposase accessible chromatin) assays, coupled with whole genome sequencing is essential to identify the target genes and the regulatory landscapes governed by the patterning transcription factors. In addition, live imaging techniques would provide novel information of the morphogenetic processes and dynamics during appendage specification and formation.

Funding: This work was supported by grants from the Secretaría de Estado de Investigación, Desarrollo e Innovación (Ministerio de Economía y Competitividad) (No. BFU2015-65728-P to C.E.).

Acknowledgments: We specially thank Jose Felix de Celis and Antonio Baonza for helpful criticisms during manuscript preparation. We apologize to all the authors whose work has not been cited due to space limitations.

Conflicts of Interest: The authors declare no conflict of interest.

References

1. Engel, M.S. Insect evolution. *Curr. Biol.* **2015**, *25*, R868–R872. [CrossRef] [PubMed]
2. Angelini, D.R.; Kaufman, T.C. Insect appendages and comparative ontogenetics. *Dev. Biol.* **2005**, *286*, 57–77. [CrossRef] [PubMed]
3. Tabin, C.; Carroll, S.; Panganiban, G. Out on a limb: Parallels in vertebrate and invertebrate limb patterning and the origin of appendages. *Am. Zool.* **1999**, *39*, 650–663. [CrossRef]
4. Pueyo, J.I.; Couso, J.P. Parallels between the proximal-distal development of vertebrate and arthropod appendages: Homology without an ancestor? *Curr. Opin. Genet. Dev.* **2005**, *15*, 439–446. [CrossRef] [PubMed]
5. Panganiban, G.; Irvine, S.M.; Lowe, C.; Roehl, H.; Corley, L.S.; Sherbon, B.; Grenier, J.K.; Fallon, J.F.; Kimble, J.; Walker, M.; et al. The origin and evolution of animal appendages. *Proc. Natl. Acad. Sci. USA* **1997**, *94*, 5162–5166. [CrossRef] [PubMed]
6. Shubin, N.; Tabin, C.; Carroll, S. Fossils, genes and the evolution of animal limbs. *Nature* **1997**, *388*, 639–648. [CrossRef] [PubMed]
7. Shubin, N.; Tabin, C.; Carroll, S. Deep homology and the origins of evolutionary novelty. *Nature* **2009**, *457*, 818–823. [CrossRef] [PubMed]
8. Cohen, B.; Simcox, A.A.; Cohen, S.M. Allocation of the thoracic imaginal primordia in the *Drosophila embryo*. *Development* **1993**, *117*, 597–608. [PubMed]
9. Bate, M.; Arias, A.M. The embryonic origin of imaginal discs in *Drosophila*. *Development* **1991**, *112*, 755–761. [PubMed]
10. Estrada, B.; Casares, F.; Sanchez-Herrero, E. Development of the genitalia in *Drosophila melanogaster*. *Differentiation* **2003**, *71*, 299–310. [CrossRef] [PubMed]
11. Younossi-Hartenstein, A.; Tepass, U.; Hartenstein, V. Embryonic origin of the imaginal discs of the head of *Drosophila melanogaster*. *Roux. Arch. Dev. Biol.* **1993**, *203*, 60–73. [CrossRef] [PubMed]
12. Haynie, J.L.; Bryant, P.J. Development of the eye-antenna imaginal disc and morphogenesis of the adult head in *Drosophila melanogaster*. *J. Exp. Zool.* **1986**, *237*, 293–308. [CrossRef] [PubMed]
13. Requena, D.; Alvarez, J.A.; Gabilondo, H.; Loker, R.; Mann, R.S.; Estella, C. Origins and specification of the *Drosophila* wing. *Curr. Biol.* **2017**, *27*, 3826–3836. [CrossRef] [PubMed]
14. Estella, C.; McKay, D.J.; Mann, R.S. Molecular integration of wingless, decapentaplegic, and autoregulatory inputs into *Distalless* during *Drosophila* leg development. *Dev. Cell* **2008**, *14*, 86–96. [CrossRef] [PubMed]
15. Galindo, M.I.; Fernandez-Garza, D.; Phillips, R.; Couso, J.P. Control of distal-less expression in the *Drosophila* appendages by functional 3′ enhancers. *Dev. Biol.* **2011**, *353*, 396–410. [CrossRef] [PubMed]
16. McKay, D.J.; Estella, C.; Mann, R.S. The origins of the *Drosophila* leg revealed by the cis-regulatory architecture of the *Distalless* gene. *Development* **2009**, *136*, 61–71. [CrossRef] [PubMed]
17. Bieli, D.; Kanca, O.; Requena, D.; Hamaratoglu, F.; Gohl, D.; Schedl, P.; Affolter, M.; Slattery, M.; Müller, M.; Estella, C. Establishment of a developmental compartment requires interactions between three synergistic cis-regulatory modules. *PLoS Genet.* **2015**, *11*, e1005376. [CrossRef] [PubMed]
18. Emmons, R.B.; Duncan, D.; Duncan, I. Regulation of the *Drosophila* distal antennal determinant spineless. *Dev. Biol.* **2007**, *302*, 412–426. [CrossRef] [PubMed]
19. Cohen, S.M. Specification of limb development in the *Drosophila embryo* by positional cues from segmentation genes. *Nature* **1990**, *343*, 173–177. [CrossRef] [PubMed]
20. Weaver, T.A.; White, R.A. Headcase, an imaginal specific gene required for adult morphogenesis in *Drosophila melanogaster*. *Development* **1995**, *121*, 4149–4160. [PubMed]

21. Hayashi, S.; Hirose, S.; Metcalfe, T.; Shirras, A.D. Control of imaginal cell development by the escargot gene of *Drosophila*. *Development* **1993**, *118*, 105–115. [CrossRef]
22. Carroll, S.B.; Weatherbee, S.D.; Langeland, J.A. Homeotic genes and the regulation and evolution of insect wing number. *Nature* **1995**, *375*, 58–61. [CrossRef] [PubMed]
23. Vachon, G.; Cohen, B.; Pfeifle, C.; McGuffin, M.E.; Botas, J.; Cohen, S.M. Homeotic genes of the Bithorax complex repress limb development in the abdomen of the *Drosophila embryo* through the target gene *Distal-less*. *Cell* **1992**, *71*, 437–450. [CrossRef]
24. Castelli-Gair, J.; Akam, M. How the hox gene *Ultrabithorax* specifies two different segments: The significance of spatial and temporal regulation within metameres. *Development* **1995**, *121*, 2973–2982. [PubMed]
25. Hughes, C.L.; Kaufman, T.C. Hox genes and the evolution of the arthropod body plan. *Evol. Dev.* **2002**, *4*, 459–499. [CrossRef] [PubMed]
26. Whiteley, M.; Noguchi, P.D.; Sensabaugh, S.M.; Odenwald, W.F.; Kassis, J.A. The *Drosophila* gene escargot encodes a zinc finger motif found in snail-related genes. *Mech. Dev.* **1992**, *36*, 117–127. [CrossRef]
27. Fuse, N.; Hirose, S.; Hayashi, S. Determination of wing cell fate by the escargot and snail genes in *Drosophila*. *Development* **1996**, *122*, 1059–1067. [PubMed]
28. Williams, J.A.; Bell, J.B.; Carroll, S.B. Control of *Drosophila* wing and haltere development by the nuclear vestigial gene product. *Genes Dev.* **1991**, *5*, 2481–2495. [CrossRef] [PubMed]
29. Schock, F.; Purnell, B.A.; Wimmer, E.A.; Jackle, H. Common and diverged functions of the *Drosophila* gene pair D-Sp1 and buttonhead. *Mech. Dev.* **1999**, *89*, 125–132. [CrossRef]
30. Estella, C.; Rieckhof, G.; Calleja, M.; Morata, G. The role of buttonhead and Sp1 in the development of the ventral imaginal discs of *Drosophila*. *Development* **2003**, *130*, 5929–5941. [CrossRef] [PubMed]
31. Panganiban, G.; Rubenstein, J.L. Developmental functions of the *Distal-less*/Dlx homeobox genes. *Development* **2002**, *129*, 4371–4386. [PubMed]
32. Cohen, S.M.; Jurgens, G. Proximal-distal pattern formation in *Drosophila*: Cell autonomous requirement for *Distal-less* gene activity in limb development. *EMBO J.* **1989**, *8*, 2045–2055. [PubMed]
33. Goto, S.; Hayashi, S. Specification of the embryonic limb primordium by graded activity of Decapentaplegic. *Development* **1997**, *124*, 125–132. [PubMed]
34. Gebelein, B.; Culi, J.; Ryoo, H.D.; Zhang, W.; Mann, R.S. Specificity of *Distalless* repression and limb primordia development by abdominal Hox proteins. *Dev. Cell* **2002**, *3*, 487–498. [CrossRef]
35. Gebelein, B.; McKay, D.J.; Mann, R.S. Direct integration of Hox and segmentation gene inputs during *Drosophila* development. *Nature* **2004**, *431*, 653–659. [CrossRef] [PubMed]
36. Uhl, J.D.; Zandvakili, A.; Gebelein, B. A Hox transcription factor collective binds a highly conserved distal-less cis-regulatory module to generate robust transcriptional outcomes. *PLoS Genet.* **2016**, *12*, e1005981. [CrossRef] [PubMed]
37. Wieschaus, E.; Gehring, W. Clonal analysis of primordial disc cells in the early embryo of *Drosophila melanogaster*. *Dev. Biol.* **1976**, *50*, 249–263. [CrossRef]
38. Wieschaus, E.; Gehring, W. Gynandromorph analysis of the thoracic disc primordia in *Drosophila melanogaster*. *Wilhelm Roux. Arch. Dev. Biol.* **1976**, *180*, 31–46. [CrossRef] [PubMed]
39. Lawrence, P.A.; Morata, G. The early development of mesothoracic compartments in *Drosophila*. An analysis of cell lineage and fate mapping and an assessment of methods. *Dev. Biol.* **1977**, *56*, 40–51. [CrossRef]
40. Kubota, K.; Goto, S.; Eto, K.; Hayashi, S. EGF receptor attenuates Dpp signaling and helps to distinguish the wing and leg cell fates in *Drosophila*. *Development* **2000**, *127*, 3769–3776. [PubMed]
41. Kubota, K.; Goto, S.; Hayashi, S. The role of Wg signaling in the patterning of embryonic leg primordium in *Drosophila*. *Dev. Biol.* **2003**, *257*, 117–126. [CrossRef]
42. Hamaguchi, T.; Yabe, S.; Uchiyama, H.; Murakami, R. *Drosophila* Tbx6-related gene, *Dorsocross*, mediates high levels of Dpp and Scw signal required for the development of amnioserosa and wing disc primordium. *Dev. Biol.* **2004**, *265*, 355–368. [CrossRef] [PubMed]
43. Reim, I.; Lee, H.H.; Frasch, M. The T-box-encoding *Dorsocross* genes function in amnioserosa development and the patterning of the dorsolateral germ band downstream of Dpp. *Development* **2003**, *130*, 3187–3204. [CrossRef] [PubMed]
44. Niwa, N.; Akimoto-Kato, A.; Niimi, T.; Tojo, K.; Machida, R.; Hayashi, S. Evolutionary origin of the insect wing via integration of two developmental modules. *Evol. Dev.* **2010**, *12*, 168–176. [CrossRef] [PubMed]

45. Clark-Hachtel, C.M.; Tomoyasu, Y. Exploring the origin of insect wings from an evo-devo perspective. *Curr. Opin. Insect Sci.* **2016**, *13*, 77–85. [CrossRef] [PubMed]
46. Crampton, G. The phylogenetic origin and the nature of the wings of insects according to the paranotal theory. *J. N. Y. Entomol. Soc.* **1916**, *24*, 1–39.
47. Kukalova-Peck, J. Origin and evolution of insect wings and their relation to metamorphosis, as documented by the fossil record. *J. Morphol.* **1978**, *156*, 53–125. [CrossRef]
48. Kukalova-Peck, J. Origin of the insect wing and wing articulation from the arthropodan leg. *Can. J. Zool.* **1983**, *61*, 1618–1669. [CrossRef]
49. Averof, M.; Cohen, S.M. Evolutionary origin of insect wings from ancestral gills. *Nature* **1997**, *385*, 627–630. [CrossRef] [PubMed]
50. Linz, D.M.; Tomoyasu, Y. Dual evolutionary origin of insect wings supported by an investigation of the abdominal wing serial homologs in *Tribolium*. *Proc. Natl. Acad. Sci. USA* **2018**, *115*, E658–E667. [CrossRef] [PubMed]
51. Clark-Hachtel, C.M.; Linz, D.M.; Tomoyasu, Y. Insights into insect wing origin provided by functional analysis of vestigial in the red flour beetle, *Tribolium castaneum*. *Proc. Natl. Acad. Sci. USA* **2013**, *110*, 16951–16956. [CrossRef] [PubMed]
52. Elias-Neto, M.; Belles, X. Tergal and pleural structures contribute to the formation of ectopic prothoracic wings in cockroaches. *R. Soc. Open Sci.* **2016**, *3*, 160347. [CrossRef] [PubMed]
53. Medved, V.; Marden, J.H.; Fescemyer, H.W.; Der, J.P.; Liu, J.; Mahfooz, N.; Popadic, A. Origin and diversification of wings: Insights from a neopteran insect. *Proc. Natl. Acad. Sci. USA* **2015**, *112*, 15946–15951. [CrossRef] [PubMed]
54. Estella, C.; Mann, R.S. Non-redundant selector and growth-promoting functions of two sister genes, buttonhead and Sp1, in *Drosophila* leg development. *PLoS Genet.* **2010**, *6*, e1001001. [CrossRef] [PubMed]
55. Prokop, J.; Pecharova, M.; Nel, A.; Hornschemeyer, T.; Krzeminska, E.; Krzeminski, W.; Engel, M.S. Paleozoic nymphal wing pads support dual model of insect wing origins. *Curr. Biol.* **2017**, *27*, 263–269. [CrossRef] [PubMed]
56. Couso, J.P.; Bate, M.; Martinez-Arias, A. A wingless-dependent polar coordinate system in *Drosophila* imaginal discs. *Science* **1993**, *259*, 484–489. [CrossRef] [PubMed]
57. Williams, J.A.; Paddock, S.W.; Carroll, S.B. Pattern formation in a secondary field: A hierarchy of regulatory genes subdivides the developing *Drosophila* wing disc into discrete subregions. *Development* **1993**, *117*, 571–584. [PubMed]
58. Ng, M.; Diaz-Benjumea, F.J.; Vincent, J.P.; Wu, J.; Cohen, S.M. Specification of the wing by localized expression of wingless protein. *Nature* **1996**, *381*, 316–318. [CrossRef] [PubMed]
59. Haro, E.; Delgado, I.; Junco, M.; Yamada, Y.; Mansouri, A.; Oberg, K.C.; Ros, M.A. Sp6 and sp8 transcription factors control AER formation and dorsal-ventral patterning in limb development. *PLoS Genet.* **2014**, *10*, e1004468. [CrossRef] [PubMed]
60. Kawakami, Y.; Esteban, C.R.; Matsui, T.; Rodriguez-Leon, J.; Kato, S.; Izpisua Belmonte, J.C. Sp8 and Sp9, two closely related buttonhead-like transcription factors, regulate Fgf8 expression and limb outgrowth in vertebrate embryos. *Development* **2004**, *131*, 4763–4774. [CrossRef] [PubMed]
61. Bell, S.M.; Schreiner, C.M.; Waclaw, R.R.; Campbell, K.; Potter, S.S.; Scott, W.J. Sp8 is crucial for limb outgrowth and neuropore closure. *Proc. Natl. Acad. Sci. USA* **2003**, *100*, 12195–12200. [CrossRef] [PubMed]
62. Beermann, A.; Aranda, M.; Schroder, R. The Sp8 zinc-finger transcription factor is involved in allometric growth of the limbs in the beetle *Tribolium castaneum*. *Development* **2004**, *131*, 733–742. [CrossRef] [PubMed]
63. Setton, E.V.W.; Sharma, P.P. Cooption of an appendage-patterning gene cassette in the head segmentation of arachnids. *Proc. Natl. Acad. Sci. USA* **2018**, *115*, E3491–E3500. [CrossRef] [PubMed]
64. Schaeper, N.D.; Prpic, N.M.; Wimmer, E.A. A clustered set of three Sp-family genes is ancestral in the Metazoa: Evidence from sequence analysis, protein domain structure, developmental expression patterns and chromosomal location. *BMC Evol. Biol.* **2010**, *10*, 88. [CrossRef] [PubMed]
65. Galindo, M.I.; Bishop, S.A.; Couso, J.P. Dynamic EGFR-Ras signalling in *Drosophila* leg development. *Dev. Dyn.* **2005**, *233*, 1496–1508. [CrossRef] [PubMed]
66. Snodgrass, R. *Principles of Insect Morphology*; McGraw-Hill: New York, NY, USA, 1935; pp. 83–99.
67. Bolinger, R.A.; Boekhoff-Falk, G. *Distal-less* functions in subdividing the *Drosophila* thoracic limb primordium. *Dev. Dyn.* **2005**, *232*, 801–816. [CrossRef] [PubMed]

68. Campbell, G.; Tomlinson, A. The roles of the homeobox genes *Aristaless* and *Distal-less* in patterning the legs and wings of *Drosophila*. *Development* **1998**, *125*, 4483–4493. [PubMed]
69. Mandaravally Madhavan, M.; Schneiderman, H.A. Histological analysis of the dynamics of growth of imaginal discs and histoblast nests during the larval development of *Drosophila melanogaster*. *Wilhelm Roux. Arch. Dev. Biol.* **1977**, *183*, 269–305. [CrossRef] [PubMed]
70. Gonzalez-Crespo, S.; Abu-Shaar, M.; Torres, M.; Martinez, A.C.; Mann, R.S.; Morata, G. Antagonism between extradenticle function and Hedgehog signalling in the developing limb. *Nature* **1998**, *394*, 196–200. [CrossRef] [PubMed]
71. González-Crespo, S.; Morata, G. Genetic evidence for the subdivision of the arthropod limb into coxopodite and telopodite. *Development* **1996**, *122*, 3921–3928. [PubMed]
72. Irvine, K.D.; Rauskolb, C. Boundaries in development: Formation and function. *Annu. Rev. Cell Dev. Biol.* **2001**, *17*, 189–214. [CrossRef] [PubMed]
73. Struhl, G. Genes controlling segmental specification in the *Drosophila* thorax. *Proc. Natl. Acad. Sci. USA* **1982**, *79*, 7380–7384. [CrossRef] [PubMed]
74. Basler, K.; Struhl, G. Compartment boundaries and the control of *Drosophila* limb pattern by hedgehog protein. *Nature* **1994**, *368*, 208–214. [CrossRef] [PubMed]
75. Brook, W.J.; Cohen, S.M. Antagonistic interactions between wingless and decapentaplegic responsible for dorsal-ventral pattern in the *Drosophila* leg. *Science* **1996**, *273*, 1373–1377. [CrossRef] [PubMed]
76. Jiang, J.; Struhl, G. Complementary and mutually exclusive activities of decapentaplegic and wingless organize axial patterning during *Drosophila* leg development. *Cell* **1996**, *86*, 401–409. [CrossRef]
77. Johnston, L.A.; Schubiger, G. Ectopic expression of wingless in imaginal discs interferes with decapentaplegic expression and alters cell determination. *Development* **1996**, *122*, 3519–3529. [PubMed]
78. Morimura, S.; Maves, L.; Chen, Y.; Hoffmann, F.M. Decapentaplegic overexpression affects *Drosophila* wing and leg imaginal disc development and wingless expression. *Dev. Biol.* **1996**, *177*, 136–151. [CrossRef] [PubMed]
79. Penton, A.; Hoffmann, F.M. Decapentaplegic restricts the domain of wingless during *Drosophila* limb patterning. *Nature* **1996**, *382*, 162–164. [CrossRef] [PubMed]
80. Theisen, H.; Haerry, T.E.; O'Connor, M.B.; Marsh, J.L. Developmental territories created by mutual antagonism between wingless and decapentaplegic. *Development* **1996**, *122*, 3939–3948. [PubMed]
81. Svendsen, P.C.; Formaz-Preston, A.; Leal, S.M.; Brook, W.J. The Tbx20 homologs midline and H15 specify ventral fate in the *Drosophila melanogaster* leg. *Development* **2009**, *136*, 2689–2693. [CrossRef] [PubMed]
82. Held, L.I., Jr.; Heup, M.A. Genetic mosaic analysis of decapentaplegic and wingless gene function in the *Drosophila* leg. *Dev. Genes Evol.* **1996**, *206*, 180–194. [CrossRef] [PubMed]
83. Campbell, G.; Weaver, T.; Tomlinson, A. Axis specification in the developing *Drosophila* appendage: The role of wingless, decapentaplegic, and the homeobox gene aristaless. *Cell* **1993**, *74*, 1113–1123. [CrossRef]
84. Diaz-Benjumea, F.J.; Cohen, B.; Cohen, S.M. Cell interaction between compartments establishes the proximal-distal axis of *Drosophila* legs. *Nature* **1994**, *372*, 175–179. [CrossRef] [PubMed]
85. Lecuit, T.; Cohen, S.M. Proximal-distal axis formation in the *Drosophila* leg. *Nature* **1997**, *388*, 139–145. [CrossRef] [PubMed]
86. Estella, C.; Mann, R.S. Logic of wg and dpp induction of distal and medial fates in the *Drosophila* leg. *Development* **2008**, *135*, 627–636. [CrossRef] [PubMed]
87. Giorgianni, M.W.; Mann, R.S. Establishment of medial fates along the proximodistal axis of the *Drosophila* leg through direct activation of dachshund by *Distalless*. *Dev. Cell* **2011**, *20*, 455–468. [CrossRef] [PubMed]
88. Wu, J.; Cohen, S.M. Proximodistal axis formation in the *Drosophila* leg: Subdivision into proximal and distal domains by homothorax and distal-less. *Development* **1999**, *126*, 109–117. [PubMed]
89. Abu-Shaar, M.; Mann, R.S. Generation of multiple antagonistic domains along the proximodistal axis during *Drosophila* leg development. *Development* **1998**, *125*, 3821–3830. [PubMed]
90. Weihe, U.; Dorfman, R.; Wernet, M.F.; Cohen, S.M.; Milan, M. Proximodistal subdivision of *Drosophila* legs and wings: The *elbow-no ocelli* gene complex. *Development* **2004**, *131*, 767–774. [CrossRef] [PubMed]
91. Dong, P.D.; Chu, J.; Panganiban, G. Coexpression of the homeobox genes distal-less and homothorax determines *Drosophila* antennal identity. *Development* **2000**, *127*, 209–216. [PubMed]
92. Casares, F.; Mann, R.S. A dual role for homothorax in inhibiting wing blade development and specifying proximal wing identities in *Drosophila*. *Development* **2000**, *127*, 1499–1508. [PubMed]

93. Casares, F.; Mann, R.S. The ground state of the ventral appendage in *Drosophila*. *Science* **2001**, *293*, 1477–1480. [CrossRef] [PubMed]
94. Emerald, B.S.; Cohen, S.M. Spatial and temporal regulation of the homeotic selector gene antennapedia is required for the establishment of leg identity in *Drosophila*. *Dev. Biol.* **2004**, *267*, 462–472. [CrossRef] [PubMed]
95. Galindo, M.I.; Bishop, S.A.; Greig, S.; Couso, J.P. Leg patterning driven by proximal-distal interactions and EGFR signaling. *Science* **2002**, *297*, 256–259. [CrossRef] [PubMed]
96. Campbell, G. Distalization of the *Drosophila* leg by graded EGF-receptor activity. *Nature* **2002**, *418*, 781–785. [CrossRef] [PubMed]
97. Kojima, T. The mechanism of *Drosophila* leg development along the proximodistal axis. *Dev. Growth Differ.* **2004**, *46*, 115–129. [CrossRef] [PubMed]
98. Kojima, T. Developmental mechanism of the tarsus in insect legs. *Curr. Opin. Insect Sci.* **2017**, *19*, 36–42. [CrossRef] [PubMed]
99. Rauskolb, C. The establishment of segmentation in the *Drosophila* leg. *Development* **2001**, *128*, 4511–4521. [PubMed]
100. Cordoba, S.; Requena, D.; Jory, A.; Saiz, A.; Estella, C. The evolutionarily conserved transcription factor Sp1 controls appendage growth through notch signaling. *Development* **2016**, *143*, 3623–3631. [CrossRef] [PubMed]
101. De Celis, J.F.; Tyler, D.M.; de Celis, J.; Bray, S.J. Notch signalling mediates segmentation of the *Drosophila* leg. *Development* **1998**, *125*, 4617–4626. [PubMed]
102. Rauskolb, C.; Irvine, K.D. Notch-mediated segmentation and growth control of the *Drosophila* leg. *Dev. Biol.* **1999**, *210*, 339–350. [CrossRef] [PubMed]
103. Kerber, B.; Monge, I.; Mueller, M.; Mitchell, P.J.; Cohen, S.M. The AP-2 transcription factor is required for joint formation and cell survival in *Drosophila* leg development. *Development* **2001**, *128*, 1231–1238. [PubMed]
104. Mirth, C.; Akam, M. Joint development in the *Drosophila* leg: Cell movements and cell populations. *Dev. Biol.* **2002**, *246*, 391–406. [CrossRef] [PubMed]
105. Tajiri, R.; Misaki, K.; Yonemura, S.; Hayashi, S. Joint morphology in the insect leg: Evolutionary history inferred from Notch loss-of-function phenotypes in *Drosophila*. *Development* **2011**, *138*, 4621–4626. [CrossRef] [PubMed]
106. Hao, I.; Green, R.B.; Dunaevsky, O.; Lengyel, J.A.; Rauskolb, C. The odd-skipped family of zinc finger genes promotes *Drosophila* leg segmentation. *Dev. Biol.* **2003**, *263*, 282–295. [CrossRef] [PubMed]
107. Cordoba, S.; Estella, C. The bHLH-PAS transcription factor dysfusion regulates tarsal joint formation in response to notch activity during *Drosophila* leg development. *PLoS Genet.* **2014**, *10*, e1004621. [CrossRef] [PubMed]
108. Suzanne, M. Molecular and cellular mechanisms involved in leg joint morphogenesis. *Semin. Cell Dev. Biol.* **2016**, *55*, 131–138. [CrossRef] [PubMed]
109. Paul, L.; Wang, S.H.; Manivannan, S.N.; Bonanno, L.; Lewis, S.; Austin, C.L.; Simcox, A. Dpp-induced Egfr signaling triggers postembryonic wing development in *Drosophila*. *Proc. Natl. Acad. Sci. USA* **2013**, *110*, 5058–5063. [CrossRef] [PubMed]
110. Rafel, N.; Milan, M. Notch signalling coordinates tissue growth and wing fate specification in *Drosophila*. *Development* **2008**, *135*, 3995–4001. [CrossRef] [PubMed]
111. Wang, S.H.; Simcox, A.; Campbell, G. Dual role for *Drosophila* epidermal growth factor receptor signaling in early wing disc development. *Genes Dev.* **2000**, *14*, 2271–2276. [CrossRef] [PubMed]
112. Baonza, A.; Roch, F.; Martin-Blanco, E. Der signaling restricts the boundaries of the wing field during *Drosophila* development. *Proc. Natl. Acad. Sci. USA* **2000**, *97*, 7331–7335. [CrossRef] [PubMed]
113. Kim, J.; Sebring, A.; Esch, J.J.; Kraus, M.E.; Vorwerk, K.; Magee, J.; Carroll, S.B. Integration of positional signals and regulation of wing formation and identity by *Drosophila* vestigial gene. *Nature* **1996**, *382*, 133–138. [CrossRef] [PubMed]
114. Azpiazu, N.; Morata, G. Function and regulation of homothorax in the wing imaginal disc of *Drosophila*. *Development* **2000**, *127*, 2685–2693. [PubMed]
115. Zirin, J.D.; Mann, R.S. Differing strategies for the establishment and maintenance of teashirt and homothorax repression in the *Drosophila* wing. *Development* **2004**, *131*, 5683–5693. [CrossRef] [PubMed]

116. Wu, J.; Cohen, S.M. Repression of teashirt marks the initiation of wing development. *Development* **2002**, *129*, 2411–2418. [PubMed]
117. Blair, S.S.; Brower, D.L.; Thomas, J.B.; Zavortink, M. The role of apterous in the control of dorsoventral compartmentalization and ps integrin gene expression in the developing wing of *Drosophila*. *Development* **1994**, *120*, 1805–1815. [PubMed]
118. Diaz-Benjumea, F.J.; Cohen, S.M. Interaction between dorsal and ventral cells in the imaginal disc directs wing development in *Drosophila*. *Cell* **1993**, *75*, 741–752. [CrossRef]
119. Cohen, B.; McGuffin, M.E.; Pfeifle, C.; Segal, D.; Cohen, S.M. Apterous, a gene required for imaginal disc development in *Drosophila* encodes a member of the LIM family of developmental regulatory proteins. *Genes Dev.* **1992**, *6*, 715–729. [CrossRef] [PubMed]
120. Couso, J.P.; Knust, E.; Martinez Arias, A. Serrate and wingless cooperate to induce vestigial gene expression and wing formation in *Drosophila*. *Curr. Biol.* **1995**, *5*, 1437–1448. [CrossRef]
121. Klein, T.; Arias, A.M. Different spatial and temporal interactions between notch, wingless, and vestigial specify proximal and distal pattern elements of the wing in *Drosophila*. *Dev. Biol.* **1998**, *194*, 196–212. [CrossRef] [PubMed]
122. Klein, T.; Arias, A.M. Interactions among delta, serrate and fringe modulate notch activity during *Drosophila* wing development. *Development* **1998**, *125*, 2951–2962. [PubMed]
123. Letizia, A.; Barrio, R.; Campuzano, S. Antagonistic and cooperative actions of the EGFR and Dpp pathways on the iroquois genes regulate *Drosophila* mesothorax specification and patterning. *Development* **2007**, *134*, 1337–1346. [CrossRef] [PubMed]
124. Zecca, M.; Struhl, G. Subdivision of the *Drosophila* wing imaginal disc by EGFR-mediated signaling. *Development* **2002**, *129*, 1357–1368. [PubMed]
125. Zecca, M.; Struhl, G. Control of growth and patterning of the *Drosophila* wing imaginal disc by EGFR-mediated signaling. *Development* **2002**, *129*, 1369–1376. [PubMed]
126. Cavodeassi, F.; Rodriguez, I.; Modolell, J. Dpp signalling is a key effector of the wing-body wall subdivision of the *Drosophila* mesothorax. *Development* **2002**, *129*, 3815–3823. [PubMed]
127. Calleja, M.; Herranz, H.; Estella, C.; Casal, J.; Lawrence, P.; Simpson, P.; Morata, G. Generation of medial and lateral dorsal body domains by the pannier gene of *Drosophila*. *Development* **2000**, *127*, 3971–3980. [PubMed]
128. Minami, M.; Kinoshita, N.; Kamoshida, Y.; Tanimoto, H.; Tabata, T. Brinker is a target of dpp in *Drosophila* that negatively regulates Dpp-dependent genes. *Nature* **1999**, *398*, 242–246. [CrossRef] [PubMed]
129. Jazwinska, A.; Kirov, N.; Wieschaus, E.; Roth, S.; Rushlow, C. The *Drosophila* gene brinker reveals a novel mechanism of Dpp target gene regulation. *Cell* **1999**, *96*, 563–573. [CrossRef]
130. Campbell, G.; Tomlinson, A. Transducing the Dpp morphogen gradient in the wing of *Drosophila*: Regulation of Dpp targets by brinker. *Cell* **1999**, *96*, 553–562. [CrossRef]
131. Blair, S.S. Wing vein patterning in *Drosophila* and the analysis of intercellular signaling. *Annu. Rev. Cell Dev. Biol.* **2007**, *23*, 293–319. [CrossRef] [PubMed]
132. De Celis, J.F. Pattern formation in the *Drosophila* wing: The development of the veins. *BioEssays* **2003**, *25*, 443–451. [CrossRef] [PubMed]
133. Schneuwly, S.; Klemenz, R.; Gehring, W.J. Redesigning the body plan of *Drosophila* by ectopic expression of the homoeotic gene antennapedia. *Nature* **1987**, *325*, 816–818. [CrossRef] [PubMed]
134. Postlethwait, J.H.; Schneiderman, H.A. Pattern formation and determination in the antenna of the homoeotic mutant antennapedia of *Drosophila melanogaster*. *Dev. Biol.* **1971**, *25*, 606–640. [CrossRef]
135. Kumar, J.P.; Moses, K. EGF receptor and notch signaling act upstream of Eyeless/Pax6 to control eye specification. *Cell* **2001**, *104*, 687–697. [CrossRef]
136. Kenyon, K.L.; Ranade, S.S.; Curtiss, J.; Mlodzik, M.; Pignoni, F. Coordinating proliferation and tissue specification to promote regional identity in the *Drosophila* head. *Dev. Cell* **2003**, *5*, 403–414. [CrossRef]
137. Wang, C.W.; Sun, Y.H. Segregation of eye and antenna fates maintained by mutual antagonism in *Drosophila*. *Development* **2012**, *139*, 3413–3421. [CrossRef] [PubMed]
138. Kumar, J.P.; Moses, K. Eye specification in *Drosophila*: Perspectives and implications. *Semin. Cell Dev. Biol.* **2001**, *12*, 469–474. [CrossRef] [PubMed]
139. Bryant, P.J. Determination and pattern formation in the imaginal discs of *Drosophila*. *Curr. Top. Dev. Biol.* **1974**, *8*, 41–80. [PubMed]

140. Morata, G.; Lawrence, P.A. Development of the eye-antenna imaginal disc of *Drosophila*. *Dev. Biol.* **1979**, *70*, 355–371. [CrossRef]
141. Cho, K.O.; Chern, J.; Izaddoost, S.; Choi, K.W. Novel signaling from the peripodial membrane is essential for eye disc patterning in *Drosophila*. *Cell* **2000**, *103*, 331–342. [CrossRef]
142. Baonza, A.; Freeman, M. Control of *Drosophila* eye specification by wingless signalling. *Development* **2002**, *129*, 5313–5322. [CrossRef] [PubMed]
143. Dominguez, M.; Casares, F. Organ specification-growth control connection: New in-sights from the *Drosophila* eye-antennal disc. *Dev. Dyn.* **2005**, *232*, 673–684. [CrossRef] [PubMed]
144. Dong, P.D.; Chu, J.; Panganiban, G. Proximodistal domain specification and interactions in developing *Drosophila* appendages. *Development* **2001**, *128*, 2365–2372. [PubMed]
145. Dong, P.D.; Dicks, J.S.; Panganiban, G. Distal-less and homothorax regulate multiple targets to pattern the *Drosophila* antenna. *Development* **2002**, *129*, 1967–1974. [PubMed]
146. Casares, F.; Mann, R.S. Control of antennal versus leg development in *Drosophila*. *Nature* **1998**, *392*, 723–726. [CrossRef] [PubMed]
147. Duncan, D.; Kiefel, P.; Duncan, I. Control of the spineless antennal enhancer: Direct repression of antennal target genes by antennapedia. *Dev. Biol.* **2010**, *347*, 82–91. [CrossRef] [PubMed]
148. Suzanne, M.; Estella, C.; Calleja, M.; Sanchez-Herrero, E. The hernandez and fernandez genes of *Drosophila* specify eye and antenna. *Dev. Biol.* **2003**, *260*, 465–483. [CrossRef]
149. Emerald, B.S.; Curtiss, J.; Mlodzik, M.; Cohen, S.M. Distal antenna and distal antenna related encode nuclear proteins containing pipsqueak motifs involved in antenna development in *Drosophila*. *Development* **2003**, *130*, 1171–1180. [CrossRef] [PubMed]

6

Insights into the Etiology of Mammalian Neural Tube Closure Defects from Developmental, Genetic and Evolutionary Studies

Diana M. Juriloff * and Muriel J. Harris

Department of Medical Genetics, University of British Columbia, 2350 Health Sciences Mall, Vancouver, BC V6T 1Z3, Canada; muriel.harris@ubc.ca
* Correspondence: juriloff@mail.ubc.ca

Abstract: The human neural tube defects (NTD), anencephaly, spina bifida and craniorachischisis, originate from a failure of the embryonic neural tube to close. Human NTD are relatively common and both complex and heterogeneous in genetic origin, but the genetic variants and developmental mechanisms are largely unknown. Here we review the numerous studies, mainly in mice, of normal neural tube closure, the mechanisms of failure caused by specific gene mutations, and the evolution of the vertebrate cranial neural tube and its genetic processes, seeking insights into the etiology of human NTD. We find evidence of many regions along the anterior–posterior axis each differing in some aspect of neural tube closure—morphology, cell behavior, specific genes required—and conclude that the etiology of NTD is likely to be partly specific to the anterior–posterior location of the defect and also genetically heterogeneous. We revisit the hypotheses explaining the excess of females among cranial NTD cases in mice and humans and new developments in understanding the role of the folate pathway in NTD. Finally, we demonstrate that evidence from mouse mutants strongly supports the search for digenic or oligogenic etiology in human NTD of all types.

Keywords: neural tube; neural folds; anencephaly; exencephaly; spina bifida; craniorachischisis; folate; epigenetics; developmental genetics

---

1. Introduction

The mechanisms of closure of the neural tube and the development of neural tube closure defects (NTD) are genetically complex and diverse. With an incidence of one to ten per thousand births, NTD are among the most common severe birth defects [1]. Defects in cranial and spinal neural tube closure (anencephaly and spina bifida) seem to occur at almost equal frequencies in human embryos [1,2]. In the almost 300 mouse mutants with NTD, cranial defects strongly predominate [3,4]. However, much of the molecular genetic understanding of mammalian neural tube development is based on studies of the spinal region in mouse embryos. Many aspects have been comprehensively reviewed previously [4–9]. This review gives an overview of the developmental process and the genetic underpinnings of neural tube closure in mammals, and then focuses on some aspects that have potential to bring new insight.

The formation of a neural tube by the mechanism of rolling up a neural plate into a tube evolved before the evolutionary emergence of vertebrates, and some mammalian molecular mechanisms likely reflect this ancient origin. We briefly consider the process of neural tube formation in the animal usually used as a proxy for the ancestral chordate, Amphioxus, and in Ascidian embryos, representative of another ancient pre-vertebrate lineage, the Tunicates. There are extensive differences in the physical process of neural tube closure among modern species and even among mammals; we shall briefly also consider these to point to the need for assessment of these differences before using observations from model animals to predict the etiologies of human NTD.

We give considerable emphasis to the anterior–posterior regional differences in the mechanisms of neural tube closure at the morphological and molecular levels, underlining the concept that defects of closure of the neural tube are mechanistically heterogeneous with distinct or partially overlapping etiologies. This heterogeneity predicts genetic differences in the etiologies of human NTD located at different sites along the anterior–posterior axis, as well as genetic heterogeneity at any given site. We focus on cranial neural tube development, its genetic mechanisms of failure to close, and the evolutionary New Head hypothesis that predicts consequential differences between the cranial and spinal regions in the mechanisms of neural tube closure. These differences should inform genetic studies of human NTD.

We consider recent advances in understanding of the epigenetic mechanism by which the sex chromosome complement of mammalian embryos influences risk of cranial NTD, seeking new insight into the neural tube closure process. Similarly, we focus on the phenomenon of the preventative effect of maternal folic acid supplementation on occurrence of NTD in humans and in mice and recent developments in understanding of folate mechanisms, seeking possible insights into the etiologies of NTD.

That there is a genetic etiology for human NTD is clear, with 30× elevated risk in siblings of NTD cases, but it is complex. Typical recent studies are based on exon sequencing of candidate genes in cohorts of NTD cases, including a variety of locations and types of lesion. This approach for several planar cell polarity (PCP) genes has detected heterozygous rare putatively deleterious variants in a small number of cases [10].

Despite many genetic studies, the etiology is not well understood [11], and likely involves combinations of deleterious versions of genes in affected individuals, with the combinations differing between individuals [12]. We discuss examples of similar genetic complexity in mouse mutant models.

Failure to close the neural tube leads to three principal open neural tube defects: craniorachischisis, in which the entire spinal region and the hindbrain fail to close; spina bifida aperta, in which a limited region of the spinal neural tube fails to close, and anencephaly or exencephaly, in which the midbrain and sometimes also the hindbrain fail to close. There are other defects that arise subsequently from defects in the tissue of the closed neural tube or surrounding mesenchyme, such as spina bifida occulta (abnormal formation of the vertebral arches from mesenchyme) and encephalocele (extrusion of neural tissue through a defect in the formation of the roof of the closed neural tube). Semantically or clinically, they are often included in NTD, but they are not within the scope of the etiology of failure to form a closed neural tube [6,13].

2. The Basics of Neural Tube Formation

2.1. Components

The major physical components of the complex that will comprise the neural tube (Figure 1) include: the anterior to posterior midline neural plate, comprised of neuroectoderm (neuroepithelium) that develops on the dorsal aspect of the embryo, curls from its lateral edges toward the midline and fuses its formerly lateral edges in the midline; the notochord that underlies the neural plate and produces signaling molecules that affect dorsal–ventral cell identity in the incipient neural tube; the lateral ectoderm that initially abuts the lateral edges of the neural plate, covers the lateral surfaces of the elevating neural folds and after the bending of the folds to the midline, meets and fuses to form a contiguous surface ectoderm over the tube; the paraxial mesoderm lateral to, and supporting, the neuroepithelium of the elevating folds; and the neural crest that is induced in the neuroepithelium of the lateral tips of the spinal neural plate or in the dorsal tips of the elevated cranial neural folds. There are several mechanisms affecting these components that lead to failure of neural tube closure (Table 1).

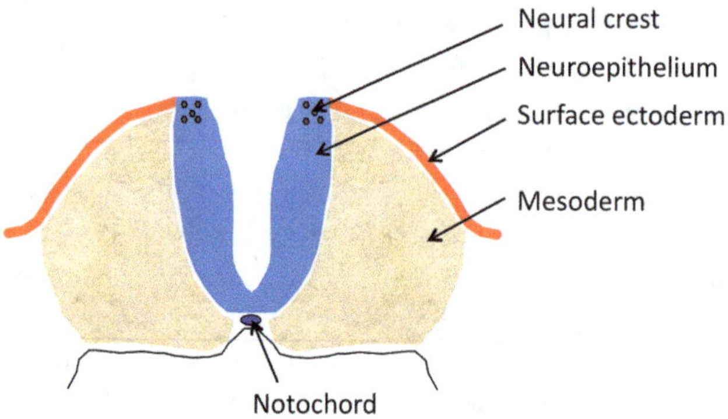

Figure 1. Diagram of a transverse section of upper spinal neural folds of an E8 mouse embryo before neural tube closure. Somites not shown.

2.2. Convergent Extension and Other Functions of the Planar Cell Polarity Pathway

Immediately preceding the initiation of neural tube formation, the vertebrate embryo undergoes convergent extension, which involves the lateral-to-medial convergence and anterior–posterior extension of axial tissues, including the neural plate. The non-canonical Wnt/PCP (planar cell polarity) pathway is crucial to this process. Cells change position in the tissue, intercalating between surrounding cells to shift toward the midline. A prerequisite is the directional orientation of cells relative to a morphogenic signal. The cells become asymmetrical in the positioning of various subcellular structures, including the core proteins of the PCP pathway [14,15]. Mutations in PCP genes interfere with the directionality of neuroepithelial cell movement in a variety of gene-specific ways [16]. The planar cell polarity (PCP) pathway is complex, comprising two main sets of interacting genes, the "core" pathway, and the Fat-Dachsous pathway, probably connected by *Prickle* [14]. The "core" pathway is highly conserved and is regulated by gradients of particular Wnt proteins, (e.g., *Wnt5a* and *Wnt11*). The "core" proteins include the transmembrane receptors *Vangl1/2*, *Celsr1/2/3*, and *Fzd* (Frizzled) and the cytosolic proteins *Dvl*, *Prickle*, Inversin (*Invs*) and *Scrib* [14]. The Fat/Dachsous pathway, acting upstream or in parallel with the PCP core pathway, includes the large membrane cadherins, *Fat* and *Dchs*, which bind to each other and are regulated by *Fjx* [5]. An additional assortment of genes serve various functions necessary to the PCP pathways; for example, *Ptk7*, *Sec24b*, *Intu* and *Fuz* [10].

In addition to its essential role in planar cell polarity required for convergent extension, the PCP pathway affects ciliogenesis, the arrangement of cytoskeletal actin and microtubules, the basal body and the centriole within cells [14,17,18]. Small Rho family-GTPases regulate the localization of actin polymerization. Microtubules are polarized, and their polarity is required for "correct axis establishment of PCP", then PCP signaling is required for stability, correct localization and remodeling of microtubule networks. It is thought that PCP, actin and microtubules together determine the asymmetrical positioning of the centriole, ciliary basal bodies, cilia and the orientation of the mitotic spindle [14]. There is also interconnection between the convergent extension system and the neural plate bending system based on actomyosin. For example, RhoA and its effector ROCK are involved in both systems [19,20].

Table 1. Ways in which the neural tube can fail to form.

Lack of convergent extension
Lack of medial hinge point (MHP) [1]
Lack of dorsolateral hinge point (DLHP) [1]
Lack of neuroepithelial bending by apical constriction (cranial region)
Lack of structural integrity of the neuroepithelium
Lack of support from surrounding mesenchymeLack of midline fusion of neuroepithelium and/or surface ectodermInterference by excessive anterior-posterior curvature of the neural plate

[1] Shown in Figure 2.

2.3. Bending and Hinge Points

Two areas of the neural plate undergo pronounced bending that strongly contributes to forming a tube; these are the medial hinge point (MHP) located at the midline base of the neural tube throughout most of its length, and the dorsolateral hinge point (DLHP) located in the dorsal third of each fold in some zones along the length of the neural tube [21] (Figure 2). In addition, the contraction of actomyosin found in the apices of all neuroepithelial cells has long been thought to contribute to bending the neural folds, and indeed, apical actomyosin is essential for cranial neural tube closure [19], but its specific role is unclear. In contrast, evidence indicates that apical actomyosin is not essential for spinal neural tube closure [19]. Instead, in the neuroepithelium of the spinal region, an orchestrated dynamic of actin turnover and actomyosin disassembly is essential for closure, and has been suggested to function in maintaining a necessary tension in the neuroepithelium (neither too floppy nor too stiff) between the MHP and DLHP sites [19]. The MHP and DLHP in the spinal region differ in their mechanisms that cause neuroepithelial bending. For the MHP to form in mouse embryos, signaling from the notochord is required [22]. At the MHP, bending is attributed to neuroepithelial cell shape. During the cell cycle, in S phase, nuclei move to the base of the cells, causing them to become wedge-shaped, with a narrow apex and wide base. Compared with the rest of the neuroepithelium, the MHP has an increased proportion of cells in S-phase, and also has cell-cycle dependent changes in apical and basolateral junctional proteins that contribute to cell shape [23,24].

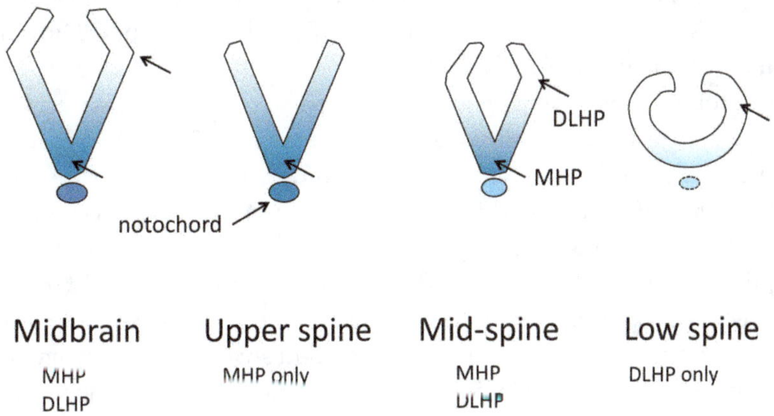

Figure 2. Conceptual views of the shape of neuroepithelium in transverse sections, at various anterior–posterior locations, of a mouse neural tube before closure, demonstrating differences in bending mechanisms. Shading represents the concentration of *Shh* signaling. DLHP, dorsolateral hinge point. MHP, medial hinge point. Modified from Figure 3 in [13].

The mouse spinal DLHPs form at the dorsoventral location where the neuroepithelium ceases to have basal contact with the paraxial mesoderm and begins basal contact with the surface ectoderm [23]; however, the significance of this relationship is unknown, given that the upper-spinal neural tube, which can form DLHP in the absence of *Shh* [22], does not have this mesoderm–ectoderm contact transition. Cell wedging related to cell cycle is not seen at the DLHP. Instead, there is cell proliferation and cell movement dorsally in the neuroepithelium, and evidence of a boundary at the DLHP between areas of different cell density. It has been proposed that the neural folds are caused to bend by the cell density difference [23]. There is an important relationship of the formation of the DLHP in the neuroepithelium with *Shh* signaling molecules from the ventral neural plate, *BMP* signaling from the surface ectoderm, and *Noggin* from the neuroepithelium [22]. The dorsal bends that appear to be DLHP in the midbrain neural folds have not been studied and it is possible that, despite appearing similar morphologically, they are driven by different cellular or molecular mechanisms.

2.4. Cilia

Most cells have a primary cilium. In the neural tube, primary cilia can be seen extending from the neuroepithelium into the incipient lumen (see Figure 6 in [25]). Vital components of signaling pathways that are important to development of the neural tube are located in cilia, and if gene mutations cause the structure/function of the cilia to be defective, these signaling pathways are dysregulated [13,26]. The best understood is *Sonic hedgehog* (*Shh*) signaling. Other pathways, such as signaling through *Wnt*, Calcium, *Pdgfr*, *Notch*, and *Tgfb*, may also function via cilia [13,26,27]. Some of the PCP genes affect the position and angle of some types of cilia on cells [28], and some types of PCP genes (the "effectors") have a role in ciliogenesis in vertebrates [13,29]. There are many components to the cilia, such as intraflagellar transport proteins and parts of the dynein motors, and when they are mutated, they lead to defects in the structure and function of cilia.

2.5. The Sonic Hedgehog Pathway Signaling in NTD

Several components of the *Sonic hedgehog* (*Shh*) signaling pathway, including positive and negative regulators, are present in the cilia [13]. Shh is a secreted glycoprotein whose binding with the transmembrane receptor Patched (Ptch1) in the cilia releases Smoothened (Smo) from repression and activates the *Shh* signaling pathway. The *Shh* signaling pathway affects the transcription of downstream target genes via the Gli transcription factors; there are many transcriptional targets and complex feedback loops [30]. An active *Shh* signaling pathway inhibits the processing of a full-length activator form of Gli3 (Gli3FL) into a repressor form (Gli3R) and the ratio of Gli3FL:Gli3R is a measure of functionality of the *Shh* signaling pathway [13,30].

During neural tube development, *Shh* is produced by the notochord and prechordal plate extending from the future forebrain area to the caudal area [22], and induces *Shh* production from the midline of the neural plate with a ventral to dorsal gradient in the developing neural tube [13]. While the notochord is required for formation of the MHP, and *Shh* signaling is required for development of the floor plate, it is not clear that *Shh* is required for formation of the MHP; the MHP is not always required for neural tube closure [6,22]. The formation of the DLHP is inhibited by *Shh* signaling [22]. Absence of *Shh* signaling in the dorsal neural folds permits a neuroepithelial response to *Noggin*, which suppresses *Bmp* signaling from the adjacent surface ectoderm, enabling formation of dorsolateral hinge points (DLHP) [31]. *Shh* is expressed in a rostral to caudal and temporal gradient in the developing neural tube from the upper spinal region to the caudal spinal region, being expressed in the caudal notochord and floor plate after neural tube closure [31]. In the cranial neural tube, no gradient has been reported and expression is seen from before closure onwards (e.g., Figure 4c in [32] and Figure 1a in [22]). The presence of DLHP in the cranial neural folds, particularly the midbrain, despite possibly robust *Shh* signaling in the ventral midline seems contradictory to the Shh concentration-dependent inhibition of DLHP in the upper spine. However, in many species, the cranial neural folds are clearly much larger than the spinal neural folds; this is particularly notable in mouse embryos. Therefore, the physical distance for diffusion of signaling molecules from the midline to the dorsal tips is greater than in the spine. We suggest that the large size of the cranial neural folds creates the necessary lack of signaling concentration in the cranial dorsal folds, enabling cranial DLHP formation.

Loss of function of the *Shh* gene, or loss of function of components of the *Shh* signaling pathway such as *Smo*, do not cause NTD; the neural tube forms a midline bend and DLHP and closes. Typically, loss of function of the *Shh* signaling pathway causes a lack of development of the cranial midline tissues that results in holoprosencephaly and related facial defects [13,33,34]. In contrast, elevated activity of the *Shh* signaling pathway downstream of *Shh* due to the loss of negative regulators such as *Ptch1*, *Rab23*, *Tulp3* or *Sufu* [13] leads to elevated signaling in the dorsal neural folds, inhibiting development of the DLHP (e.g., *Tulp3* [35]) and causing NTD (exencephaly and sometimes caudal spina bifida) in the regions that normally have DLHP (Figure 2).

Mutations in genes that affect the structure or function of cilia fall into two major types with respect to *Shh* signaling [13]. Some mutations that affect the structure/function of cilia, such as *Ttc21b* (intraflagellar transport), *Kif7* (kinesin motor protein) and *Arl13b* (cilial structure), seem to increase *Shh* signaling pathway activity based on expansion of ventral markers, and cause exencephaly, parallel to the other mutants with increased *Shh* signaling. Others, such as mutants for several intraflagellar transport (*Ift*) genes, dynein components (*Dync2h1*, *Dync2li1*), PCP effector genes (*Fuzzy*, *Inturned*), and *C2cd3* (which recruits *Ift* proteins for ciliogenesis) seem to reduce *Shh* signaling pathway activity, based on reduced ventral markers in the neural tube, but contrary to the mutations with loss of *Shh* signaling, they also cause exencephaly. Notably, it seems that cilia mutants that cause exencephaly have an increased Gli3FL:Gli3R ratio [13], but there are many other complex changes and regulatory loop alterations specific to each mutant, and possible involvement of the other cilia-based signaling pathways. Shh signaling also functions through a network of non-canonical, Gli-independent pathways, which could be involved [36]. The process of cranial neural fold formation and bending does not seem to have been examined in each mutant with exencephaly and it is unknown whether there is a shared mechanism, such as lack of DLHP formation, or a heterogeneity of mechanisms such as apoptosis in the neuroepithelium or deficient cell adhesion.

2.6. Adhesion and Fusion

To complete neural tube closure, the neuroepithelium and surface ectoderm of the neural fold must make contact with their counterparts in the apposed fold and form de novo adhesions to create a continuous surface ectoderm covering a continuous neuroepithelium. The neuroepithelia and surface ectoderms do not necessarily make contact simultaneously, and the order varies along the anterior–posterior axis. Some details have emerged from recent studies. Filopodia and ruffles project from apposed surface ectoderms into the gap between the apposed folds and are required for successful closure, although their role is unknown [37–39]. Various membrane-bound *Eph* receptor tyrosine kinases and their membrane bound ligands, the *Ephrins*, which are generally involved in cell–cell attraction and repulsion interactions, are found in the neural folds, including the dorsal tips, just prior to fusion and are possibly involved in initiation of adhesions between apposed folds [38,40].

3. Regional Differences in Closure Mechanisms

Studies of mouse embryos have demonstrated that different regions along the anterior–posterior axis have different major mechanisms for bending the neuroepithelium to meet in the midline and also differ in fusion mechanisms (Table 2). These reflect several factors [5,6] including the early spinal role of planar cell polarity, the timing of anterior–posterior development of the notochord relative to closure, a gradient of *Shh* signaling along the anterior to posterior spine [22], presence of MHP and DLHP, differences in expression domains along the anterior–posterior axis of various genes important to closure, such as the *Grhl* gene family [41] and the *Ephrin/Eph* gene families [40,42,43], types of cell projections, cell types that initiate fusion, and factors in the molecular genetics of closure that are unique to the cranial region, such as the dynamics of the actomyosin cytoskeleton. Given the distinct regional differences in mechanism and genes involved in neurulation, studies to identify causative genetic variants in human NTD would be strengthened by studying pools of cases that share a similar location of lesion.

3.1. Planar Cell Polarity, MHP and DLHP

The upper spine region depends on planar cell polarity for cell movements and convergent extension that narrows the midline gap between the future neural folds, and on cell cycle dynamics in the neural plate midline that cause cell wedging to create a medial hinge point (MHP). The mid-spinal region also depends on the effects of convergent extension, and has the MHP, but additionally has DLHPs. The most caudal area of primary neurulation, the posterior neuropore (PNP), likely also depends on initial convergent extension to narrow the neural plate, but unlike the rest of the neural tube, the notochord is not present before closure [22]. The MHP does not form, but DLHPs bend the folds to the midline; there is also a greater dependence on actomyosin contraction for bending than in the more rostral spinal region [44]. Note: the PNP is sometimes used to refer to an extended region including part of the mid-lower spine to the most caudal end of primary neurulation; in this review we refer to only the most caudal part of this region as the PNP.

Caudal to the PNP, neural folds do not form and a tube is hollowed out from the caudal mesenchyme in a process termed "secondary neurulation". Defects in secondary neurulation per se would not cause closure defects, but there is evidence that mis-regulation of the development of the junction between the PNP and secondary neurulation may cause some lumbosacral open neural tube defects [45,46].

Overall, given the various overlapping regional differences in molecular genetic mechanisms and closure mechanisms from head to tail, it is likely that NTD in different anterior–posterior regions have region-specific causes that can be predicted. For example, in families with recurrence of NTDs, in about 70% of families, the second cases have the same defect as the first case, but in about 30% of families, one has spina bifida and one has anencephaly [47]. Given that failure of DLHP is a mechanism of NTD shared by anencephaly and lumbosacral spina bifida aperta, we suggest these latter families may be enriched for defects in genes that affect DLHP, which can inform the genetic approaches to unraveling some of the genetic complexity of human NTD.

Table 2. Regional differences in mechanisms for neural tube closure.

Mechanism	Difference between Regions of Neural Tube
Convergent extension	Absent in forebrain
Notochord in head	Absent in forebrain
Notochord in spinal region	Absent at posterior neuropore during closure
Level of *Shh* expression	Gradient; lowest in posterior neuropore
Medial Hinge Point	Absent in posterior neuropore
Dorsolateral Hinge Point	Absent in upper spinal region
Convex mesenchymal expansion	Midbrain only
Grhl2, *Grhl3* expression	Cranial, caudal and upper spine differ
Ephrins and Ephrin receptors	Different members of gene family regionally
Hox gene expression	Absent in midbrain and forebrain regions
Closure initiation site spacing	Caudal closure furthest from an initiation site
"Zipping" vs. simultaneous closure	Regional differences in axial bending
Meeting of neural folds in midline	Contact of neuroepithelium vs. surface ectoderm
Ruffles versus filopodia	Forebrain, midbrain, hindbrain, spine differ
Apical actomyosin contractility	Required for cranial closure, not spinal
Neural crest emigration	Spinal after closure; cranial before closure
Apoptosis	Required for cranial closure, not caudal

3.2. Closure Initiation Sites

At several specific locations along the anterior–posterior axis the neural folds achieve close apposition and initiate fusion. Closure then is thought to "zip" along the neural folds until it meets a closed region "zipped" from a different initiation site—an intermittent pattern rather than a single "zipper". The mechanisms that determine the locations of closure initiation sites are not known. The locations of cranial closure initiation sites differ somewhat between normal mouse strains [48]; the possibility of similar variation within other species should be considered. The locations of cranial closure initiation sites and patterns of closure vary greatly between mammalian species as summarized in Figure 3 (mouse [49–51], rat [52,53], hamster [54,55], rabbit [56], pig [57], human [58–62]). The pattern of cranial closure initiation sites differs greatly between humans and mice, rats or hamsters, particularly with respect to the presence of an initiation site near the forebrain–midbrain junction that is not seen in human embryos (Figure 3). The closure process in rabbit and pig seem more similar to human. In rabbit, pig and human, instead of "zipping", extended regions in the head and spinal region come into close apposition and then fuse simultaneously. Greater axial curvature seems to reduce the rate of closure along the anterior–posterior axis and between species [63]; flat areas tend to close simultaneously rather than "zip". Otherwise, the location of the Closure 1 initiation site at the future neck (cervical) region and pattern of expansion of closure from it to the posterior neuropore seems generally similar across mammalian species.

The relevance to NTD of species differences in locations of closure initiation sites is debatable. Generally, the locations of open cranial NTD and open spinal NTD in mice and humans appear to be the same, i.e., the same parts of the neural tube have failed to close. As we have said previously " . . . if the defect in both species is a failure of elevation itself, a species difference in the location of first contact of already elevated folds would be subsequent to, and unrelated to, the cause of NTDs" [64]. Furthermore, the exact positions of open defects may not be interpretable with respect to inferences about locations of closure initiation sites. In mouse models, embryos sharing the same genetic etiology often differ in the exact location of the open neural tube. For example, in the SELH/Bc model, the rostral edge of exencephaly may be located anywhere between the caudal limit of the forebrain and the middle of the midbrain and, independently, the caudal edge of the exencephaly may be located anywhere between the caudal limit of the midbrain and the caudal limit of the hindbrain (Figure 4 in [65]). In the SELH model, we have shown that, in the absence of a midbrain initiation site, the normal closure initiated in the forebrain can compensate by advancing caudally to close the entire midbrain, and that closure of the hindbrain can also compensate by continuing rostrally to close part of the caudal midbrain. The variation in location of exencephaly seems to reflect stochastic variation in the efficiencies of these compensatory processes. We suggest that this type of compensatory mechanism might explain part of the documented variation in locations of human anencephaly [66].

3.3. Cell Type of Initial Contact

The tips of the neural folds expose the border between the neuroepithelium and surface ectoderm. The cell type involved in initial contact between apposed folds varies regionally. In the mouse forebrain region, first contact across the gap is between the apposed neuroepithelia, whereas in the midbrain and hindbrain, the initial contact and apparent adhesion is between cells of the surface ectoderm, which in the midbrain wraps around the neuroectoderm as the folds meet so that surface ectoderm faces across the closing gap [37,67]. In the posterior neuropore, the surface ectoderm may make first contact, based on locations of cellular projections [5].

The fusion of the neuroepithelium and the surface ectoderm may be independent; a closed cerebellar defect arising in an unfused neuroepithelium beneath a fused surface ectoderm has been described in mice [65].

3.4. Cell Projection Types

Fusions between epithelia in various tissues and species are typically preceded by various cellular projections such as ruffles, filopodia and lamellipodia [38]). Dynamic cell behaviors and interactions, such as the interdigitation of ruffles across the gap between the closely apposed neural folds of the closing neural tube, are seen and the projections are required for neural tube closure [38,55,68]. The density and type of cell projections vary along the anterior–posterior axis of the neural folds [38,55]. The forebrain, midbrain and hindbrain differ in the presence of filopodia and ruffles. In the forebrain region, live imaging of surface ectoderm has shown that there are a few small narrow projections close to the area of contact, but the majority of the surface ectoderm does not have projections [39]. In the midbrain region as it approaches closure, there are many surface ectoderm projections that appear to be filopodia [39]. In the closing hindbrain region there are many long round projections from the surface ectoderm, some of which may be filopodia, and others a relatively narrow form of ruffles [39]. The upper spine Closure 1 initiation region has wide membranous ruffles or lamellipodia and narrow bulbous projections, both extending from the surface ectoderm towards the gap at the midline and increasing as the folds move closer together; these projections are the first points of contact across the gap [39]. The caudally advancing closure region of the upper spine region has filopodia [37]. The mid-spinal region has both filopodia and ruffles in the closing area. The most caudal region of primary neurulation has only ruffles [38,55].

The regional differences in the abundance of filopodia and ruffles reflect, at least in part, differences in regulators of the actin cytoskeleton, *Rac* and *Cdc42* [38]. These GTPases are molecular switches that link extracellular signals to specific downstream effectors. In general, *Rac1* induces the formation of branched actin networks needed for lamellipodia and ruffles, whereas *Cdc42* leads to assembly of unbranched actin bundles that form filopodia [38].

3.5. Regional Differences in Ephrins and Ephrin Receptors

The gene families of membrane-bound Ephrin receptor tyrosine kinases (*Ephs*) and their membrane-bound ligands, the *Ephrins*, are involved in cell–cell interactions throughout development. Interaction between *Ephrins* and their receptors may cause cell adhesion or repulsion. They are involved in the adhesion of apposed neural folds for fusion and may have a role in the etiology of NTD. This area has not been deeply studied and roles of more members of these gene families are likely to be identified in future studies. Expression of some combinations of *Ephrins* and *Ephs* may differ between cranial regions and also differ from the caudal region. *EphA4* is expressed in the midbrain region in early neural fold stages [69]; *EphA2* and *EphA4* are expressed in the hindbrain neural folds before closure [70]. *EphrinA5* and its receptor *EphA7* splice forms are expressed in the dorsal cranial neural folds (midbrain and hindbrain) before and during closure, but not in the spinal neural folds [40]. In the posterior neuropore, *EphA1* is expressed in the neural plate and other tissues, *EphA2* in the neural plate and tips of the apposed neural folds, *EphA4* in the mesoderm and tips of the apposed neural folds, and *EphA5* in the neural plate [43]. Further study of *EphA2* in the tips of the neural folds located its expression to surface ectoderm cells and their lamellipodia-like protrusions that extend towards the apposed neural fold [43].

3.6. Requirement for Grhl Gene Family Expression

The regionality of the requirement for expression of different members of the *Grhl* gene family along the anterior–posterior axis of the neural folds in mice is an important demonstration that the locations of NTD reflect the regional requirement for expression of specific genes. Loss of function of *Grhl3* causes thoraco-lumbo-sacral spina bifida and curled tail (a microform of spina bifida) in all homozygotes and concomitant exencephaly in a few [71]. Loss of function of *Grhl2* causes split face and open cranial neural tube extending through the hindbrain to the cervical region and, depending on the mutant allele and genetic background, spina bifida [37,41]. Heterozygotes at either locus are normal, but some digenic heterozygotes (*Grhl2*$^{+/-}$;*Grhl3*$^{+/-}$) have lumbo-sacral spina bifida with curled tail or exencephaly and/or curled tail. *Grhl2* null and *Grhl3* null embryos fail to form DLHP in the regions that will have failed neural tube closure [41].

Further study [41] indicates that *Grhl2* is essential to closure of the forebrain and cooperates with *Grhl3*, whereas *Grhl3* is essential to closure posterior to the mid-thoracic region, and cooperates with *Grlh2* for closure of the posterior neuropore (PNP). Neither *Grhl* gene is required for closure of the cervical region to the thoracic region, the rostral part of the Closure 1 zone; this is the only region of the neural tube that does not form a concave shape of neural folds before closure, and may therefore be the only region that does not have DLHPs. DLHPs are lacking in the PNP of both *Grhl2* and *Grhl3* mutants; other regions were not studied. In summary, the *Grhl2* and *Grhl3* genes may be respectively essential for DLHP formation in separate zones of the developing neural tube, and may also function cooperatively for DLHP formation in other zones. The *Grhl* mutants did not alter the expression of some genes known to be important in DLHP formation, such as *Bmp2*, *Noggin*, and *Zic2* [41]. The failure to form DLHP may be explained by the observation that *Grhl2*, which is specifically expressed in the surface ectoderm of the cranial neural folds, is required to suppress transformation of ectoderm to mesenchyme [72]. The acquisition of mesenchymal characteristics and loss of integrity of the surface ectoderm likely causes the inability to bend the dorsal neural fold. A similar disruptive effect on the development of DLHPs or the bending to concave shape is suggested for the homolog, *Grhl3*. In the caudal region *Grhl3*, is expressed in surface ectoderm during neural tube closure, and lack of *Grhl3* seems to cause delay of specification of surface ectoderm [73], a situation somewhat like the findings for *Grhl2* [72]. Later, at the most caudal end of primary neurulation, a second mechanism is observed. *Grhl3* is expressed in the gut endoderm that underlies the neural tube, and lack of *Grhl3* expression there is associated with defective hindgut cell proliferation that causes excessive axial bending that mechanically interferes with closure of the posterior neuropore [73].

3.7. Dynamics of the Actomyosin Cytoskeleton of the Neuroepithelium

Actomyosin "machinery" (actin filaments, phosphorylated myosin light chain (pMLC) and RhoGTPases) is located circumferentially in the apices and apical junctions of all neuroepithelial cells. Its contraction is thought to be responsible for apical constriction, contributing to neural fold bending, and is required for cranial neural tube closure but not spinal neural tube closure. Mice with mutations affecting actin-associated proteins develop cranial, but not spinal, NTDs. Conversely, the caudal spinal region as well as the cranial region requires actin turnover for neural tube closure and abnormal accumulation of apical actomyosin does cause failure of neural fold elevation in both regions [19]. RhoA and its effector ROCK are required in the neuroepithelium, to maintain balanced apical actomyosin accumulation, to regulate actin turnover, as well as for convergent extension [19,20]. A related caudal difference is the contribution of bipotential neuromesodermal progenitors to the caudal neuroectoderm and somatic mesoderm, and the influence of PCP (*Vangl2*) on their differentiation as well as on neuroepithelial actomyosin contractility [44].

3.8. Other Examples: Hox, Neural Crest and Apoptosis

Other regional differences in the neural tube also have an unknown or uncertain role in closure mechanisms. For example, numerous *Hox* genes, which convey anterior–posterior positional information, are expressed in many different combinations along the midline axis before neural tube formation, with the earliest and fewest expressed in the hindbrain region and the latest and most numerous expressed in the caudal spinal region; in contrast, none are expressed in the cranial region rostral to the hindbrain [74,75]. Another example is the timing of emigration from the tips of the neural folds by the neural crest, which in mouse emigrates after closure in the spinal region and before closure in the head, and whose role in neural tube closure is uncertain [42,76]. There also may be a requirement for normal apoptosis in the cranial neural folds to enable closure, but having less effect in the spinal region [42,77].

4. Evolutionary Aspects

4.1. Modern Vertebrates

Most of our understanding of neural tube development is based on a small number of species: mouse, chick, frog (Xenopus), rat, and zebrafish. Among these, there are very important differences in the basic mechanisms by which the neural tube is formed. All likely share the basic planar cell polarity/convergent extension mechanism [5]. Details of convergent extension in development of the neural tube have been studied in frog (Xenopus) and zebrafish, but their cellular process is expected to differ from mammals and chick because their organization of neuroepithelial cells is known to differ [16]. For the hinge point mechanism, Xenopus lacks the MHP and DLHP [20]. Overall, the studied species differ in various mechanisms of closure, such as the relationship of the neuroepithelium to the surface ectoderm, the presence of neural folds, the presence of hinge points and the degree to which closure is intermittent or simultaneous along the whole neural tube [5,78–80]. Of course, a species that does not normally undergo the cellular/tissue process whose failure causes human NTD cannot model the mechanism of human NTD. Among the well-studied models for neural tube development, mouse, chick and rat appear to be most similar to human, but little is known about human embryos except histological and morphological studies. Based on morphological studies, the little-studied species, rabbit and pig, may be most similar to human (Figure 3).

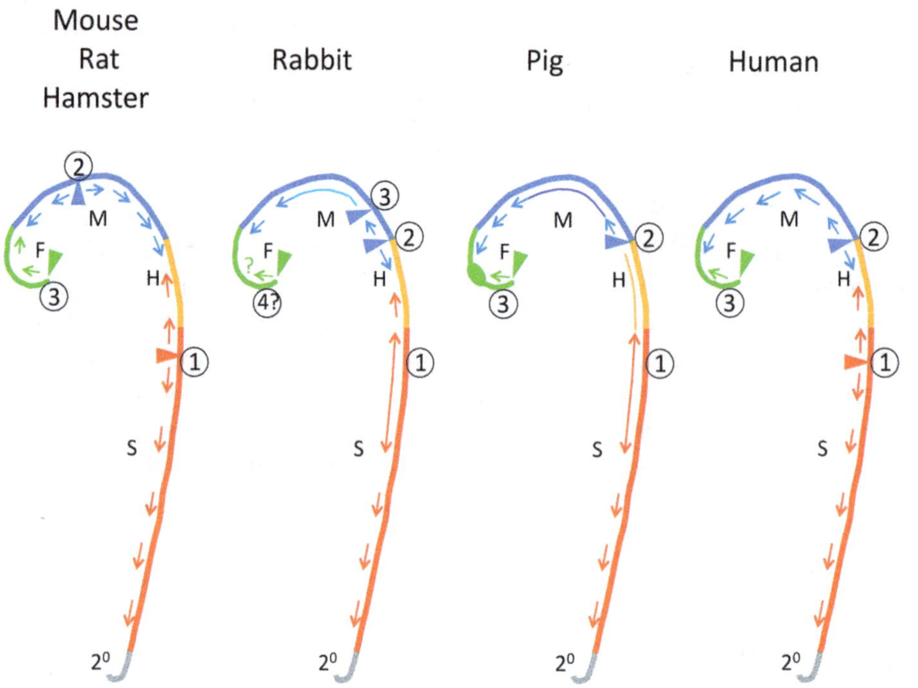

Figure 3. An interpretation of the patterns of neural tube closure in various mammalian species based on published images and studies. Colors denote future anterior–posterior fates of the neural tube. F, forebrain; M, midbrain; H, hindbrain; S, spine. Short arrows indicate direction of zipping. Long stems on arrows and lines lacking arrowheads denote areas that appose and then fuse simultaneously. Triangles and circled numbers indicate closure initiation sites. The forebrain oblong denotes a region that appears to close from all sides, rather than apposition or zipping. 2°, region of secondary neurulation.

4.2. Proxies for the Ancestral Pre-Vertebrate

The Cephalochordate, Amphioxus (lancelet), whose lineage diverged from vertebrates 550–600 million years ago, is used as a proxy for the ancestral chordate [81,82]. Like vertebrates, Amphioxus forms a neural plate, with a neural plate border region. Whereas in the representative mammals studied for neural tube development, the border region remains intact and is carried to the midline with the medial folding of the neural folds, in Amphioxus, the neural plate border cells, using lamellipodia, migrate medially to cover the neural plate as continuous surface ectoderm and beneath it, the neuroectoderm bends to form a tube [81,83]. The Amphioxus genome has been sequenced [83]. *Noggin/BMP* signaling and parts of the *Wnt*/PCP signaling pathway are present [81] but a detailed analysis of the pathways known to be important in neural tube formation does not seem to have been done.

The lineage of Tunicate chordates, the Ascidians (sea squirts), diverged from the vertebrate lineage around the same time as Cephalochordates [84]. The larval sea squirt forms a neural plate and neural folds in a manner similar to the representative mammals. The neural plate curls towards the midline to form a tube, pulling with it the adjacent surface ectoderm to the midline, which meets and fuses to form the epidermis that covers the neural tube. The neural plate cells elongate and change from columnar to wedge-shaped during this process, with narrow apices. Closure progresses from posterior to anterior [85]. Nodal signaling may be required for neurulation. Convergent extension is present [86]. It is not clear that the notochord has a role in patterning of the neural tube, unlike in vertebrates [87]. As in vertebrate neuroepithelium, cell cycle alterations in both the surface ectoderm and the neural plate facilitate the bending required for closure. Movement of the midline epidermal cells requires Rho/ROCK signaling and medial actin filament accumulation [87].

The similarities of neural tube development in comparisons of mammals and Amphioxus and Ascidian embryos suggests that the rolling up of the neural plate into a tube is the ancient mechanism, and from Ascidian embryos that the formation of neural folds is also the ancient mechanism. The divergences from this pattern in species such as zebrafish are therefore new evolutionary products. Some of the ancient processes, such as migration of the neural plate border cells to form the surface ectoderm over the tube in Amphioxus suggest the possibility of atavistic capabilities lurking in the mammalian genome, potentially available in some circumstances.

5. The Cranial Neural Tube

5.1. The New Head and the Cranial Neural Crest

An important context for understanding cranial neural tube closure is that the vertebrate head and its neural tube are considered by the "New Head" hypothesis to be a new structure that was added by evolution, stepwise, onto the basic chordate ancestor which already had the spinal neural tube [88,89]. The genetic basis used to create the New Head includes conservation of early-expressed genes, co-option of other existing genes, and evolution of novel functions for the newly evolved duplicate orthologs of developmental genes [81,82,90]. Consequently, cranial neural tube developmental processes use a blend of old genes and pathways interacting with more recently evolved genes and pathways; processes in the head that may appear to be similar to those in the spinal neural tube may be different at the molecular genetic level.

The ancestral pre-vertebrate chordate representative, Amphioxus, is a filter-feeder with tentacle-like structures around a small mouth, pharyngeal gill slits, a notochord that extends all the way to the front of the head, a neural tube, and a post-anal tail; it does not have jaws, skull, neural crest or ectodermal placodes (optic, otic, olfactory). It has a very simple tiny brain with rudiments of hindbrain, diencephalon, and perhaps some midbrain, but no telencephalon [81]. Amphioxus has some basic ancient conserved gene networks used in development of the neural plate, neural crest and placodes, but not the more downstream components. It does not have a neural crest per se, because genes required for the early development of neural crest cells, although present, are not expressed at the edges of the neural plate [90].

Evolution of a neural crest set the stage for the development of the New Head [91]. Most of the vertebrate face and jaw complex is derived from cranial neural crest cells [92], which arise from the neuroectoderm at the edge of the neural plate and migrate subectodermally ventrally and rostrally. The sensory placodes, another aspect of the New Head, also arise at the edge of the neural plate, induced in the adjacent non-neural ectoderm [89]. In the mammalian head, the neural crest cells emigrate from the neural folds before the folds have elevated; in the spine, they emigrate after closure [60,76]. The neural crest functions differently in the New Head than in the spinal region. For example, in the New Head, cartilage is derived from neural crest cells; this seems to have been enabled by the acquisition of a new cis regulator on the *SoxE* gene that causes its expression in cranial neural crest cells [93]. A consideration for cranial neural tube closure is whether there are differences in gene expression in the cranial neural crest versus the spinal neural crest that precede neural tube closure—if so, these differences have potential to affect the process of closure.

Unlike the rest of the vertebrate neural tube, the anterior forebrain neural folds do not generate neural crest cells, and the neural crest cells of the forebrain mesenchyme migrate in from the caudal forebrain and the midbrain [60,94–96]. In mouse embryos, the transcription factor *Tcf7l1* expressed in the anterior forebrain neural folds acts to repress the *Wnt/B-catenin* pathway that otherwise would induce neuroectoderm cells to become neural crest cells [97]. Loss of *Tcf7l1* leads to conversion of the neuroectoderm of the most rostral neural folds to neural crest cells and a subsequent reduction of the rostral forebrain, which if severe is accompanied by exencephaly [97]. Interestingly, in the images published, the midbrain also appears to be reduced.

5.2. Absence of Hox Gene Expression

Hox genes, some of which are present in Amphioxus, are expressed in the neuroectoderm and paraxial mesoderm of vertebrates and impart anterior–posterior positional identity. Under the regulation of *Fgf*, *Wnt* and other genes, various combinations of *Hox* genes are expressed in the newly formed axial tissues from the primitive streak stage onward [75,98]. In vertebrates, *Hox* expression extends from the embryonic hindbrain through the tailbud; there is no cranial *Hox* gene expression rostral of the border between rhombomere 1 and 2 of the hindbrain [98]. An interesting question is whether the absence of *Hox* gene expression affects the subsequent genetic program determining cranial neural tube closure.

5.3. The Prechordal Plate and Cranial Flexure

The prechordal plate [34,99] underlies only the rostral forebrain (the telencephalon and the diencephalon region that will become the optic primordium). Histologically, the prechordal and notochordal plates appear continuous, but are distinguishable [99]. Located immediately below the midline of the neural plate, the prechordal plate functions in similar ways to the notochord, which underlies the rest of the neural plate, producing essential regulatory factors such as *Shh*, but it also differs from the notochord, by producing factors such as *Goosecoid* (*Gsc*), an inhibitor of convergent extension [100] and by not producing others such as *cNot1* [34,101,102]. Thus the basic regulatory mechanism underlying the forebrain neural tube differs from the rest of the neural tube.

Although the prechordal plate is initially located in line with the notochord, it is bent to lie at right angles to the notochord just before and during the time the mesencephalic folds are elevating [34,60,99,103]. Concurrently, and centered on the mid-mesencephalon, the angle of the cranial flexure narrows from 150 degrees to 100 degrees in human, and from 100 degrees to 70 degrees in mouse [60,103]. The increased bend of the mesencephalic neural tube at the time of closure can be hypothesized to be a physical force working in opposition to the elevation of the mesencephalic neural folds, and the relative acuteness of the angle in mouse is notable in this respect. The prechordal plate also takes part in the formation of the oropharygeal membrane, which, beneath the neural tube, temporarily separates the foregut from the stomodeum. The oropharyngeal membrane ruptures around the time the midbrain neural tube is closing [60,61]. It would be interesting to discover whether the rupture has any mechanical effect that offsets the narrowing of the angle of the flexure.

5.4. Optic Sulci

A further complexity in the mechanical context of closure of the cranial neural tube is the development of the optic sulci in the forebrain. These begin as indentations in the neuroepithelium and balloon outward toward the surface ectoderm as the forebrain neural folds elevate toward each other, so that the apposed forebrain folds resemble a pair of castanets [104] (Figure 5c in [105] and Figures 19 and 20 in [106]). This contrasts with the relative simplicity of the shape of the spinal neural folds during elevation and leads to the question whether the process of shaping the neuroepithelium to form optic sulci affects the process of neural fold elevation that occurs concurrently.

5.5. Sonic Hedgehog and Cranial DLHP

A *Sonic hedgehog* expression gradient in neuroepithelium from upper spine to lower spine accounts for a correlated inverse gradient of dorsal bending of spinal neural folds during closure, as high levels of *Shh* signaling in the upper spine inhibit formation of DLHP, as discussed in Section 2.5. The levels of *Shh* in the ventral neural tube of the head do not seem to have been compared to those of the upper spine; in published whole mount in situ hybridization images, the intensity in the cranial region seems as high as the upper spine (Figure 4C in [32]), but this is not a quantitative method.

The *BMP-Noggin*-DLHP relationship and cell behavior at the cranial DLHP does not seem to have been studied, and it is possible that the mechanisms are not entirely the same as in the spine. Conversely, the induction of exencephaly in response to elevated *Shh* signaling suggests that the relationship of cranial DLHP to *Shh* is similar to that in the spinal region and modulated by the large size of the cranial neural folds, as discussed in Section 2.5.

5.6. The Actomyosin Cytoskeleton of the Cranial Neuroepithelium

Mammalian cranial neural fold bending for closure requires the actomyosin cytoskeleton, and mutations in its components or chemical disruption of apical actin microfilaments cause exencephaly (summarized in [19]). Thus, mutations in components such as *palladin, vinculin, cofilin 1* (*Cfl1*) and *Marcks*, or in genes for actin regulatory proteins such as *Mena, Vasp* and *Evl*, or in genes for protein kinases with cytoskeletal influence such as *Abl1, Abl2, Mapk8* and *Mapk9* all cause exencephaly in mice [19]. In contrast, closure of the spinal region does not require these genes. Therefore, these genes important to the actomyosin cytoskeleton are candidate genes specific to human anencephaly etiology.

5.7. Apoptosis in Cranial Neural Folds

Programmed cell death, apoptosis, normally occurs in the neuroepithelium during development of the neural tube [42], with a pattern of areas of particular abundance associated with bending and fusion [107]. Various mouse mutants that demonstrate decreased or increased apoptosis in the neural folds develop exencephaly but not spinal NTD [42]. For example, in mouse embryos that constitutively lack apoptosis, the *Casp3* and *Apaf1* null mutants, the midbrain and hindbrain fail to close but the rest of the neural tube closes normally. Live imaging of the midbrain–hindbrain neural folds undergoing bending and closure in normal embryos show conventional and unconventional apoptotic cells [77]. Conventional apoptotic cells (having caspase activation, followed by shrinkage and fragmentation) are in the boundary region and surface ectoderm at the tip of the neural folds and at the midline before and after the completion of closure. The unconventional apoptotic cells (having caspase activation, but not shrinkage and fragmentation) emerge from the neuroepithelium of the boundary domain and dorsal neural fold and accumulate as the fold bends; they are round, protrude from the surface, remain attached for a time, and then tumble along the surface of the neuroepithelium into the lumen of the neural tube. In *Casp3* and *Apaf1* null mutants, neither type of apoptotic cell is found and the apical bending of the folds towards the midline is severely reduced. The mechanism of the effect of apoptosis on bending is not known.

5.8. Mutations in Genes Expressed in the Cranial Neural Crest That Are Associated with NTD

Emigration of neural crest cells from the midbrain and hindbrain open neural folds has been thought to be necessary for cranial neural tube closure [42], but the evidence from mouse mutants is limited. Two genes that cause NTD and are expressed in the cranial neural crest are the transcription factor *Tcfap2a* and the cell adhesion factor *Cdh2* (N-cadherin). In mice on E8.5, *Tcfap2a* is expressed in the cranial neuroepithelium and cranial neural crest; *Cdh2* is expressed in the cranial neural plate. Conditional mutants in which either *Tcfap2a* or *Cdh2* is deleted in the neural crest cells in which *Wnt1* is normally expressed, suggest that some aspect of cranial neural crest abnormality causes midbrain exencephaly (15–20% in *Tcfap2a*, [108]; 100% in *Cdh2*; [109]). Although for both mutants, it appears that exencephaly is due to the absence of the targeted gene's expression in the cranial neural crest cells, for neither mutant is the mechanism of interference with neural tube closure understood. In the *Tcfap2a* mutants, it may involve the positioning of the neural plate border [108]. A complication of these studies is that *Wnt1*-Cre is expressed in both the lateral non-neural ectoderm of E8.5 cranial neural folds and the neural crest [110]. The potential role of the lateral non-neural ectoderm cells in neural tube closure or its failure has yet to be explored.

It is interesting that whereas 100% of *Tcfap2a* null embryos have split face and forebrain/midbrain/hindbrain exencephaly [111,112], the cranial neural crest conditional null mutants have failure of closure of only the midbrain. Further evidence for neural crest involvement emerges from study of *Cdh2* null mutants rescued from early lethality [113,114], which are 100% exencephalic. On E8.5, the rescued embryos have increased apoptosis in the tips of the cranial neural folds, in what appear to be neural crest cells starting to migrate from the caudal midbrain region [114]. In summary, although abnormality in cranial neural crest cells appears to be the cause of exencephaly in the *Wnt1*-Cre conditional null *Tcfap2a* and *Cdh2* mutants, the mechanisms are unclear. We have found no definitive evidence from mouse mutants for delayed or abnormal neural crest cell emigration as a cause for exencephaly.

5.9. Role of Cranial Mesoderm in Neural Fold Elevation

The vertebrate cranial mesoderm has an important role in cranial neural fold elevation. The vertebrate cranial mesoderm, created by gastrulation [115], may have accrued differences from spinal mesoderm as part of the evolution of the New Head. Comparisons of mesodermal gene expression between Amphioxus and vertebrates indicate that expression in late gastrula of the mesodermal patterning genes found in both groups is reorganized in vertebrates, limiting expression of some to the head and others to the body, thus creating a novel type of mesoderm in the vertebrate head [116].

The role of cranial mesenchyme (comprised of cranial mesoderm and neural crest cells after their emigration from the neuroectoderm) in the cranial neural tube has been studied in mouse, rat and chick embryos [117]. During cranial neural tube formation, the shape of the folds progresses from biconvex, with the mid-lateral folds cushioned on a large supporting cranial mesenchyme, to biconcave (Figure 4). The neural folds of the spinal region do not have the convex stage. During the process of cranial neural fold elevation, the mesenchyme expands by both increased numbers of cells and increased space between the cells [117,118] and this mesenchymal expansion and reorganization is thought to be important in causing elevation. During the biconvex stage, the neural crest cells leave the neuroectoderm of the lateral tips of the neural folds and begin to migrate subectodermally through the mesenchyme to the first branchial arch and facial prominence areas. As the neural folds elevate, becoming biconcave, and the dorsolateral hinge points form, the mesodermal cells become more widely spaced in the medial region and more closely packed in the lateral region. The extracellular matrix undergoes changes in hyaluronic acid concentration that may provide a mechanism for elevation by mesenchyme expansion, as hydration of hyaluronic acid is known to cause expansion of mesenchyme [117,119].

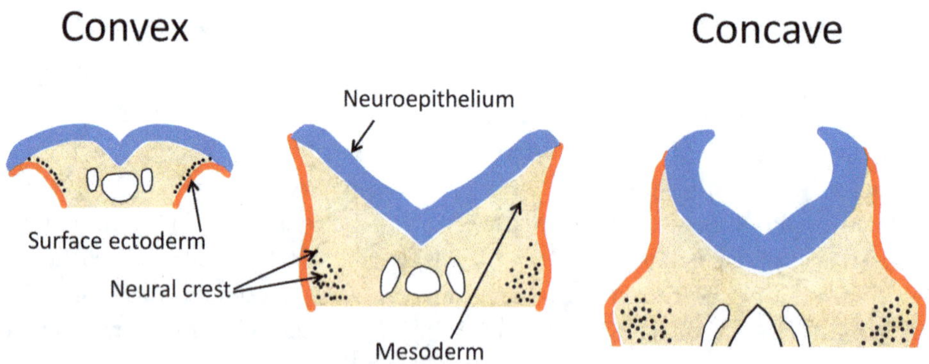

Figure 4. Diagrams of transverse sections of midbrain neural folds during their elevation, showing the morphological change from convex to concave shape. Based on Figure 1 in [117].

Several mouse mutants demonstrate the importance of the cranial mesenchyme to cranial neural fold elevation. Loss of function of the transcription factor *twist* causes severe craniofacial defects, including midbrain exencephaly; the neural folds fail to become concave and remain everted [120]. *twist* is highly expressed in the cranial mesenchyme just before and during the time of neural tube closure [121,122]. At the normal time of cranial neural fold elevation, mutant embryos have a morphologically normal neuroepithelium, but have abnormal morphology of mesenchyme cells in the forebrain and midbrain, which are rounded, lack normal stellate shape and have greater intercellular space than normal [120]. The exencephaly is attributable to a defective cranial mesenchyme and not the neuroepithelium, based on chimera studies [120] and on conditional mutants. Ablation of *twist* in the precursors of the cranial mesoderm causes reduced proliferation and abnormal clustering of mesoderm cells in the lateral regions of the head folds and subsequent exencephaly [121]. No evidence of increased cell death has been found [121]. The neural crest seems to form normally [123] and the neural crest cell-based facial abnormalities are considered to be a secondary effect of the abnormal mesenchyme failing to provide essential signals to the neural crest cells as they migrate to the face [121]. Based on studies later in development, the loss of expression of *Cart1* (*Alx1*) and *Alx3* in *twist* null mutant embryos suggests that they are in the same regulatory framework [123], and it is interesting that they also express exencephaly in mutants.

The *Cart1* (*Alx1*) transcription factor is expressed in mesenchyme cells of the forebrain but not the midbrain, and not in neuroepithelium, during neural tube development [124]. Null mutant embryos for *Cart1* (*Alx1*) have increased apoptosis and transient deficiency of mesenchyme in the forebrain, with delayed closure of the forebrain folds; subsequently, the midbrain neural folds fail to elevate, leading to exencephaly [124]. Whereas *Cart1* (*Alx1*) early expression may be limited to the forebrain mesenchyme, a homolog, *Alx3*, appears to be expressed in paraxial mesoderm of the body and the mesenchyme of the head, but also not in the neuroectoderm, at the time the cranial folds are elevating [125]. Although some authors assume *Cart1 (Alx1)* and *Alx3* expression in the neural fold mesenchyme is in the neural crest cell component, this seems unproven. About 10% of mutants lacking *Alx3* develop midbrain exencephaly and/or split face (failure of closure of the forebrain neural folds); the neural folds of these embryos fail to undergo the shift from convex to concave or development of a dorsolateral hinge point and therefore fail to meet in the midline [126]. Increased apoptosis and decreased mesenchymal cell density in mutants are present in later developmental stages [126], but is not known if any abnormalities in apoptosis and mesenchymal cell density are present earlier when they could contribute to the failure of neural tube closure. Interestingly, *Alx3* expression is lost during folic acid deficiency at the time of cranial neural tube closure, whereas expression of other neural tube genes such as *Cart1 (Alx1)*, *twist*, and *Cited2* is not affected [126].

Inka1 is first expressed in the cranial mesenchyme prior to and during neural tube formation and then in the cranial neural crest during its migration in the mesenchyme [127]. It is not expressed in the neuroepithelium. The *Inka1* protein inhibits the *Pak4* kinase that may function in cell–cell contacts and adherent junction formation [128]. *Inka1* is regulated by *Tcfap2a* in non-mammalian vertebrates, but not in mice [127]. Loss of function of *Inka1* causes a predisposition to midbrain exencephaly (5–7%); otherwise, individuals are normal and fertile. The mechanism by which the neural tube fails to close has not been demonstrated, and the mechanistic role of the mesoderm or neural crest or both is unknown.

Loss of function of the transcriptional regulator, *Ski*, [129] which causes exencephaly or split face [130], is sometimes listed as having abnormal cranial mesenchyme [117] but cranial tissues do not seem to have been studied during the period of cranial fold elevation [130,131], and an etiology of exencephaly based on effects on cranial mesenchyme seems unproven.

Null mutant embryos for *Smarca4* (*Brg1* or *SW1/SNF*), involved in chromatin remodeling, die before implantation. Heterozygotes often have midbrain exencephaly, and no other defects [132]. *Smarca4* is expressed strongly in the cranial mesenchyme before elevation [133]. The mechanism causing the failure of neural tube closure has not been studied.

Mutations causing loss of function of the E3 ubiquitin ligase *Hectd1* (*opm*) lead to midbrain exencephaly in all homozygotes and some heterozygotes [134]. *Hectd1* is ubiquitously expressed in embryos, including the head mesenchyme, and the cause of exencephaly is attributable to the effects of *Hectd1* deficiency on the cranial mesoderm cells, which are abnormally shaped and packed more densely than normal near the neural tube at the normal time of elevation, and which fail to orient parallel to the neuroepithelium [117,134]. The neural folds lack mesenchyme expansion and remain flat or convex and fail to form the dorsolateral hinge points. The neural crest cells' migration pattern and differentiation are normal; mesenchymal cell proliferation and apoptosis are normal [134]. Before the normal time of elevation and closure, *twist* has an expanded area of expression in the midbrain folds in the *Hectd1* mutants, and after failure of full elevation, *Snail* and *Pdgfr* are abnormally expressed in the dorsal area of the midbrain folds [134]. One of the *Hectd1* substrates, heat shock protein 90 (*Hsp90*), is known to enhance migration in various cell type, is secreted at elevated rates from *Hectd1* mutant cranial mesenchyme cells which are abnormally migratory in vitro [135]. It seems that the effects of the lack of *Hectd1* in the midbrain neural folds lead to excess *Hsp90* secretion, which in turn alters the behavior of the mesoderm cells, leading to their failure to contribute to elevation of the folds.

Loss-of-function mutants for the *Cecr2* chromatin-remodeling factor develop exencephaly [136]. *Cecr2* is expressed in neuroepithelium and mesenchyme during cranial neural tube development [137]. Studies of gene expression arrays from mutant embryos show reduced expression of *Cart1/Alx1* transcription factor [138]. As this mesenchymally-expressed gene is required for cranial neural tube closure [124], it appears that the defect in *Cecr2* mutants may be attributable to defective cranial mesenchyme. The observation that in *Cecr2* mutants, the cranial neural folds do not elevate sufficiently for contact between the apposed folds, despite forming DLHP [137], is consistent with the hypothesis that the NTD is caused by an insufficient expansion of the cranial mesenchyme to assist elevation.

Exencephaly in mouse mutants is usually caused by a failure of midbrain neural fold elevation and it appears that defects in cranial mesoderm are a common etiology for this defect. Exencephaly becomes anencephaly, owing to degradation of exposed neural tissue during later gestation, and therefore is seen as anencephaly at birth in humans [6]. Interestingly, as in the mesodermal mutants in mice, unelevated everted midbrain neural folds have been observed in some early NTD human embryos [139,140], suggesting that lack of elevation owing to mesodermal defects also could be a common mechanism in human anencephaly.

5.10. Neuroepithelium-Expressed Genes and Mechanisms That Cause Exencephaly

The actomyosin-dependent shaping and bending of the cranial neural folds is based in the neuroepithelium, and it seems predictable that the many actomyosin-related genes that cause NTD would show neuroepithelium-specific expression, but data to demonstrate this is lacking. Many studies report whole mount in situ hybridization in the cranial neural folds without demonstration of the cell type involved (neuroectoderm, mesenchyme, surface ectoderm). A few genes that cause cranial neural tube defects when dysfunctional have been reported to be expressed in the neuroepithelium.

Cfl1 (*Cofilin 1*, *Cofilin-n*) is expressed in the pre-closure cranial and caudal neuroepithelium and neural crest [141,142]. *Cfl1* enables the dynamic reorganization of the actin cytoskeleton by its function in F-actin, severing and depolymerization and by promoting the recycling of monomeric actin [141]. Mutants have exencephaly and delayed closure of the posterior neuropore (PNP) [19,141]. Originally, the exencephaly was ascribed to a demonstrated lack of neural crest cell migration, but no mechanism for the lack of neural fold elevation was demonstrated [142]. More recent work has shown that the mechanism causing the exencephaly is a lack of activation of apical actomyosin in neuroepithelial cells and abnormal basal accumulation of F-actin [141]. Consistent with the different roles of actomyosin in cranial versus caudal neural folds, the delayed PNP closure of *Cfl1* mutants is owing to actomyosin accumulation, lack of F-actin turnover and actomyosin disassembly [19].

Fat1 cadherin, a large cell adhesion molecule, is expressed in the early anterior neural plate in the forebrain and midbrain neural folds during neural fold elevation and in the caudal neural folds [143]. It seems to be located in the neuroepithelium and neural crest. *Fat1* is thought to regulate actin cytoskeletal organization at cell peripheries. In cell cultures, it is found in filopodia and cell–cell boundaries, overlapping with dynamic actin structures; it regulates actin polymerization [144]. Therefore, the cause of the NTD is likely a deficiency of neural fold bending. On some strain backgrounds, *Fat1* null mutants have exencephaly, but not spina bifida. Although mutations of the family member *Fat4* do not cause NTD, and *Fat4* binds different actin-regulating and junctional proteins than *Fat1*, they form heterodimers, and on some strain backgrounds, the combination of homozygous null mutations at both *Fat1* and *Fat4* cause increased frequencies of exencephaly compared with *Fat1* alone [145]; this is an example of the genetic complexity of NTD etiology.

Cited2 (*Mrg1*) is a transcription regulator that serves as a co-activator for several transcription factors such as *Tfcap2*, *Smad4*, *Lhx2*, *Nanog* and *Tbx3* [146–148]. It is expressed during neural fold elevation in the midbrain neuroepithelium and in migrating cranial neural crest [146,149]. Null mutants have exencephaly. The mechanism leading to exencephaly is likely the disrupted integrity of the neuroepithelium, owing to the observed intense localized apoptosis in the midbrain neuroepithelium, particularly at the midbrain–forebrain boundary, at the time of normal midbrain-fold bending [146].

Members of the *Sall* gene family *Sall1*, *Sall2*, and *Sall4* are transcription repressors that interact with chromatin remodeling complexes, can form heterodimers [150] and are all expressed in the cranial neuroepithelium during neural fold elevation [151,152]. Loss-of-function mutants for each of these *Sall* genes cause exencephaly, depending on strain background [150,152,153], and digenic heterozygous mutant combinations of *Sall1* with *Sall4* or *Sall2* with *Sall4* also cause exencephaly. Based on embryos with a combination of null *Sall1* and hypomorphic *Sall4*, it appears that the cause of exencephaly is a deficiency of elevation of the cranial neural folds, which do elevate part way but do not reach approximation in order to fuse [152].

Members of the *Nuak* gene family, *Nuak1* and *Nuak2*, are serine-threonine kinases that have many functions [154]. *Nuak1* is expressed in many embryonic tissues including the neuroepithelium, but null mutants do not have NTD; *Nuak2* is specifically expressed in the entire neuroepithelium during neural tube closure, and some of the null mutants have midbrain and hindbrain exencephaly; digenic null mutants for both genes all have exencephaly of the forebrain, midbrain and hindbrain [154]. The cranial neural folds in the digenic mutants seem to lack DLHP, and the cause seems to be a significant lack of apical constriction of the cranial neuroepithelial cells during the normal elevation stage [154], which would be expected to cause a lack of bending.

Shroom3 (*Shrm*), encodes a cytoskeletal protein that binds F-actin and regulates its distribution within the cell. During neural fold elevation, *Shroom3* is strongly expressed in the cranial neuroepithelium, rostral to the otic vesicle, but not in the adjacent cranial mesenchyme [155]. Later, it is expressed in the caudal neural folds at the posterior neuropore during elevation (gene k8220b03, Appendix, and Figure 1 in [156]). All *Shroom3* mutant homozygotes have midbrain and hindbrain exencephaly, most also have split face (failure of forebrain neural tube closure), and some also have spina bifida aperta. The morphology of the neuroepithelium of the closed spinal neural tube is also abnormal [155]. In *Shroom3*-mutant embryonic neuroepithelial cells, F-actin is shifted away from the apical surface, whereas in wild-type cells, it is predominantly localized at the apical surface of the cells. This pattern suggests that the mechanism of the neural tube closure defects in *Shroom3* mutants is the lack of actomyosin-dependent apical constriction in the neuroepithelium. There is evidence, based on various digenic heterozygous mutants involving *Vangl2* (*Lp*) and *Shroom3*, that the actomyosin system is linked to the PCP-convergent extension system by *Shroom3* [157]. For example, whereas neither *Shroom3/+* nor *Lp/+* mutants have spina bifida aperta, the digenic mutants (*Shroom3/+*, *Lp/+*) have almost 40% spina bifida aperta. At the molecular level, there is evidence that that a PCP factor (*Dvl2*) determines the location of *Shroom3* at mediolateral cell junctions along with F-actin, *Rock1* and *Myosin IIb*, components of the actinomyosin pathway [157], suggesting that *Shroom3* and *Dvl2* may interact to link PCP signaling to actomyosin contractility. This is an example of how simultaneous reduction in a genetic component of each of two pathways can cause NTD.

5.11. Cranial Ephrins and Ephrin Receptors

An example of the potential molecular complexity underlying neural tube fusion and leading to a cranial-specific etiology of an NTD is an apparent regulatory relationship between the transcription factors *Dlx5* and *Msx2* with each other and with the downstream membrane-bound cellular adhesion/repulsion factors *EphrinA5* and *EphA7*. *Dlx5* is expressed in the dorsal tips of the neural fold before, during, and after neural tube closure, extending from the forebrain to the posterior neuropore [158,159], and null mutants have some (19–28%) exencephaly [159,160]. Similarly, *Msx2* is expressed before closure in the dorsal edges of the neural tube, except in the forebrain region [161,162], but null mutants do not have NTD [160,163]. In contrast, double mutant embryos null for *Msx2* and *Dlx5* have a high frequency (73%) of exencephaly that seems to be caused by their joint regulatory effect on the splicing of *EphA7* transcripts [160]. The ligand *EphrinA5* and its receptor *EphA7* splice forms are expressed in the dorsal cranial neural folds (midbrain and hindbrain) before and during closure [40]. The interaction of *EphrinA5* with full length *EphA7* causes cellular repulsion, but a co-expressed alternative splice form of truncated *EphA7* converts the effect of the interaction to adhesion [40]. *EphrinA5* null mutants sometimes develop exencephaly (17%), a few of which have failed forebrain closure as well as midbrain and hindbrain, having failed to fuse despite elevating to become juxtaposed at the midline; some (25%) *EphA7* null mutants also have exencephaly [40]. Null mutants for either *Msx2* or *Dlx5* have fairly normal expression of *EphrinA5*, full-length *EphA7* and truncated *EphA7* in the cranial neural tube. In contrast, in *Dlx5/Msx2* digenic null mutants during the stage of neural tube closure, the expression of *EphrinA5* at the apex of the neural folds is decreased, full length *EphA7* expression is unchanged but the truncated form of *EphA7* is greatly reduced, a situation that is expected to cause repulsion, rather than adhesion of the apposed midbrain and hindbrain neural folds. It appears, therefore, that the mechanism leading to exencephaly is likely lack of adhesion of neural folds in each of the *EphrinA5*, *EphA7* and digenic null *Msx/Dlx5* mutants, and the limitation of the NTD to the midbrain and hindbrain reflects the limited expression domain of *EphrinA5* and *EphA7*.

Demonstrating another kind of complexity, null mutation carriers for the X-linked gene, *EphrinB1*, express exencephaly at a higher rate in heterozygous females than in homozygotes and null males [164]. Embryos have not been studied during the period of cranial neural tube closure, but the mechanism may be segregation of *EphrinB1*-positive and -negative cells in the neuroepithelium, affecting its structural integrity [164].

5.12. Cranial Neural Fold Projections

In general, it seems likely that genetic defects whose mechanisms cause cranial neural tube defects through failure of fusion of apposed neural folds, would demonstrate defects in the surface ectoderm projections normally present at the fusion sites. However, it appears that data is lacking in this area.

Loss of function of Myosin-X (*Myo10*), an unconventional myosin known for its roles in formation of filopodia and localization to the tips of filopodia, causes exencephaly, along with other non-neural-tube defects [165].

5.13. The Phenomenon of Sex Ratio Distortions in Cranial NTD

A striking predominance of females in cranial neural tube closure defects has been recognized for many years; about two-thirds of human anencephalics and mouse exencephalics are female (reviewed in [166]). Human craniorachischisis may also exhibit an excess of females; no sex data are available for mouse mutants with craniorachischisis [166]. Spina bifida as a whole generally does not exhibit female excess in humans or mice [166], although the subtype located in the upper spine may do so [167]. Previously, 12 mouse mutants and strains for which sex had been reported showed an excess of females among exencephalics [166]. There is similar data for two additional mutants: *Cecr2* with 66% females among exencephalics [168], and *Scarb1 (srb1)* with 72% females among exencephalics [169]. The 14 mutants and strains represent a diversity of functions and pathways, suggesting that the increased liability of females to cranial NTD extends across a wide range of causal molecular mechanisms. There may be exceptions. For example, the lack of female predominance in NTD caused by mutations at *Gldc* [170] and *Cited2* [146] requires further confirmation of sex ratios with larger samples. If other exceptional mutants that lack female excess in exencephalics can be found, a comparison of the function of genes with female excess with those without might point to the mechanism responsible for elevated female susceptibility to cranial NTD.

The causes of the female excess among cranial neural tube closure defects in humans are difficult to identify, but in mice, several of the possible explanations have been ruled out by the data. For example, the neural tube develops long before gonadal hormones can be present, the developmental rate of the cranial neural tube during closure does not differ between males and females, and there is no evidence of preferential death of affected males during gestation [166]. In mice, genetic experiments have demonstrated that the higher susceptibility to cranial NTD is caused by the presence of two X chromosomes in females, not the absence of the Y (reviewed in [166]).

There is an intriguing phenomenon of methylation differences correlated to number of X chromosomes in mouse and bovine embryos, during the early developmental period when both X's are normally active, such as in blastocysts or embryonic stem cells. The global DNA is hypomethylated and the *Dnmt3b* (*DNA methyltransferase 3 beta*) transcript and protein levels are reduced in XX embryos relative to embryos with one X [171,172]. Whether this phenomenon has later consequences affecting gene expression in neurulation stage embryos is unknown.

In mammals such as humans and mice, males have one large X chromosome with over a thousand genes and one small Y chromosome, which contains, amongst a few genes, the *Sry* testis-determining gene; females have two copies of the large X, of which one copy has been mostly silenced beginning from about the time of implantation [173] by a combination of epigenetic changes including DNA methylation, histone modifications and chromatin remodeling [174]. Chromatin (DNA and its attached histones) that has undergone transcriptional silencing and remodeling to become densely packed is termed "heterochromatin".

A mechanism that has been suggested to explain a variety of disorders that occur at higher frequencies in females than in males is that some genes on the inactive X variably escape from silencing and cause a gene expression imbalance; there is evidence in support of this mechanism for the autoimmune disease lupus erythematosus [174].

Another hypothesis to explain the female excess is that the use of cellular resources to inactivate a large X chromosome in females after every mitosis would competitively reduce the availability of those resources for gene regulation on the autosomes, weakening the process of neural tube closure [64]; updated in [166]. A similar hypothesis has emerged from genetic studies in Drosophila, the "heterochromatin sink" [175]. In Drosophila, a chromosome with a large heterochromatin component can sequester various limited factors required to regulate genes, and this can alter the expression of genes in the rest of the genome [175]. Some observations suggest that the large inactive heterochromatic X chromosome in female mammals can act as a sink that sequesters epigenetic factors that are present in limited amounts (e.g., DNA methyl transferases and histone modifying enzymes), thus altering their availability in the rest of the genome. A mutation at *Smchd1*, a gene involved in the methylation of DNA (formerly *Momme D1*) causes reduced methylation and silencing of a reporter transgene and autosomal retrotransposons, but only in females. The female-specific effects led Blewitt et al. [176] to suggest that the inactive X chromosome could be sequestering proteins that are involved in controlling gene expression at autosomal loci. An example of this phenomenon has been demonstrated by the "four core genotypes" system, where the male sex determining factor *Sry* has been moved to an autosome, comparing XX females, XX males, XY males, and XY females. It shows that the expression level of hundreds of genes in mouse thymus cells is determined by the number of X chromosomes present and not by gender [177]. Genes differ by whether they are sensitive to the number of X chromosome and in degree of sensitivity [177]. The sensitive genes have lower expression in the presence of two X chromosomes [177]. If some of the important genes that have cranial-specific neural tube expression (Section 5) are relatively sensitive, and if there are not sensitive genes among the important spinal-specific neural tube genes, this would explain the female excess in cranial, but not spinal, NTD. Perhaps examination of expression levels of genes in the neural folds of embryos from this "four core genotypes" system would yield insight into the causes of female excess in cranial NTD.

There is an interesting interaction between embryo gender and the preventative effects of folate on human NTD. In at least three populations from Chile, Argentina and Canada, the female excess among anencephalic births disappeared soon after the start of folic acid fortification of wheat flour [167,178]. Prior to folate supplementation, anencephalic births in each country exhibited the typical excess of females, where almost two-thirds of anencephalic births were female. Fortification with folic acid reduced the occurrence of anencephaly in both males and females, but had a greater effect in females; for example, 50% prevention in males versus 70% in females [167]. One way to interpret the pattern would be that the additional folic acid compensates for the "heterochromatin sink" in females and restores availability of essential chromatin regulatory factors, reducing female risk of NTD to the same level as males. Then, through perhaps a different mechanism, the abundant folate interacts with a particular type of genetic weakness present in a subset of cases to reduce NTD risk in that subset equally in both sexes.

6. The Spinal Region, PCP and Craniorachischisis

Much of the understanding of basic mechanisms of neural tube development is based on studies of the spinal region, and spinal neural tube closure has been well reviewed by others [5,6]. We have considered the basic mechanisms (Section 2) and differences between cranial and spinal regions (Section 3). In this section we consider the NTD that is caused by failure of closure of the entire spinal region. The first initiation site for neural tube closure is in the upper spine. There the folds are brought together by, in part, the combined effect of convergent extension and the medial hinge point (MHP, Section 2.3).

The failure of convergent extension (Section 2.2) results in a short body axis and broad neural plate, with no medial hinge point (MHP), placing the lateral neural folds too far apart to meet in the midline and fuse [179,180]. This results in the NTD, craniorachischisis, in which all but the most rostral zone of primary neurulation has failed and the neural tube remains open from the midbrain or hindbrain, along the length of the spine to the lumbar region in humans and to the sacral region in mice. It is interesting to note that the caudal limit of primary neurulation is more rostral in humans than in mice [46]. The constant feature of craniorachischisis, a closed neural tube in the most rostral, forebrain region, is likely explained by Ulmer et al. [100], who show that *Goosecoid* (*Gsc*), present in the prechordal plate and not in the notochord, appears to repress convergent extension; i.e., because of *Goosecoid*, normally there is no convergent extension in the forebrain. The role of a disruption in the planar cell polarity (PCP) pathway and the consequent lack of convergent extension in causing craniorachischisis have been well reviewed recently [5].

In mice, among the numerous mouse gene mutations that cause NTDs, those that produce craniorachischisis are all components of the planar cell polarity pathway, including the PCP core factors, *Vangl2*, *Celsr1*, *Dvl1/2/3* and *Scrib*; and the less central factors, *Ptk7*, *Cdx1/2*, *Sec24b*, *Sfrp1/2/5* and *Smurf1/2* [10].

In humans, craniorachischisis is a relatively rare NTD for which the only known genetic cause is mutations in PCP genes [181]. The developmental morphology and resultant NTD (wide neural plate and wide open neural tube, except for the closed forebrain) is similar in humans and in mice. Histological study of human embryos with craniorachischisis revealed that they have cervical notochord duplication, side by side, beneath the open neural tube, and a broad *Sonic hedgehog* expression domain; if the neural tube in these embryos appeared closed caudal to the notochord duplication, two parallel neural tubes side by side were found housed in a single dural tube [182]. In the homozygous *Lp* mouse mutant at the *Vangl2* gene (*Vangl2-Lp/Lp*), the notochord at the time of initiation of neural tube closure is loosely organized and comprised of two or three parallel longitudinal structures [179]. The process of convergent extension has been shown to be cell-autonomously defective in both the notochord and neural plate of these homozygous embryos [180].

It is becoming clear that PCP genes can have roles other than in convergent extension that can influence neural tube closure. One possibility (for spina bifida) is that caudal neural tube closure is delayed in some PCP-mutant heterozygotes, as has been observed in the mouse heterozygous *Vangl2-Lp* mutant that has 2% spina bifida [183]. A second possibility, which applies more to the cranial region than the spinal region, is a relationship between PCP and cilia that can cause exencephaly in some PCP mutants. Numerous mouse mutants for genes involved in ciliogenesis have exencephaly [13], including mutants for the PCP factor *Inturned* (*Intu*) with 30% exencephaly [18]. A third possibility arises from observations of interaction between the PCP pathway and the regulation of actomyosin dynamics. In several mutants, the PCP gene involved has been shown to have a role in both convergent extension and actomyosin dynamics [19]. Conversely, the actin-binding protein, *Shroom3*, has been found to interact with PCP pathway proteins during neural tube morphogenesis in mice [157]. PCP gene mutations probably contribute to risk of human anencephaly and spina bifida, albeit in a very small proportion of cases [10].

7. Studies of Digenic Mechanisms of NTD Involving PCP Gene Mutations

In mice, a large and growing number of digenic mutants, heterozygous for mutations at two gene loci, have been observed to have NTD. The molecular mechanisms by which these digenic or oligogenic combinations lead to NTD are largely unknown, but likely include both cumulative effects along the same pathway or deficiencies in interacting pathways. The known digenics involve at least twenty different kinds of genes [3,10,12], but the extensive study of the planar cell polarity (PCP) system has generated numerous well-studied examples, which provide insight into mechanisms of the genetic complexity of human NTD. Most of the examples at hand involve two genes in the PCP pathway, many involving interactions between heterozygosity for both the *Vangl2* mutant, *Lp*, and one of several other PCP genes. In addition to the PCP–PCP gene interactions, there are digenic mutant examples of PCP genes interacting with genes in actomyosin dynamics, as in *Vangl2;Cfl1* [184] and *Vangl2;Shroom3* [44,157].

Mouse PCP-mutant studies focused on the NTD craniorachischisis demonstrate the genetic complexity of the role of PCP genes in craniorachischisis [10]. In some examples, mutants are homozygous at only one PCP gene (e.g., *Vangl2*, *Celsr1*, *Scrib* or *Sec24b*) and all embryos have craniorachischisis; heterozygotes do not. Although virtually all mouse mutants with craniorachischisis have mutations in genes of the PCP pathway, often mutations at two or more PCP genes are involved. In one type of digenic example, mutants are homozygous at both of two homologous PCP genes with functional redundancy (e.g., *Dvl1;Dvl2* or *Fzd3;Fzd6*) and nearly all embryos have craniorachischisis; homozygotes for any one of these genes do not. In yet another type of digenic cause, high frequencies of craniorachischisis are found in mutants that are double heterozygotes for mutations at two PCP genes of different gene families (e.g., *Vangl2;Celsr1*, *Vangl2;Dvl2*, *Vangl2;Scrib*, or *Celsr1;Scrib)* [10,185]. It is notable that several of the digenic mutant combinations that cause craniorachischisis also produce some normal, fertile, double heterozygotes.

Craniorachischisis is not the only NTD caused by PCP mutations in mice. Mutations of various single PCP genes or digenic combinations of PCP genes do not cause craniorachischisis, but do cause the more common types of NTD (exencephaly or spina bifida [10]); several others cause craniorachischisis or exencephaly or spina bifida in different individuals. For example, only exencephaly is found in homozygous single mutants at *Fzd3*, *Intu* or *Fat1*, or digenic mutants at *Vangl2;Fzd1* (both heterozygous) or *Vangl2;Cthrc1* (heterozygous; homozygous). Only spina bifida is found in digenic heterozygotes at *Vangl2;Ptk7*, *Vangl2;Grhl3*, *Vangl2;Sec24b*, and *Celsr1;Ptk7* [10]. Finally, different individuals can have craniorachischisis or exencephaly or spina bifida in digenics heterozygous at *Vangl2;Scrib* or *Celsr1;Scrib* [185].

The mechanisms by which these non-craniorachischisis defects of neural tube closure are caused by PCP pathway mutations are largely unknown. However, the clear differences known for neural tube formation between anterior–posterior regions (discussed above) raise the possibility that the causal mechanisms involve the ciliogenesis, actomyosin, and microtubule functions of PCP genes, in addition to their convergent extension role (Sections 2.2 and 6). Effects of the *Wnt*-PCP pathway on the caudal neural tube that cause spina bifida include effects on cell differentiation, cell shape, and actomyosin contractility [44].

Mouse mutants support the digenic (or oligogenic) hypothesis for human craniorachischisis and, further, the hypothesis that digenic heterozygotes for PCP genes can also cause risk for human exencephaly and spina bifida. Human NTD cases of various types have been screened for rare deleterious variants in several PCP genes. Human craniorachischisis is relatively rare and few studies are of craniorachischisis only. Most variant screening studies include exencephaly, spina bifida and other neural tube abnormalities [10]. Among craniorachischisis cases, about 20% have potentially deleterious variants in single PCP genes (*VANGL2*, *CELSR1*, *SCRIB* or *DACT1*) [181,186,187]. The mouse digenic PCP mutants support the suggestion that these cases may be undetected double heterozygotes for PCP mutations. Recently, six human NTD cases digenic for PCP variants were reported as follows: three spina bifida and one anencephaly digenic at *CELSR1;SCRIB*; one anencephaly digenic at *CELSR1;DVL3* and one spina bifida digenic at *PTK7;SCRIB* [188].

Overall, the mouse digenic PCP mutants support the strategy of screening cases with anencephaly, spina bifida or craniorachischisis for PCP gene mutations and, given the usual observation of human heterozygosity for the PCP mutations found, strongly encourage the search for second gene locus mutations in these NTD cases.

8. Folate and NTD

8.1. Human NTD Prevention by Folate

In addition to genetic susceptibility, environmental factors contribute to risk of developing NTD, including various aspects of diet and maternal health [11,189]. One of the important factors is supplementary maternal peri-conceptional intake of folic acid, which has been demonstrated to reduce the occurrence of human NTD by as much as 70% [190,191] and has led to government policies of fortification of various foods with folic acid in some countries [192]. There is a significant residual, perhaps a third, of NTD cases that appear to be "folate unresponsive" [193]. The mechanism of beneficial effects of supplemental folate on human NTD occurrence is unknown. A paradigm of "folate deficiency" is widely assumed, but the data suggest this concept may be an over-simplification [7]. Although in the absence of folic acid supplementation, risk of NTD appears to increase with lower maternal blood folate levels on average [194], the levels are generally in the normal range, not deficient [195]. In animal models, dietary folate deficiency seldom, if ever, causes NTD [189,196,197]. A hypothesis that supplemental folate compensates for a genetic weakness in the folate metabolic pathway is long standing [194,195]. The folate metabolic pathway is well known and frequently reviewed [7,198]. The response to folic acid has directed the search for genes involved in the etiology of human NTD to focus on the genes that affect folate availability to cells and the genes in the known folate metabolic pathways, such as those involved in DNA and protein methylation and in DNA and RNA synthesis [11,199,200]. Despite extensive studies, in general this approach has not discovered genetic variants with major roles in the etiology of human NTD.

8.2. Mouse NTD Prevention by Folate

In parallel to humans, in mice, mutations in most folate metabolic pathway genes that have been tested do not cause NTD [201]. A small proportion of the almost 300 mouse gene mutations that cause NTD have been tested for response to folate supplementation. There are a few mouse mutants whose NTD frequency (usually exencephaly) is reduced by maternal folic acid, and most have no known defect in folate metabolism [201]. Examples are the loss of function mutants for transcription factors *Cart1* and *Cited2*, the histone actetyltransferase *Gcn5*, and the "*Crooked*" gain-of-function mutation of the transmembrane co-receptor for *Wnt*, the *Lrp6* gene [202]. Interestingly, the loss-of-function mutation of *Lrp6* also causes exencephaly and spina bifida but responds to maternal folate supplementation with *increased* frequency of these NTDs [202]. These mutants indicate that the NTD preventative effects of folate are not necessarily through compensation for a deficiency in the folate metabolic pathway [201].

The mutant genes in folate-responsive mouse NTD models may have unexpected connections to folate pathways. An example is the *Pax3* transcription factor mutations [203,204], which have several effects on neural tube cell functions and whose homozygous "*Splotch*" mutants have spina bifida, exencephaly, or both. These mutants have a defect in the folate metabolic pathway that leads to deficiency of de novo pyrimidine biosynthesis; this defect seems to be ameliorated by folate supplementation, which reduces the frequency of NTD by about 40% [203,204].

8.3. A Possible Effect of Folate on Cilia

A recent study has suggested that folate deficiency or lack of the folate-dependent methylation process may exert adverse effects on neural tube development through an effect on formation of primary cilia [205]. In Xenopus embryos, mouse embryos and various types of cell cultures, folate deficiency, inhibition of methylation, or loss of the reduced folate carrier *Slc19a1* each causes reduced cilia formation, indicating that the folate and methylation pathways are required for cilium formation. The lack of cilia is attributable to the demonstrated deficiency of methylation of *Septin2*, a cytoskeletal protein whose methylation state affects whether it contributes to cilia formation [205]. Cilia are required for normal functioning of the *Hedgehog* signaling pathway [13]. An indicator of the functioning of the *Hedgehog* pathway, the truncated form of Gli3 (Gli3R), is decreased in mouse embryonic fibroblast cells lacking folate or methylation, indicating that the lack of cilia in these cells leads to dysregulation of the *Hedgehog* pathway [205]. Cilial structural defects often cause over-expression of the *Sonic hedgehog* signaling pathway in the developing neural tube, which in turn interferes with dorsal bending of the neural folds and causes exencephaly or spina bifida [13] (Section 2.5). In summary, folate status may affect cytoskeletal regulation, cilia formation and *Hedgehog* signaling.

Folr1 (*Fbp1*) is a membrane protein that takes folate into cells. Normally, *Folr1* in mouse embryos is expressed in the neural folds of the upper spine just before and during closure and expression expands bidirectionally along the anterior–posterior axis with advancing closure; it is also expressed in the cranial neural folds just before and during closure, similarly expanding with the advancing closure; it is not detected in the forebrain region and posterior neuropore [206,207]. The *Folr1* null mutant is lethal before neural tube closure, but with maternal folate supplementation, null *Folr1* mutant embryos survive to express NTD, particularly exencephaly [208]. *FOLR1* gene variants do not seem to be important in human anencephaly [209] nor in spina bifida, although *FOLR2* and *FOLR3* may have a small role [210]. A functionally related gene that also causes exencephaly in mice is *Lrp2*, a multi-ligand membrane receptor that seems to mediate uptake of folic acid into cells by *Folr1* at the apical surface of the developing neuroepithelium [211]. Maternal injection with folic acid during early gestation reduces the NTD frequency in *Lrp2* mutants by half; dietary supplementation with folic acid is not effective [212].

Cellular folate deficiency may not be the mechanism by which mutants for *Folr1* and *Lrp2* cause exencephaly. There are other mechanisms of uptake of folate into cells that may be more efficient than *Folr1*, such as the reduced folate carriers [200]. In this light, it is notable that there is evidence that *Folr1* can also function as a transcription factor in mammalian cells, with a large number of developmentally important genes as its target [213]. Furthermore, recent observations in *Xenopus* have pointed to a novel mechanism for the role of *Folr1* and folate in neural tube closure [214]. Knockdown of *Folr1* during neurulation in Xenopus causes failure to develop a neural tube. In Xenopus, *Folr1* is expressed at the apical surface of neural plate cells, and interaction between *Folr1* and folate in the medial neural plate is required for these cells' apical constriction and the bending of the neural plate. *Folr1* co-localizes and interacts with *C-cadherin* and *beta-catenin* in the medial neural plate cells. Apical constriction requires internalization of apically localized *C-cadherin* by endocytosis; in *Folr1* knockdown neural plates, there is a decrease in *C-cadherin*–containing endosomes. Overall, the observations indicate that the primary function of *Folr1* interacting with folate in Xenopus neural tube development may not be related to its functions in metabolism, but rather as a regulator of the cytoskeleton and cell adhesion remodeling [200,214]. If a similar link of the *Folr1*-folate interaction to cytoskeletal remodeling can be found in mammals, then a fresh interpretation of the folate-responsiveness of human and mouse NTD occurrences should be undertaken. Perhaps supplemental availability of folate to the *Folr1* receptor can maximize its activity, and can also compensate for weaknesses in other unrelated modules that also regulate the cytoskeleton.

8.4. Heterogeneity of Effects of Folate on Neural Tube Closure?

Overall, the mouse NTD mutants that interact with folate present a complex story that suggests that some folate effects on NTD risk may function through connections to the known metabolic roles of folate; other folate effects on NTD may be through previously unknown roles in ciliogenesis and regulation of the cytoskeleton. Given the great variety of cellular functions essential for successful neural tube closure, abundant levels of folic acid may be important via different aspects of its function in multiple aspects of the closure process, such as cell proliferation, regulation of a complex repertoire of developmental gene signaling via DNA and histone methylation, *Hedgehog* signaling, and cytoskeletal-driven bending of the neural plate. Given the genetic heterogeneity of NTD, individual cases may have different types of weakness in the closure process and interact with the aspect of supplemental folate that matches the functional deficit.

8.5. Mouse NTD Prevention by Formate

A new set of NTD mutants in folate-pathway genes is emerging. The genes are all involved in folate one-carbon metabolism in the mitochondria, the process that supplies one-carbon formate molecules to the cytosol for nucleotide biosynthesis and DNA methylation [6]. The reported mutants, which have between 25% and 100% exencephaly, are in genes *Amt*, *Gldc*, *Mthfd1l* and *Slc25a32* [170,215–217]. It appears that, whereas most of the non-mitochondrial folate-pathway mutants previously reported [201] do not have NTDs (the exception being *Folr1*), almost all of the mitochondrial folate-pathway mutants reported to date have exencephaly (with occasional craniorachischisis); none have spina bifida. The exception among the mitochondrial folate-pathway mutants that does not cause NTD is *Mthfd2*, which has redundancy with *Mthfd2l* [218].

Interestingly, in three of the mitochondrial mutants (*Gldc*, *Mthfd1l*, *Slc25a32*), the frequency of NTDs is reduced by between 60% and 100% by sodium formate supplementation (in drinking water) to the pregnant dams [170,216,217]; testing of the *Amt* mutant with formate has not been reported. In none of these mutants has folate/folic acid treatment reduced the NTD frequency.

A further example is the curly tail mouse strain, in which 10–20% of embryos have spina bifida. Folic acid has no preventative effect in curly tail strain embryos, but nucleotide precursors reduce spina bifida by 85% [219]. Recently, expression of the mitochondrial folate-pathway gene, *Mthfd1l*, was found to be 50% deficient in curly tail strain embryos and, as with the *Mthfd1l* null mutant (above), treatment of curly tail strain dams with prenatal sodium formate was found to reduce the spina bifida frequency by 75% [220].

Because the main source of one-carbon units for cytoplasmic processes comes from the mitochondria, any failure in a step of mitochondrial folate one-carbon metabolism results in a deficiency of one-carbon components (formate) for essential cellular processes, such as DNA synthesis and methylation. Momb et al. [216] suggest that a metabolic block in the mitochondria creates a "formyl trap", starving the cytoplasm of formate, even in the presence of plentiful folate. It may be that formate, rather than folate, supplementation could be helpful in a similar category of human NTDs in which folate supplementation has no preventative effect.

9. An Overview of the Genetics of NTD

9.1. Genetic Architecture of Human NTD

Understanding of the genetic architecture of human NTD is largely derived from two approaches: family-based data and population-based data. The family-based approach led to the multifactorial model that explains the pattern of recurrence of NTD in families. Rates of recurrence of NTD in various types of relatives of NTD cases demonstrate that NTD have a strong genetic etiology, with a 30- to 50-fold increased rate of occurrence in siblings of NTD cases, compared to a population rate of about one per 1000 births [11,47]. The pattern of recurrence rates of NTD in other types of relatives does not fit simple single gene locus inheritance, but generally fits predictions for multiple gene locus inheritance, where the effects of variants at the multiple genes add together with each other and with environmental influences. We have summarized the foundational studies previously [12]. Conceptually, the genes involved may be unequal in effect [221,222] and the various genes contributing to risk of NTD in an individual case can be imagined to differ in impact, some having fairly large effects on risk, some having moderate effects, and some, perhaps many common polymorphisms, having small effects. A derivative of quantitative genetics, the multifactorial model assumes that individual cases, even with the same type of NTD, will have different subsets of the risk genes.

Population-based studies aimed at finding statistical associations with single nucleotide polymorphisms (SNPs) in over 200 candidate genes have found a few weak associations often lacking statistical significance [1,223,224]. Association studies likely have been hampered by the heterogeneity of etiology of NTD at different locations along the neural tube, mismatch of candidate genes to the types of NTD in the pool of cases studied, dilution of power by use of pools of multiple types of NTD, and lack of adequate sample size for statistical significance of modest associations.

Studies based on sequencing of coding regions of candidate genes usually are done as a population-based approach, often using pools of various types of NTD cases [10]. This approach does not depend on statistical association, only on the ability to identify sequence changes that affect gene function. As the multifactorial model for NTD predicts that individual cases have different combinations in subsets of the gene variants that cause risk of NTD, sequencing a pool of cases may detect few mutations at any given NTD risk gene sequenced. The pattern of findings fits this prediction. A few putative mutations have been found at each of some of the PCP system loci that have been sequenced in several pools of various types of NTD [10]. When examined, the mutations are usually inherited from a normal parent. The multifactorial concept predicts that the NTD cases have also inherited additional variants or mutations at other genes, probably from the other parent. Sequencing of coding regions of five PCP genes in a pool of 36 human craniorachischisis cases has yielded putative mutations at either of two genes, in about 13% of cases [181]. Similarly, exon sequencing of 191 candidate genes in a pool of 94 cases, most of which were anencephaly, has yielded 21 putative mutations affecting 19 genes [209] in about 22% of cases. In this study, consistent with multifactorial etiology, more of the additional rare/novel sequence variants that might have deleterious effects are found in each of the NTD cases than in the controls. The good yields from these two studies [181,209] perhaps demonstrate that increased effectiveness can be conferred by use of pools of only one type of NTD at a time. It should be noted that the usual sequencing approach cannot detect mutations in the more distant regulatory regions that direct the expression of genes to specific tissues (enhancers), and many such regulatory mutations at previously studied candidate genes may yet be discovered.

A current hypothesis about the genetic architecture of various multifactorial, post-reproductive diseases ("complex traits") [225] suggests that a large proportion of the genome can have some impact on a "complex trait" disease. This concept is also a long-standing component of genetic selection theory based on empirical data for quantitative traits [226]. The genetic architecture likely differs between NTD and "complex traits" or quantitative traits because of the effects of genetic lethality of NTD on gene frequencies. For NTD, perhaps the important goal is to identify the variants of major effect whose combinations account for most of the risk of NTD, not all of the variants that can add smaller effects. There are probably many variants of major effect.

The common forms of NTD that are explained by the multifactorial model are non-syndromic, meaning the only primary defect is the defect in neural tube closure.

There are some rare human syndromes of multiple unrelated defects that include anencephaly or spina bifida, and they are caused by chromosomal defects or Mendelian mutations [227]. They are not included in studies of common, non-syndromic NTD.

9.2. The Role of Environmental Effects

For a multifactorial trait, it is expected that environmental factors contribute to the trait to some degree. Theoretically, the effects of environmental factors add to the genetic risk factors in individuals. Individual embryos with a load of genetic risk factors but marginal ability to close the neural tube may be hampered enough by the additional effects of environmental factors to fail to close, whereas an exposed embryo with fewer genetic risk factors would still be able to successfully close. Folate supplementation (Section 8) is an environmental factor with a large effect that apparently compensates for the genetic risk factors in some way. Some other environmental factors are deleterious, with much weaker but statistically detectable effects. They include maternal diabetes, maternal obesity, maternal hyperthermia during pregnancy, parental occupations and exposure to chemicals, and a few pharmaceuticals [1].

9.3. The Role of Mouse Mutants in NTD Genetics

The almost 300 mouse NTD mutants cannot model the human genetic architecture of NTD. The mouse mutants are maintained on inbred strain backgrounds, within which there is essentially no genetic variation between individuals. Very few strain backgrounds are used, and the strains themselves are a very limited sample of the variation in the mouse species' genome. Some mutations show different penetrance or expressivity of NTD on different strain backgrounds, demonstrating their potential for multifactorial genetics. In the studies of digenic effects, to test the combined effects of two different genes on NTD risk, the specific digenic combination of interest is created by crosses between mutant stocks. Unexpected genetic interactions are unlikely to be discovered in this hypothesis-driven system.

The mouse mutations are usually highly penetrant recessive "knockouts" that cause loss of function of the gene, and for many of the genes, these cause syndromes of multiple defects. In general, gene loci that cause syndromic phenotypes from null (knockout) mutations may present non-syndromic phenotypes from hypomorphic (reduced function) mutations. An example in mice is the kidney defect and cleft lip syndrome caused by a null mutation at *Wnt9b* versus the non-syndromic cleft lip phenotype from a hypomorphic mutation at *Wnt9b* [228]. Therefore, the mouse mutants serve to identify candidate loci for human NTD despite the severity of the null mutant phenotypes.

The use of candidate genes from mice should be tempered by awareness of the differences in the morphological process between human and mouse neural tube development (Section 4.1), and ideally should be informed by gene expression data from other species whose neural tube closure is morphologically more similar to human, such as pig and rabbit.

Many of the candidate genes that have been tested in human NTD cases are based on various hypotheses, such as roles related to folate metabolism or folate transport, and the lack of finding gene variants associated with human NTD concurs with the mouse mutants; i.e., they do not cause NTD in mouse mutants.

The pools of human cases studied usually focus on spina bifida. The majority of the NTD genes in mouse mutants cause cranial NTD, and based on the regional heterogeneity of etiology discussed in Sections 3 and 5, these genes would not be expected to participate in human spina bifida. This mismatch may contribute to the disappointing results from some studies.

Two successful sequencing studies are discussed in Section 9.1. Both demonstrate the value of mouse NTD mutants as candidate genes. The study of a few of the PCP genes in craniorachischisis cases [181], based on the observation that nearly all genetic craniorachischisis in mice is caused by PCP gene mutations [12] has been fruitful, yielding putative mutations from 13% of cases. In a similar study [209] based on mostly anencephalic cases, about half of the 19 genes with putative mutations have mouse NTD mutants at the same gene or a member of the same gene family.

10. Concluding Remarks

Various insights emerge from an overview of the many studies of neural tube closure. The mammalian way of forming a neural tube by elevating and bending the lateral edges of a neural plate to meet in the midline is an ancient mechanism that evolved before vertebrates. Several systems contribute to forming the neural tube. Among them are the planar cell polarity/convergent extension system; the medial hinge point and related cell cycle phenomena; the dorsolateral hinge point and the *Sonic hedgehog* cilia system on which it depends; the actomyosin and cytoskeletal system responsible for some bending of the neuroepithelium; the various types of cell projections and the *Ephrin* system that facilitate contact and fusion. Each of these systems involves many genes. Despite the understanding of many details of the genetic determination of neural tube development, the genetic causes of human NTD are mostly undiscovered.

The evidence that the planar cell polarity system is involved in various aspects of neural tube closure in addition to convergent extension, such as the cytoskeleton, ciliogenesis, and the actomyosin system, offers an explanation for the observed role of planar cell polarity genes in exencephaly/anencephaly and spina bifida as well as in craniorachischisis.

Although the mouse has been used extensively as the mammalian model for human NTD, the development of the neural tube in other mammals, such as pig and rabbit, is more similar to human in morphology. Gene expression arrays of regions of neural folds from these species during elevation and closure might point to the subset of genes and mechanisms previously identified in mice that are likely to be involved in human neural tube closure.

The vertebrate head evolved as an "add-on" to the pre-vertebrate body plan and its neural tube. The formation of the cranial neural tube consequently differs in several ways from that of the spinal region. For example, the mouse cranial mesoderm has a much greater role in neural tube closure and etiology of NTD. The neuroepithelial actomyosin system is also more important in the cranial region than in the spinal region. The large number of genes involved in these two systems may account for the predominance of genes causing exencephaly among mouse NTD mutants. The genes identified by these mutants may be good candidates for examination in the etiology of human anencephaly.

Cranial NTD occur more frequently in females than in males. This phenomenon is seen in humans and in nearly all mouse NTD mutants examined. Recent advances in the understanding of epigenetics suggest the hypothesis that the second, inactive X chromosome of female embryos is a "heterochromatin sink" that sequesters essential epigenetic factors, altering expression of genes in the rest of the genome and thereby increasing the risk of NTD. The limitation of the effect to cranial NTD suggests that specific genes are involved.

It is well established that peri-conceptional dietary folic acid supplementation reduces the risk of occurrence of NTD in humans and in some mouse mutants, but the genetic risk of folate-responsive NTD does not seem to involve genes in the folate metabolic pathway in either species. There is recent evidence for a function of folate-pathway genes that is separate from the established metabolic functions in DNA methylation and nucleotide synthesis. In Xenopus, the *Folr1* membrane protein that takes folate into cells seems to act also as a regulator of the cytoskeleton and of cell adhesion remodeling in the neural plate. Secondly, independent evidence in Xenopus demonstrates a requirement for folate in ciliogenesis, which is known to be involved in NTD. These observations may shift the focus of candidate gene research in folate-responsive human NTD away from the metabolic functions of folate.

There are many examples in mouse mutants of NTD caused by combinations of mutant genes at two or more loci from the same or different genetic pathways. Usually, the molecular mechanism of interaction between the mutant genes is unknown. The mechanism of the interaction effect can be surprising, as in the case of the *Msx2/Dlx5* double mutant, where the combined effect of the two mutant genes alters the balance of splice forms of a third gene, *EphA7*. The extensive study of the planar cell polarity system has generated several examples of digenic combinations of different planar cell polarity genes that cause a variety of types of NTD. The observations in mice suggest that many of the cases of human NTD in which a putative mutation has been detected at one planar cell polarity gene locus may well harbor an undetected mutation at a second planar cell polarity gene locus or other type of gene.

Overall, the cumulative data show that whereas the neural tube appears to be a single anterior–posterior structure, in contradiction, the gene expression, cell behavior, and closure systems occupy a variety of different regions. The neural tube along its length is neither homogeneous nor segmented. Each gene and mechanism is expressed in areas along the length that are different from the expression patterns for many of the other genes or mechanisms so that each linear area is the result of a unique combination. NTD may be a series of genetically distinct defects occurring along a contiguous physical structure. Setting aside the planar cell polarity system mutations, the remainder of human NTD may have genetic etiology that varies by anterior–posterior location of the lesion, a concept that could inform human genetic studies in search of causative variants. For example, families within which there are both anencephalic and spina bifida cases may tend to have mutations in the DLHP system. Perhaps the most important conclusion from this review is recognition of the degree to which the neural tube is regionalized along the anterior–posterior axis.

Funding: This research received no external funding.

Conflicts of Interest: The authors declare no conflict of interest.

References

1. Au, K.S.; Ashley-Koch, A.; Northrup, H. Epidemiologic and Genetic Aspects of Spina Bifida and Other Neural Tube Defects. *Dev. Disabil. Res. Rev.* **2010**, *16*, 6–15. [CrossRef] [PubMed]
2. McBride, M.L. Sib Risks of Anencephaly and Spina Bifida in British Columbia. *Am. J. Med. Genet.* **1979**, *3*, 377–387. [CrossRef] [PubMed]
3. Harris, M.J.; Juriloff, D.M. An Update to the List of Mouse Mutants with Neural Tube Closure Defects and Advances toward a Complete Genetic Perspective of Neural Tube Closure. *Birth Defects Res. A Clin. Mol. Teratol.* **2010**, *88*, 653–669. [CrossRef] [PubMed]
4. Wilde, J.J.; Petersen, J.R.; Niswander, L. Genetic, Epigenetic, and Environmental Contributions to Neural Tube Closure. *Annu. Rev. Genet.* **2014**, *48*, 583–611. [CrossRef] [PubMed]
5. Nikolopoulou, E.; Galea, G.L.; Rolo, A.; Greene, N.D.; Copp, A.J. Neural Tube Closure: Cellular, Molecular and Biomechanical Mechanisms. *Development* **2017**, *144*, 552–566. [CrossRef] [PubMed]
6. Greene, N.D.; Copp, A.J. Neural Tube Defects. *Annu. Rev. Neurosci.* **2014**, *37*, 221–242. [CrossRef] [PubMed]
7. Copp, A.J.; Stanier, P.; Greene, N.D. Neural Tube Defects: Recent Advances, Unsolved Questions, and Controversies. *Lancet Neurol.* **2013**, *12*, 799–810. [CrossRef]
8. Wallingford, J.B.; Niswander, L.A.; Shaw, G.M.; Finnell, R.H. The Continuing Challenge of Understanding, Preventing, and Treating Neural Tube Defects. *Science* **2013**, *339*, 1222002. [CrossRef] [PubMed]
9. Yamaguchi, Y.; Miura, M. How to Form and Close the Brain: Insight into the Mechanism of Cranial Neural Tube Closure in Mammals. *Cell Mol. Life Sci.* **2013**, *70*, 3171–3186. [CrossRef] [PubMed]
10. Juriloff, D.M.; Harris, M.J. A Consideration of the Evidence that Genetic Defects in Planar Cell Polarity Contribute to the Etiology of Human Neural Tube Defects. *Birth Defects Res. A Clin. Mol. Teratol.* **2012**, *94*, 824–840. [CrossRef] [PubMed]
11. Au, K.S.; Findley, T.O.; Northrup, H. Finding the Genetic Mechanisms of Folate Deficiency and Neural Tube Defects-Leaving no Stone Unturned. *Am. J. Med. Genet. A* **2017**, *173*, 3042–3057. [CrossRef] [PubMed]
12. Harris, M.J.; Juriloff, D.M. Mouse Mutants with Neural Tube Closure Defects and their Role in Understanding Human Neural Tube Defects. *Birth Defects Res. A Clin. Mol. Teratol.* **2007**, *79*, 187–210. [CrossRef] [PubMed]
13. Murdoch, J.N.; Copp, A.J. The Relationship between Sonic Hedgehog Signaling, Cilia, and Neural Tube Defects. *Birth Defects Res. A Clin. Mol. Teratol.* **2010**, *88*, 633–652. [CrossRef] [PubMed]
14. Carvajal-Gonzalez, J.M.; Mulero-Navarro, S.; Mlodzik, M. Centriole Positioning in Epithelial Cells and its Intimate Relationship with Planar Cell Polarity. *Bioessays* **2016**, *38*, 1234–1245. [CrossRef] [PubMed]

15. Montcouquiol, M.; Sans, N.; Huss, D.; Kach, J.; Dickman, J.D.; Forge, A.; Rachel, R.A.; Copeland, N.G.; Jenkins, N.A.; Bogani, D.; et al. Asymmetric Localization of Vangl2 and Fz3 Indicate Novel Mechanisms for Planar Cell Polarity in Mammals. *J. Neurosci.* **2006**, *26*, 5265–5275. [CrossRef] [PubMed]
16. Williams, M.; Yen, W.; Lu, X.; Sutherland, A. Distinct Apical and Basolateral Mechanisms Drive Planar Cell Polarity-Dependent Convergent Extension of the Mouse Neural Plate. *Dev. Cell.* **2014**, *29*, 34–46. [CrossRef] [PubMed]
17. Tissir, F.; Qu, Y.; Montcouquiol, M.; Zhou, L.; Komatsu, K.; Shi, D.; Fujimori, T.; Labeau, J.; Tyteca, D.; Courtoy, P.; et al. Lack of Cadherins Celsr2 and Celsr3 Impairs Ependymal Ciliogenesis, Leading to Fatal Hydrocephalus. *Nat. Neurosci.* **2010**, *13*, 700–707. [CrossRef] [PubMed]
18. Zeng, H.; Hoover, A.N.; Liu, A. PCP Effector Gene Inturned is an Important Regulator of Cilia Formation and Embryonic Development in Mammals. *Dev. Biol.* **2010**, *339*, 418–428. [CrossRef] [PubMed]
19. Escuin, S.; Vernay, B.; Savery, D.; Gurniak, C.B.; Witke, W.; Greene, N.D.; Copp, A.J. Rho-Kinase-Dependent Actin Turnover and Actomyosin Disassembly are Necessary for Mouse Spinal Neural Tube Closure. *J. Cell Sci.* **2015**, *128*, 2468–2481. [CrossRef] [PubMed]
20. Rolo, A.; Escuin, S.; Greene, N.D.E.; Copp, A.J. Rho GTPases in Mammalian Spinal Neural Tube Closure. *Small GTPases* **2018**, *9*, 283–289. [CrossRef] [PubMed]
21. Shum, A.S.; Copp, A.J. Regional Differences in Morphogenesis of the Neuroepithelium Suggest Multiple Mechanisms of Spinal Neurulation in the Mouse. *Anat. Embryol.* **1996**, *194*, 65–73. [CrossRef] [PubMed]
22. Ybot-Gonzalez, P.; Cogram, P.; Gerrelli, D.; Copp, A.J. Sonic Hedgehog and the Molecular Regulation of Mouse Neural Tube Closure. *Development* **2002**, *129*, 2507–2517. [PubMed]
23. McShane, S.G.; Mole, M.A.; Savery, D.; Greene, N.D.; Tam, P.P.; Copp, A.J. Cellular Basis of Neuroepithelial Bending during Mouse Spinal Neural Tube Closure. *Dev. Biol.* **2015**, *404*, 113–124. [CrossRef] [PubMed]
24. Amarnath, S.; Agarwala, S. Cell-Cycle-Dependent TGFbeta-BMP Antagonism Regulates Neural Tube Closure by Modulating Tight Junctions. *J. Cell Sci.* **2017**, *130*, 119–131. [CrossRef] [PubMed]
25. Cui, C.; Chatterjee, B.; Francis, D.; Yu, Q.; SanAgustin, J.T.; Francis, R.; Tansey, T.; Henry, C.; Wang, B.; Lemley, B.; et al. Disruption of Mks1 Localization to the Mother Centriole Causes Cilia Defects and Developmental Malformations in Meckel-Gruber Syndrome. *Dis. Model. Mech.* **2011**, *4*, 43–56. [CrossRef] [PubMed]
26. Wallingford, J.B. Planar Cell Polarity and the Developmental Control of Cell Behavior in Vertebrate Embryos. *Annu. Rev. Cell Dev. Biol.* **2012**, *28*, 627–653. [CrossRef] [PubMed]
27. Pala, R.; Alomari, N.; Nauli, S.M. Primary Cilium-Dependent Signaling Mechanisms. *Int. J. Mol. Sci.* **2017**, *18*, 18. [CrossRef] [PubMed]
28. May-Simera, H.; Kelley, M.W. Planar Cell Polarity in the Inner Ear. *Curr. Top. Dev. Biol.* **2012**, *101*, 111–140. [PubMed]
29. Adler, P.N.; Wallingford, J.B. From Planar Cell Polarity to Ciliogenesis and Back: The Curious Tale of the PPE and CPLANE Proteins. *Trends Cell Biol.* **2017**, *27*, 379–390. [CrossRef] [PubMed]
30. Hui, C.C.; Angers, S. Gli Proteins in Development and Disease. *Annu. Rev. Cell Dev. Biol.* **2011**, *27*, 513–537. [CrossRef] [PubMed]
31. Ybot-Gonzalez, P.; Gaston-Massuet, C.; Girdler, G.; Klingensmith, J.; Arkell, R.; Greene, N.D.; Copp, A.J. Neural Plate Morphogenesis during Mouse Neurulation is Regulated by Antagonism of Bmp Signalling. *Development* **2007**, *134*, 3203–3211. [CrossRef] [PubMed]
32. Xue, Y.; Wang, X.; Li, Z.; Gotoh, N.; Chapman, D.; Skolnik, E.Y. Mesodermal Patterning Defect in Mice Lacking the Ste20 NCK Interacting Kinase (NIK). *Development* **2001**, *128*, 1559–1572. [PubMed]
33. Chiang, C.; Litingtung, Y.; Lee, E.; Young, K.E.; Corden, J.L.; Westphal, H.; Beachy, P.A. Cyclopia and Defective Axial Patterning in Mice Lacking Sonic Hedgehog Gene Function. *Nature* **1996**, *383*, 407–413. [CrossRef] [PubMed]
34. Aoto, K.; Shikata, Y.; Imai, H.; Matsumaru, D.; Tokunaga, T.; Shioda, S.; Yamada, G.; Motoyama, J. Mouse Shh is Required for Prechordal Plate Maintenance during Brain and Craniofacial Morphogenesis. *Dev. Biol.* **2009**, *327*, 106–120. [CrossRef] [PubMed]
35. Patterson, V.L.; Damrau, C.; Paudyal, A.; Reeve, B.; Grimes, D.T.; Stewart, M.E.; Williams, D.J.; Siggers, P.; Greenfield, A.; Murdoch, J.N. Mouse Hitchhiker Mutants have Spina Bifida, Dorso-Ventral Patterning Defects and Polydactyly: Identification of Tulp3 as a Novel Negative Regulator of the Sonic Hedgehog Pathway. *Hum. Mol. Genet.* **2009**, *18*, 1719–1739. [CrossRef] [PubMed]

36. Robbins, D.J.; Fei, D.L.; Riobo, N.A. The Hedgehog Signal Transduction Network. *Sci. Signal.* **2012**, *5*, re6. [CrossRef] [PubMed]
37. Pyrgaki, C.; Trainor, P.; Hadjantonakis, A.K.; Niswander, L. Dynamic Imaging of Mammalian Neural Tube Closure. *Dev. Biol.* **2010**, *344*, 941–947. [CrossRef] [PubMed]
38. Rolo, A.; Savery, D.; Escuin, S.; de Castro, S.C.; Armer, H.E.; Munro, P.M.; Mole, M.A.; Greene, N.D.; Copp, A.J. Regulation of Cell Protrusions by Small GTPases during Fusion of the Neural Folds. *Elife* **2016**, *5*, e13273. [CrossRef] [PubMed]
39. Ray, H.J.; Niswander, L.A. Dynamic Behaviors of the Non-Neural Ectoderm during Mammalian Cranial Neural Tube Closure. *Dev. Biol.* **2016**, *416*, 279–285. [CrossRef] [PubMed]
40. Holmberg, J.; Clarke, D.L.; Frisen, J. Regulation of Repulsion versus Adhesion by Different Splice Forms of an Eph Receptor. *Nature* **2000**, *408*, 203–206. [CrossRef] [PubMed]
41. Rifat, Y.; Parekh, V.; Wilanowski, T.; Hislop, N.R.; Auden, A.; Ting, S.B.; Cunningham, J.M.; Jane, S.M. Regional Neural Tube Closure Defined by the Grainy Head-Like Transcription Factors. *Dev. Biol.* **2010**, *345*, 237–245. [CrossRef] [PubMed]
42. Copp, A.J.; Greene, N.D.; Murdoch, J.N. The Genetic Basis of Mammalian Neurulation. *Nat. Rev. Genet.* **2003**, *4*, 784–793. [CrossRef] [PubMed]
43. Abdul-Aziz, N.M.; Turmaine, M.; Greene, N.D.; Copp, A.J. EphrinA-EphA Receptor Interactions in Mouse Spinal Neurulation: Implications for Neural Fold Fusion. *Int. J. Dev. Biol.* **2009**, *53*, 559–568. [CrossRef] [PubMed]
44. Lopez-Escobar, B.; Caro-Vega, J.M.; Vijayraghavan, D.S.; Plageman, T.F.; Sanchez-Alcazar, J.A.; Moreno, R.C.; Savery, D.; Marquez-Rivas, J.; Davidson, L.A.; Ybot-Gonzalez, P. The Non-Canonical Wnt-PCP Pathway Shapes the Mouse Caudal Neural Plate. *Development* **2018**, *145*, 145. [CrossRef] [PubMed]
45. Saitsu, H.; Shiota, K. Involvement of the Axially Condensed Tail Bud Mesenchyme in Normal and Abnormal Human Posterior Neural Tube Development. *Congenit. Anom.* **2008**, *48*, 1–6. [CrossRef] [PubMed]
46. Dady, A.; Havis, E.; Escriou, V.; Catala, M.; Duband, J.L. Junctional Neurulation: A Unique Developmental Program Shaping a Discrete Region of the Spinal Cord Highly Susceptible to Neural Tube Defects. *J. Neurosci.* **2014**, *34*, 13208–13221. [CrossRef] [PubMed]
47. Detrait, E.R.; George, T.M.; Etchevers, H.C.; Gilbert, J.R.; Vekemans, M.; Speer, M.C. Human Neural Tube Defects: Developmental Biology, Epidemiology, and Genetics. *Neurotoxicol. Teratol.* **2005**, *27*, 515–524. [CrossRef] [PubMed]
48. Juriloff, D.M.; Harris, M.J.; Tom, C.; MacDonald, K.B. Normal Mouse Strains Differ in the Site of Initiation of Closure of the Cranial Neural Tube. *Teratology* **1991**, *44*, 225–233. [CrossRef] [PubMed]
49. Kaufman, M. Cephalic Neurulation and Optic Vesicle Formation in the Early Mouse Embryo. *Am. J. Anat.* **1979**, *155*, 425–443. [CrossRef] [PubMed]
50. Sakai, Y. Neurulation in the Mouse: Manner and Timing of Neural Tube Closure. *Anat. Rec.* **1989**, *223*, 194–203. [CrossRef] [PubMed]
51. Golden, J.A.; Chernoff, G.F. Intermittent Pattern of Neural Tube Closure in Two Strains of Mice. *Teratology* **1993**, *47*, 73–80. [CrossRef] [PubMed]
52. Morriss, G.M.; Solursh, M. Regional Differences in Mesenchymal Cell Morphology and Glycosaminoglycans in Early Neural-Fold Stage Rat Embryos. *J. Embryol. Exp. Morphol.* **1978**, *46*, 37–52. [PubMed]
53. Christie, G.A. Developmental Stages in Somite and Post-Somite Rat Embryos, Based on External Appearance, and Including some Features of the Macroscopic Development of the Oral Cavity. *J. Morphol.* **1964**, *114*, 263–283. [CrossRef] [PubMed]
54. Keyser, A. The Development of the Diencephalon of the Chinese Hamster. An Investigation of the Validity of the Criteria of Subdivision of the Brain. *Acta Anat. Suppl.* **1972**, *59*, 1–178. [PubMed]
55. Waterman, R.E. Topographical Changes Along the Neural Fold Associated with Neurulation in the Hamster and Mouse. *Am. J. Anat.* **1976**, *146*, 151–171. [CrossRef] [PubMed]
56. Peeters, M.C.; Viebahn, C.; Hekking, J.W.; van Straaten, H.W. Neurulation in the Rabbit Embryo. *Anat. Embryol.* **1998**, *197*, 167–175. [CrossRef] [PubMed]
57. Van Straaten, H.W.; Peeters, M.C.; Hekking, J.W.; van der Lende, T. Neurulation in the Pig Embryo. *Anat. Embryol.* **2000**, *202*, 75–84. [CrossRef] [PubMed]
58. Hamilton, W.J.; Mossman, H.W. *Human Embryology: Prenatal Development of Form and Function*, 4th ed.; Williams and Wilkins: Baltimore, MD, USA, 1972; pp. 1–646.

59. Jirasek, J.E. *Atlas of Human Prenatal Morphogenesis*, 1st ed.; Springer: Dordrecht, the Netherlands, 1983; p. 168.
60. Muller, F.; O'Rahilly, R. The First Appearance of the Neural Tube and Optic Primordium in the Human Embryo at Stage 10. *Anat. Embryol.* **1985**, *172*, 157–169. [CrossRef] [PubMed]
61. Muller, F.; O'Rahilly, R. The Development of the Human Brain and the Closure of the Rostral Neuropore at Stage 11. *Anat. Embryol.* **1986**, *175*, 205–222. [CrossRef] [PubMed]
62. Nakatsu, T.; Uwabe, C.; Shiota, K. Neural Tube Closure in Humans Initiates at Multiple Sites; Evidence from Human Embryos and Implications for the Pathogenesis of Neural Tube Defects. *Anat. Embryol.* **2000**, *201*, 455–466. [CrossRef] [PubMed]
63. Peeters, M.C.; Hekking, J.W.; Shiota, K.; Drukker, J.; Van Straaten, H.W. Differences in Axial Curvature Correlate with Species-Specific Rate of Neural Tube Closure in Embryos of Chick, Rabbit, Mouse, Rat and Human. *Anat. Embryol.* **1998**, *198*, 185–194. [CrossRef] [PubMed]
64. Juriloff, D.M.; Harris, M.J. Mouse Models for Neural Tube Closure Defects. *Hum. Mol. Genet.* **2000**, *9*, 993–1000. [CrossRef] [PubMed]
65. Harris, M.J.; Juriloff, D.M.; Gunn, T.M.; Miller, J.E. Development of the Cerebellar Defect in Ataxic SELH/Bc Mice. *Teratology* **1994**, *50*, 63–73. [CrossRef] [PubMed]
66. Golden, J.A.; Chernoff, G.F. Multiple Sites of Anterior Neural Tube Closure in Humans: Evidence from Anterior Neural Tube Defects (Anencephaly). *Pediatrics* **1995**, *95*, 506–510. [PubMed]
67. Geelen, J.A.; Langman, J. Closure of the Neural Tube in the Cephalic Region of the Mouse Embryo. *Anat. Rec.* **1977**, *189*, 625–640. [CrossRef] [PubMed]
68. Geelen, J.A.; Langman, J. Ultrastructural Observations on Closure of the Neural Tube in the Mouse. *Anat. Embryol.* **1979**, *156*, 73–88. [CrossRef] [PubMed]
69. Nieto, M.A.; Gilardi-Hebenstreit, P.; Charnay, P.; Wilkinson, D.G. A Receptor Protein Tyrosine Kinase Implicated in the Segmental Patterning of the Hindbrain and Mesoderm. *Development* **1992**, *116*, 1137–1150. [PubMed]
70. Niederreither, K.; Vermot, J.; Schuhbaur, B.; Chambon, P.; Dolle, P. Retinoic Acid Synthesis and Hindbrain Patterning in the Mouse Embryo. *Development* **2000**, *127*, 75–85. [PubMed]
71. Ting, S.B.; Wilanowski, T.; Auden, A.; Hall, M.; Voss, A.K.; Thomas, T.; Parekh, V.; Cunningham, J.M.; Jane, S.M. Inositol- and Folate-Resistant Neural Tube Defects in Mice Lacking the Epithelial-Specific Factor Grhl-3. *Nat. Med.* **2003**, *9*, 1513–1519. [CrossRef] [PubMed]
72. Ray, H.J.; Niswander, L.A. Grainyhead-Like 2 Downstream Targets Act to Suppress Epithelial-to-Mesenchymal Transition during Neural Tube Closure. *Development* **2016**, *143*, 1192–1204. [CrossRef] [PubMed]
73. De Castro, S.C.P.; Hirst, C.S.; Savery, D.; Rolo, A.; Lickert, H.; Andersen, B.; Copp, A.J.; Greene, N.D.E. Neural Tube Closure Depends on Expression of Grainyhead-Like 3 in Multiple Tissues. *Dev. Biol.* **2018**, *435*, 130–137. [CrossRef] [PubMed]
74. Deschamps, J.; van den Akker, E.; Forlani, S.; De Graaff, W.; Oosterveen, T.; Roelen, B.; Roelfsema, J. Initiation, Establishment and Maintenance of Hox Gene Expression Patterns in the Mouse. *Int. J. Dev. Biol.* **1999**, *43*, 635–650. [PubMed]
75. Deschamps, J.; van Nes, J. Developmental Regulation of the Hox Genes during Axial Morphogenesis in the Mouse. *Development* **2005**, *132*, 2931–2942. [CrossRef] [PubMed]
76. Morriss-Kay, G.; Tan, S. Mapping Cranial Neural Crest Cell Migration Pathways in Mammalian Embryos. *TIG* **1987**, *3*, 257–261. [CrossRef]
77. Yamaguchi, Y.; Shinotsuka, N.; Nonomura, K.; Takemoto, K.; Kuida, K.; Yosida, H.; Miura, M. Live Imaging of Apoptosis in a Novel Transgenic Mouse Highlights its Role in Neural Tube Closure. *J. Cell Biol.* **2011**, *195*, 1047–1060. [CrossRef] [PubMed]
78. Lowery, L.A.; Sive, H. Strategies of Vertebrate Neurulation and a Re-Evaluation of Teleost Neural Tube Formation. *Mech. Dev.* **2004**, *121*, 1189–1197. [CrossRef] [PubMed]
79. Schroeder, T.E. Neurulation in Xenopus Laevis. An Analysis and Model Based upon Light and Electron Microscopy. *J. Embryol. Exp. Morphol.* **1970**, *23*, 427–462. [PubMed]
80. Eagleson, G.W. Developmental Neurobiology of the Anterior Areas in Amphibians: Urodele Perspectives. *Int. J. Dev. Biol.* **1996**, *40*, 735–743. [PubMed]
81. Holland, L.Z. The Origin and Evolution of Chordate Nervous Systems. *Philos. Trans. R. Soc. Lond. B Biol. Sci.* **2015**, *370*, 370. [CrossRef] [PubMed]

82. Shimeld, S.M.; Donoghue, P.C. Evolutionary Crossroads in Developmental Biology: Cyclostomes (Lamprey and Hagfish). *Development* **2012**, *139*, 2091–2099. [CrossRef] [PubMed]
83. Holland, N.D.; Holland, L.Z. The Ups and Downs of Amphioxus Biology: A History. *Int. J. Dev. Biol.* **2017**, *61*, 575–583. [CrossRef] [PubMed]
84. Putnam, N.H.; Butts, T.; Ferrier, D.E.; Furlong, R.F.; Hellsten, U.; Kawashima, T.; Robinson-Rechavi, M.; Shoguchi, E.; Terry, A.; Yu, J.K.; et al. The Amphioxus Genome and the Evolution of the Chordate Karyotype. *Nature* **2008**, *453*, 1064–1071. [CrossRef] [PubMed]
85. Satoh, N. Cellular Morphology and Architecture during Early Morphogenesis of the Ascidian Egg: An SEM Study. *Biol. Bull.* **1978**, *155*, 608–614. [CrossRef] [PubMed]
86. Jiang, D.; Munro, E.M.; Smith, W.C. Ascidian Prickle Regulates both Mediolateral and Anterior-Posterior Cell Polarity of Notochord Cells. *Curr. Biol.* **2005**, *15*, 79–85. [CrossRef] [PubMed]
87. Sasakura, Y.; Mita, K.; Ogura, Y.; Horie, T. Ascidians as Excellent Chordate Models for Studying the Development of the Nervous System during Embryogenesis and Metamorphosis. *Dev. Growth Differ.* **2012**, *54*, 420–437. [CrossRef] [PubMed]
88. Northcutt, R.G. The New Head Hypothesis Revisited. *J. Exp. Zool. B Mol. Dev. Evol.* **2005**, *304*, 274–297. [CrossRef] [PubMed]
89. Bronner-Fraser, M. On the Trail of the 'New Head' in Les Treilles. *Development* **2008**, *135*, 2995–2999. [CrossRef] [PubMed]
90. Holland, L.Z. Chordate Roots of the Vertebrate Nervous System: Expanding the Molecular Toolkit. *Nat. Rev. Neurosci.* **2009**, *10*, 736–746. [CrossRef] [PubMed]
91. Bronner, M.E.; LeDouarin, N.M. Development and Evolution of the Neural Crest: An Overview. *Dev. Biol.* **2012**, *366*, 2–9. [CrossRef] [PubMed]
92. Santagati, F.; Rijli, F.M. Cranial Neural Crest and the Building of the Vertebrate Head. *Nat. Rev. Neurosci.* **2003**, *4*, 806–818. [CrossRef] [PubMed]
93. Jandzik, D.; Garnett, A.T.; Square, T.A.; Cattell, M.V.; Yu, J.K.; Medeiros, D.M. Evolution of the New Vertebrate Head by Co-Option of an Ancient Chordate Skeletal Tissue. *Nature* **2015**, *518*, 534–537. [CrossRef] [PubMed]
94. Tan, S.S.; Morriss-Kay, G. The Development and Distribution of the Cranial Neural Crest in the Rat Embryo. *Cell Tissue Res.* **1985**, *240*, 403–416. [CrossRef] [PubMed]
95. Osumi-Yamashita, N.; Ninomiya, Y.; Doi, H.; Eto, K. The Contribution of both Forebrain and Midbrain Crest Cells to the Mesenchyme in the Frontonasal Mass of Mouse Embryos. *Dev. Biol.* **1994**, *164*, 409–419. [CrossRef] [PubMed]
96. Ezin, M.; Barembaum, M.; Bronner, M.E. Stage-Dependent Plasticity of the Anterior Neural Folds to Form Neural Crest. *Differentiation* **2014**, *88*, 42–50. [CrossRef] [PubMed]
97. Masek, J.; Machon, O.; Korinek, V.; Taketo, M.M.; Kozmik, Z. Tcf7l1 Protects the Anterior Neural Fold from Adopting the Neural Crest Fate. *Development* **2016**, *143*, 2206–2216. [CrossRef] [PubMed]
98. Nolte, C.; Krumlauf, R. Expression of Hox genes in the nervous system of vertebrates. In *Hox Gene Expression*; Papageorgiou, S., Ed.; Springer: New York, NY, USA, 2007; pp. 14–41.
99. Muller, F.; O'Rahilly, R. The Prechordal Plate, the Rostral End of the Notochord and Nearby Median Features in Staged Human Embryos. *Cells Tissues Organs* **2003**, *173*, 1–20. [CrossRef] [PubMed]
100. Ulmer, B.; Tingler, M.; Kurz, S.; Maerker, M.; Andre, P.; Monch, D.; Campione, M.; Deissler, K.; Lewandoski, M.; Thumberger, T.; et al. A Novel Role of the Organizer Gene Goosecoid as an Inhibitor of Wnt/PCP-Mediated Convergent Extension in Xenopus and Mouse. *Sci. Rep.* **2017**, *7*, 43010. [CrossRef] [PubMed]
101. Shimamura, K.; Rubenstein, J.L. Inductive Interactions Direct Early Regionalization of the Mouse Forebrain. *Development* **1997**, *124*, 2709–2718. [PubMed]
102. Belo, J.A.; Leyns, L.; Yamada, G.; De Robertis, E.M. The Prechordal Midline of the Chondrocranium is Defective in Goosecoid-1 Mouse Mutants. *Mech. Dev.* **1998**, *72*, 15–25. [CrossRef]
103. Muller, F.; O'Rahilly, R. The First Appearance of the Major Divisions of the Human Brain at Stage 9. *Anat. Embryol.* **1983**, *168*, 419–432. [CrossRef] [PubMed]
104. Jacobson, A.G.; Tam, P.P. Cephalic Neurulation in the Mouse Embryo Analyzed by SEM and Morphometry. *Anat. Rec.* **1982**, *203*, 375–396. [CrossRef] [PubMed]
105. Takeuchi, T.; Kojima, M.; Nakajima, K.; Kondo, S. Jumonji Gene is Essential for the Neurulation and Cardiac Development of Mouse Embryos with a C3H/He Background. *Mech. Dev.* **1999**, *86*, 29–38. [CrossRef]

106. Kaufman, M.H. The Role of Embryology in Teratological Research, with Particular Reference to the Development of the Neural Tube and Heart. *J. Reprod. Fertil.* **1981**, *62*, 607–623. [CrossRef] [PubMed]
107. Massa, V.; Savery, D.; Ybot-Gonzalez, P.; Ferraro, E.; Rongvaux, A.; Cecconi, F.; Flavell, R.; Greene, N.D.; Copp, A.J. Apoptosis is Not Required for Mammalian Neural Tube Closure. *Proc. Natl. Acad. Sci. USA* **2009**, *106*, 8233–8238. [CrossRef] [PubMed]
108. Brewer, S.; Feng, W.; Huang, J.; Sullivan, S.; Williams, T. Wnt1-Cre-Mediated Deletion of AP-2alpha Causes Multiple Neural Crest-Related Defects. *Dev. Biol.* **2004**, *267*, 135–152. [CrossRef] [PubMed]
109. Luo, Y.; High, F.A.; Epstein, J.A.; Radice, G.L. N-Cadherin is Required for Neural Crest Remodeling of the Cardiac Outflow Tract. *Dev. Biol.* **2006**, *299*, 517–528. [CrossRef] [PubMed]
110. Breau, M.A.; Pietri, T.; Stemmler, M.P.; Thiery, J.P.; Weston, J.A. A Nonneural Epithelial Domain of Embryonic Cranial Neural Folds Gives Rise to Ectomesenchyme. *Proc. Natl. Acad. Sci. USA* **2008**, *105*, 7750–7755. [CrossRef] [PubMed]
111. Schorle, H.; Meier, P.; Buchert, M.; Jaenisch, R.; Mitchell, P.J. Transcription Factor AP-2 Essential for Cranial Closure and Craniofacial Development. *Nature* **1996**, *381*, 235–238. [CrossRef] [PubMed]
112. Zhang, J.; Hagopian-Donaldson, S.; Serbedzija, G.; Elsemore, J.; Plehn-Dujowich, D.; McMahon, A.P.; Flavell, R.A.; Williams, T. Neural Tube, Skeletal and Body Wall Defects in Mice Lacking Transcription Factor AP-2. *Nature* **1996**, *381*, 238–241. [CrossRef] [PubMed]
113. Radice, G.L.; Rayburn, H.; Matsunami, H.; Knudsen, K.A.; Takeichi, M.; Hynes, R.O. Developmental Defects in Mouse Embryos Lacking N-Cadherin. *Dev. Biol.* **1997**, *181*, 64–78. [CrossRef] [PubMed]
114. Luo, Y.; Ferreira-Cornwell, M.; Baldwin, H.; Kostetskii, I.; Lenox, J.; Lieberman, M.; Radice, G. Rescuing the N-Cadherin Knockout by Cardiac-Specific Expression of N- Or E-Cadherin. *Development* **2001**, *128*, 459–469. [PubMed]
115. Kinder, S.J.; Tsang, T.E.; Quinlan, G.A.; Hadjantonakis, A.K.; Nagy, A.; Tam, P.P. The Orderly Allocation of Mesodermal Cells to the Extraembryonic Structures and the Anteroposterior Axis during Gastrulation of the Mouse Embryo. *Development* **1999**, *126*, 4691–4701. [PubMed]
116. Onai, T.; Aramaki, T.; Inomata, H.; Hirai, T.; Kuratani, S. Ancestral Mesodermal Reorganization and Evolution of the Vertebrate Head. *Zool. Lett.* **2015**, *1*, 29. [CrossRef] [PubMed]
117. Zohn, I.E.; Sarkar, A.A. Does the Cranial Mesenchyme Contribute to Neural Fold Elevation during Neurulation? *Birth Defects Res. A Clin. Mol. Teratol.* **2012**, *94*, 841–848. [CrossRef] [PubMed]
118. Copp, A.J. Neurulation in the Cranial Region—Normal and Abnormal. *J. Anat.* **2005**, *207*, 623–635. [CrossRef] [PubMed]
119. Morris-Wiman, J.; Brinkley, L.L. Changes in Mesenchymal Cell and Hyaluronate Distribution Correlate with in vivo Elevation of the Mouse Mesencephalic Neural Folds. *Anat. Rec.* **1990**, *226*, 383–395. [CrossRef] [PubMed]
120. Chen, Z.F.; Behringer, R.R. Twist is Required in Head Mesenchyme for Cranial Neural Tube Morphogenesis. *Genes Dev.* **1995**, *9*, 686–699. [CrossRef] [PubMed]
121. Bildsoe, H.; Loebel, D.A.; Jones, V.J.; Hor, A.C.; Braithwaite, A.W.; Chen, Y.T.; Behringer, R.R.; Tam, P.P. The Mesenchymal Architecture of the Cranial Mesoderm of Mouse Embryos is Disrupted by the Loss of Twist1 Function. *Dev. Biol.* **2013**, *374*, 295–307. [CrossRef] [PubMed]
122. Wolf, C.; Thisse, C.; Stoetzel, C.; Thisse, B.; Gerlinger, P.; Perrin-Schmitt, F. The M-Twist Gene of Mus is Expressed in Subsets of Mesodermal Cells and is Closely Related to the Xenopus X-Twi and the Drosophila Twist Genes. *Dev. Biol.* **1991**, *143*, 363–373. [CrossRef]
123. Soo, K.; O'Rourke, M.P.; Khoo, P.L.; Steiner, K.A.; Wong, N.; Behringer, R.R.; Tam, P.P. Twist Function is Required for the Morphogenesis of the Cephalic Neural Tube and the Differentiation of the Cranial Neural Crest Cells in the Mouse Embryo. *Dev. Biol.* **2002**, *247*, 251–270. [CrossRef] [PubMed]
124. Zhao, Q.; Behringer, R.R.; de Crombrugghe, B. Prenatal Folic Acid Treatment Suppresses Acrania and Meroanencephaly in Mice Mutant for the Cart1 Homeobox Gene. *Nat. Genet.* **1996**, *13*, 275–283. [CrossRef] [PubMed]
125. ten Berge, D.; Brouwer, A.; el Bahi, S.; Guenet, J.L.; Robert, B.; Meijlink, F. Mouse Alx3: An Aristaless-Like Homeobox Gene Expressed during Embryogenesis in Ectomesenchyme and Lateral Plate Mesoderm. *Dev. Biol.* **1998**, *199*, 11–25. [CrossRef] [PubMed]
126. Lakhwani, S.; Garcia-Sanz, P.; Vallejo, M. Alx3-Deficient Mice Exhibit Folic Acid-Resistant Craniofacial Midline and Neural Tube Closure Defects. *Dev. Biol.* **2010**, *344*, 869–880. [CrossRef] [PubMed]

127. Reid, B.S.; Sargent, T.D.; Williams, T. Generation and Characterization of a Novel Neural Crest Marker Allele, Inka1-LacZ, Reveals a Role for Inka1 in Mouse Neural Tube Closure. *Dev. Dyn.* **2010**, *239*, 1188–1196. [CrossRef] [PubMed]
128. Baskaran, Y.; Ang, K.C.; Anekal, P.V.; Chan, W.L.; Grimes, J.M.; Manser, E.; Robinson, R.C. An in Cellulo-Derived Structure of PAK4 in Complex with its Inhibitor Inka1. *Nat. Commun.* **2015**, *6*, 8681. [CrossRef] [PubMed]
129. Baranek, C.; Atanasoski, S. Modulating Epigenetic Mechanisms: The Diverse Functions of Ski during Cortical Development. *Epigenetics* **2012**, *7*, 676–679. [CrossRef] [PubMed]
130. Berk, M.; Desai, S.Y.; Heyman, H.C.; Colmenares, C. Mice Lacking the Ski Proto-Oncogene have Defects in Neurulation, Craniofacial, Patterning, and Skeletal Muscle Development. *Genes Dev.* **1997**, *11*, 2029–2039. [CrossRef] [PubMed]
131. Lyons, G.E.; Micales, B.K.; Herr, M.J.; Horrigan, S.K.; Namciu, S.; Shardy, D.; Stavnezer, E. Protooncogene c-Ski is Expressed in both Proliferating and Postmitotic Neuronal Populations. *Dev. Dyn.* **1994**, *201*, 354–365. [CrossRef] [PubMed]
132. Bultman, S.; Gebuhr, T.; Yee, D.; La Mantia, C.; Nicholson, J.; Gilliam, A.; Randazzo, F.; Metzger, D.; Chambon, P.; Crabtree, G.; et al. A Brg1 Null Mutation in the Mouse Reveals Functional Differences among Mammalian SWI/SNF Complexes. *Mol. Cell* **2000**, *6*, 1287–1295. [CrossRef]
133. Randazzo, F.M.; Khavari, P.; Crabtree, G.; Tamkun, J.; Rossant, J. Brg1: A Putative Murine Homologue of the Drosophila Brahma Gene, a Homeotic Gene Regulator. *Dev. Biol.* **1994**, *161*, 229–242. [CrossRef] [PubMed]
134. Zohn, I.E.; Anderson, K.V.; Niswander, L. The Hectd1 Ubiquitin Ligase is Required for Development of the Head Mesenchyme and Neural Tube Closure. *Dev. Biol.* **2007**, *306*, 208–221. [CrossRef] [PubMed]
135. Sarkar, A.A.; Zohn, I.E. Hectd1 Regulates Intracellular Localization and Secretion of Hsp90 to Control Cellular Behavior of the Cranial Mesenchyme. *J. Cell Biol.* **2012**, *196*, 789–800. [CrossRef] [PubMed]
136. Banting, G.S.; Barak, O.; Ames, T.M.; Burnham, A.C.; Kardel, M.D.; Cooch, N.S.; Davidson, C.E.; Godbout, R.; McDermid, H.E.; Shiekhattar, R. CECR2, a Protein Involved in Neurulation, Forms a Novel Chromatin Remodeling Complex with SNF2L. *Hum. Mol. Genet.* **2005**, *14*, 513–524. [CrossRef] [PubMed]
137. Dawe, C.E.; Kooistra, M.K.; Fairbridge, N.A.; Pisio, A.C.; McDermid, H.E. Role of Chromatin Remodeling Gene Cecr2 in Neurulation and Inner Ear Development. *Dev. Dyn.* **2011**, *240*, 372–383. [CrossRef] [PubMed]
138. Fairbridge, N.A.; Dawe, C.E.; Niri, F.H.; Kooistra, M.K.; King-Jones, K.; McDermid, H.E. Cecr2 Mutations Causing Exencephaly Trigger Misregulation of mesenchymal/ectodermal Transcription Factors. *Birth Defects Res. A Clin. Mol. Teratol.* **2010**, *88*, 619–625. [CrossRef] [PubMed]
139. Nishi, T.; Nakano, R. First-Trimester Diagnosis of Exencephaly by Transvaginal Ultrasonography. *J. Ultrasound Med.* **1994**, *13*, 149–151. [CrossRef] [PubMed]
140. Chatzipapas, I.K.; Whitlow, B.J.; Economides, D.L. The 'Mickey Mouse' Sign and the Diagnosis of Anencephaly in Early Pregnancy. *Ultrasound Obstet. Gynecol.* **1999**, *13*, 196–199. [CrossRef] [PubMed]
141. Grego-Bessa, J.; Hildebrand, J.; Anderson, K.V. Morphogenesis of the Mouse Neural Plate Depends on Distinct Roles of Cofilin 1 in Apical and Basal Epithelial Domains. *Development* **2015**, *142*, 1305–1314. [CrossRef] [PubMed]
142. Gurniak, C.B.; Perlas, E.; Witke, W. The Actin Depolymerizing Factor n-Cofilin is Essential for Neural Tube Morphogenesis and Neural Crest Cell Migration. *Dev. Biol.* **2005**, *278*, 231–241. [CrossRef] [PubMed]
143. Cox, B.; Hadjantonakis, A.K.; Collins, J.E.; Magee, A.I. Cloning and Expression Throughout Mouse Development of mfat1, a Homologue of the Drosophila Tumour Suppressor Gene Fat. *Dev. Dyn.* **2000**, *217*, 233–240. [CrossRef]
144. Tanoue, T.; Takeichi, M. Mammalian Fat1 Cadherin Regulates Actin Dynamics and Cell-Cell Contact. *J. Cell Biol.* **2004**, *165*, 517–528. [CrossRef] [PubMed]
145. Badouel, C.; Zander, M.A.; Liscio, N.; Bagherie-Lachidan, M.; Sopko, R.; Coyaud, E.; Raught, B.; Miller, F.D.; McNeill, H. Fat1 Interacts with Fat4 to Regulate Neural Tube Closure, Neural Progenitor Proliferation and Apical Constriction during Mouse Brain Development. *Development* **2015**, *142*, 2781–2791. [CrossRef] [PubMed]
146. Barbera, J.P.; Rodriguez, T.A.; Greene, N.D.; Weninger, W.J.; Simeone, A.; Copp, A.J.; Beddington, R.S.; Dunwoodie, S. Folic Acid Prevents Exencephaly in Cited2 Deficient Mice. *Hum. Mol. Genet.* **2002**, *11*, 283–293. [CrossRef] [PubMed]

147. Bamforth, S.D.; Braganca, J.; Eloranta, J.J.; Murdoch, J.N.; Marques, F.I.; Kranc, K.R.; Farza, H.; Henderson, D.J.; Hurst, H.C.; Bhattacharya, S. Cardiac Malformations, Adrenal Agenesis, Neural Crest Defects and Exencephaly in Mice Lacking Cited2, a New Tfap2 Co-Activator. *Nat. Genet.* **2001**, *29*, 469–474. [CrossRef] [PubMed]
148. Kranc, K.R.; Oliveira, D.V.; Armesilla-Diaz, A.; Pacheco-Leyva, I.; Catarina Matias, A.; Luisa Escapa, A.; Subramani, C.; Wheadon, H.; Trindade, M.; Nichols, J.; et al. Acute Loss of Cited2 Impairs Nanog Expression and Decreases Self-Renewal of Mouse Embryonic Stem Cells. *Stem Cells* **2015**, *33*, 699–712. [CrossRef] [PubMed]
149. Dunwoodie, S.L.; Rodriguez, T.A.; Beddington, R.S. Msg1 and Mrg1, Founding Members of a Gene Family, show Distinct Patterns of Gene Expression during Mouse Embryogenesis. *Mech. Dev.* **1998**, *72*, 27–40. [CrossRef]
150. Sakaki-Yumoto, M.; Kobayashi, C.; Sato, A.; Fujimura, S.; Matsumoto, Y.; Takasato, M.; Kodama, T.; Aburatani, H.; Asashima, M.; Yoshida, N.; et al. The Murine Homolog of SALL4, a Causative Gene in Okihiro Syndrome, is Essential for Embryonic Stem Cell Proliferation, and Cooperates with Sall1 in Anorectal, Heart, Brain and Kidney Development. *Development* **2006**, *133*, 3005–3013. [CrossRef] [PubMed]
151. Buck, A.; Kispert, A.; Kohlhase, J. Embryonic Expression of the Murine Homologue of SALL1, the Gene Mutated in Townes–Brocks Syndrome. *Mech. Dev.* **2001**, *104*, 143–146. [CrossRef]
152. Bohm, J.; Buck, A.; Borozdin, W.; Mannan, A.U.; Matysiak-Scholze, U.; Adham, I.; Schulz-Schaeffer, W.; Floss, T.; Wurst, W.; Kohlhase, J.; et al. Sall1, sall2, and sall4 are Required for Neural Tube Closure in Mice. *Am. J. Pathol.* **2008**, *173*, 1455–1463. [CrossRef] [PubMed]
153. Kiefer, S.M.; Ohlemiller, K.K.; Yang, J.; McDill, B.W.; Kohlhase, J.; Rauchman, M. Expression of a Truncated Sall1 Transcriptional Repressor is Responsible for Townes-Brocks Syndrome Birth Defects. *Hum. Mol. Genet.* **2003**, *12*, 2221–2227. [CrossRef] [PubMed]
154. Ohmura, T.; Shioi, G.; Hirano, M.; Aizawa, S. Neural Tube Defects by NUAK1 and NUAK2 Double Mutation. *Dev. Dyn.* **2012**, *241*, 1350–1364. [CrossRef] [PubMed]
155. Hildebrand, J.D.; Soriano, P. Shroom, a PDZ Domain-Containing Actin-Binding Protein, is Required for Neural Tube Morphogenesis in Mice. *Cell* **1999**, *99*, 485–497. [CrossRef]
156. Sousa-Nunes, R.; Rana, A.A.; Kettleborough, R.; Brickman, J.M.; Clements, M.; Forrest, A.; Grimmond, S.; Avner, P.; Smith, J.C.; Dunwoodie, S.L.; et al. Characterizing Embryonic Gene Expression Patterns in the Mouse using Nonredundant Sequence-Based Selection. *Genome Res.* **2003**, *13*, 2609–2620. [CrossRef] [PubMed]
157. McGreevy, E.M.; Vijayraghavan, D.; Davidson, L.A.; Hildebrand, J.D. Shroom3 Functions Downstream of Planar Cell Polarity to Regulate Myosin II Distribution and Cellular Organization during Neural Tube Closure. *Biol. Open* **2015**, *4*, 186–196. [CrossRef] [PubMed]
158. Yang, L.; Zhang, H.; Hu, G.; Wang, H.; Abate-Shen, C.; Shen, M.M. An Early Phase of Embryonic Dlx5 Expression Defines the Rostral Boundary of the Neural Plate. *J. Neurosci.* **1998**, *18*, 8322–8330. [CrossRef] [PubMed]
159. Depew, M.J.; Liu, J.K.; Long, J.E.; Presley, R.; Meneses, J.J.; Pedersen, R.A.; Rubenstein, J.L. Dlx5 Regulates Regional Development of the Branchial Arches and Sensory Capsules. *Development* **1999**, *126*, 3831–3846. [PubMed]
160. Lee, J.; Corcoran, A.; Han, M.; Gardiner, D.M.; Muneoka, K. Dlx5 and Msx2 Regulate Mouse Anterior Neural Tube Closure through ephrinA5-EphA7. *Dev. Growth Differ.* **2013**, *55*, 341–349. [CrossRef] [PubMed]
161. Ishii, M.; Han, J.; Yen, H.Y.; Sucov, H.M.; Chai, Y.; Maxson, R.E., Jr. Combined Deficiencies of Msx1 and Msx2 Cause Impaired Patterning and Survival of the Cranial Neural Crest. *Development* **2005**, *132*, 4937–4950. [CrossRef] [PubMed]
162. Trainor, P.A.; Sobieszczuk, D.; Wilkinson, D.; Krumlauf, R. Signalling between the Hindbrain and Paraxial Tissues Dictates Neural Crest Migration Pathways. *Development* **2002**, *129*, 433–442. [PubMed]
163. Han, J.; Ishii, M.; Bringas, P., Jr.; Maas, R.L.; Maxson, R.E., Jr.; Chai, Y. Concerted Action of Msx1 and Msx2 in Regulating Cranial Neural Crest Cell Differentiation during Frontal Bone Development. *Mech. Dev.* **2007**, *124*, 729–745. [CrossRef] [PubMed]
164. Arvanitis, D.N.; Behar, A.; Tryoen-Toth, P.; Bush, J.O.; Jungas, T.; Vitale, N.; Davy, A. Ephrin B1 Maintains Apical Adhesion of Neural Progenitors. *Development* **2013**, *140*, 2082–2092. [CrossRef] [PubMed]

165. Heimsath, E.G., Jr.; Yim, Y.I.; Mustapha, M.; Hammer, J.A.; Cheney, R.E. Myosin-X Knockout is Semi-Lethal and Demonstrates that Myosin-X Functions in Neural Tube Closure, Pigmentation, Hyaloid Vasculature Regression, and Filopodia Formation. *Sci. Rep.* **2017**, *7*, 17354. [CrossRef] [PubMed]
166. Juriloff, D.M.; Harris, M.J. Hypothesis: The Female Excess in Cranial Neural Tube Defects Reflects an Epigenetic Drag of the Inactivating X Chromosome on the Molecular Mechanisms of Neural Fold Elevation. *Birth Defects Res. A Clin. Mol. Teratol.* **2012**, *94*, 849–855. [CrossRef] [PubMed]
167. Poletta, F.A.; Rittler, M.; Saleme, C.; Campana, H.; Gili, J.A.; Pawluk, M.S.; Gimenez, L.G.; Cosentino, V.R.; Castilla, E.E.; Lopez-Camelo, J.S. Neural Tube Defects: Sex Ratio Changes After Fortification with Folic Acid. *PLoS ONE* **2018**, *13*, e0193127. [CrossRef] [PubMed]
168. Davidson, C.E.; Li, Q.; Churchill, G.A.; Osborne, L.R.; McDermid, H.E. Modifier Locus for Exencephaly in Cecr2 Mutant Mice is Syntenic to the 10q25.3 Region Associated with Neural Tube Defects in Humans. *Physiol. Genom.* **2007**, *31*, 244–251. [CrossRef] [PubMed]
169. Santander, N.G.; Contreras-Duarte, S.; Awad, M.F.; Lizama, C.; Passalacqua, I.; Rigotti, A.; Busso, D. Developmental Abnormalities in Mouse Embryos Lacking the HDL Receptor SR-BI. *Hum. Mol. Genet.* **2013**, *22*, 1086–1096. [CrossRef] [PubMed]
170. Pai, Y.J.; Leung, K.Y.; Savery, D.; Hutchin, T.; Prunty, H.; Heales, S.; Brosnan, M.E.; Brosnan, J.T.; Copp, A.J.; Greene, N.D. Glycine Decarboxylase Deficiency Causes Neural Tube Defects and Features of Non-Ketotic Hyperglycinemia in Mice. *Nat. Commun.* **2015**, *6*, 6388. [CrossRef] [PubMed]
171. Zvetkova, I.; Apedaile, A.; Ramsahoye, B.; Mermoud, J.E.; Crompton, L.A.; John, R.; Feil, R.; Brockdorff, N. Global Hypomethylation of the Genome in XX Embryonic Stem Cells. *Nat. Genet.* **2005**, *37*, 1274–1279. [CrossRef] [PubMed]
172. Dobbs, K.B.; Rodriguez, M.; Sudano, M.J.; Ortega, M.S.; Hansen, P.J. Dynamics of DNA Methylation during Early Development of the Preimplantation Bovine Embryo. *PLoS ONE* **2013**, *8*, e66230. [CrossRef] [PubMed]
173. Vallot, C.; Ouimette, J.F.; Rougeulle, C. Establishment of X Chromosome Inactivation and Epigenomic Features of the Inactive X Depend on Cellular Contexts. *Bioessays* **2016**, *38*, 869–880. [CrossRef] [PubMed]
174. Carrel, L.; Brown, C.J. When the Lyon(Ized Chromosome) Roars: Ongoing Expression from an Inactive X Chromosome. *Philos. Trans. R. Soc. Lond. B Biol. Sci.* **2017**, *372*, 372. [CrossRef] [PubMed]
175. Wijchers, P.J.; Festenstein, R.J. Epigenetic Regulation of Autosomal Gene Expression by Sex Chromosomes. *Trends Genet.* **2011**, *27*, 132–140. [CrossRef] [PubMed]
176. Blewitt, M.E.; Vickaryous, N.K.; Hemley, S.J.; Ashe, A.; Bruxner, T.J.; Preis, J.I.; Arkell, R.; Whitelaw, E. An *N*-Ethyl-*N*-Nitrosourea Screen for Genes Involved in Variegation in the Mouse. *Proc. Natl. Acad. Sci. USA* **2005**, *102*, 7629–7634. [CrossRef] [PubMed]
177. Wijchers, P.J.; Yandim, C.; Panousopoulou, E.; Ahmad, M.; Harker, N.; Saveliev, A.; Burgoyne, P.S.; Festenstein, R. Sexual Dimorphism in Mammalian Autosomal Gene Regulation is Determined Not Only by Sry but by Sex Chromosome Complement as Well. *Dev. Cell.* **2010**, *19*, 477–484. [CrossRef] [PubMed]
178. Evans, J.A. Comment on Changes in Sex Ratio in Neural Tube Defects since Food Fortification with Folic Acid: Re "Hypothesis: The Female Excess in Cranial Neural Tube Defects Reflects an Epigenetic Drag of the Inactivating X Chromosome on the Molecular Mechanisms of Neural Tube Fold Elevation". *Birth Defects Res. A Clin. Mol. Teratol.* **2012**, *94*, 958. [CrossRef] [PubMed]
179. Greene, N.D.; Gerrelli, D.; van Straaten, H.W.; Copp, A.J. Abnormalities of Floor Plate, Notochord and Somite Differentiation in the Loop-Tail (Lp) Mouse: A Model of Severe Neural Tube Defects. *Mech. Dev.* **1998**, *73*, 59–72. [CrossRef]
180. Ybot-Gonzalez, P.; Savery, D.; Gerrelli, D.; Signore, M.; Mitchell, C.E.; Faux, C.H.; Greene, N.D.; Copp, A.J. Convergent Extension, Planar-Cell-Polarity Signalling and Initiation of Mouse Neural Tube Closure. *Development* **2007**, *134*, 789–799. [CrossRef] [PubMed]
181. Robinson, A.; Escuin, S.; Doudney, K.; Vekemans, M.; Stevenson, R.E.; Greene, N.D.; Copp, A.J.; Stanier, P. Mutations in the Planar Cell Polarity Genes CELSR1 and SCRIB are Associated with the Severe Neural Tube Defect Craniorachischisis. *Hum. Mutat.* **2012**, *33*, 440–447. [CrossRef] [PubMed]
182. Kirillova, I.; Novikova, I.; Auge, J.; Audollent, S.; Esnault, D.; Encha-Razavi, F.; Lazjuk, G.; Attie-Bitach, T.; Vekemans, M. Expression of the Sonic Hedgehog Gene in Human Embryos with Neural Tube Defects. *Teratology* **2000**, *61*, 347–354. [CrossRef]

183. Murdoch, J.N.; Doudney, K.; Paternotte, C.; Copp, A.J.; Stanier, P. Severe Neural Tube Defects in the Loop-Tail Mouse Result from Mutation of Lpp1, a Novel Gene Involved in Floor Plate Specification. *Hum. Mol. Genet.* **2001**, *10*, 2593–2601. [CrossRef] [PubMed]
184. Mahaffey, J.P.; Grego-Bessa, J.; Liem, K.F., Jr.; Anderson, K.V. Cofilin and Vangl2 Cooperate in the Initiation of Planar Cell Polarity in the Mouse Embryo. *Development* **2013**, *140*, 1262–1271. [CrossRef] [PubMed]
185. Murdoch, J.N.; Damrau, C.; Paudyal, A.; Bogani, D.; Wells, S.; Greene, N.D.; Stanier, P.; Copp, A.J. Genetic Interactions between Planar Cell Polarity Genes Cause Diverse Neural Tube Defects in Mice. *Dis. Model. Mech.* **2014**, *7*, 1153–1163. [CrossRef] [PubMed]
186. Doudney, K.; Ybot-Gonzalez, P.; Paternotte, C.; Stevenson, R.E.; Greene, N.D.; Moore, G.E.; Copp, A.J.; Stanier, P. Analysis of the Planar Cell Polarity Gene Vangl2 and its Co-Expressed Paralogue Vangl1 in Neural Tube Defect Patients. *Am. J. Med. Genet. A* **2005**, *136*, 90–92. [CrossRef] [PubMed]
187. Shi, Y.; Ding, Y.; Lei, Y.P.; Yang, X.Y.; Xie, G.M.; Wen, J.; Cai, C.Q.; Li, H.; Chen, Y.; Zhang, T.; et al. Identification of Novel Rare Mutations of DACT1 in Human Neural Tube Defects. *Hum. Mutat.* **2012**, *33*, 1450–1455. [CrossRef] [PubMed]
188. Wang, L.; Xiao, Y.; Tian, T.; Jin, L.; Lei, Y.; Finnell, R.H.; Ren, A. Digenic Variants of Planar Cell Polarity Genes in Human Neural Tube Defect Patients. *Mol. Genet. Metab.* **2018**, *124*, 94–100. [CrossRef] [PubMed]
189. Scott, J.M.; Kirke, P.N.; Weir, D.G. The Role of Nutrition in Neural Tube Defects. *Annu. Rev. Nutr.* **1990**, *10*, 277–295. [CrossRef] [PubMed]
190. MRC Vitamin Study Research Group. Prevention of Neural Tube Defects: Results of the Medical Research Council Vitamin Study. MRC Vitamin Study Research Group. *Lancet* **1991**, *338*, 131–137. [CrossRef]
191. Czeizel, A.E.; Dudas, I. Prevention of the First Occurrence of Neural-Tube Defects by Periconceptional Vitamin Supplementation. *N. Engl. J. Med.* **1992**, *327*, 1832–1835. [CrossRef] [PubMed]
192. Garrett, G.S.; Bailey, L.B. A Public Health Approach for Preventing Neural Tube Defects: Folic Acid Fortification and Beyond. *Ann. N. Y. Acad. Sci.* **2018**, *1414*, 47–58. [CrossRef] [PubMed]
193. Greene, N.D.; Leung, K.Y.; Copp, A.J. Inositol, Neural Tube Closure and the Prevention of Neural Tube Defects. *Birth Defects Res.* **2017**, *109*, 68–80. [CrossRef] [PubMed]
194. Kirke, P.N.; Molloy, A.M.; Daly, L.E.; Burke, H.; Weir, D.G.; Scott, J.M. Maternal Plasma Folate and Vitamin B12 are Independent Risk Factors for Neural Tube Defects. *Q. J. Med.* **1993**, *86*, 703–708. [PubMed]
195. Scott, J.M. Folate and Vitamin B12. *Proc. Nutr. Soc.* **1999**, *58*, 441–448. [CrossRef] [PubMed]
196. Heid, M.K.; Bills, N.D.; Hinrichs, S.H.; Clifford, A.J. Folate Deficiency Alone does Not Produce Neural Tube Defects in Mice. *J. Nutr.* **1992**, *122*, 888–894. [CrossRef] [PubMed]
197. Burgoon, J.M.; Selhub, J.; Nadeau, M.; Sadler, T.W. Investigation of the Effects of Folate Deficiency on Embryonic Development through the Establishment of a Folate Deficient Mouse Model. *Teratology* **2002**, *65*, 219–227. [CrossRef] [PubMed]
198. Blom, H.J.; Shaw, G.M.; den Heijer, M.; Finnell, R.H. Neural Tube Defects and Folate: Case Far from Closed. *Nat. Rev. Neurosci.* **2006**, *7*, 724–731. [CrossRef] [PubMed]
199. Doudney, K.; Grinham, J.; Whittaker, J.; Lynch, S.A.; Thompson, D.; Moore, G.E.; Copp, A.J.; Greene, N.D.; Stanier, P. Evaluation of Folate Metabolism Gene Polymorphisms as Risk Factors for Open and Closed Neural Tube Defects. *Am. J. Med. Genet. A* **2009**, *149A*, 1585–1589. [CrossRef] [PubMed]
200. Balashova, O.A.; Visina, O.; Borodinsky, L.N. Folate Action in Nervous System Development and Disease. *Dev. Neurobiol.* **2018**, *78*, 391–402. [CrossRef] [PubMed]
201. Harris, M.J. Insights into Prevention of Human Neural Tube Defects by Folic Acid Arising from Consideration of Mouse Mutants. *Birth Defects Res. A Clin. Mol. Teratol.* **2009**, *85*, 331–339. [CrossRef] [PubMed]
202. Gray, J.D.; Nakouzi, G.; Slowinska-Castaldo, B.; Dazard, J.E.; Rao, J.S.; Nadeau, J.H.; Ross, M.E. Functional Interactions between the LRP6 WNT Co-Receptor and Folate Supplementation. *Hum. Mol. Genet.* **2010**, *19*, 4560–4572. [CrossRef] [PubMed]
203. Fleming, A.; Copp, A.J. Embryonic Folate Metabolism and Mouse Neural Tube Defects. *Science* **1998**, *280*, 2107–2109. [CrossRef] [PubMed]
204. Greene, N.D.; Massa, V.; Copp, A.J. Understanding the Causes and Prevention of Neural Tube Defects: Insights from the Splotch Mouse Model. *Birth Defects Res. A Clin. Mol. Teratol.* **2009**, *85*, 322–330. [CrossRef] [PubMed]
205. Toriyama, M.; Toriyama, M.; Wallingford, J.B.; Finnell, R.H. Folate-Dependent Methylation of Septins Governs Ciliogenesis during Neural Tube Closure. *FASEB J.* **2017**, *31*, 3622–3635. [CrossRef] [PubMed]

206. Saitsu, H.; Ishibashi, M.; Nakano, H.; Shiota, K. Spatial and Temporal Expression of Folate-Binding Protein 1 (Fbp1) is Closely Associated with Anterior Neural Tube Closure in Mice. *Dev. Dyn.* **2003**, *226*, 112–117. [CrossRef] [PubMed]
207. Saitsu, H. Folate Receptors and Neural Tube Closure. *Congenit. Anom.* **2017**, *57*, 130–133. [CrossRef] [PubMed]
208. Tang, L.S.; Finnell, R.H. Neural and Orofacial Defects in Folp1 Knockout Mice [Corrected]. *Birth Defects Res. A Clin. Mol. Teratol.* **2003**, *67*, 209–218. [CrossRef] [PubMed]
209. Ishida, M.; Cullup, T.; Boustred, C.; James, C.; Docker, J.; English, C.; GOSgene; Lench, N.; Copp, A.J.; Moore, G.E.; et al. A Targeted Sequencing Panel Identifies Rare Damaging Variants in Multiple Genes in the Cranial Neural Tube Defect, Anencephaly. *Clin. Genet.* **2018**, *93*, 870–879. [CrossRef] [PubMed]
210. O'Byrne, M.R.; Au, K.S.; Morrison, A.C.; Lin, J.I.; Fletcher, J.M.; Ostermaier, K.K.; Tyerman, G.H.; Doebel, S.; Northrup, H. Association of Folate Receptor (FOLR1, FOLR2, FOLR3) and Reduced Folate Carrier (SLC19A1) Genes with Meningomyelocele. *Birth Defects Res. A Clin. Mol. Teratol.* **2010**, *88*, 689–694. [CrossRef] [PubMed]
211. Kur, E.; Mecklenburg, N.; Cabrera, R.M.; Willnow, T.E.; Hammes, A. LRP2 Mediates Folate Uptake in the Developing Neural Tube. *J. Cell Sci.* **2014**, *127*, 2261–2268. [CrossRef] [PubMed]
212. Sabatino, J.A.; Stokes, B.A.; Zohn, I.E. Prevention of Neural Tube Defects in Lrp2 Mutant Mouse Embryos by Folic Acid Supplementation. *Birth Defects Res.* **2017**, *109*, 16–26. [CrossRef] [PubMed]
213. Boshnjaku, V.; Shim, K.W.; Tsurubuchi, T.; Ichi, S.; Szany, E.V.; Xi, G.; Mania-Farnell, B.; McLone, D.G.; Tomita, T.; Mayanil, C.S. Nuclear Localization of Folate Receptor Alpha: A New Role as a Transcription Factor. *Sci. Rep.* **2012**, *2*, 980. [CrossRef] [PubMed]
214. Balashova, O.A.; Visina, O.; Borodinsky, L.N. Folate Receptor 1 is Necessary for Neural Plate Cell Apical Constriction during Xenopus Neural Tube Formation. *Development* **2017**, *144*, 1518–1530. [CrossRef] [PubMed]
215. Narisawa, A.; Komatsuzaki, S.; Kikuchi, A.; Niihori, T.; Aoki, Y.; Fujiwara, K.; Tanemura, M.; Hata, A.; Suzuki, Y.; Relton, C.L.; et al. Mutations in Genes Encoding the Glycine Cleavage System Predispose to Neural Tube Defects in Mice and Humans. *Hum. Mol. Genet.* **2012**, *21*, 1496–1503. [CrossRef] [PubMed]
216. Momb, J.; Lewandowski, J.P.; Bryant, J.D.; Fitch, R.; Surman, D.R.; Vokes, S.A.; Appling, D.R. Deletion of Mthfd1l Causes Embryonic Lethality and Neural Tube and Craniofacial Defects in Mice. *Proc. Natl. Acad. Sci. USA* **2013**, *110*, 549–554. [CrossRef] [PubMed]
217. Kim, J.; Lei, Y.; Guo, J.; Kim, S.E.; Wlodarczyk, B.J.; Cabrera, R.M.; Lin, Y.L.; Nilsson, T.K.; Zhang, T.; Ren, A.; et al. Formate Rescues Neural Tube Defects Caused by Mutations in Slc25a32. *Proc. Natl. Acad. Sci. USA* **2018**, *115*, 4690–4695. [PubMed]
218. Di Pietro, E.; Sirois, J.; Tremblay, M.L.; MacKenzie, R.E. Mitochondrial NAD-Dependent Methylenetetrahydrofolate Dehydrogenase-Methenyltetrahydrofolate Cyclohydrolase is Essential for Embryonic Development. *Mol. Cell. Biol.* **2002**, *22*, 4158–4166. [CrossRef] [PubMed]
219. Leung, K.Y.; De Castro, S.C.; Savery, D.; Copp, A.J.; Greene, N.D. Nucleotide Precursors Prevent Folic Acid-Resistant Neural Tube Defects in the Mouse. *Brain* **2013**, *136*, 2836–2841. [CrossRef] [PubMed]
220. Sudiwala, S.; De Castro, S.C.; Leung, K.Y.; Brosnan, J.T.; Brosnan, M.E.; Mills, K.; Copp, A.J.; Greene, N.D. Formate Supplementation Enhances Folate-Dependent Nucleotide Biosynthesis and Prevents Spina Bifida in a Mouse Model of Folic Acid-Resistant Neural Tube Defects. *Biochimie* **2016**, *126*, 63–70. [CrossRef] [PubMed]
221. Lalouel, J.M.; Rao, D.C.; Morton, N.E.; Elston, R.C. A Unified Model for Complex Segregation Analysis. *Am. J. Hum. Genet.* **1983**, *35*, 816–826. [PubMed]
222. Morton, N.E.; MacLean, C.J. Analysis of Family Resemblance. III. Complex Segregation of Quantitative Traits. *Am. J. Hum. Genet.* **1974**, *26*, 489–503. [PubMed]
223. Greene, N.D.E.; Stanier, P.; Copp, A.J. Genetics of Human Neural Tube Defects. *Hum. Mol. Genet.* **2009**, *18*, R113–R129. [CrossRef] [PubMed]
224. Lu, W.; Guzman, A.R.; Yang, W.; Chapa, C.J.; Shaw, G.M.; Greene, R.M.; Pisano, M.M.; Lammer, E.J.; Finnell, R.H.; Zhu, H. Genes Encoding Critical Transcriptional Activators for Murine Neural Tube Development and Human Spina Bifida: A Case-Control Study. *BMC Med. Genet.* **2010**, *11*, 141. [CrossRef] [PubMed]
225. Boyle, E.A.; Li, Y.; Pritchard, J.K. An Expanded View of Complex Traits: From Polygenic to Omnigenic. *Cell* **2017**, *169*, 1177–1186. [CrossRef] [PubMed]

226. Hill, W.G. Understanding and Using Quantitative Genetic Variation. *Philos. Trans. R. Soc. Lond. B Biol. Sci.* **2010**, *365*, 73–85. [CrossRef] [PubMed]
227. Lynch, S.A. Non-multifactorial Neural Tube Defects. *Am. J. Med. Genet. C Semin. Med. Genet.* **2005**, *135C*, 69–76. [CrossRef] [PubMed]
228. Juriloff, D.M.; Harris, M.J.; McMahon, A.P.; Carroll, T.J.; Lidral, A.C. Wnt9b is the Mutated Gene Involved in Multifactorial Nonsyndromic Cleft Lip with or without Cleft Palate in A/WySn Mice, as Confirmed by a Genetic Complementation Test. *Birth Defects Res. A Clin. Mol. Teratol.* **2006**, *76*, 574–579. [CrossRef] [PubMed]

7

It's about Time: Ossification Center Formation in C57BL/6 Mice from E12–E16

Kevin Flaherty [1,2,*] and Joan T. Richtsmeier [1]

[1] Department of Anthropology, Pennsylvania State University, University Park, PA 16802, USA; jta10@psu.edu
[2] Department of Biology, Augustana College, Rock Island, IL 61201, USA
[*] Correspondence: kevin.v.flaherty@gmail.com

Abstract: The establishment of precise, high-resolution temporal sequences for morphogenetic events in laboratory mice remains a vexing issue in developmental biology. Mouse embryos collected at the same period of gestation, even those from the same litter, show wide variation in individual levels of progress along their developmental trajectory. Therefore, age at harvest does not provide sufficient information about developmental progress to serve as the basis for forming samples for the study of rapidly or near-simultaneously occurring events such as the sequence of ossification center formation. Here, we generate two measures of individual developmental progress (developmental age) for a large sample of mouse embryos using crown–rump lengths that measures size, and limbstaging ages produced by the embryonic Mouse Ontogenetic Staging System (eMOSS) that measure shape. Using these measures, we establish fine-grained sequences of ossification center appearance for mouse embryos. The two measures of developmental progress generate slightly different sequences of ossification center formation demonstrating that despite their tight correlation throughout the developmental period, size and shape are aspects of form that are at least partially dissociated in development.

Keywords: ossification center; staging system; heterochrony; morphometrics; OSX; embryogenesis; mouse models; eMOSS

"Time is the longest distance between two places."
—Tennessee Williams, The Glass Menagerie

1. Introduction

The establishment of precise, high-resolution temporal sequences for morphogenetic events in laboratory mice, currently the most extensively used experimental mammal for studying human development and disease, remains a vexing issue in developmental biology. One of the main obstacles to establishing a precise developmental timeline for any developmental event is the high level of variability in developmental progress exhibited by embryos collected at similar days of gestation. Mouse embryos collected at the same embryonic age, even littermates, show disparities in morphological progress indicating variation in the rate of development [1–3]. As a result, for any given developmental process (e.g., palate elevation, heart septation, limb bud development, organogenesis, etc.), each mouse within a litter will exhibit a different amount of progress along the trajectory towards completion of that process. Similar developmental differences appear across litters collected at the same time, with some litters appearing older or younger in aggregate than other litters. Consequently, selecting specimens to analyze molecular or cellular details of rapidly occurring developmental phenomena based solely on age information determined from timed matings is problematic. Inter-individual differences in rates of development make it difficult to select individuals to represent precise stages of development, even when matings are conducted on short-duration timelines and when time-consuming and often subjective examination of developmental morphology is used to stage individuals.

One example of the difficulty in establishing precise temporal sequence is the initial establishment of the mineralized skeleton. Dermal bone is not preformed in cartilage, but forms directly through intramembranous ossification [4,5] that involves the differentiation and proliferation of osteoprogenitor cells directly from mesenchyme. The initiation of skeletogenesis of dermal bones of the cranial vault and facial skeleton (the dermatocranium) is marked by ossification center differentiation, while skeletogenesis of endochondrally forming bone (the chondrocranium and the majority of post cranial skeletal elements) is marked by the initiation of mineralization of cartilage models. Despite nearly a century of research into bone development using laboratory mice as a model, few researchers have precisely characterized the developmental timing and sequence of ossification center appearance in mice. Even the meticulous atlases of mouse development compiled by Theiler [6] and Kaufman [7] do not attempt to establish an absolute sequence of ossification center development in the mouse.

The production of an embryo represents an assembly of diverse developmental events within any specific developmental process whose sequence is obscured due to the rapidity and variability of mouse development. Skeletogenesis is a developmental process marked by a series of developmental events: the proliferation, condensation, and differentiation of mesenchymal cells to a chondrogenic or osteogenic fate [8], continued proliferation of those cells, and the production and mineralization of osteoid by osteoblasts. The appearance of ossification centers for cranial bones can occur rapidly, being separated by less than two hours of embryonic development, which confounds even short duration (commonly two hour) mating schemes (e.g., [9]). Moreover, mice within a litter can display a range of developmental morphologies that exceed what can be achieved during two hours of gestational development. Consequently, since each embryo grows at its own rate and represents a slightly different stage of development, we cannot determine the precise sequence and timing of many developmental events strictly by using timed conceptions [10]. Establishing a reliable read-out of the sequence of ossification center appearance requires a method for sorting embryos according to the relative level of development [2,3].

One potential method for sorting embryos is the use of crown–rump length (CRL), a simple and easy-to-collect metric of the maximum total length of an embryo, to order specimens according to size (Figure 1). Ferguson [11] found that CRL predicted the level of palatal development in rats better than embryo weight and other linear measurements. The precise nature of the relationship between size measured by CRL during embryogenesis and developmental progress is not clear, however. It is possible, despite the high correlation between size and developmental progress, that these two variables are autonomous for some developmental processes, and that CRL measurements cannot be used to generate relative sequences for some aspects of development. Despite the fact that embryos increase in size as their development progresses, the advancement of any individual developmental process (e.g., palate development) may not inherently depend on or be correlated with overall size increase.

Another solution to determining the precise sequence and timing of specific developmental events is the implementation of staging systems that use developmental morphology to establish a metric of the relative level of development of a mouse embryo (developmental age). Developmental age estimates provide information about the stage of development of an individual that cannot be determined from the amount of time elapsed between conception and embryo collection (age at harvest). Staging systems enable researchers to estimate a developmental age for each individual, sort individuals on the basis of developmental ages, and determine the precise sequence of a set of developmental events. There is no single method for determining developmental age for a mouse embryo, as any aspect of embryonic morphology that changes in synchrony with the level of overall development provides a potential metric for measuring developmental age. Thus, if methods are precise, staging systems based on the developmental morphology of the head [1] and those based on the limb [2,12,13] are equally valid for assigning developmental ages to embryos. However, some may be more or less appropriate for particular uses (e.g., head morphology-based staging systems may prove to be better predictors of palatal development relative to limb-based systems).

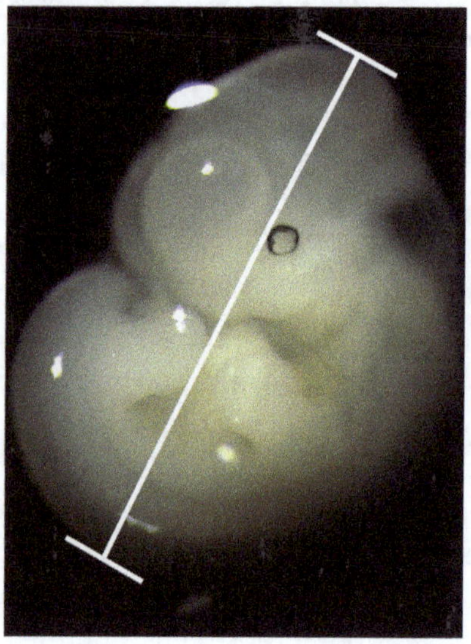

Figure 1. Crown-rump length measures the maximum total length of an embryo using a sagittal view.

Though many systems for staging mouse embryos exist [1,6,12,13], they are not routinely used because they are not quantitative and thus contain implicit bias, are low resolution (grouping embryos by embryonic days or half days) providing no better precision than time elapsed since fertilization, or are too time-consuming to be of practical use for most research programs. The embryonic Mouse Ontogenetic Staging System (eMOSS) [3] is a high-resolution staging method that uses the two-dimensional outline of an embryo's hindlimb to provide an estimate of the relative level of development for embryos collected on embryonic days 10 through 15 (E10–E15). eMOSS age estimates (hereafter called limbstaging ages) are scaled to developmental timing, so that limbstaging ages are provided as point estimates in embryonic days and hours, with an associated confidence interval given as a range of hours of development. eMOSS estimates can be generated rapidly, since the method is scale-invariant and only a picture of the hindlimb is required to establish a limbstaging age estimate. eMOSS is particularly well-suited to the task of establishing a sequence for ossification center appearance, as the process of ossification begins and a large number of ossification centers appear during the embryonic period covered by eMOSS.

The distinction between these two classes of systems for ordering embryos is that one measures the extent of development on the basis of a size metric (CRL) while the other uses a metric designed to represent shape (limb outline) to track development. In evolutionary developmental biology, changes in the timing or rate of developmental events, known as heterochrony [14–17], lead to changes in size and shape that affect the overall morphology of the organism. Previous investigations into evolution by heterochrony assume that growth and development, represented mathematically by change in size and change in shape, respectively, are evolutionarily dissociable processes [15–17]. A large literature rests on this assumption, though little evidence has been gathered on the dissociability of size and shape (as proxies for growth and development) among individuals of the same species.

Here, we use precise data collected from laboratory mice to determine the magnitude of variation exhibited by mouse embryos of the same age at harvest to determine variability in growth and development across our sample. We stage our samples of embryos using information on change in size (using crown–rump length) and change in shape (using eMOSS limbstaging ages) to establish a detailed sequence of cranial ossification center appearance in C57BL/6 and in Osx-GFP mice in which the expression machinery of the Osx (Osterix) gene drives expression of a green fluorescent protein (GFP) reporter gene. We examine the difference in sequences of ossification center appearance in mouse embryos using these two systems. Our use of these two systems enables the establishment of concrete timelines for the appearance of ossification centers using two different aspects of the initiation of ossification centers (osteoblast differentiation and bone mineralization), and provides an indication of the delay between these two events in the development of an ossification center. Our data also provide the first experimental evidence of the dissociability of size and shape in developing mice.

2. Results

2.1. *Analysis of Size Indicates Variation in Growth*

Mouse embryos ranging in age from E12 to E16 exhibit considerable variation in overall size. The mean, standard deviation, and range of CRL for each embryonic day (Table 1) are similar to those obtained by Otis and Brent [18], though the ranges of our data are larger, particularly for the oldest specimens. Analysis of variance (ANOVA) showed a significant litter effect across all ages, as well as significant associations of CRL with both harvesting age and limbstaging age (Table 2). The relationships between CRL and both age descriptors is expected, given the inevitable correspondence between the increase in size and increasing complexity of developmental anatomy. More surprising is the significant overlap in CRL ranges between consecutive embryonic days (Figure 2). The range for each day overlaps with the range of the preceding and following day, and average CRL for each day corresponds roughly with the minimum CRL for the following day (e.g., the average for E13 is approximately the minimum for E14). This overlap in size measurements between embryos collected 24 h apart illustrates that the relationship between size and age is highly variable during this period of embryonic development.

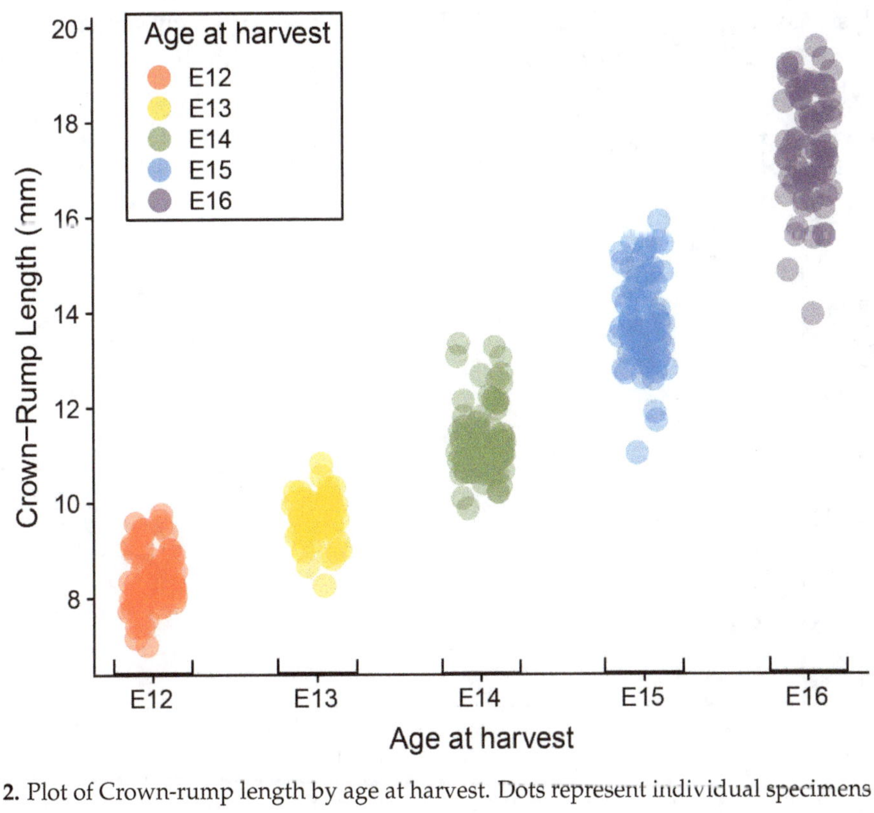

Figure 2. Plot of Crown-rump length by age at harvest. Dots represent individual specimens and are colored according to age at harvest.

Table 1. Summary statistics for crown–rump length.

Harvesting Age	Mean	SD	Min	Max	Range
E12	8.41	0.61	7.00	9.75	2.75
E13	9.61	0.43	8.26	10.33	2.07
E14	11.29	0.71	9.87	13.33	3.46
E15	13.70	0.94	11.03	15.46	4.43
E16	17.49	1.19	13.94	19.57	5.63

Table 2. ANOVA for Crown-Rump Length. Statistically significant *p*-values **bolded**.

	Df	Sum Sq	Mean Sq	F Value	Pr (>F)
Limbstaging Age	1	1062.36	1062.36	10261.27	**0.0000**
Harvesting Age	1	1.21	1.21	11.71	**0.0007**
Litter	38	70.83	1.86	18.00	**0.0000**
Uterus Location	1	0.02	0.02	0.22	0.6375
Residuals	238	24.64	0.10		

2.2. Analysis of Shape Indicates Variation in Development

Developmental shape, measured by the limbstaging ages generated by eMOSS, shows substantial variation during this period. This means that mice collected at the same period of gestation exhibit a substantial range of variation in limbstaging scores. Individual embryos show differences between limbstaging age and their expected age based on age at harvest greater than 24 h (Figures 3 and 4). For instance, one E12 specimen had a limbstaging age that was 24.5 h later than the expected age based on age at harvest, indicating that the individual had developed at such an advanced rate relative to a typically developing individual that its hindlimb shape was in the normal range for a mouse collected on E13. Analysis of variance of our shape metric (Table 3) shows a strikingly similar pattern to that of size, with age, size, and litter all showing significant relationships with shape.

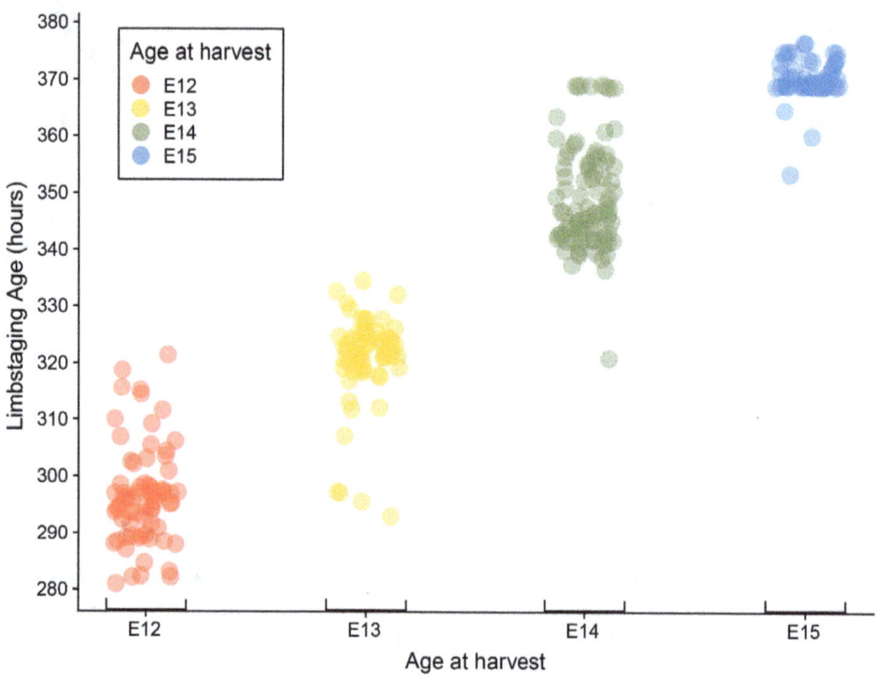

Figure 3. Plot of limbstaging age by age at harvest. Dots represent individual specimens and are colored according to age at harvest.

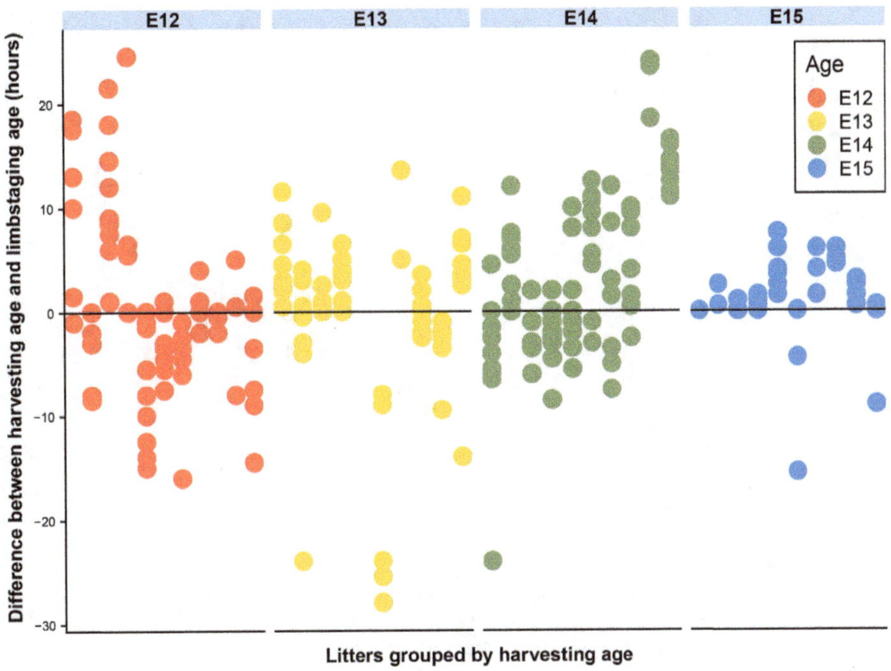

Figure 4. Deviation between age at harvest and limbstaging age for E12–E15. Each column represents a litter, with each data point representing an individual embryo. Colors represent age at harvest. Scores on the *y*-axis represent the difference between the expected age of an embryo based on age at harvest and limbstaging age for each individual.

Table 3. ANOVA for limbstaging Age. Statistically significant *p*-values **bolded**.

	Df	Sum Sq	Mean Sq	F Value	Pr (>F)
Limbstaging Age	1	8703.86	8703.86	612.08	**0.0000**
CRL	1	214,058.47	214,058.47	15,053.19	**0.0000**
Litter	38	7342.67	193.23	13.59	**0.0000**
Uterus Location	1	54.67	54.67	3.84	0.0511
Residuals	238	3384.39	14.22		

2.3. Relationship between Size and Shape

There is a generally strong positive relationship between CRL and limbstaging age when all data are considered ($R^2 = 0.92$), though the correlation between these variables for each embryonic day is often much lower and exhibits significant daily variation, with R^2 values ranging between 0.53 and 0.8 (Table 4). It is notable, however, that the spread of variation for CRL and limbstaging age appears to move in opposite directions, with limbstaging age being relatively variable for younger mice in the sample and diminishing on days E14 and E15. CRLs, on the other hand, are more tightly clustered in the younger mice, but show increased variation in the older mice (Figure 5).

Table 4. Pearson's correlation values for Crown-Rump Length and Developmental Age.

Harvesting Age	Correlation	R^2
E12	0.89	0.79
E13	0.73	0.53
E14	0.90	0.80
E15	0.84	0.71
Overall	0.96	0.92

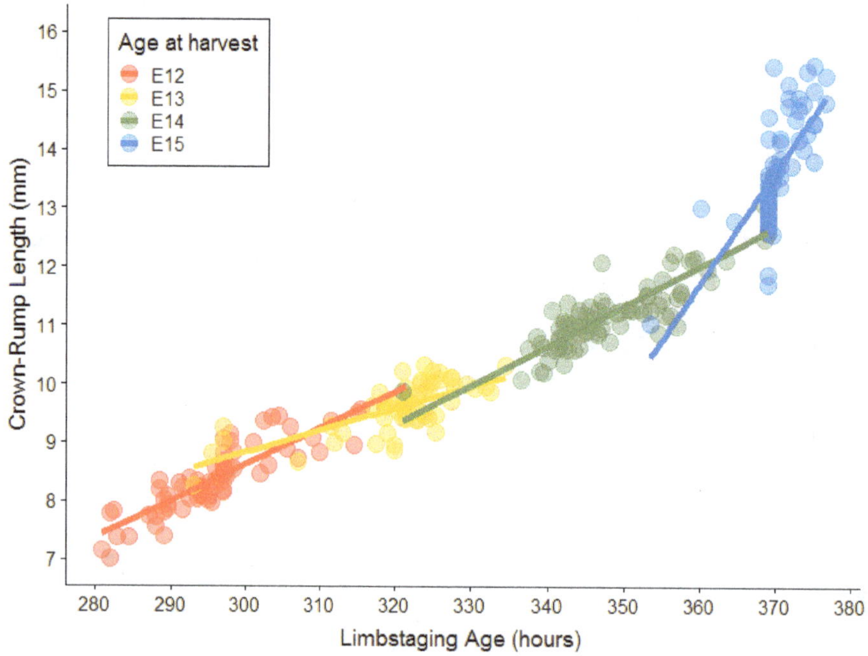

Figure 5. Graph of crown–rump length and limbstaging age. Dots represent individual specimens and are colored according to age at harvest.

2.4. Size, Shape, Age, and Bone Development

We collected Osx-GFP mice at six-hour intervals from E12.0–E15.0 to establish high-resolution timelines for the appearance of ossification centers indicated by osteoblast differentiation. C57BL/calcein mice were collected every six hours from E14.0–E16.0 to examine the timeline for the appearance of ossification centers based on bone mineralization. Figure 6 shows that arranging mice based on increasing CRL or limbstaging age produces a significantly different sequence of the appearance of ossification centers relative to the sequence based on age at harvest. This results from the fact that each embryo in a litter is by definition assigned the same age at harvest but may exhibit differences in CRL and limbstaging age. Of the three metrics (age at harvest, CRL, and limbstaging age), limbstaging age appears to provide the most reasonable sequence of ossification center expansion, as mice with less developed hindlimbs (younger limbstaging ages) show smaller ossification centers than those with more developed hindlimbs (older limbstaging ages). When specimens are ordered according to limbstaging age, ossification centers expand in an orderly sequence, with developmentally younger mice having smaller ossification centers and developmentally older mice having larger ones. CRL also provides a logical sequence of ossification center expansion, with smaller mice tending to have smaller ossification centers and larger mice having larger ones. Age at harvest, on the other hand, provides a poor sequence of ossification center expansion, as several mice appear to be developing significantly faster or slower than would be expected based on their age at harvest. In Figure 6, mice collected at E13.25 appear to be significantly younger than those collected at E13.00, as well as some of those collected at E12.75. It is notable that age at harvest appears to be a relatively coarse measure for development in terms of ossification center size compared to either limbstaging age or CRL. Age at harvest is the same for each embryo in a litter but does not reflect the fact that some litters may develop faster or slower than others. Limbstaging age and CRL reveal individual and inter-litter variation and order specimens accordingly. Thus, age at harvest cannot account for individual and inter-litter variation, resulting in the absence of an orderly sequence of ossification center expansion where ossification center size gets progressively larger over time.

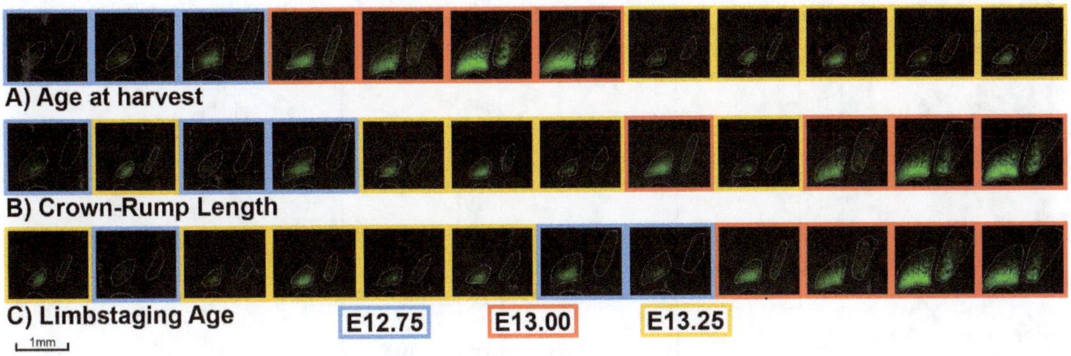

Figure 6. Appearance of frontal and parietal bones in Osx-GFP mice ordered by age at harvest, CRL, and limbstaging age. Each image shows a lateral view of the cranial vault of a left-facing Osx-GFP embryo, with Osx expression revealed by green fluorescent protein. The frontal and parietal bones and the eye are outlined in white to show the extent of Osx expression in each bone. The boxes outlining each specimen are colored according to age at harvest. Each age group is represented by a single litter. (**A**) specimens ordered according to age at harvest; (**B**) specimens ordered according to CRL; (**C**) specimens ordered according to limbstaging age.

Ordering mice based on CRL and based on limbstaging age produce similar, but not identical, ossification center appearance sequences for the Osx-GFP mice (Figure 7) and the C57BL/calcein mice (Figure 8). This suggests that size and shape reflect different aspects of developmental progress, despite their strong correlation. In addition, both CRL and limbstaging age provide higher resolution timelines for bone development than age at harvest. Embryos collected on each embryonic day during this period exhibit wide variation in the rate of development which limbstaging age and CRL can capture but age at harvest cannot.

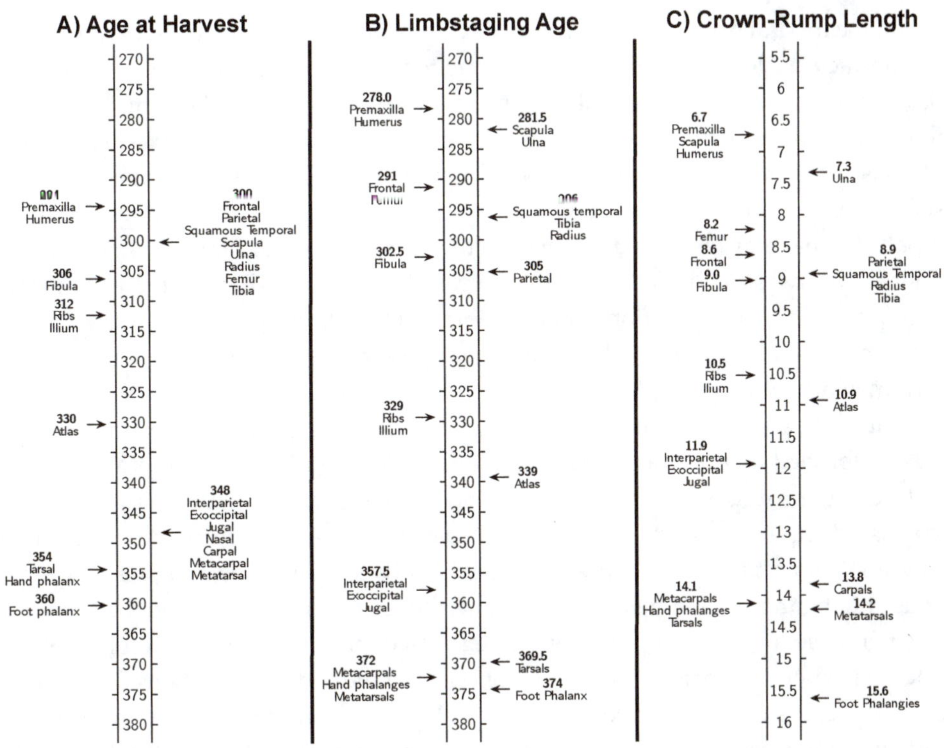

Figure 7. Timeline of ossification center appearance marked by osteoblast differentiation (expression of Osx) generated using Osx-GFP mice ordered according to: (**A**) age at harvest; (**B**) limbstaging age; and (**C**) CRL.

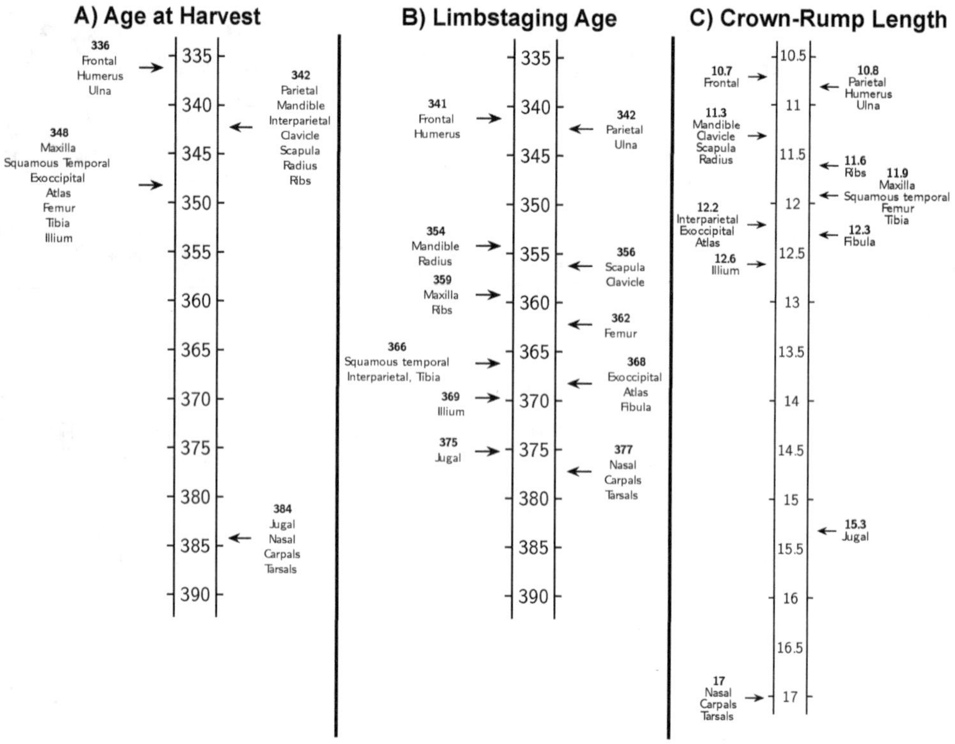

Figure 8. Timeline of ossification center appearance marked by mineralization of osteoid (marked by calcein staining) generated using C57BL/6-calcein mice ordered according to: (**A**) age at harvest; (**B**) limbstaging age; and (**C**) CRL.

3. Discussion

Establishing temporal priority for a series of developmental events using laboratory mice presents a significant obstacle in developmental research due to variation in the rate of development among individuals. A synchrony of developmental events in mice contributes to the complexity. Unravelling the complex nature of sequential and causally-linked events such as gene expression and cellular differentiation is crucial to understanding the developmental processes that comprise morphogenesis. To understand the initiation of a developmental event, it is often necessary to detect changes in gene expression patterns that may be limited in time and therefore difficult to capture using standard mating schemes. Here, we have shown that measures of growth (CRL) and development (limbstaging age) provide significantly higher resolution timelines for a relatively obvious and binary developmental event (ossification center appearance) than the age at harvest for mouse embryos. Moreover, our data establish a framework for breeding for selected timepoints for research into osteoblast differentiation and mineralization by providing more precise timelines of when specific ossification centers emerge. The methods established here are easily transferable for characterizing the temporal sequence of other events of interest. Limbstaging methodology is readily applicable to establishing temporal priority of gene expression sequences, patterns that shift rapidly during developmental events [19], yet are critical to understanding the nature of the relationship between genotype and phenotype.

Previous staging methodologies have not generally been used to provide the type of variation and developmental sequence data presented here for two main reasons. First, many of the staging systems lack the temporal precision needed to properly sort individual specimens. The Theiler staging system, for instance, describes the major developmental events on each embryonic day, but does not provide sufficient information to sort embryos according to specific levels of development within embryonic days [6]. Second, while other staging systems possess the temporal resolution to correctly sort specimens according to size or developmental morphology, they lack the ease of collecting CRL or eMOSS data. Most prominent among these is the craniofacial development staging system outlined by Miyake et al. [1] that established substages within the Theiler system for E11–E14. Over this range,

Miyake et al. found similarly high levels of variation for embryos collected at the same harvesting age. However, scoring the level of development for individual embryos using this system requires high-resolution images of the craniofacial region of embryos as well as the knowledge that enables a researcher to discriminate among subtle differences in craniofacial developmental anatomy. While CRL data has always been available to researchers using inbred mice, it has not been used to establish temporal sequences and variation data outside of a few notable exceptions [11,18].

In a broader context, we have demonstrated that, although size and shape are coupled during development, this relationship is not fixed. Size and shape are highly correlated over this sample whose chronological ages span E12–E15 ($R^2 = 0.92$), but the correlation varies across harvesting age groups (ranging from $R^2 = 0.53–0.8$). Additionally, we have shown that size and shape show contrasting patterns of variation during this age period for our metrics. Variation in CRL increases with age, while variation in shape (limbstaging age) decreases with age, supporting previous findings of other researchers who studied variation in CRL [18] and in craniofacial development [1].

Finally, the sequence of appearance of ossification centers summarizing data collected from Osx-GFP and C57BL/calcein mice establish slightly different ontogenetic trajectories based on whether the specimens are ordered according to size or shape, though both measures present a more reasonable sequence than that obtained by using harvesting age. This indicates that for research projects seeking to establish temporal priority in developmental events, the metric used to sort individual embryos can seriously impact the results and their interpretation. It is evident that no single "best metric" exists for sorting individuals according to their developmental morphology. Sorting our data according to both limbstaging age (shape) and CRL (size) resulted in highly detailed and reasonable timelines of ossification center appearance, but the timelines are not the same. There may be no unique way to precisely measure the level of development of an individual embryo, and developmental processes may each bear a certain level of dissociation. The detected pattern of dissociability may reflect underlying modular patterns of development that facilitate flexibility that contributes to morphological evolution (see an example in opercle development [20]). In short, each developmental event has its own timeline which may or may not correspond with the pace of other developmental events. Consequently, developmental events used to mark developmental progress of specific processes or of varying anatomical systems may not be appropriate across developmental time or across anatomical systems. Whether an internal (pertaining to the development of a particular system) or external (pertaining to general developmental principles) system is appropriate depends largely upon the research question being posed.

4. Methods

Litters of C57BL/6J embryos were generated by timed, overnight matings. The mice were bred twice a week, with males being introduced into the cages housing females at 3:00 p.m. and removed the following morning at 9:00 a.m. Pregnancy was verified by visually confirming the presence of a copulatory plug and weighing breeder females daily. Pregnant females were sacrificed at 9:00 a.m. via CO_2 exposure followed by cervical dislocation. Embryos were removed from their placental sac, washed with saline, and fixed in 4% PFA for 24 h. The embryos were then transferred to 0.01% Sodium azide solution in PBS for preservation. All mouse procedures employed in this research were approved by the Institutional Animal Care and Use Committee of the Pennsylvania State University (IACUC46558, IBC46590).

To test the dissociability of size and shape, a total of 356 C57BL/6J embryos were collected at 24-h intervals for embryonic days 12–16. Ten litters were bred for each embryonic day, with sample sizes for each day ranging between 65 and 86 individual embryos (Table 5).

Table 5. Sample sizes for each embryological age. Each embryo was collected at 9:00 a.m. on the designated embryological day.

Age	E12	E13	E14	E15	E16
N=	71	65	86	67	67

To establish temporal sequence for ossification center development, two lines were used. Litters of Osx-GFP mice [21] were collected at six-hour intervals from E12.0 through E15.0 to study the appearance of osteoblasts aggregated in developing ossification centers. Osx marks the differentiation of cells into osteoblasts [22]. To study the appearance of mineralized ossification centers, we used C57BL6/J embryos whose mothers were injected with (20 mg/kg of body weight) calcein one hour prior to sacrifice. Litters of C57BL/calcein mice were collected at 6-hour intervals from E14.0 through E16.0 from the injected C57BL6/J dams. Mice from these two lines will be referred to as Osx-GFP mice and C57BL/calcein mice, respectively. Embryos were imaged using a light microscope with an attached Retiga 2000R monochrome camera (QImaging: Surry, BC, Canada) with an attached X-Cite 120Q to detect fluorescence.

Crown–rump length (CRL) was recorded for each specimen using ImageJ [23]. CRL is defined in this study as the maximum cranial-caudal dimension of an embryo, which typically runs from the most superior/posterior portion of the head to the base of the tail in the sagittal plane (Figure 1). CRL measurements were acquired for each embryo twice on different days and had to meet an intra-observer measurement error standard of no greater than 0.1 mm to be included in the study. The two CRL measurements were then averaged to establish a single CRL score for each embryo.

2D outlines of the hindlimb of each embryo were used to estimate limb staging ages. Briefly, a photograph of the hindlimb was uploaded to the embryonic Mouse Ontogenetic Staging System (eMOSS) [3]. The 2D outline of the limb of each embryo is traced and then compared with a library of mean hindlimb shapes for each hour of development between E10–E15 to provide an estimate of an embryo's limbstaging age [3]. Whenever possible, both the left and right hindlimbs were staged and their average was used to estimate limbstaging age. All statistical analyses for this work were performed using R [24]. Determination of the average time of appearance for each ossification center in the Osx-GFP and C57BL/calcein samples were estimated using the R package OptimalCutpoints [25]. OptimalCutpoints uses receiver operating characteristic (ROC) curves to establish boundaries between binary variables (presence/absence of ossification centers) based on their relationship to another variable (limbstaging age or CRL) and accuracy measures selected by the researcher (sensitivity/specificity ratio).

The use of two different samples of mice to compare the level of bone development with crown–rump length and limbstaging age provides two advantages. First, osteoblast differentiation and bone mineralization are distinct developmental processes with osteoblast differentiation preceding mineralization, and it is possible that each process has a unique individual relationship with size and shape. Second, the two samples cover two embryonic time periods, which allows examination of the relationship between size and shape and of the progress of bone development over a larger proportion of developmental time.

Author Contributions: Conceptualization, K.F.; Data Curation, K.F.; Formal Analysis, K.F.; Funding Acquisition, K.F. and J.T.R.; Investigation, J.T.R.; Methodology, K.F.; Project Administration, J.T.R.; Resources, J.T.R.; Visualization, K.F.; Writing—Original Draft, K.F.; Writing—Review and Editing, J.T.R.

Funding: This research was funded by NIH/NIDCR R01 DE022988, NIH/NICHD P01 HD078233, NIH/NIDCR R01 DE027677, NSF BCS-1650824 and a collaborative research travel grant from the Burroughs Wellcome Fund (2013).

Acknowledgments: Expert breeding of the Osx mice was conducted at Pennsylvania State University by Mizuho Kawasaki. Kazuhiko Kawasaki was consulted for his expertise on multiple aspects of this project. We thank Patrick Drew for access to equipment for visualization of mouse embryos. Kate Lesciotto generously assisted with data collection during this project.

Conflicts of Interest: The authors declare no conflict of interest. The funders had no role in the design of the study; in the collection, analyses, or interpretation of data; in the writing of the manuscript, or in the decision to publish the results.

References

1. Miyake, T.; Cameron, A.M.; Hall, B.K. Detailed staging of inbred C57BL/6 mice between Theiler's [1972] stages 18 and 21 (11–13 days of gestation) based on craniofacial development. *J. Craniofac. Genet. Dev. Biol.* **1995**, *16*, 1–31.
2. Boehm, B.; Rautschka, M.; Quintana, L.; Raspopovic, J.; Jan, Z.; Sharpe, J. A landmark-free morphometric staging system for the mouse limb bud. *Development* **2011**, *138*, 1227–1234. [CrossRef] [PubMed]
3. Musy, M.; Flaherty, K.; Raspopovic, J.; Robert-Moreno, A.; Richtsmeier, J.T.; Sharpe, J. A quantitative method for staging mouse embryos based on limb morphometry. *Development* **2018**, *145*. [CrossRef] [PubMed]
4. Percival, C.J.; Richtsmeier, J.T. Angiogenesis and intramembranous osteogenesis. *Dev. Dyn.* **2013**, *242*, 909–922. [CrossRef] [PubMed]
5. Kawasaki, K.; Richtsmeier, J.T. Association of the chondrocranium and dermatocranium in early skull formation. In *Building Bones: Bone Formation and Development in Anthropology*; Cambridge University Press: Cambridge, UK, 2017; pp. 52–78.
6. Theiler, K. *The House Mouse: Atlas of Embryonic Development*; Springer: New York, NY, USA, 1989.
7. Kaufman, M.H. *The Atlas of Mouse Development*; Academic Press: London, UK, 1992.
8. Hall, B.K.; Miyake, T. Divide, accumulate, differentiate: cell condensation in skeletal development revisited. *Int. J. Dev. Biol.* **1995**, *39*, 881–894. [PubMed]
9. Heyne, G.W.; Plisch, E.H.; Melberg, C.G.; Sandgren, E.P.; Peter, J.A.; Lipinski, R.J. A simple and reliable method for early pregnancy detection in inbred mice. *J. Am. Assoc. Lab. Anim. Sci.* **2015**, *54*, 368–371. [PubMed]
10. Flaherty, K.V.; Halmi, C.; Richtsmeier, J. Using developmental rate variation to establish a fine-grained sequence of ossification center differentiation. *FASEB J.* **2018**, *32*, 9650.
11. Ferguson, M.W. Palatal shelf elevation in the Wistar rat fetus. *J. Anat.* **1978**, *125*, 555. [PubMed]
12. Wanek, N.; Muneoka, K.; Holler-Dinsmore, G.; Burton, R.; Bryant, S.V. A staging system for mouse limb development. *J. Exp. Zool.* **1989**, *249*, 41–49. [CrossRef] [PubMed]
13. Geyer, S.H.; Reissig, L.; Rose, J.; Wilson, R.; Prin, F.; Szumska, D.; Ramirez-Solis, R.; Tudor, C.; White, J.; Mohun, T.J.; et al. A staging system for correct phenotype interpretation of mouse embryos harvested on embryonic day 14 (E14.5). *J. Anat.* **2017**, *230*, 710–719. [CrossRef] [PubMed]
14. De Beer, G.R. *Embryos and Ancestors*, 2nd ed.; Oxford University Press: London, UK, 1951.
15. Gould, S.J. *Ontogeny and Phylogeny*; Harvard University Press: Cambridge, MA, USA, 1977.
16. McKinney, M.L.; McNamara, K.J. *Heterochrony: The Evolution of Ontogeny*; Springer: Berlin, Germany, 1991.
17. Klingenberg, C.P. Heterochrony and allometry: the analysis of evolutionary change in ontogeny. *Biol. Rev.* **1998**, *73*, 79–123. [CrossRef] [PubMed]
18. Otis, E.M.; Brent, R. Equivalent ages in mouse and human embryos. *Anat. Rec.* **1954**, *120*, 33–63. [CrossRef] [PubMed]
19. Martinez-Abadias, N.; Estivill, R.M.; Tomas, J.S.; Perrine, S.M.; Yoon, M.; Robert-Moreno, A.; Swoger, J.; Russo, L.; Kawasaki, K.; Richtsmeier, J.; et al. Quantification of gene expression patterns to reveal the origins of abnormal morphogenesis. *bioRxiv* **2018**, 246256. [CrossRef] [PubMed]
20. Kimmel, C.B.; Hohenlohe, P.A.; Ullmann, B.; Currey, M.; Cresko, W.A. Developmental dissociation in morphological evolution of the stickleback opercle: Dissociation in opercle evolution. *Evol. Dev.* **2012**, *14*, 326–337. [CrossRef] [PubMed]
21. Rodda, S.J. Distinct roles for Hedgehog and canonical Wnt signaling in specification, differentiation and maintenance of osteoblast progenitors. *Development* **2006**, *133*, 3231–3244. [CrossRef] [PubMed]
22. Nakashima, K.; Zhou, X.; Kunkel, G.; Zhang, Z.; Deng, J.M.; Behringer, R.R.; de Crombrugghe, B. The novel zinc finger-containing transcription factor osterix is required for osteoblast differentiation and bone formation. *Cell* **2002**, *108*, 17–29. [CrossRef]
23. Abràmoff, M.D.; Magalhães, P.J.; Ram, S.J. Image processing with ImageJ. *Biophotonics Int.* **2004**, *11*, 36–42.

8

Sonic Hedgehog—'Jack-of-All-Trades' in Neural Circuit Formation

Nikole R. Zuñiga and Esther T. Stoeckli *

Department of Molecular Life Sciences and Neuroscience Center Zurich, University of Zurich, Winterthurerstrasse 190, 8057 Zurich, Switzerland
* Correspondence: esther.stoeckli@imls.uzh.ch

Academic Editors: Henk Roelink and Simon J. Conway

Abstract: As reflected by the term morphogen, molecules such as Shh and Wnts were identified based on their role in early development when they instruct precursor cells to adopt a specific cell fate. Only much later were they implicated in neural circuit formation. Both in vitro and in vivo studies indicated that morphogens direct axons during their navigation through the developing nervous system. Today, the best understood role of Shh and Wnt in axon guidance is their effect on commissural axons in the spinal cord. Shh was shown to affect commissural axons both directly and indirectly via its effect on Wnt signaling. In fact, throughout neural circuit formation there is cross-talk and collaboration of Shh and Wnt signaling. Thus, although the focus of this review is on the role of Shh in neural circuit formation, a separation from Wnt signaling is not possible.

Keywords: axon guidance; commissural axons; midline crossing; synaptogenesis

1. Introduction

The best known function of Shh (Sonic hedgehog) is its role as a morphogen in the developing spinal cord [1–6]. Cells in the developing neural tube adopt a different cell fate depending on their position along the dorsoventral axis. Each position is specified by opposing gradients of morphogens. In the dorsal spinal cord, neurons are mainly influenced by Bone Morphogenetic Proteins (BMPs) and members of the Wnt family, secreted by the roof plate [7–10]. In the ventral spinal cord, the floor plate is the source of the ventralizing morphogen Shh.

Shh signaling affects cell differentiation by binding to the surface receptor Patched1 (Ptc1). In the absence of Shh, Ptc1 represses Shh signaling by keeping Smoothened (Smo) out of the primary cilium. Upon Shh binding, Smo translocates to the cilium, allowing for the processing of Gli from a repressor to an activator form, which in turn regulates the transcriptional program of the cell (Figure 1). Canonical Shh and Wnt signaling pathways converge in this process, as Gli expression and activation is also affected by Wnt signaling [11]. Thus, Shh and Wnt are important for the differentiation of neurons early on.

The role of Shh in axon guidance was identified much later than its function in cell differentiation and tissue patterning [12]. Initially, Shh was shown to inhibit the axon growth of retinal ganglion cells [13]. Later, Shh was implicated in commissural axon guidance in the spinal cord [14–18]. This role was extended to other populations of neurons [19–23].

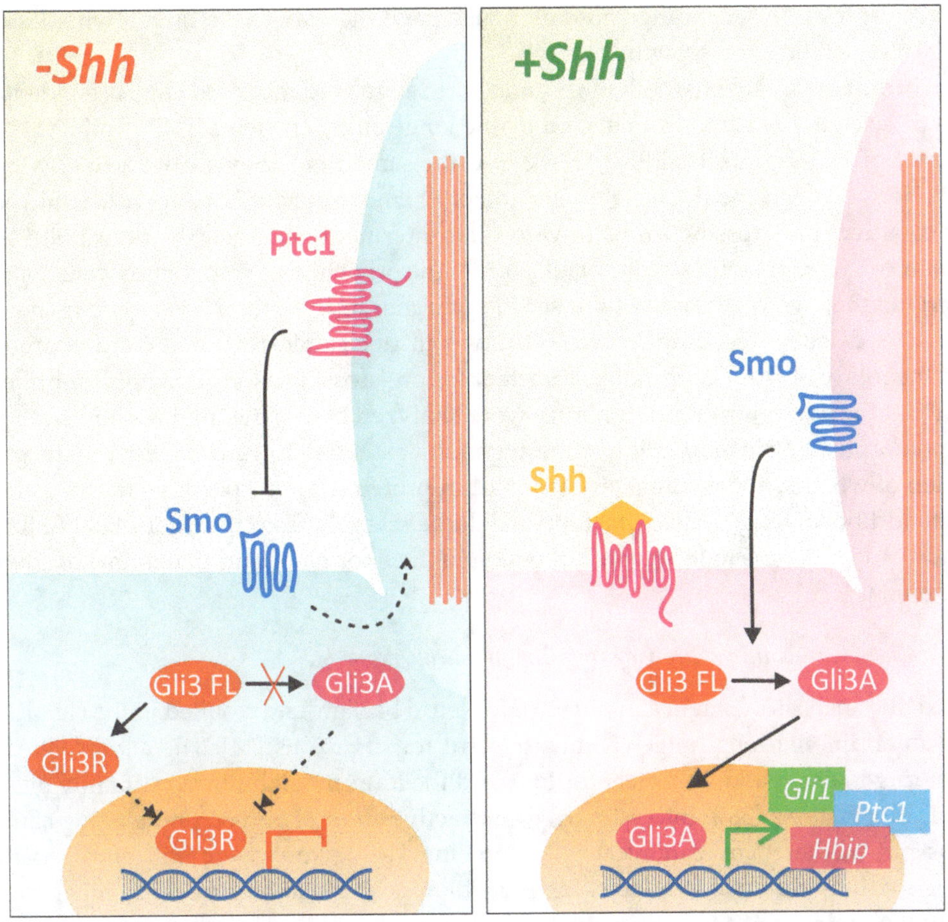

Figure 1. Canonical Shh signaling requires translocation of Smo into the primary cilium. Canonical Shh signaling depends on the de-repression of the G-protein-coupled receptor-like molecule Smoothend (Smo) by a mechanism involving the multiple-pass transmembrane receptor Patched1 (Ptc1). In the absence of Shh, Ptc1 represses Smo activity by preventing its translocation to the cilium. Shh binding to Ptc1 relieves this repression, allowing Smo to translocate to the primary cilium, where it activates intracellular signaling pathways. This activation of Smo culminates in transcriptional activity, via its regulation of Gli transcription factors that can activate or repress Shh target genes. Gli1 is always an activator of gene expression, whereas Gli2 and Gli3 (Gli3FL) need to be processed from a repressor (Gli3R) to an activator form (Gli3A) in a Smo-dependent manner.

2. Shh in Commissural Axon Guidance

Commissural neurons in the dorsal spinal cord have been used as a model to study molecular mechanisms of axon guidance mainly due to their accessibility to experimental manipulations and their stereotypic trajectory [24–27]. The dI1 subpopulation of commissural neurons [28] extends their axons towards the floor plate, which forms the ventral midline of the developing spinal cord. Axons cross the midline and turn rostrally towards the brain [29] along the contralateral floor-plate border. Axons navigate through the developing neural tissue based on the interaction of receptors expressed on the growth cone surface with guidance molecules in the environment [30]. Guidance cues can be classified into four different groups: long- and short-range cues, which can either be attractive or repulsive. By definition, long-range cues are released from a source and either attract growing axons towards the source (long-range attractants) or repel them from the source (long-range repellents). Due to their diffusible nature, long-range guidance cues outline the overall direction towards the target of the extending axons but they cannot specify the actual pathway. This is done by the integration of signals derived from the interaction between short-range guidance cues and receptors on the growth cone surface. Attractive short-range guidance cues pave the way to the target, whereas repulsive short-range cues inhibit axon growth.

For dI1 commissural axons, both long- and short-range guidance cues have been identified [24]. Axons are directed to the ventral spinal cord by the long-range repellents BMP7 and Draxin derived from the roof plate, the dorsal midline [31–33]. At the same time, they are attracted by long-range attractants derived from the floor plate. The major chemoattractant is Netrin1 [34,35]. Netrin1 is sufficient to attract pre-crossing axons toward their intermediate target, the floor plate, as shown recently in mice that express no Netrin1 at all [36,37]. The originally analyzed mice were hypomorphic for Netrin1, as they were generated by insertion of a marker into the *Netrin* gene (gene trap line) rather than its deletion resulting in reduced but still detectable Netrin protein expression [35]. However, the role of Shh as a long-range attractant for pre-crossing axons was only identified thanks to the fact that the hypomorphic Netrin mice were analyzed before the full knockout was available. In the hypomorphic *Netrin* mouse line, some axons still extended toward the floor plate, prompting the question of whether the residual levels of Netrin protein were responsible for their navigation, or whether additional long-range attractants would be released from the floor plate. Shh appeared to be a good candidate due to its strong expression in the floor plate during the time of commissural axon navigation.

2.1. *Shh Is a Chemoattractant for Pre-Crossing Commissural Axons*

Indeed, the analysis of Shh in a hypomorphic Netrin background resulted in the characterization of Shh as an additional long-range chemoattractant for dI1 axons [14]. The axons were attracted by Shh binding to its well-known receptor Ptc1 and the immunoglobulin superfamily member Boc (Brother of Cdo; Figure 2 [38]). Interestingly, the attractive effect of Shh on pre-crossing commissural axons was not mediated by canonical Shh signaling [16]. Axons were still attracted to Shh in the presence of transcriptional inhibitors. Instead, Shh was found to activate Src family kinases in dI1 neurons. In the presence of a Shh gradient, Src and Fyn are rapidly phosphorylated and distributed asymmetrically in the growth cone. Since Src family kinases can regulate cytoskeletal dynamics [39,40], the polarized distribution of phosphorylated Src and Fyn affects growth cone turning locally without affecting transcription.

2.2. *Shh Is a Repellent for Post-Crossing Commissural Axons*

Interestingly, in vivo studies in chicken embryos indicated a role of Shh in the guidance of post-crossing axons [15]. Because chicken embryos are readily accessible for manipulations in vivo in a precisely temporally controlled manner, Shh's role on pre- and post-crossing axons can be analyzed separately [25–27]. Silencing Shh after its role in cell differentiation and its effect on pre-crossing axons indicated that Shh was a repellent for post-crossing commissural axons [15,18]. Initially, these results seemed to be at odds with the finding that Shh can attract pre-crossing commissural axons. However, the difference between the responses of pre- versus post-crossing commissural axons was shown to be the level of 14-3-3 protein expression [41] and the expression of Hhip (Hedgehog-interacting protein) as a Shh receptor [15,18]. In vitro experiments with rat commissural neurons exposed to a Shh gradient indicated a requirement of increasing levels of 14-3-3 proteins to explain the observed repulsion of axons by high Shh levels [41].

In vivo experiments in chicken embryos indicated that Shh itself triggered the expression of Hhip, its receptor mediating the repulsive response, by binding to Glypican1 expressed on pre-crossing axons [18]. As Glypican1 is not a transmembrane protein but rather a glycosylphosphatidyl-inositol anchored proteoglycan, it is unclear how Shh binding to Glypican1 triggers a change in transcription (Figure 2). The most likely explanation is that Glypican1 interacts with another molecule in the plane of the growth cone membrane. Whether this molecule is one of the known Shh-binding molecules Ptc1 or Boc remains to be shown. The induction of Hhip expression requires canonical Shh signaling [18]. Additional experiments will be required to identify the regulatory mechanisms in more detail and to show if and how the induction of Hhip and the increase in protein 14-3-3 levels are linked. The analysis of Hhip knockout mice did not indicate a requirement for Hhip in post-crossing commissural axons [41]. However, genetic redundancy can obscure the role of a particular protein in axon guidance, as has been shown in other cases. For instance, a role of Neuropilin2 in post-crossing commissural axon guidance was identified in mice, which could explain why post-crossing axons still turn rostrally in *Hhip* knockout mice [41,42].

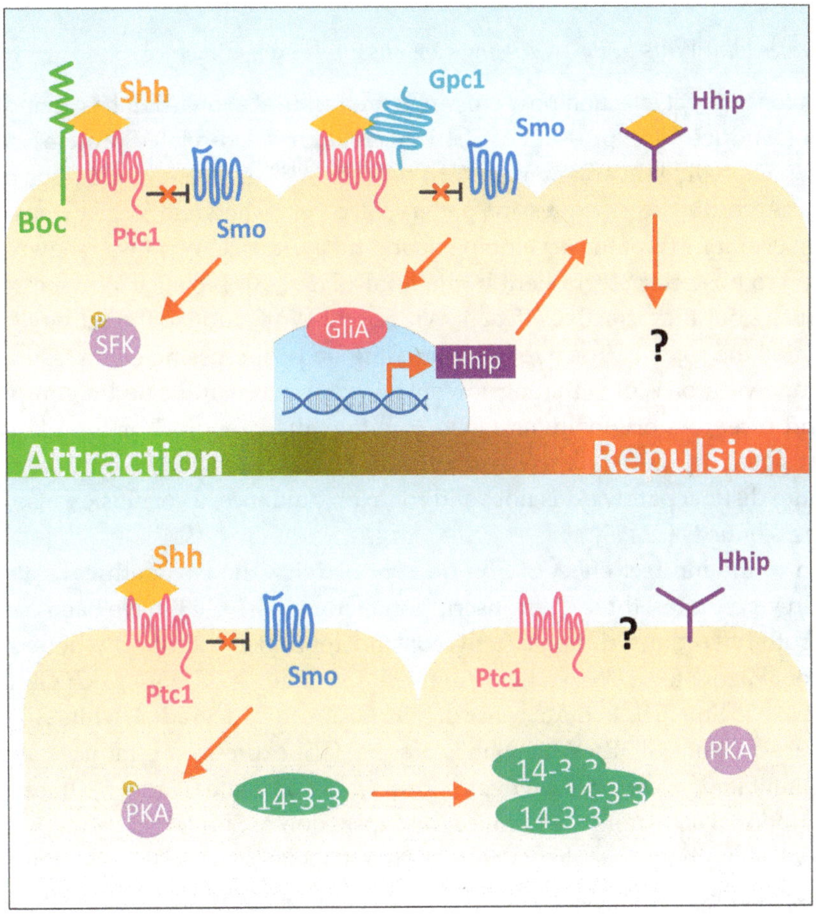

Figure 2. Non-canonical Shh signaling in commissural axon guidance. In vivo and in vitro studies have implicated Shh in commissural axon guidance. Shh is attractive for pre-crossing but repulsive for post-crossing commissural axons. On pre-crossing commissural axons Shh binds to Boc (Brother of Cdo) and Ptc1 (Patched1), thus de-repressing Smo (Smoothened). However, in contrast to canonical signaling (see Figure 1), binding of Shh results in the phosphorylation of Src family kinases (SFKs). Shh itself induces expression of Hhip (Hedgehog-interacting protein), its receptor for post-crossing axon navigation, by binding to Glypican1 (Gpc1) and Ptc1. The de-repression of Smo results in transcription of Hhip via the canonical signaling pathway. How Shh binding to Hhip on post-crossing commissural axons mediates repulsion is not yet known. In vitro (lower panel), protein 14-3-3 levels have been linked to the transition from attraction to repulsion. Pre-crossing axons initially express low levels of protein 14-3-3 (left). High levels of 14-3-3 were required for the repulsive response to Shh (right). It is unknown how 14-3-3 is linked to the precise switch from attraction to repulsion. Protein 14-3-3 can reduce the activity of PKA (Protein Kinase A), but no changes in the levels of cAMP were found between pre- and post-crossing axons. Furthermore, it is unknown if (and which) surface receptors could be linked to the function of protein 14-3-3.

At present it is unknown how Hhip transmits the Shh signal that results in repulsive axon guidance. Although Hhip is a type I transmembrane protein that can bind directly to all vertebrate hedgehog proteins [43], it does not contain an intracellular domain. Thus, it is unclear how Hhip would affect intracellular signaling. The most straightforward explanation would be that Hhip could mediate repulsive axon guidance by acting as a Shh-binding unit in a receptor complex containing an unidentified signal-transducing component. Alternatively, Hhip might modulate Shh signaling by sequestering it away from other receptors [44]. Consistent with this idea, Hhip has been reported as a negative regulator of Shh signaling [43], by competing with other Hedgehog receptors for Shh binding, including Cdo [45] and possibly Ptc [46].

2.3. Shh Collaborates with Wnts in Post-Crossing Commissural Axon Guidance

In addition to its direct effect on post-crossing commissural axons, Shh has an indirect effect [47]. Shh was shown to induce the expression of Sfrp1 (Secreted frizzled-related protein1), an endogenous antagonist of Wnt signaling, in an anterior^{low} to posterior^{high} gradient in the floor plate of chicken embryos in vivo. According to its expression pattern, Sfrp1 was shown to form a Wnt activity gradient along the anteroposterior axis of the developing neural tube. In mice, Wnt4 was shown to be expressed in an anterior^{high} to posterior^{low} gradient in the floor plate [48]. No such Wnt expression gradient was found in chicks, but the attractive effect of Wnt5a and Wnt7a due to the activity gradient shaped by Sfrp1 confirmed the role of attractive Wnt signaling on post-crossing commissural axons [47,48]. More detailed analyses of Wnt signaling in commissural axons indicated a complex network of components and receptors originally assigned to either the canonical or the planar cell polarity pathway of Wnt signaling [49–51]. Based on these studies, it is clear that the traditional segregation of Wnt signaling into distinct pathways is not valid for axon guidance, a conclusion also drawn for Wnt signaling in other contexts [52–55].

In addition to the indirect effect of Shh on Wnt activity, the two pathways also converge on Gli activation. In vertebrates, three Gli transcription factors (Gli1-Gli3) have been described [56–62]. Gli2 and Gli3 both have transcriptional activator and repressor activities, whereas Gli1 acts only as a transcriptional activator (reviewed by Ruiz i Altaba et al., 2007 [63]). Each Gli family member responds differently to Shh [64]. In the absence of Shh, Gli2 is fully degraded, while Gli3 is converted to a repressor of transcription (Gli3R). When Shh is present, Gli1 expression is induced and proteolysis of Gli2 and Gli3 is inhibited, allowing Gli activator forms to accumulate (Figure 1). In turn, in a feedback loop, Gli transcriptional activators then induce the expression of Shh target genes [65–67].

Shh and Wnt signaling cooperate via regulation of Gli3 [68]. Gli3R, the transcriptional repressor form of Gli3, is generated in the absence of Shh. Gli3R was shown to interact directly with β-Catenin and thus influence its role in Tcf3/4-mediated control of transcription, although the precise mechanism is currently unknown [69].

In addition to its direct effect on axon guidance and its indirect effect as a modulator of the axon guidance activity of Wnts, Shh was also suggested to regulate growth cone responsiveness via local intracellular mechanisms. Shh was shown to influence levels of cyclic nucleotides, which are important for signaling in response to guidance cues [13,70]. The ratio of cAMP to cGMP determines attractive versus repulsive axonal responses [71]. High levels of cAMP and low levels of cGMP favor attraction, whereas the opposite ratio of cAMP to cGMP favors repulsion. Since Shh can influence cyclic nucleotide levels by reducing the activity of protein kinase A (PKA), it may influence the responsiveness of axons to additional guidance cues, such as semaphorins [13,17].

3. The Role of Shh in Other Neuronal Populations

In contrast to the role of Shh in commissural axon guidance in the spinal cord, much less is known about the effect of Shh on other populations of neurons. Shh was also shown to be repulsive for serotonergic neurons forming the raphespinal tract in a Ptc1- and Smo-dependent manner [23] and for axons of enteric neurons in a Gas1-dependent manner [21].

Other effects of Shh were explained by an effect on axon growth rather than guidance. For instance, Shh was shown to inhibit the growth of retinal ganglion cell axons [13]. This effect was shown to involve intracellular levels of cAMP. A more recent study revisiting the role of Shh on retinal ganglion cell axons concluded that the effect was mediated by protein kinase Cα and integrin-linked kinase (ILK) [72].

In an in vitro model, Shh was found to positively affect regenerative neurite growth [73]. Shh added to the medium enhanced neurite lengths of cortical neurons exposed to reactive astrocytes. Similar to its effect on pre-crossing commissural axons [16], Shh was concluded to activate Src family kinases [73]. This is in contrast to a report on the effect of Shh on hippocampal neurons [74]. Hippocampal axon elongation was found to be stimulated by Shh in an Smo- and Gli1-dependent manner. Interestingly, Shh did not act directly on the axon but rather via the somatodendritic compartment, where the Shh-mediated Gli1 activation induced higher intracellular Profilin activity that induced axon elongation.

4. Shh at the Synapse

Upon reaching their appropriate target cell, axons initiate and stabilize contacts, by remodeling the growth cone into synaptic boutons and assembling pre-synaptic vesicle release machinery. Similarly, the post-synaptic cell must recruit receptors for neurotransmitters. Recent studies have demonstrated contributions of Wnts to synapse formation, plasticity, and maintenance in both vertebrates and invertebrates (reviewed by Salinas and Zou, 2008 [75]; Budnik and Salinas, 2011 [76]). Wnt7a was demonstrated to promote synapse assembly via its effector Dishevelled-1 (Dvl1) [77,78]. In pre-synaptic hippocampal neurons, this activity of Wnt7a was mediated by the receptor Frizzled5 (Fzd5) [79]. The expression of Dvl1 has been demonstrated to increase the stability of microtubules through a signaling pathway that inhibits GSK3β but does not require transcription [80]. Inhibition of GSK3β reduces phosphorylation of the microtubule-associated protein MAP1B, thus linking Wnt signaling to the regulation of microtubule stability and growth cone remodeling [81,82].

Wnt5a was demonstrated to stimulate post-synaptic assembly [83], suggesting that different Wnts might regulate synapse formation by promoting protein assembly in either pre- or post-synaptic cells. In the post-synaptic cell, Wnt5a does not act via GSK3β, but instead leads to clustering of the scaffold molecule PSD95.

Wnt signaling has also been demonstrated to promote synaptic plasticity. The expression and release of Wnt proteins is enhanced by depolarization [84,85]. High-frequency stimulation of hippocampal neurons increased the level of cell-surface Fzd5 and its localization to synapses, in a Wnt-dependent manner [79]. In addition, the remodeling of hippocampal circuits in response to an enriched environment was repressed when Wnt signaling was blocked [86]. Together, these studies link Wnt activity to synaptic remodeling in response to behavioral experience and activity.

In contrast to these roles of Wnts at synapses, there is currently only limited evidence that Shh signaling is directly involved in synaptogenesis or synaptic plasticity [87,88]. Shh and its receptors are appropriately expressed to support such activities. In the retina, Shh-N was shown to be transported along the optic nerve to the superior colliculus in vivo [89]. Similarly, Shh is distributed intracellularly in regulated secretory vesicles and on the extracellular surface of the soma, dendrites, and axons of rat retinal ganglion cells (RGCs). Shh secretion can be induced by depolarization of RGCs, suggesting a link between neuronal activity and Shh release [90]. Ptc and Smo were localized to dendrites and spines of mature rat hippocampal neurons, suggesting that Shh signaling may influence synaptic transmission [91]. Indeed, in a follow-up study, Shh was found to increase presynaptic terminal size in hippocampal neurons in vitro [92]. Furthermore, the addition of Shh to these cultured neurons increased the frequency, but not the amplitude, of mEPSCs (miniature excitatory postsynaptic currents) [93]. In an earlier study, Smo was also localized to the presynapse [94]. These findings match what was found in the only in vivo study done for the analysis of Shh function in synaptogenesis so far [88]. In cortical circuit formation, Shh and its receptor Boc were found in a layer-specific manner. The analysis of mouse brains lacking either Shh or Boc due to gene silencing by in utero electroporation revealed that connections between layer 2/3 axons and layer 5 dendrites require Boc in the presynaptic and Shh in the postsynaptic partner. These studies indicate that Shh most likely also acts during the last steps of neural circuit formation, synaptogenesis, and synaptic plasticity. However, many more studies will be required to elucidate the molecular mechanisms of Shh's effect.

5. Conclusions

Taken together, in vitro and in vivo studies have clearly indicated the multifaceted role of Shh in development. In the nervous system, Shh and Wnt signaling are involved in all aspects of neural circuit formation, from cell differentiation to synaptic plasticity. During most of these steps, Shh and Wnt signaling cooperate and influence each other. The cross-talk occurs at different levels, from shaping extracellular activity gradients to the regulation of common downstream effectors. No other class of molecules has been identified that was shown to affect the development of neural circuits throughout the entire process from cell differentiation to the regulation of synaptic plasticity.

Acknowledgments: Work in the lab of E.T.S. is supported by the Swiss National Science Foundation. Funds for covering the costs to publish in open access have been received from the University of Zurich.

Conflicts of Interest: The authors declare no conflict of interest.

References

1. Dessaud, E.; McMahon, A.P.; Briscoe, J. Pattern formation in the vertebrate neural tube: A sonic hedgehog morphogen-regulated transcriptional network. *Development* **2008**, *135*, 2489–2503. [CrossRef] [PubMed]
2. Gouti, M.; Metzis, V.; Briscoe, J. The route to spinal cord cell types: A tale of signals and switches. *Trends Genet.* **2015**, *31*, 282–289. [CrossRef] [PubMed]
3. Jessell, T.M. Neuronal specification in the spinal cord: Inductive signals and transcriptional codes. *Nat. Rev. Genet.* **2000**, *1*, 20–29. [CrossRef] [PubMed]
4. Le Dréau, G.; Martí, E. Dorsal-ventral patterning of the neural tube: A tale of three signals. *Dev. Neurobiol.* **2012**, *72*, 1471–1481. [CrossRef] [PubMed]
5. Matise, M.P. Molecular genetic control of cell patterning and fate determination in the developing ventral spinal cord. *Wiley Interdiscip. Rev. Dev. Biol.* **2013**, *2*, 419–425. [CrossRef] [PubMed]
6. Ulloa, F.; Martí, E. Wnt won the war: Antagonistic role of Wnt over Shh controls dorso-ventral patterning of the vertebrate neural tube. *Dev. Dyn.* **2010**, *239*, 69–76. [CrossRef] [PubMed]
7. Liem, K.F.; Tremml, G.; Jessell, T.M. A role for the roof plate and its resident TGFb-related proteins in neuronal patterning in the dorsal spinal cord. *Cell* **1997**, *91*, 127–138. [CrossRef]
8. Tozer, S.; Le Dréau, G.; Marti, E.; Briscoe, J. Temporal control of BMP signalling determines neuronal subtype identity in the dorsal neural tube. *Development* **2013**, *140*, 1467–1474. [CrossRef] [PubMed]
9. Yamauchi, K.; Phan, K.D.; Butler, S.J. BMP type I receptor complexes have distinct activities mediating cell fate and axon guidance decisions. *Development* **2008**, *135*, 1119–1128. [CrossRef] [PubMed]
10. Zechner, D.; Müller, T.; Wende, H.; Walther, I.; Taketo, M.M.; Crenshaw, E.B.; Treier, M.; Birchmeier, W.; Birchmeier, C. Bmp and Wnt/β-catenin signals control expression of the transcription factor Olig3 and the specification of spinal cord neurons. *Dev. Biol.* **2007**, *303*, 181–190. [CrossRef] [PubMed]
11. Alvarez-Medina, R.; Cayuso, J.; Okubo, T.; Takada, S.; Martí, E. Wnt canonical pathway restricts graded Shh/Gli patterning activity through the regulation of Gli3 expression. *Development* **2008**, *135*, 237–247. [CrossRef] [PubMed]
12. Wilson, N.H.; Stoeckli, E.T. Sonic Hedgehog regulates Wnt activity during neural circuit formation. *Vitam. Horm.* **2012**, *88*, 173–209. [CrossRef] [PubMed]
13. Trousse, F.; Martí, E.; Gruss, P.; Torres, M.; Bovolenta, P. Control of retinal ganglion cell axon growth: A new role for Sonic hedgehog. *Development* **2001**, *128*, 3927–3936. [PubMed]
14. Charron, F.; Stein, E.; Jeong, J.; McMahon, A.P.; Tessier-Lavigne, M. The morphogen sonic hedgehog is an axonal chemoattractant that collaborates with Netrin-1 in midline axon guidance. *Cell* **2003**, *113*, 11–23. [CrossRef]
15. Bourikas, D.; Pekarik, V.; Baeriswyl, T.; Grunditz, A.; Sadhu, R.; Nardó, M.; Stoeckli, E.T. Sonic hedgehog guides commissural axons along the longitudinal axis of the spinal cord. *Nat. Neurosci.* **2005**, *8*, 297–304. [CrossRef] [PubMed]

16. Yam, P.T.; Langlois, S.D.; Morin, S.; Charron, F. Sonic Hedgehog Guides Axons through a Noncanonical, Src-Family-Kinase-Dependent Signaling Pathway. *Neuron* **2009**, *62*, 349–362. [CrossRef] [PubMed]
17. Parra, L.M.; Zou, Y. Sonic hedgehog induces response of commissural axons to Semaphorin repulsion during midline crossing. *Nat. Neurosci.* **2010**, *13*, 29–35. [CrossRef] [PubMed]
18. Wilson, N.H.; Stoeckli, E.T. Sonic Hedgehog regulates its own receptor on postcrossing commissural axons in a glypican1-dependent manner. *Neuron* **2013**, *79*, 478–491. [CrossRef] [PubMed]
19. Fabre, P.J.; Shimogori, T.; Charron, F. Segregation of ipsilateral retinal ganglion cell axons at the optic chiasm requires the Shh receptor Boc. *J. Neurosci.* **2010**, *30*, 266–275. [CrossRef] [PubMed]
20. Hammond, R.; Blaess, S.; Abeliovich, A. Sonic hedgehog is a chemoattractant for midbrain dopaminergic axons. *PLoS ONE* **2009**, *4*, e7007. [CrossRef] [PubMed]
21. Jin, S.; Martinelli, D.C.; Zheng, X.; Tessier-Lavigne, M.; Fan, C.M. Gas1 is a receptor for sonic hedgehog to repel enteric axons. *Proc. Natl. Acad. Sci. USA* **2015**, *112*, E73–E80. [CrossRef] [PubMed]
22. Liu, F.; Placzek, M. Axon guidance effects of classical morphogens Shh and BMP7 in the hypothalamo-pituitary system. *Neurosci. Lett.* **2014**, *562*, 108–113. [CrossRef] [PubMed]
23. Song, L.; Liu, Y.; Yu, Y.; Duan, X.; Qi, S.; Liu, Y. Shh signaling guides spatial pathfinding of raphespinal tract axons by multidirectional repulsion. *Cell Res.* **2012**, *22*, 697–716. [CrossRef] [PubMed]
24. De Ramon Francàs, G.; Zuñiga, N.R.; Stoeckli, E.T. The spinal cord shows the way—How axons navigate intermediate targets. *Dev. Biol.* **2017**. [CrossRef] [PubMed]
25. Andermatt, I.; Wilson, N.; Stoeckli, E.T. In ovo electroporation of miRNA-based-plasmids to investigate gene function in the developing neural tube. *Methods Mol. Biol.* **2014**, *1101*, 353–368. [CrossRef] [PubMed]
26. Wilson, N.H.; Stoeckli, E.T. Cell type specific, traceable gene silencing for functional gene analysis during vertebrate neural development. *Nucleic Acids. Res.* **2011**, *39*, e133. [CrossRef] [PubMed]
27. Wilson, N.H.; Stoeckli, E.T. In ovo electroporation of miRNA-based plasmids in the developing neural tube and assessment of phenotypes by DiI injection in open-book preparations. *J. Vis. Exp.* **2012**, *68*, 4384. [CrossRef]
28. Helms, A.W.; Johnson, J.E. Specification of dorsal spinal cord interneurons. *Curr. Opin. Neurobiol.* **2003**, *13*, 42–49. [CrossRef]
29. Sakai, N.; Insolera, R.; Sillitoe, R.V.; Shi, S.H.; Kaprielian, Z. Axon sorting within the spinal cord marginal zone via Robo-mediated inhibition of N-cadherin controls spinocerebellar tract formation. *J. Neurosci.* **2012**, *32*, 15377–15387. [CrossRef] [PubMed]
30. Kolodkin, A.L.; Tessier-Lavigne, M. Mechanisms and molecules of neuronal wiring: A primer. *Cold Spring Harb. Perspect. Biol.* **2011**, *3*, a001727. [CrossRef] [PubMed]
31. Augsburger, A.; Schuchardt, A.; Hoskins, S.; Dodd, J.; Butler, S. BMPs as mediators of roof plate repulsion of commissural neurons. *Neuron* **1999**, *24*, 127–141. [CrossRef]
32. Butler, S.J.; Dodd, J. A role for BMP heterodimers in roof plate-mediated repulsion of commissural axons. *Neuron* **2003**, *38*, 389–401. [CrossRef]
33. Islam, S.M.; Shinmyo, Y.; Okafuji, T.; Su, Y.; Naser, I.B.; Ahmed, G.; Zhang, S.; Chen, S.; Ohta, K.; Kiyonari, H.; et al. Draxin, a repulsive guidance protein for spinal cord and forebrain commissures. *Science* **2009**, *323*, 388–393. [CrossRef] [PubMed]
34. Kennedy, T.E.; Serafini, T.; de la Torre, J.; Tessier-Lavigne, M. Netrins are diffusible chemotropic factors for commissural axons in the embryonic spinal cord. *Cell* **1994**, *78*, 425–435. [CrossRef]
35. Serafini, T.; Colamarino, S.A.; Leonardo, E.D.; Wang, H.; Beddington, R.; Skarnes, W.C.; Tessier-Lavigne, M. Netrin-1 is required for commissural axon guidance in the developing vertebrate nervous system. *Cell* **1996**, *87*, 1001–1014. [CrossRef]
36. Bin, J.M.; Han, D.; Lai Wing Sun, K.; Croteau, L.P.; Dumontier, E.; Cloutier, J.F.; Kania, A.; Kennedy, T.E. Complete Loss of Netrin-1 Results in Embryonic Lethality and Severe Axon Guidance Defects without Increased Neural Cell Death. *Cell Rep.* **2015**, *12*, 1099–1106. [CrossRef] [PubMed]
37. Yung, A.R.; Nishitani, A.M.; Goodrich, L.V. Phenotypic analysis of mice completely lacking netrin. *Development* **2015**, *142*, 3686–3691. [CrossRef] [PubMed]
38. Okada, A.; Charron, F.; Morin, S.; Shin, D.S.; Wong, K.; Fabre, P.J.; Tessier-Lavigne, M.; McConnell, S.K. Boc is a receptor for sonic hedgehog in the guidance of commissural axons. *Nature* **2006**, *444*, 369–373. [CrossRef] [PubMed]

39. Suter, D.M.; Forscher, P. Substrate-cytoskeletal coupling as a mechanism for the regulation of growth cone motility and guidance. *J. Neurobiol.* **2000**, *44*, 97–113. [CrossRef]
40. Robles, E.; Woo, S.; Gomez, T.M. Src-dependent tyrosine phosphorylation at the tips of growth cone filopodia promotes extension. *J. Neurosci.* **2005**, *25*, 7669–7681. [CrossRef] [PubMed]
41. Yam, P.T.; Kent, C.B.; Morin, S.; Farmer, W.T.; Alchini, R.; Lepelletier, L.; Colman, D.R.; Tessier-Lavigne, M.; Fournier, A.E.; Charron, F. 14-3-3 Proteins Regulate a Cell-Intrinsic Switch from Sonic Hedgehog-Mediated Commissural Axon Attraction to Repulsion after Midline Crossing. *Neuron* **2012**, *76*, 735–749. [CrossRef] [PubMed]
42. Zou, Y.; Stoeckli, E.; Chen, H.; Tessier-Lavigne, M. Squeezing axons out of the gray matter: A role for slit and semaphorin proteins from midline and ventral spinal cord. *Cell* **2000**, *102*, 363–375. [CrossRef]
43. Chuang, P.T.; McMahon, A.P. Vertebrate Hedgehog signalling modulated by induction of a Hedgehog-binding protein. *Nature* **1999**, *397*, 617–621. [CrossRef] [PubMed]
44. Capurro, M.I.; Xu, P.; Shi, W.; Li, F.; Jia, A.; Filmus, J. Glypican-3 inhibits Hedgehog signaling during development by competing with patched for Hedgehog binding. *Dev. Cell* **2008**, *14*, 700–711. [CrossRef] [PubMed]
45. McLellan, J.S.; Zheng, X.; Hauk, G.; Ghirlando, R.; Beachy, P.A.; Leahy, D.J. The mode of Hedgehog binding to Ihog homologues is not conserved across different phyla. *Nature* **2008**, *455*, 979–983. [CrossRef] [PubMed]
46. Bosanac, I.; Maun, H.R.; Scales, S.J.; Wen, X.; Lingel, A.; Bazan, J.F.; de Sauvage, F.J.; Hymowitz, S.G.; Lazarus, R.A. The structure of SHH in complex with HHIP reveals a recognition role for the Shh pseudo active site in signaling. *Nat. Struct. Mol. Biol.* **2009**, *16*, 691–697. [CrossRef] [PubMed]
47. Domanitskaya, E.; Wacker, A.; Mauti, O.; Baeriswyl, T.; Esteve, P.; Bovolenta, P.; Stoeckli, E.T. Sonic Hedgehog Guides Post-Crossing Commissural Axons Both Directly and Indirectly by Regulating Wnt Activity. *J. Neurosci.* **2010**, *30*, 11167–11176. [CrossRef] [PubMed]
48. Lyuksyutova, A.I.; Lu, C.-C.; Milanesio, N.; King, L.A.; Guo, N.; Wang, Y.; Nathans, J.; Tessier-Lavigne, M.; Zou, Y. Anterior-posterior guidance of commissural axons by Wnt-frizzled signaling. *Science* **2003**, *302*, 1984–1988. [CrossRef] [PubMed]
49. Shafer, B.; Onishi, K.; Lo, C.; Colakoglu, G.; Zou, Y. Vangl2 Promotes Wnt/Planar Cell Polarity-like Signaling by Antagonizing Dvl1-Mediated Feedback Inhibition in Growth Cone Guidance. *Dev. Cell* **2011**, *20*, 177–191. [CrossRef] [PubMed]
50. Onishi, K.; Shafer, B.; Lo, C.; Tissir, F.; Goffinet, A.M.; Zou, Y. Antagonistic Functions of Dishevelleds Regulate Frizzled3 Endocytosis via Filopodia Tips in Wnt-Mediated Growth Cone Guidance. *J. Neurosci.* **2013**, *33*, 19071–19085. [CrossRef] [PubMed]
51. Avilés, E.C.; Stoeckli, E.T. Canonical Wnt signaling is required for commissural axon guidance. *Dev. Neurobiol.* **2016**, *76*, 190–208. [CrossRef] [PubMed]
52. Avilés, E.C.; Wilson, N.H.; Stoeckli, E.T. Sonic hedgehog and Wnt: Antagonists in morphogenesis but collaborators in axon guidance. *Front. Cell. Neurosci.* **2013**, *7*, 86. [CrossRef] [PubMed]
53. Van Amerongen, R.; Nusse, R. Towards an integrated view of Wnt signaling in development. *Development* **2009**, *136*, 3205–3214. [CrossRef] [PubMed]
54. Van Amerongen, R. Alternative Wnt pathways and receptors. *Cold Spring Harb. Perspect. Biol.* **2012**, *4*. [CrossRef] [PubMed]
55. Niehrs, C. The complex world of WNT receptor signalling. *Nat. Rev. Mol. Cell Biol.* **2012**, *13*, 767–779. [CrossRef] [PubMed]
56. Hui, C.C.; Slusarski, D.; Platt, K.A.; Holmgren, R.; Joyner, A.L. Expression of three mouse homologs of the Drosophila segment polarity gene cubitus interruptus, Gli, Gli-2, and Gli-3, in ectoderm- and mesoderm-derived tissues suggests multiple roles during postimplantation development. *Dev. Biol.* **1994**, *162*, 402–413. [CrossRef] [PubMed]
57. Lee, J.; Platt, K.A.; Censullo, P.; Ruiz i Altaba, A. Gli1 is a target of Sonic hedgehog that induces ventral neural tube development. *Development* **1997**, *124*, 2537–2552. [PubMed]
58. Ruiz i Altaba, A. Combinatorial Gli gene function in floor plate and neuronal inductions by Sonic hedgehog. *Development* **1998**, *125*, 2203–2212. [PubMed]
59. Bai, C.B.; Auerbach, W.; Lee, J.S.; Stephen, D.; Joyner, A.L. Gli2, but not Gli1, is required for initial Shh signaling and ectopic activation of the Shh pathway. *Development* **2002**, *129*, 4753–4761. [PubMed]

60. Jacob, J.; Briscoe, J. Gli proteins and the control of spinal-cord patterning. *EMBO Rep.* **2003**, *4*, 761–765. [CrossRef] [PubMed]
61. Cayuso, J.; Ulloa, F.; Cox, B.; Briscoe, J.; Martí, E. The Sonic hedgehog pathway independently controls the patterning, proliferation and survival of neuroepithelial cells by regulating Gli activity. *Development* **2006**, *133*, 517–528. [CrossRef] [PubMed]
62. Persson, M.; Stamataki, D.; te Welscher, P.; Andersson, E.; Böse, J.; Rüther, U.; Ericson, J.; Briscoe, J. Dorsal-ventral patterning of the spinal cord requires Gli3 transcriptional repressor activity. *Genes Dev.* **2002**, *16*, 2865–2878. [CrossRef] [PubMed]
63. Ruiz i Altaba, A.; Mas, C.; Stecca, B. The Gli code: An information nexus regulating cell fate, stemness and cancer. *Trends Cell Biol.* **2007**, *17*, 438–447. [CrossRef] [PubMed]
64. Stamataki, D.; Ulloa, F.; Tsoni, S.V.; Mynett, A.; Briscoe, J. A gradient of Gli activity mediates graded Sonic Hedgehog signaling in the neural tube. *Genes Dev.* **2005**, *19*, 626–641. [CrossRef] [PubMed]
65. Dai, P.; Akimaru, H.; Tanaka, Y.; Maekawa, T.; Nakafuku, M.; Ishii, S. Sonic Hedgehog-induced activation of the Gli1 promoter is mediated by GLI3. *J. Biol. Chem.* **1999**, *274*, 8143–8152. [CrossRef] [PubMed]
66. Ruiz i Altaba, A. Gli proteins encode context-dependent positive and negative functions: Implications for development and disease. *Development* **1999**, *126*, 3205–3216. [PubMed]
67. Pan, Y.; Bai, C.B.; Joyner, A.L.; Wang, B. Sonic hedgehog signaling regulates Gli2 transcriptional activity by suppressing its processing and degradation. *Mol. Cell. Biol.* **2006**, *26*, 3365–3377. [CrossRef] [PubMed]
68. Ulloa, F.; Itasaki, N.; Briscoe, J. Inhibitory Gli3 activity negatively regulates Wnt/beta-catenin signaling. *Curr. Biol.* **2007**, *17*, 545–550. [CrossRef] [PubMed]
69. Alvarez-Medina, R.; Le Dreau, G.; Ros, M.; Martí, E. Hedgehog activation is required upstream of Wnt signalling to control neural progenitor proliferation. *Development* **2009**, *136*, 3301–3309. [CrossRef] [PubMed]
70. Song, H.J.; Ming, G.L.; Poo, M.M. cAMP-induced switching in turning direction of nerve growth cones. *Nature* **1997**, *388*, 275–279. [CrossRef] [PubMed]
71. Song, H.; Ming, G.; He, Z.; Lehmann, M.; McKerracher, L.; Tessier-Lavigne, M.; Poo, M. Conversion of neuronal growth cone responses from repulsion to attraction by cyclic nucleotides. *Science* **1998**, *281*, 1515–1518. [CrossRef] [PubMed]
72. Guo, D.; Standley, C.; Bellve, K.; Fogarty, K.; Bao, Z.Z. Protein kinase Cα and integrin-linked kinase mediate the negative axon guidance effects of Sonic hedgehog. *Mol. Cell. Neurosci.* **2012**, *50*, 82–92. [CrossRef] [PubMed]
73. Berretta, A.; Gowing, E.K.; Jasoni, C.L.; Clarkson, A.N. Sonic hedgehog stimulates neurite outgrowth in a mechanical stretch model of reactive-astrogliosis. *Sci. Rep.* **2016**, *6*, 21896. [CrossRef] [PubMed]
74. Yao, P.J.; Petralia, R.S.; Ott, C.; Wang, Y.X.; Lippincott-Schwartz, J.; Mattson, M.P. Dendrosomatic Sonic Hedgehog Signaling in Hippocampal Neurons Regulates Axon Elongation. *J. Neurosci.* **2015**, *35*, 16126–16141. [CrossRef] [PubMed]
75. Salinas, P.C.; Zou, Y. Wnt signaling in neural circuit assembly. *Annu. Rev. Neurosci.* **2008**, *31*, 339–358. [CrossRef] [PubMed]
76. Budnik, V.; Salinas, P.C. Wnt signaling during synaptic development and plasticity. *Curr. Opin. Neurobiol.* **2011**, *21*, 151–159. [CrossRef] [PubMed]
77. Hall, A.C.; Lucas, F.R.; Salinas, P.C. Axonal remodeling and synaptic differentiation in the cerebellum is regulated by WNT-7a signaling. *Cell* **2000**, *100*, 525–535. [CrossRef]
78. Ahmad-Annuar, A.; Ciani, L.; Simeonidis, I.; Herreros, J.; Fredj, N.B.; Rosso, S.B.; Hall, A.; Brickley, S.; Salinas, P.C. Signaling across the synapse: A role for Wnt and Dishevelled in presynaptic assembly and neurotransmitter release. *J. Cell Biol.* **2006**, *174*, 127–139. [CrossRef] [PubMed]
79. Sahores, M.; Gibb, A.; Salinas, P.C. Frizzled-5, a receptor for the synaptic organizer Wnt7a, regulates activity-mediated synaptogenesis. *Development* **2010**, *137*, 2215–2225. [CrossRef] [PubMed]
80. Ciani, L.; Krylova, O.; Smalley, M.J.; Dale, T.C.; Salinas, P.C. A divergent canonical WNT-signaling pathway regulates microtubule dynamics: Dishevelled signals locally to stabilize microtubules. *J. Cell Biol.* **2004**, *164*, 243–253. [CrossRef] [PubMed]
81. Lucas, F.R.; Goold, R.G.; Gordon-Weeks, P.R.; Salinas, P.C. Inhibition of GSK-3beta leading to the loss of phosphorylated MAP-1B is an early event in axonal remodelling induced by WNT-7a or lithium. *J. Cell Sci.* **1998**, *111*, 1351–1361. [PubMed]

82. Meli, R.; Weisová, P.; Propst, F. Repulsive axon guidance by Draxin is mediated by protein Kinase B (Akt), glycogen synthase kinase-3β (GSK-3β) and microtubule-associated protein 1B. *PLoS ONE* **2015**, *10*, e0119524. [CrossRef] [PubMed]
83. Farías, G.G.; Alfaro, I.E.; Cerpa, W.; Grabowski, C.P.; Godoy, J.A.; Bonansco, C.; Inestrosa, N.C. Wnt-5a/JNK signaling promotes the clustering of PSD-95 in hippocampal neurons. *J. Biol. Chem.* **2009**, *284*, 15857–15866. [CrossRef] [PubMed]
84. Wayman, G.A.; Impey, S.; Marks, D.; Saneyoshi, T.; Grant, W.F.; Derkach, V.; Soderling, T.R. Activity-dependent dendritic arborization mediated by CaM-kinase I activation and enhanced CREB-dependent transcription of Wnt-2. *Neuron* **2006**, *50*, 897–909. [CrossRef] [PubMed]
85. Ataman, B.; Ashley, J.; Gorczyca, M.; Ramachandran, P.; Fouquet, W.; Sigrist, S.J.; Budnik, V. Rapid activity-dependent modifications in synaptic structure and function require bidirectional Wnt signaling. *Neuron* **2008**, *57*, 705–718. [CrossRef] [PubMed]
86. Gogolla, N.; Galimberti, I.; Deguchi, Y.; Caroni, P. Wnt signaling mediates experience-related regulation of synapse numbers and mossy fiber connectivities in the adult hippocampus. *Neuron* **2009**, *62*, 510–525. [CrossRef] [PubMed]
87. Yao, P.J.; Petralia, R.S.; Mattson, M.P. Sonic Hedgehog Signaling and Hippocampal Neuroplasticity. *Trends Neurosci.* **2016**, *39*, 840–850. [CrossRef] [PubMed]
88. Harwell, C.C.; Parker, P.R.; Gee, S.M.; Okada, A.; McConnell, S.K.; Kreitzer, A.C.; Kriegstein, A.R. Sonic hedgehog expression in corticofugal projection neurons directs cortical microcircuit formation. *Neuron* **2012**, *73*, 1116–1126. [CrossRef] [PubMed]
89. Traiffort, E.; Moya, K.L.; Faure, H.; Hässig, R.; Ruat, M. High expression and anterograde axonal transport of aminoterminal sonic hedgehog in the adult hamster brain. *Eur. J. Neurosci.* **2001**, *14*, 839–850. [CrossRef] [PubMed]
90. Beug, S.T.; Parks, R.J.; McBride, H.M.; Wallace, V.A. Processing-dependent trafficking of Sonic hedgehog to the regulated secretory pathway in neurons. *Mol. Cell. Neurosci.* **2011**, *46*, 583–596. [CrossRef] [PubMed]
91. Petralia, R.S.; Schwartz, C.M.; Wang, Y.-X.; Mattson, M.P.; Yao, P.J. Subcellular localization of patched and smoothened, the receptors for sonic hedgehog signaling, in the hippocampal neuron. *J. Comp. Neurol.* **2011**, *519*, 3684–3699. [CrossRef] [PubMed]
92. Petralia, R.S.; Wang, Y.X.; Mattson, M.P.; Yao, P.J. Subcellular distribution of patched and smoothened in the cerebellar neurons. *Cerebellum* **2012**, *11*, 972–981. [CrossRef] [PubMed]
93. Mitchell, N.; Petralia, R.S.; Currier, D.G.; Wang, Y.X.; Kim, A.; Mattson, M.P.; Yao, P.J. Sonic hedgehog regulates presynaptic terminal size, ultrastructure and function in hippocampal neurons. *J. Cell Sci.* **2012**, *125*, 4207–4213. [CrossRef] [PubMed]
94. Masdeu, C.; Bernard, V.; Faure, H.; Traiffort, E.; Ruat, M. Distribution of Smoothened at hippocampal mossy fiber synapses. *Neuroreport* **2007**, *18*, 395–399. [CrossRef] [PubMed]

9

Regulation of Hedgehog Signalling Inside and Outside the Cell

Simon A. Ramsbottom [1,*] and Mary E. Pownall [2]

1. Institute of Genetic Medicine, International Centre for Life, Newcastle University, NE1 3BZ Newcastle upon Tyne, UK
2. Biology Department, University of York, YO10 5YW York, UK; betsy.pownall@york.ac.uk

* Correspondence: simon.ramsbottom@newcastle.ac.uk

Academic Editors: Henk Roelink and Simon J. Conway

Abstract: The hedgehog (Hh) signalling pathway is conserved throughout metazoans and plays an important regulatory role in both embryonic development and adult homeostasis. Many levels of regulation exist that control the release, reception, and interpretation of the hedgehog signal. The fatty nature of the Shh ligand means that it tends to associate tightly with the cell membrane, and yet it is known to act as a morphogen that diffuses to elicit pattern formation. Heparan sulfate proteoglycans (HSPGs) play a major role in the regulation of Hh distribution outside the cell. Inside the cell, the primary cilium provides an important hub for processing the Hh signal in vertebrates. This review will summarise the current understanding of how the Hh pathway is regulated from ligand production, release, and diffusion, through to signal reception and intracellular transduction.

Keywords: signalling; regulation; heparan sulfate; Hedgehog

---

1. Introduction

Hedgehog (*hh*) was originally identified as a segment polarity gene in *Drosophila* [1]. Its name comes from the mutant phenotype of the larvae, where the neatly patterned rows of denticles are disrupted and instead a lawn of denticles form, resembling the spines of a hedgehog. *Drosophila* only has one *hh* gene, whereas vertebrates have three (or more) and, of these, Sonic hedgehog (*Shh*) has garnered the greatest interest due to its beautiful and specific expression in major signalling centres of the vertebrate embryo [2–4]. Since the initial description of Shh, it has been found to be an essential regulator in a wide range of contexts during vertebrate development, and was the first morphogen proven to act in vertebrates, where it diffuses to pattern the different neural cell types in the ventral neural tube in a concentration-dependent manner [5].

The discoveries made while investigating the molecular and cellular basis of the hedgehog signalling pathway have revealed novel and unpredicted mechanisms. For instance, the recognition that primary cilia serve as a signalling hub was an unexpected finding from a forward genetic screen for embryonic patterning defects similar to the *Shh* mutant phenotype [6]. While this finding was made in mice, the complex machinery of kinases and microtubule-associated proteins involved in the intracellular relay of the hedgehog signal were largely dissected genetically in *Drosophila*; again, an unbiased genetic approach was essential for the discovery of many components of the Hh signal transduction pathway. Biochemical analyses [7–9] have also been crucial in discovering the extensive processing of the hedgehog ligand, revealing the fatty nature of the ligand that was difficult to reconcile with its long range morphogen functions in the *Drosophila* wing disc and the vertebrate neural tube and limb bud [10]. In addition to what is termed the "canonical" signalling pathway, Hh signalling can be elicited outside of this pathway to affect a number of diverse roles, including regulation of proliferation, chemotaxis, and cell motility using a "non-canonical" pathway that has been reviewed elsewhere [11,12]. This review will focus on the regulation of canonical hedgehog signalling, from expression of the ligand through to the activation of target genes.

2. Hedgehog Expression

Regulation of hedgehog gene expression was first identified in *Drosophila*, where the expression of *engrailed* (*en*) within the posterior compartment of the wing disc specifies posterior identity and drives *hh* expression in the same compartment [13]. Ectopic En protein in the anterior compartment results in ectopic *hh* activation, while loss of *en* in the posterior compartment results in a loss of *hh*. In the eye imaginal disc, however, *en* is not required for *hh* expression [14], suggesting a role for multiple regulatory factors. Additional transcriptional regulators have been identified including *Master of thickveins* (*mtv*), and *Serpent* (*Srp*), which act to restrict the expression of *hh* in anterior cells and activate *hh* during haematopoiesis, respectively [15,16]. In vertebrates, a screen identified Hepatocyte nuclear factor 3 (Hnf3)-class transcriptional regulators, which bind some but not all *Shh* enhancers [17], driving expression of *Shh* in the notochord and floorplate, where it acts to pattern the neural tube [18]. The homeobox transcription factor *Aristaless related* (ARX) similarly acts to promote *Shh* within the neural tube, working in conjunction with FOXA2 (previously known as HNF3β) within the ventral neural tube [19]. Ectopic expression of HNF3β (FoxA2) results in ectopic activation of *Shh* [20], and misexpression of proneural transcription factors. *Shh* is also expressed within the posterior margin of the developing tetrapod limb bud, in a region known as the Zone of Polarising activity (ZPA), where it acts to specify positional identity along the anterior–posterior axis. Expression of *Shh* within the ZPA is controlled by a cis-enhancer located a megabase upstream of *Shh* within intron five of the *Lmbr1* gene [21]. This ZPA regulatory sequence (ZRS) contains binding sites for the ETS transcription factors *Ets1*, *Ets2 Etv4*, *Etv5*, *Elf1* and *Gabpα*, which act to delineate the expression boundary of *Shh* while simultaneously repressing ectopic expression of *Shh* outside the ZPA [22].

Hedgehog signalling is crucial for the development of a number of different embryonic systems, and consequently its expression is tightly regulated by the combined action of multiple transcription factors. Mutations within transcription factor binding elements, or of the transcription factors themselves, can therefore significantly affect hedgehog expression, resulting in a number of conditions associated with aberrant hedgehog signalling including polydactyly, tibial hypoplasia, X-linked lissencephaly, and epilepsy [21–26].

3. Ligand Synthesis

The hedgehog ligand is initially synthesised as a 46 kDa precursor [7,8], with two distinct domains: the N-terminal "hedge" domain is processed to a 19 kDa fragment (Hh-N) following proteolytic cleavage that is executed by the C-terminal "hog" domain [10] within the endoplasmic reticulum [27] (Figure 1). The C-terminus acts as a cholesterol transferase to covalently attach a cholesterol group to the carboxy end of the Hh amino terminal fragment, Hh-N [8,28]. The nascent Hh-N is further modified by the subsequent addition of a palmitoyl group at Cys-24 [9], resulting in an extremely hydrophobic molecule that is referred to as Hh-Np for Hh-N-processed. The processing of Hh-N takes place in the secretory pathway and is mediated by a palmitoylacyltransferase which is coded for by the Skinny hedgehog gene (*Ski/Skn*) [29]. This gene is also known as *sightless* (*sit*) [30], *rasp* [31], and hedgehog aceyltransferase (Hhat) [32]. Mice deficient in *Skn* exhibit similar defects to *Shh* mutants: they lack a differentiated floor plate and display neural tube defects. These data indicate that the palmitoyl addition is essential for SHH function [33]. The addition of cholesterol and palmitate increases the efficacy of SHH-Np [33,34], while addition of hydrophilic adducts to the N terminus reduces the activity of SHH [35]. Overall, the consensus is that lipidation of the Hh signalling protein is an essential step in producing a functional signalling molecule.

4. Ligand Release and Dispersal

The dual lipidation of Hedgehog results in a very hydrophobic molecule that should find it difficult to move very far from the cell membrane. This is evident in tissue culture experiments: when the lipid-modified form of SHH (SHH-Np) is expressed in cells, it is found associated with cells, and is not released into the surrounding medium [7,36,37]. In contrast, when a form of Hedgehog that has no cholesterol adduct (termed Hh-Nu) is expressed, it is released by cells in the media [37]. Hh-Nu has also been shown to have an increased range compared with Hh-Np [38], further supporting the idea the cholesterol has a role in Hedgehog tethering. These experiments are important to note because they are in sharp contrast to the well-documented ability of SHH to act as a morphogen in vivo. A morphogen is a signalling molecule that can elicit different responses depending on its concentration. Therefore, a morphogen is present as a gradient. Consistent with this, SHH has been shown to display the characteristics of a freely-diffusible molecule, and is able to act from a distance in a morphogenic manner [39]. Due to the extensive lipid modification of the Hh ligand, the a priori assumption would be that Hh should be tethered to the membrane, thus precluding its ability to diffuse and act as a morphogen. The ability of Hh to signal at a distance, however, suggests that modification of the protein is not for the purpose of protein restriction but instead forms part of a regulated process, tightly controlling the release and subsequent diffusion of Hh ligands.

Evidence for the regulated release of Hh ligands was found when the membrane protein *Dispatched* (*Disp*) was shown to be required for the release of Hh-Np [40]. Disp contains a sterol sensing domain (SSD), which can be found in other proteins that directly bind cholesterol [41]; Disp binds SHH in a cholesterol-dependent manner [42], and loss of Disp leads to retention of Hh-Np in producing cells [40]. Hh-Nu, however (which is not cholesterol modified), is not retained following loss of Disp, suggesting that addition of cholesterol during ligand processing is responsible for Hh tethering at the membrane, and that this is overcome by the action of Disp [40,42]. Mice homozygous null for *Disp* show the characteristic phenotype resulting from a lack of hedgehog signalling; they exhibit cyclopia and holoprosencephaly and display reduced expression of *Shh* target genes [43]. The failure of SHH and Hh to be released from its site of synthesis in the absence of Disp in both flies and mice indicates a conserved mechanism for Hedgehog release. DISP alone, however, is not sufficient for the release of cholesterol-modified SHH from cells, but instead requires the action of a secreted protein—SCUBE2. The interaction of SCUBE2 is also dependent on cholesterol, and the differential affinity of SCUBE2 and DISP for different regions of the cholesterol molecule, suggest a hand-over mechanism between the two proteins, which promotes SHH release from the membrane [42]. SCUBE2 also forms an interaction with sheddases [44] which, with the support of heparan sulfate proteoglycans (HSPGs) [45], trim off the lipid groups with which hedgehog proteins interact with the plasma membrane, and so promote Hh release into the extracellular space (Figure 1-1).

SHH can be released from cells in multimeric complexes, the formation of which depends on N-terminal palmitoylation and the addition of cholesterol [33,34,46,47]. HSPGs present at the cell surface provide a scaffold to allow the formation of Hedgehog multimers, presenting a platform to promote the accumulation of SHH on producing cells, thereby facilitating SHH oligomerization and release from the plasma membrane [47]. The multimeric complex of SHH is more stable [47] and more active [33] than its monomeric form. A form of SHH lacking palmitate, SHH-N, is unable to form multimeric complexes while retaining the ability to activate downstream signalling, indicating that this modification is an important level of regulation. When expressed in the developing limb bud, unmodified SHH-N is unable to signal effectively over any significant distance, underlining the observation that long-range signalling requires lipid modifications that promote the formation of multimers [48]. When presented at the cell surface, multimeric Hh complexes may be internalised by ESCRT (endosomal sorting complexes required for transport) proteins [49], and packaged into intraluminal vesicles (ILVs), which subsequently fuse with the plasma membrane and are released as exovesicles [50]; evidence in *Drosophila* suggests that this is a Disp-dependent process [51] (Figure 1-2). More evidence for multimeric Hh complexes come from the identification of exovesicles containing Hedgehog proteins in *C. Elegans*, *D. Melanogaster* and *Mus musculus*, suggesting this is an evolutionarily-conserved release mechanism [50,52,53]. Release of multimeric hedgehog into the extracellular space requires the action of DISP, SCUBE2, and sheddases [54]. The route of export may therefore be determined by the availability of sheddases at the cell surface; if sheddases are present, hedgehog is released in a multimeric lipid-free form into the extracellular space (Figure 1-1), whereas if they are not, Hh is internalised and exported via exosomes (Figure 1-2).

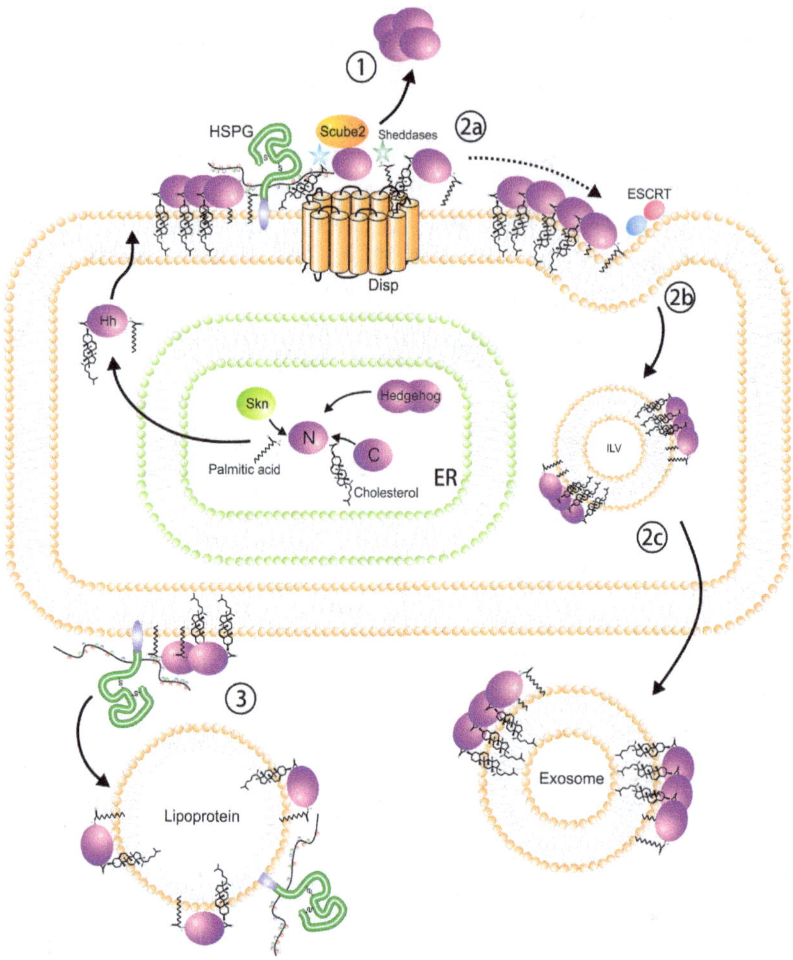

Figure 1. The sonic hedgehog (Shh) ligand is initially synthesised as a precursor with two distinct domains: the N-terminal "hedge" and C-terminal "hog" domain, which undergoes autoproteolysis to give N- and C-terminal fragments. The C-terminal fragment acts as a cholesterol transferase to attach cholesterol to the N-terminal fragment. Skinny hedgehog attaches palmitate to the N-terminus of hedgehog (Hh) to give rise to the fully processed form. Following synthesis, Hh is delivered from the endoplasmic reticulum to the cell membrane, where it is released from the cell via a number of different mechanisms. (**1**) Hh binds Disp in a cholesterol-dependent manner, and through the combined action of Disp and Scube2, is released from the cell. Heparan sulfate proteoglycans (HSPGs) act as assembly points for multiple components. Hh monomers are able to form multimeric complexes aided by association with HSPGs and Scube2. Hh is released following proteolytic processing by sheddases, which remove cholesterol and palmitate. These complexes are more soluble than the monomeric form and so are able to diffuse away from the cell; (**2**) Unprocessed Hh may re-enter the cell in a Disp-dependent fashion (**2a**) and be internalised by endosomal sorting complexes required for transport (ESCRT) proteins, which sort Hh proteins into intra-luminal vesicles (**2b**); These vesicles subsequently fuse with the plasma membrane and are released from the cell (**2c**); (**3**) Association of Hh with HSPGs results in loading of Hh into lipoprotein particles. Glypicans (Glycosylphosphatidylinositol (GPI)-linked HSPGs) may be cleaved and released along with Hh.

5. Heparan Sulfate Proteoglycans

Heparan sulfate proteoglycans (HSPGs) are large glycoproteins that are ubiquitously expressed and can be found both at the cell surface and within the extracellular matrix [55,56]. HSPGs are comprised of a core protein covalently linked to long, unbranched glycosaminoglycan (GAG) chains [57]. The heparan sulfate (HS) chains can be modified by sulfation during their synthesis in the golgi, giving rise to a high degree of heterogeneity along their length. The first evidence demonstrating an essential role for HSPGs in the regulation of hedgehog signalling came from *Drosophila* mutants. Mutation of any of the genes coding for the enzymes required for HS synthesis (the UDP glucose dehydrogenase *sugarless*, the N-deacetylase/N-sulfotransferase *sulfateless*, the glycosyltransferases *tout-velu, brother of tout-velu, sister of tout-velu*), or the proteoglycans themselves (the glypicans *dally* and *dally-like protein*) results in disrupted hedgehog signalling [58]. Glypicans, Glycosylphosphatidylinositol (GPI)-linked HSPGs, are able to bind the lipoprotein lipophorin, and its recruitment to Hh signalling cells promotes the loading of Hh into soluble lipoproteins [59] (Figure 1-3). Furthermore, this association of Hh with lipophorin has been shown to be required for the full signalling activity of Hh [60]. Vyas, et al. (2008) visualised nanoscale oligomers of Hh that are associated with HSPGS at the surface of ligand producing cells [61]. The relevance of HSPGs in hedgehog signalling was brought into question however when a mouse carrying a mutation in the Cardin–Weintraub (CW) motif of *Shh*, a domain important for the ligand interaction with HSPGs, showed decreased SHH-HSPG interaction, but displayed no patterning defects associated with *Shh* disruption [62]. These data suggest that the interaction of the CW domain of SHH with HSPGs is not important for its ability to pattern the neural tube. However, a study [63] examining the crystal structure of SHH complexed with heparin found that a SHH dimer forms a continuous stretch of positive amino acids that interact with the heparin chain, and concluded that both the CW motif and the newly described "core GAG-binding site" contribute to SHH interactions with HSPGs.

Some evidence suggests that HSPGs can act as co-receptors for Hh signal transduction; a role for HS as a co-receptor is already well-established for other signalling pathways, such as Fibroblast growth factor (FGF) signalling (reviewed by [64]). The important role for glypicans in hedgehog signalling in *Drosophila* and vertebrates has been recently reviewed [65]; in short, glypicans can act to either sequester Hh ligand to prevent signalling or, in other cases, glypicans can stabilise the association of the ligand with the Ptc receptor to promote signalling. Opposite effects of glypicans in modulating hedgehog signalling have been described in mice and *Drosophila*, where GPC3 acts as a negative regulator of SHH activity by competing with PTC1 for SHH binding, while dally-like protein (Dlp) allows Hh to interact with Ptc1 [66,67]. These differences could be species- or context-dependent; alternatively, it is possible that different structural modification of the HS chain could underlie these contrasting findings. Recently, GPC5 carrying a specific sulfation modification (2OS-Iro) to the GAG chain has also been identified as a *Shh* co-receptor, as it is enriched in cells responsive to SHH and can promote downstream signalling [68].

6. Regulation by SULF1

Structural modification of heparan sulfate chains by the cell surface endosulfatase SULF1 plays a significant role in regulating the establishment of a gradient of Hedgehog ligand during the development of the *Drosophila* imaginal disc [69] and the vertebrate neural tube [70–72]. SULF1 activity hydrolyses a sulfate group from the 6O position of glucosamine sugars in highly sulfated HS chains [73]. The notable expression of *Sulf1* in the floorplate points to a possible role in *Shh* signalling, and Cathy Soula's lab has used several vertebrate models to demonstrate that *Sulf1* expression is essential to the enhancement of *Shh* signalling in a cell-autonomous manner within the ventral spinal cord, and that this happens in precursor cells that require an increase in *Shh* signalling at a specific time in order to form oligodendrocytes [69–71]. *Sulf1*$^{-/-}$ mice had fewer cells expressing the oligodendrocyte transcription factor *Olig2* in the mantle zone compared to wild type, due to a failure of motor neuron precursors to make the switch to the oligodendrocyte lineage in the absence of *Sulf1*. These authors suggest that SULF1 acts to promote the production of active forms of SHH at the right time and place to induce specific neuronal cell types—an idea that is supported by the observation that HSPGs can promote the processing (removal of lipid groups) of active SHH at the cell surface [45].

A role for Sulf1 in modulating the diffusion of Hh in the *Drosophila* imaginal disc was demonstrated by Wojcinski, et al. [74], while prior to this, it was shown that Hh is unable to cross a field of cells lacking HSPGs [75,76]. These data suggest that HSPGs are essential for Hh diffusion, and that the nature of the HS structure is relevant for regulating this diffusion. Further evidence to support this notion came from studies in *Xenopus* explants, where the expression of *sulf1* leads to the retention of fluorescently-tagged SHH close to the cell membrane [72]. This same study showed that in the developing neural tube, the absence of *sulf1* resulted in a loss of *foxA2*-positive floorplate cells, which require a high level of SHH. Instead, *nkx2.2*-positive cells—which require lower concentrations of SHH—were found in the most ventral position. Further dorsally within the neural tube of *Xenopus* embryos, the *nkx6.1*-positive domain is extended, suggesting flattening and dorsal extension of the SHH gradient in the absence of *sulf1*. Consistent with this, direct visualisation of the endogenous ligand reveals an altered distribution, with ventral levels reduced and dorsal levels increased. These experiments demonstrate that SULF1 can limit the distance over which the Shh ligand can move, and thus increases the local concentration of the signal [72].

Further evidence of a central role for HSPGs in the recruitment of factors that facilitate release of the Shh ligand has recently been demonstrated [54]. The model supported by this data is that glypicans act as centres for the assembly of proteins that include SHH and adaptors such as SCUBE2, as well as proteolytic enzymes like sheddases; these cell-surface hubs promote the proteolytic release of SHH (Figure 1-1). The notion that proteolytic cleavage of the fatty side chains acts to promote Hh signalling seems at odds with previous research, as the palmitate group has been shown to be essential for Hh signalling [29] and long-range activity [33,47]. However, there is evidence that the proteolytic processing that releases the ligand is dependent on the N-palmitate [44,77], which anchors the protein while it is processed. An important observation from this same work [54] was that different cell types (which will be decorated with distinct HS chains with different levels of sulfation) vary in respect to how well they support the GPC/SCUBE2-dependent release of SHH. This is interesting, considering the reported differences in the release and diffusion of SHH from cells depending on whether they express the HS-editing enzyme Sulf1 [69–72]. It may be that SULF1 acts to influence the formation of these extracellular signalosomes to affect the release/distribution of Shh ligand.

HSPGs, and by extension, genes which act to modify HSPG structure, have a significant role to play in the release, dispersal, and reception of Hh. Manipulation of HSPGs can result in ligand trapping at the cell surface of cells producing the signal, or rapid diffusion away from the signalling cell. The gradient of ligand across a field of cells can be regulated, while reception of the signal at a receiving cell can be modified. In this way, HSPGs can be utilized to precisely define the range and level of activity of hedgehog signalling.

7. Regulation at the Membrane

The main receptor for the Hh ligand is the 12-pass membrane protein Patched (Ptc) [78]. In vertebrates, there are two patched genes, *Patched1 (Ptc1)* and *Patched2 (Ptc2)*, which are differentially expressed during development [79]. *Ptc1* null mice do not survive past embryogenesis, while *Ptc2* mutants are viable; *Ptc2* mutants develop alopecia and epidermal hypoplasia, however, indicating that neither protein can fully compensate for the loss of the other [80]. As well as the main receptor(s), several co-receptors for Hh signalling have been identified: Cell adhesion-associated, oncogene regulated *(Cdo)/interference hedgehog* (*ihog*) and Brother of Cdo (*Boc*)/*brother of ihog* (*boi*) are conserved between vertebrates and *Drosophila*, while a third, Growth arrest-specific 1(*Gas1*) is found only in vertebrates. Loss of *Cdo/ihog* or *Boc/boi* alone leads to minor changes in signalling, while inactivation of both results in loss of hedgehog signalling similar to that seen in the absence of Smoothened (Smo) [81,82]. In vertebrates, the membrane proteins CDO, BOC, and GAS1 interact with SHH and positively regulate signalling [82], while the co-receptor HHIP binds to SHH to negatively regulate signalling [83]. Recent evidence suggests that HHIP is not membrane associated, but rather a diffusible, extracellular regulator of Shh signalling, analogous to the well-documented bone morphogenetic protein BMP regulators like Chordin (reviewed by [84]). In the study that showed Hhip1 to have non-cell-autonomous effects in antagonising Hh signalling, it was interesting to note that the secretion of Hhip1 depended on its HS-binding motif, suggesting that its activity is modulated by HSPGs [85,86]—which, as discussed above, can also act as co-receptors for Hh signalling. A nice example of how hedgehog co-receptors can modulate SHH activity occurs during the development of the chick CNS, where SHH acts first as a chemoattractant and then as a repulsive cue as commissural axons descend to cross the ventral midline of the neural tube before they turn to continue their migration in an anterior direction. This navigation requires the expression of the glypican *Gpc1* in the commissural neurons, which, acting with SHH, promotes the expression of *Hhip1* to mediate a repulsive effect, directing an anterior migration of postcommissural axons [87,88].

Hedgehog signalling is unusual in its mode of activating downstream signalling; this is because its membrane-associated receptor Patched (Ptc) is an inhibitor of the pathway; so, rather than activating the Ptc receptor, the binding of the Hh ligand represses it. In the absence of ligand, Ptc acts to suppress the activity of a G-protein coupled receptor, Smoothened (Smo) [89]. The presence of ligand relieves the inhibition of the pathway by blocking the function of Ptc, allowing downstream signal transduction to occur. It is interesting, therefore, to observe that the depletion of the Hh receptor Ptc leads to ectopic activation of the pathway, while the lack of Smo results in a total loss of hedgehog signalling [89–92].

The inhibition of Ptc in response to ligand binding is generally agreed to be as a result of internalisation and degradation of the Ptc-Hh complex [93], however, the mechanism by which Ptc acts to inhibit Smo has yet to be fully elucidated. Originally, it was hypothesised that Ptc inhibited Smo via a direct physical interaction; Ptch and Smo display a high degree of co-localization in the absence of ligand [93,94], which is significantly reduced in Shh-responding cells [93,95]. Smo and Ptc are internalized together, but subsequently segregated within the endocytic pathway, allowing Smo to be recycled to a Ptc-negative compartment [94]. Ptc, however, acts in a substoichiometric way, able to repress signalling by half even when Smo is present in a 50-fold molar excess, suggesting a catalytic nature in its repressive activity [95], rather than direct physical inhibition. Ptc may therefore act by regulating the availability and/or activity of a Smo inhibitor or activator. Smo is known to be influenced by small molecules; for instance, the steroidal alkaloid cyclopamine is widely used to block Smo activity [96], and other drugs with this activity have been identified by screening chemical libraries [97]. The druggability of Smo, together with the structural similarity of Ptc to members of the RND (Resistance-Nodulation-Division) family of transporters, has led to the proposal that Ptc acts by importing or exporting small molecules to regulate Smo activity [98]. While the endogenous regulator of Smo is not known, a study by Bijlsma, et al. [99] showed that *Ptc*1-transfected cells display elevated levels of 3b-hydroxysteroid (pro-vitamin D3), which is released into the surrounding medium and can effectively block downstream signalling. Treatment of zebrafish with vitamin D3 leads to a loss of the *Shh* target gene *engrailed* expression in muscle pioneer cells and a down regulation of *ptc* expression, indicating reduced hedgehog signalling [99]. While vitamin D3 represses Smo, conversely, oxysterols (which are also an intermediate product in the cholesterol biosynthetic pathway) are able to activate Smo by binding a distinct region of its extracellular domain [100], resulting in the activation of Hh target genes [101]. Mutations in Smo which lead to the abolition of oxysterol binding reduce the level to which cells can respond to the Hh ligand, suggesting that these molecules may be required for endogenous signalling [100].

Other candidates for regulators of Smo activity are lipid kinases and phosphatases, given the importance of phospholipids in endosomal trafficking. Work in *Drosophila* has suggested a role for Ptc in the recruitment of lipoproteins to endosomes through which Smo is trafficked, resulting in the destabilization of Smo at the membrane [102]. Furthermore, a loss of Ptc activity leads to increased levels of the phospholipid phosphatidylinositol-4 phosphate (PI(4)P), which has been shown to promote Smo membrane localisation and activity [103], suggesting Ptc might act to control the nature of lipids in the membrane to affect Smo activity. Together, these studies point to a mechanism whereby Ptc acts to control the recruitment of lipoproteins and regulate their phospholipid composition. Recently, Khaliullina, et al. [104] undertook a screen for novel regulators of *Smo* by fractionating extracts of human very low density lipoproteins and testing the eluted fractions for the ability to repress activity of a *Shh* reporter line. A fraction showing inhibitory activity was analysed using lipid mass spectrometry and identified endocannabinoids as a class of lipids that inhibit Shh signalling, most likely at the level of SMO. Endocannabinoids are imported by lipoproteins; thus, if PTC acts to controls the level of PI(4)P in the membrane and the uptake of endocannabinoids, this would simultaneously prevent the accumulation and activity of SMO. Conversely, by blocking PTC activity, Hh promotes an increase in PI(4)P levels and a decrease in endocannabinoids, leading to the accumulation and activation of SMO. Interestingly, the finding that lipid-derived endocannabinoids regulate SMO activity means that there are likely to be additional mechanisms regulating *Shh* signal transduction through the activation or inhibition of endocannabinoid metabolism [104].

Smo is activated by phosphorylation at several sites in its C-terminal cytoplasmic tail [105]; this is thought to counteract electrostatic interactions that keep Smo in an inactive conformation [106]. Kinases involved in the hyperphosphorylation of Smo include casein kinase I (Ck1), G-protein-coupled receptor kinases (GRKs), and Protein Kinase A (Pka). However, while Smo is generally conserved, the sites phosphorylated by Pka in *Drosophila* are absent in mammalian SMO; it is, however, phosphorylated by CK1 and GRK2 [107]. In flies, Smo is initially phosphorylated by Pka [105,108,109], which promotes even tighter binding of Smo by Pka, leading to the formation of stable kinase-substrate complexes [110]. Smo is subsequently phosphorylated by CkI, CkII, and Gprk2 [105,109,111,112]. The sites on Smo targeted by GRKs are conserved throughout bilateria, suggesting that Smo regulation is fundamentally a GRK-mediated activity and GRKs are essential for activating downstream signalling [113]. However, recent work in zebrafish has indicated that while Grk2 functions as a positive regulator of Hh signalling, the genetic interactions do not support a model where Grk2 directly phosphorylates Smo, but instead regulates a step further downstream in the Shh pathway [114].

Phosphorylation of Smo by these kinases results in a conformational change in its structure. Smo contains multiple Arg clusters in its cytoplasmic tail, which together form an auto-inhibitory domain [106]. In the presence of ligand, phosphorylation of Ser/Thr residues results in the disruption of this inhibitory conformation, promoting Smo dimerisation and translocation to the cell surface. Smo activation appears to be a non-binary process, instead displaying differential levels of membrane accumulation and activation, depending on its level of phosphorylation [105,109]—a characteristic which may result from having numerous Arg clusters that are systematically disrupted by increasing levels of phosphorylation [106]. The more extensive phosphorylation of Smo promotes the expression of target genes that require high levels of Hh signalling [109]. Recently another regulatory component of this SMO activation complex has been identified. The Ser/Thr kinase CK1γ, known as *gilgamesh* (*gish*) in *Drosophila*, associates with SMO in response to Hh signalling, and phosphorylates its C-terminal tail to promote high-level pathway activation. Loss of CK1γ results in only subtle changes to Hh signalling levels, reinforcing the model of stepwise activation of SMO by kinases at multiple residues to precisely translate the concentration and longevity of the Hh ligand into a specific level of pathway activation [115].

8. Downstream of Smo

Activation of Smo leads to the up regulation of hedgehog target genes, a process mediated in *Drosophila* by cubitus interruptus (Ci). In the absence of Hh, Ci undergoes proteolytic cleavage [116] to form a truncated protein (Ci75), which acts as a transcriptional repressor [117]. In the presence of Hh, however, the full length Ci (Ci155) remains uncleaved and is able to activate Hh-responsive genes. Ci processing is regulated by a number of different proteins which act to sequester and phosphorylate Ci, controlling its activity and translocation to the nucleus [118]. A key regulator of Ci processing is the kinesin-related protein Costal-2 (Cos-2). Over-expression of Cos-2 promotes Ci cleavage and is sufficient to inhibit hedgehog signal transduction [119]. Conversely, loss of Cos-2 leads to Ci(155) accumulation, although this in itself is not sufficient to activate hedgehog signalling [119,120]. Cos-2 associates with both Ci [121] and Smo [122], as well as microtubules [123], and the affinity of Cos-2 for microtubules appears to be controlled by hedgehog, whereby its addition results in the release of Cos-2 [124]. By binding microtubules, Cos-2 provides the scaffold for a complex of proteins which contribute to the processing of Ci. This complex is comprised of four kinases: Protein Kinase A (Pka), casein kinase I (CkI), glycogen synthase kinase-3β (Gsk3β), the serine/threonine kinase Fused (Fu), as well as Ci [125]. Ci contains several Pka sites, suggesting that its cleavage may be facilitated by Pka phosphorylation, and mutation of these sites has been shown to be sufficient to inhibit Ci cleavage [126]. Similarly, loss of Pka function leads to accumulation of full-length Ci and subsequent activation of hedgehog target genes in the absence of Hh ligand [127]. Pka activity, however, does not seem to be moderated by hedgehog activity [128], suggesting that Pka activity is only a permissive factor in Ci regulation. Once phosphorylated, Ci binds to a component of the SCF (Skp, Cullin, F-box containing) ubiquitin E3 ligase complex—Supernumerary limbs (Slmb)—which acts to facilitate the processing of Ci into its repressor form [129], which then translocates to the nucleus to inhibit the transcription of hedgehog target genes.

Ci is also regulated by Suppressor of fused (Sufu). Sufu acts to sequester Ci within the cytoplasm in both its cleaved and full-length form, preventing its translocation to the nucleus [130,131]. In addition, Sufu is able to inhibit the transcriptional activity of Ci within the nucleus [131]. Therefore, in the absence of ligand, hedgehog signalling is repressed—both actively by the production of a repressor form of Ci and inhibition of transcriptional activity, and passively by the retention of unprocessed Ci within the cytoplasm. Binding of the Hedgehog ligand to Ptc leads to the release of Smo inhibition. Smo is phosphorylated by Pka, CkI, CkII, and Gprk2, and translocates to the membrane [105,108,111,112]. CkI, Pka, and Gsk3β dissociate, and the remaining parts of the signalling complex separate from microtubules [124]. Cos-2, Smo, and Fu form a complex that allows Fu to undergo autophosphorylation [132]. Subsequent CkI-dependent phosphorylation leads to full activation of Fu, which is then able promote Ci-155 stabilisation via the phosphorylation of Cos-2 and, along with CkI, inhibit the action of Sufu [132]. Activated Ci(155) translocates to the nucleus and interacts with CREB-binding protein (CBP) to activate target gene transcription [133].

In vertebrates, the Ci-related proteins GLI1, 2, and 3 [134] act as the effectors of the pathway and are differentially regulated by Shh signalling. GLI2 and GLI3 contain activator and repressor domains and can be proteolytically cleaved in a manner similar to that of Ci [135–138]. Phosphorylation of GLI2 and GLI3 promotes recognition by the F-box protein β-TRCP (homologue of Slmb), which targets them for processing [138,139]. Processing of GLI2, however, is very inefficient, and the small amount that is processed is rapidly degraded, resulting in very little cleaved GLI2 being present within the cell [135,139]. Conversely, GLI3 is efficiently processed into its repressor form [140]. This regulatory disparity is due to the fact that GLI2 and GLI3 contain a domain which confers differential susceptibility to proteolytic processing, which results in GLI3 being more efficiently processed than GLI2 [141]. Consequently, GLI2 functions primarily as an activator, while GLI3 acts predominantly in a repressive fashion. GLI1 does not contain the same processing domains as GLI2 and GLI3 and therefore acts exclusively as an activator [136,142]. Instead of undergoing cleavage to form a repressor, GLI1 is regulated by sequestration. In the absence of ligand, GLI1 is prevented from entering the nucleus by SUFU, which binds GLI1 and retains it in the cytoplasm; SUFU is also able to bind GLI within the nucleus and prevent it from activating target genes [143–145]. Upon signal activation, SUFU-GLI complexes dissociate, allowing GLI1 to enter the nucleus and activate signal transduction [146]. GLI1 can undergo alternative splicing to give rise to two variants, which either contain a truncation or a deletion within the N-terminal domain, termed GLI1ΔN and tGLI1, respectively [147,148], although only GLI1ΔN can be detected in normal (non-cancerous) tissue [148,149]. *Gli1* is transcriptionally up-regulated in response to *Shh* signalling [134,136,150], as are both of the alternative splice-forms of *Gli1* [147,148]. The presence of tGLI1 exclusively in cancerous tissue suggests that it may play a role in the transition from a healthy to a diseased state, an example of which can be observed in breast cancer, where tGLI1 is able to promote proliferation to a greater extent than GLI1 [149]. A full understanding of the expression and regulation of these isoforms is therefore crucial for the development of effective therapeutics [151].

A further difference in signal regulation between *Drosophila* and vertebrates is the respective importance of Fu and Sufu. While loss of Fu in *Drosophila* prevents downstream activation of Hh targets, there appears to be little effect in hedgehog signalling when a *fused* ortholog is knocked out in mice [152]. Alternatively, while Sufu in *Drosophila* is not a requisite part of the hedgehog pathway [153], its loss in mice leads to embryonic lethality stemming from constitutive activation of hedgehog signalling [154,155]. Another mechanistic difference between *Drosophila* and vertebrate Hh signalling can be observed at the level of the Smo–Cos-2 interaction. Requisite Cos-2 binding domains located within dSmo (*Drosophila*) are not necessary for mSmo (vertebrate) function; replacement of the C-terminal domain of mSmo with dSmo, however, renders the same Cos-2 binding sites in mSmo essential [156].

9. The Primary Cilium

While many of the mechanisms regulating hedgehog signalling are remarkably similar between vertebrates and *Drosophila*, one fundamental difference is the requirement of the primary cilium (PC) in vertebrates [6,157]. The primary cilium is an organelle that protrudes from the cell and has both a sensory and a signalling role [158]. It is comprised of a basal body formed from the mother and daughter centrioles, a central bundle of microtubules known as the axoneme, a complex of proteins at the proximal end known as the transition zone or "ciliary gate", and the ciliary membrane (Figure 2). While the ciliary membrane is essentially a continuation of the cell membrane, transport in and out of the cilial space is highly regulated [159]. The transition zone contains multiple subsets of protein complexes, each named for the diseases which arise following their disruption. These include the nephronophthisis (NPHP) module, the Joubert Syndrome (JBTS) module, and the Meckel syndrome (MKS) module [158]. While disruption of these complexes can lead to specific syndromes, mutations of some key proteins (such as Cep290) can give rise to a range of different diseases [160].

Murine models of these mutations have suggested that in addition to site-specific mutations, the severity of the phenotype may be modified by additional genetic interactions [161]. The PC acts as the hub for Hh signalling activity; receptors, repressors, and downstream activators are in continuous flux, transiting through the ciliary space in a dynamic fashion [146,162–164]. Mutations of proteins which either regulate intraflagellar transport—which is required for movement through the cilial compartment—or of the transition zone—which controls entry into the cilium—give rise to defects in hedgehog signalling [6,157,159,165].

The proteins involved in GLI processing—PKA, CKI and GSK3β—are all localised to the basal body of the primary cilia [166–168], where they act to maintain the pathway in an inactive state [169] (Figure 2). The Ser/Thr kinase CK1γ—which phosphorylates SMO to promote pathway activation—is also localised to the cilium; deletion of the C-terminal palmitoylation site of CK1γ prevents its ciliary accumulation and renders it unable to associate with SMO [115]. By controlling entry and exit of proteins, the primary cilium is able to regulate the pathway by compartmentally restricting pathway components. Furthermore, as receptors and activators of the pathway are in continuous flux, the pathway can be rapidly activated or repressed in response to ligand availability.

In the absence of ligand, signalling is repressed by PTC, which is localised within the ciliary membrane. Upon signal activation, PTC moves out of the cilium and is replaced by SMO [170,171] (Figure 2), the accumulation of which is mediated by β–arrestin and the IFT family member KIF3A [172]. Accumulation of SMO within the cilium is, however, not sufficient for pathway activation; SMO must be activated (as discussed previously) in order for it to promote downstream signalling [173]. Hedgehog signalling can therefore be regulated by either preventing accumulation of SMO within the cilium, or by inhibiting its activation, and this two-step process is thought to provide a way of preventing aberrant activation of the pathway [173]. Cilial entry of SMO also results in the removal of GPR161 from the cilium in a β–arrestin mediated manner [174]. Gpcr161 is an orphaned

G-protein-coupled receptor which acts to negatively regulate the hedgehog pathway by modulating cAMP levels, so promoting GLI processing into its repressor form by PKA [175] (Figure 2). Once SMO has entered the cilial compartment, it interacts with EVC and EVC2. These proteins are restricted to a specific region close to the basal body, and loss of this compartmental restriction leads to inhibition of the hedgehog pathway [176,177]. EVC and EVC2 act upstream of SUFU to promote GLI activation, although the precise mechanism by which this occurs is yet to be fully elucidated.

The activation and accumulation of SMO leads to increased levels of the GLI proteins within the cilium, where they localise to the distal tip [162,163,178]. Prior to translocation into the cilium, the GLI proteins form a complex with SUFU, which is co-translocated following signal activation. At the ciliary tip, this complex dissociates, allowing the GLI transcription factors to translocate to the nucleus [146].

Correct localisation of the GLI proteins within cilia is mediated by the Cos2 orthologue KIF7. The role of KIF7 appears to be to regulate cilial growth by binding to the plus end of growing microtubules, promoting catastrophe [163]. This microtubule regulation generates the compartment at the ciliary tip, allowing components of the hedgehog pathway to be appropriately localised and processed [179]. The trafficking of GLI by IFT proteins in and out of the cilial compartment is necessary to generate both activator and repressor forms of these transcription factors [180], and is therefore indispensable for the correct regulation of vertebrate hedgehog signalling.

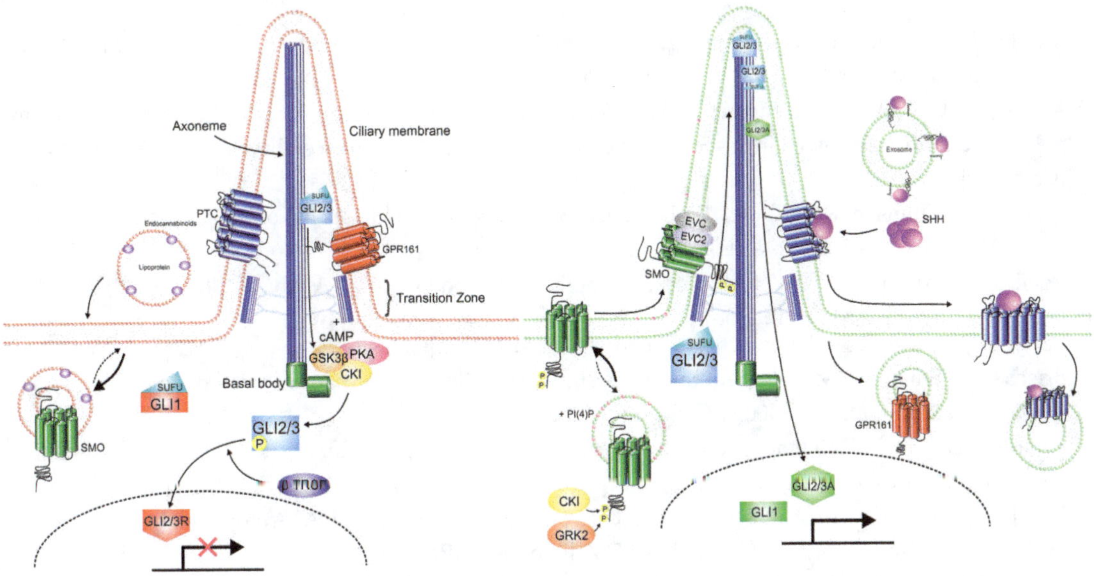

Figure 2. (**Left**) In the absence of hedgehog ligand, patched (PTC)—which is enriched in the cilial membrane—acts to repress smoothened (SMO) through the recruitment of lipoproteins and regulation of their phospholipid composition. SMO is maintained in its inactive state and sequestered within the cell. Upon exiting the cilium, GLI2 and GLI3 are phosphorylated by glycogen synthase kinsase 3β (GSK3β), casein kinase I (CKI), and protein kinase A (PKA); the activity of PKA is promoted by elevated cAMP levels due to the presence of G-protein-coupled receptor 161 GPR161. Phosphorylated GLI is recognised by β-TRCP, which promotes ubiquitylation and degradation of its C-terminal domain, giving rise to a cleaved repressor form. This cleaved repressor translocates to the nucleus and represses hedgehog target genes. GLI1 is not cleaved but is sequestered within the cytoplasm by suppressor of fused (SUFU), preventing it from activating downstream signalling; (**Right**) Upon binding of hedgehog, the hedgehog-PTC complex is internalised, and SMO inhibition by PTC is released; endocannabinoid levels are reduced, while phosphatidylinositol-4 phosphate (PI(4)P) levels increase, promoting SMO accumulation at the membrane. SMO is phosphorylated by CK1 and GRK2, leading to its activation. Activated SMO accumulates within the cilial membrane and binds EVC and EVC2. GPR161 exits the cilium and is internalised. GLI proteins within the cilial tip dissociate from SUFU and translocate the nucleus to activate Shh target genes.

10. Perspectives

This review has highlighted nodes of regulation that can enhance or inhibit output of the hedgehog signalling pathway. However, presenting one pathway in isolation does not give a true picture of cell processes in the context of a developing embryo or an adult tissue. During cellular response, input from other environmental cues and signals must be integrated, some of which can enhance the effect of Shh—such as the effect of Notch, which up-regulates SMO activity [181,182]. In addition, other signals can change the way a cell responds to Shh, as seen in neural precursors that are specified as floorplate at the intersection of FGF and Shh signalling in the posterior ventral neural plate [183]. These other pathways also include multiple levels of regulation that will also impact Shh signalling. Taken together with the mechanisms described, there is evidence for a multi-layered, complex system regulating this important signalling pathway.

Acknowledgments: S.A.R. is funded by the Medical Research Council: MR/M012212/1, and M.E.P. by the Biotechnology and Biological Sciences Research Council: BB/H010297/1.

Conflicts of Interest: The authors declare no conflict of interest. The founding sponsors had no role in the design of the study; in the collection, analyses, or interpretation of data; in the writing of the manuscript, and in the decision to publish the results.

References

1. Mohler, J. Requirements for hedgehog, a segmental polarity gene, in patterning larval and adult cuticle of drosophila. *Genetics* **1988**, *120*, 1061–1072. [PubMed]
2. Krauss, S.; Concordet, J.P.; Ingham, P.W. A functionally conserved homolog of the drosophila segment polarity gene hh is expressed in tissues with polarizing activity in zebrafish embryos. *Cell* **1993**, *75*, 1431–1444. [CrossRef]
3. Riddle, R.D.; Johnson, R.L.; Laufer, E.; Tabin, C. Sonic hedgehog mediates the polarizing activity of the ZPA. *Cell* **1993**, *75*, 1401–1416. [CrossRef]
4. Roelink, H.; Augsburger, A.; Heemskerk, J.; Korzh, V.; Norlin, S.; Ruiz i Altaba, A.; Tanabe, Y.; Placzek, M.; Edlund, T.; Jessell, T.M.; et al. Floor plate and motor neuron induction by vhh-1, a vertebrate homolog of hedgehog expressed by the notochord. *Cell* **1994**, *76*, 761–775. [CrossRef]
5. Briscoe, J.; Small, S. Morphogen rules: Design principles of gradient-mediated embryo patterning. *Development* **2015**, *142*, 3996–4009. [CrossRef] [PubMed]
6. Huangfu, D.; Liu, A.; Rakeman, A.S.; Murcia, N.S.; Niswander, L.; Anderson, K.V. Hedgehog signalling in the mouse requires intraflagellar transport proteins. *Nature* **2003**, *426*, 83–87. [CrossRef] [PubMed]
7. Lee, J.J.; Ekker, S.C.; von Kessler, D.P.; Porter, J.A.; Sun, B.I.; Beachy, P.A. Autoproteolysis in hedgehog protein biogenesis. *Science* **1994**, *266*, 1528–1537. [CrossRef] [PubMed]
8. Porter, J.A.; Young, K.E.; Beachy, P.A. Cholesterol modification of hedgehog signaling proteins in animal development. *Science* **1996**, *274*, 255–259. [CrossRef] [PubMed]
9. Pepinsky, R.B.; Zeng, C.; Wen, D.; Rayhorn, P.; Baker, D.P.; Williams, K.P.; Bixler, S.A.; Ambrose, C.M.; Garber, E.A.; Miatkowski, K.; et al. Identification of a palmitic acid-modified form of human sonic hedgehog. *J. Biol. Chem.* **1998**, *273*, 14037–14045. [CrossRef] [PubMed]
10. Ingham, P.W.; Nakano, Y.; Seger, C. Mechanisms and functions of hedgehog signalling across the metazoa. *Nat. Rev. Genet.* **2011**, *12*, 393–406. [CrossRef] [PubMed]
11. Jenkins, D. Hedgehog signalling: Emerging evidence for non-canonical pathways. *Cell. Signal.* **2009**, *21*, 1023–1034. [CrossRef] [PubMed]
12. Robbins, D.J.; Fei, D.L.; Riobo, N.A. The hedgehog signal transduction network. *Sci. Signal.* **2012**, *5*, re6. [CrossRef] [PubMed]
13. Tabata, T.; Eaton, S.; Kornberg, T.B. The drosophila hedgehog gene is expressed specifically in posterior compartment cells and is a target of engrailed regulation. *Genes Dev.* **1992**, *6*, 2635–2645. [CrossRef] [PubMed]
14. Strutt, D.I.; Mlodzik, M. The regulation of hedgehog and decapentaplegic during drosophila eye imaginal disc development. *Mech. Dev.* **1996**, *58*, 39–50. [CrossRef]
15. Bejarano, F.; Perez, L.; Apidianakis, Y.; Delidakis, C.; Milan, M. Hedgehog restricts its expression domain in the drosophila wing. *EMBO Rep.* **2007**, *8*, 778–783. [CrossRef] [PubMed]

16. Tokusumi, Y.; Tokusumi, T.; Stoller-Conrad, J.; Schulz, R.A. Serpent, suppressor of hairless and u-shaped are crucial regulators of hedgehog niche expression and prohemocyte maintenance during drosophila larval hematopoiesis. *Development* **2010**, *137*, 3561–3568. [CrossRef] [PubMed]
17. Epstein, D.J.; McMahon, A.P.; Joyner, A.L. Regionalization of sonic hedgehog transcription along the anteroposterior axis of the mouse central nervous system is regulated by hnf3-dependent and -independent mechanisms. *Development* **1999**, *126*, 281–292. [PubMed]
18. Briscoe, J.; Pierani, A.; Jessell, T.M.; Ericson, J. A homeodomain protein code specifies progenitor cell identity and neuronal fate in the ventral neural tube. *Cell* **2000**, *101*, 435–445. [CrossRef]
19. Cho, G.; Lim, Y.; Cho, I.T.; Simonet, J.C.; Golden, J.A. Arx together with foxa2, regulates shh floor plate expression. *Dev. Biol.* **2014**, *393*, 137–148. [CrossRef] [PubMed]
20. Ruiz i Altaba, A.; Jessell, T.M.; Roelink, H. Restrictions to floor plate induction by hedgehog and winged-helix genes in the neural tube of frog embryos. *Mol. Cell. Neurosci.* **1995**, *6*, 106–121. [CrossRef] [PubMed]
21. Lettice, L.A.; Heaney, S.J.; Purdie, L.A.; Li, L.; de Beer, P.; Oostra, B.A.; Goode, D.; Elgar, G.; Hill, R.E.; de Graaff, E. A long-range shh enhancer regulates expression in the developing limb and fin and is associated with preaxial polydactyly. *Hum. Mol. Genet.* **2003**, *12*, 1725–1735. [CrossRef] [PubMed]
22. Lettice, L.A.; Williamson, I.; Wiltshire, J.H.; Peluso, S.; Devenney, P.S.; Hill, A.E.; Essafi, A.; Hagman, J.; Mort, R.; Grimes, G.; et al. Opposing functions of the ets factor family define shh spatial expression in limb buds and underlie polydactyly. *Dev. Cell* **2012**, *22*, 459–467. [CrossRef] [PubMed]
23. Kitamura, K.; Yanazawa, M.; Sugiyama, N.; Miura, H.; Iizuka-Kogo, A.; Kusaka, M.; Omichi, K.; Suzuki, R.; Kato-Fukui, Y.; Kamiirisa, K.; et al. Mutation of arx causes abnormal development of forebrain and testes in mice and x-linked lissencephaly with abnormal genitalia in humans. *Nat. Genet.* **2002**, *32*, 359–369. [CrossRef] [PubMed]
24. Hill, R.E.; Lettice, L.A. Alterations to the remote control of shh gene expression cause congenital abnormalities. *Philos. Trans. R. Soc. Lond. B Biol. Sci.* **2013**, *368*, 20120357. [CrossRef] [PubMed]
25. Johnson, E.J.; Neely, D.M.; Dunn, I.C.; Davey, M.G. Direct functional consequences of zrs enhancer mutation combine with secondary long range shh signalling effects to cause preaxial polydactyly. *Dev. Biol.* **2014**, *392*, 209–220. [CrossRef] [PubMed]
26. Marsh, E.; Fulp, C.; Gomez, E.; Nasrallah, I.; Minarcik, J.; Sudi, J.; Christian, S.L.; Mancini, G.; Labosky, P.; Dobyns, W.; et al. Targeted loss of arx results in a developmental epilepsy mouse model and recapitulates the human phenotype in heterozygous females. *Brain* **2009**, *132*, 1563–1576. [CrossRef] [PubMed]
27. Chen, X.; Tukachinsky, H.; Huang, C.H.; Jao, C.; Chu, Y.R.; Tang, H.Y.; Mueller, B.; Schulman, S.; Rapoport, T.A.; Salic, A. Processing and turnover of the hedgehog protein in the endoplasmic reticulum. *J. Cell Biol.* **2011**, *192*, 825–838. [CrossRef] [PubMed]
28. Briscoe, J.; Therond, P.P. The mechanisms of hedgehog signalling and its roles in development and disease. *Nat. Rev. Mol. Cell. Biol.* **2013**, *14*, 416–429. [CrossRef] [PubMed]
29. Chamoun, Z.; Mann, R.K.; Nellen, D.; von Kessler, D.P.; Bellotto, M.; Beachy, P.A.; Basler, K. Skinny hedgehog, an acyltransferase required for palmitoylation and activity of the hedgehog signal. *Science* **2001**, *293*, 2080–2084. [CrossRef] [PubMed]
30. Lee, J.D.; Treisman, J.E. Sightless has homology to transmembrane acyltransferases and is required to generate active hedgehog protein. *Curr. Biol.* **2001**, *11*, 1147–1152. [CrossRef]
31. Micchelli, C.A.; The, I.; Selva, E.; Mogila, V.; Perrimon, N. Rasp, a putative transmembrane acyltransferase, is required for hedgehog signaling. *Development* **2002**, *129*, 843–851. [PubMed]
32. Buglino, J.A.; Resh, M.D. Hhat is a palmitoylacyltransferase with specificity for n-palmitoylation of sonic hedgehog. *J. Biol. Chem.* **2008**, *283*, 22076–22088. [CrossRef] [PubMed]
33. Chen, M.H.; Li, Y.J.; Kawakami, T.; Xu, S.M.; Chuang, P.T. Palmitoylation is required for the production of a soluble multimeric hedgehog protein complex and long-range signaling in vertebrates. *Genes Dev.* **2004**, *18*, 641–659. [CrossRef] [PubMed]
34. Zeng, X.; Goetz, J.A.; Suber, L.M.; Scott, W.J., Jr.; Schreiner, C.M.; Robbins, D.J. A freely diffusible form of sonic hedgehog mediates long-range signalling. *Nature* **2001**, *411*, 716–720. [CrossRef] [PubMed]
35. Taylor, F.R.; Wen, D.; Garber, E.A.; Carmillo, A.N.; Baker, D.P.; Arduini, R.M.; Williams, K.P.; Weinreb, P.H.; Rayhorn, P.; Hronowski, X.; et al. Enhanced potency of human sonic hedgehog by hydrophobic modification. *Biochemistry* **2001**, *40*, 4359–4371. [CrossRef] [PubMed]

36. Bumcrot, D.A.; Takada, R.; McMahon, A.P. Proteolytic processing yields two secreted forms of sonic hedgehog. *Mol. Cell. Biol.* **1995**, *15*, 2294–2303. [CrossRef] [PubMed]
37. Porter, J.A.; von Kessler, D.P.; Ekker, S.C.; Young, K.E.; Lee, J.J.; Moses, K.; Beachy, P.A. The product of hedgehog autoproteolytic cleavage active in local and long-range signalling. *Nature* **1995**, *374*, 363–366. [CrossRef] [PubMed]
38. Porter, J.A.; Ekker, S.C.; Park, W.J.; von Kessler, D.P.; Young, K.E.; Chen, C.H.; Ma, Y.; Woods, A.S.; Cotter, R.J.; Koonin, E.V.; et al. Hedgehog patterning activity: Role of a lipophilic modification mediated by the carboxy-terminal autoprocessing domain. *Cell* **1996**, *86*, 21–34. [CrossRef]
39. Briscoe, J.; Chen, Y.; Jessell, T.M.; Struhl, G. A hedgehog-insensitive form of patched provides evidence for direct long-range morphogen activity of sonic hedgehog in the neural tube. *Mol. Cell* **2001**, *7*, 1279–1291. [CrossRef]
40. Burke, R.; Nellen, D.; Bellotto, M.; Hafen, E.; Senti, K.A.; Dickson, B.J.; Basler, K. Dispatched, a novel sterol-sensing domain protein dedicated to the release of cholesterol-modified hedgehog from signaling cells. *Cell* **1999**, *99*, 803–815. [CrossRef]
41. Radhakrishnan, A.; Sun, L.P.; Kwon, H.J.; Brown, M.S.; Goldstein, J.L. Direct binding of cholesterol to the purified membrane region of scap: Mechanism for a sterol-sensing domain. *Mol. Cell* **2004**, *15*, 259–268. [CrossRef] [PubMed]
42. Tukachinsky, H.; Kuzmickas, R.P.; Jao, C.Y.; Liu, J.; Salic, A. Dispatched and scube mediate the efficient secretion of the cholesterol-modified hedgehog ligand. *Cell Rep.* **2012**, *2*, 308–320. [CrossRef] [PubMed]
43. Kawakami, T.; Kawcak, T.; Li, Y.J.; Zhang, W.; Hu, Y.; Chuang, P.T. Mouse dispatched mutants fail to distribute hedgehog proteins and are defective in hedgehog signaling. *Development* **2002**, *129*, 5753–5765. [CrossRef] [PubMed]
44. Jakobs, P.; Exner, S.; Schurmann, S.; Pickhinke, U.; Bandari, S.; Ortmann, C.; Kupich, S.; Schulz, P.; Hansen, U.; Seidler, D.G.; et al. Scube2 enhances proteolytic shh processing from the surface of shh-producing cells. *J. Cell Sci.* **2014**, *127*, 1726–1737. [CrossRef] [PubMed]
45. Ortmann, C.; Pickhinke, U.; Exner, S.; Ohlig, S.; Lawrence, R.; Jboor, H.; Dreier, R.; Grobe, K. Sonic hedgehog processing and release are regulated by glypican heparan sulfate proteoglycans. *J. Cell Sci.* **2015**, *128*, 2374–2385. [CrossRef] [PubMed]
46. Gallet, A.; Ruel, L.; Staccini-Lavenant, L.; Therond, P.P. Cholesterol modification is necessary for controlled planar long-range activity of hedgehog in drosophila epithelia. *Development* **2006**, *133*, 407–418. [CrossRef] [PubMed]
47. Goetz, J.A.; Singh, S.; Suber, L.M.; Kull, F.J.; Robbins, D.J. A highly conserved amino-terminal region of sonic hedgehog is required for the formation of its freely diffusible multimeric form. *J. Biol. Chem.* **2006**, *281*, 4087–4093. [CrossRef] [PubMed]
48. Lewis, P.M.; Dunn, M.P.; McMahon, J.A.; Logan, M.; Martin, J.F.; St-Jacques, B.; McMahon, A.P. Cholesterol modification of sonic hedgehog is required for long-range signaling activity and effective modulation of signaling by ptc1. *Cell* **2001**, *105*, 599–612. [CrossRef]
49. Matusek, T.; Wendler, F.; Poles, S.; Pizette, S.; D'Angelo, G.; Furthauer, M.; Therond, P.P. The escrt machinery regulates the secretion and long-range activity of hedgehog. *Nature* **2014**, *516*, 99–103. [CrossRef] [PubMed]
50. Parchure, A.; Vyas, N.; Ferguson, C.; Parton, R.G.; Mayor, S. Oligomerization and endocytosis of hedgehog is necessary for its efficient exovesicular secretion. *Mol. Biol. Cell* **2015**, *26*, 4700–4717. [CrossRef] [PubMed]
51. D'Angelo, G.; Matusek, T.; Pizette, S.; Therond, P.P. Endocytosis of hedgehog through dispatched regulates long-range signaling. *Dev. Cell* **2015**, *32*, 290–303. [CrossRef] [PubMed]
52. Tanaka, Y.; Okada, Y.; Hirokawa, N. Fgf-induced vesicular release of sonic hedgehog and retinoic acid in leftward nodal flow is critical for left-right determination. *Nature* **2005**, *435*, 172–177. [CrossRef] [PubMed]
53. Liegeois, S.; Benedetto, A.; Garnier, J.M.; Schwab, Y.; Labouesse, M. The v0-atpase mediates apical secretion of exosomes containing hedgehog-related proteins in caenorhabditis elegans. *J. Cell Biol.* **2006**, *173*, 949–961. [CrossRef] [PubMed]
54. Jakobs, P.; Schulz, P.; Ortmann, C.; Schurmann, S.; Exner, S.; Rebollido-Rios, R.; Dreier, R.; Seidler, D.G.; Grobe, K. Bridging the gap: Heparan sulfate and scube2 assemble sonic hedgehog release complexes at the surface of producing cells. *Sci. Rep.* **2016**, *6*, 26435. [CrossRef] [PubMed]
55. Bernfield, M.; Gotte, M.; Park, P.W.; Reizes, O.; Fitzgerald, M.L.; Lincecum, J.; Zako, M. Functions of cell surface heparan sulfate proteoglycans. *Annu. Rev. Biochem.* **1999**, *68*, 729–777. [CrossRef] [PubMed]
56. Lamanna, W.C.; Kalus, I.; Padva, M.; Baldwin, R.J.; Merry, C.L.; Dierks, T. The heparanome—The enigma of encoding and decoding heparan sulfate sulfation. *J. Biotechnol.* **2007**, *129*, 290–307. [CrossRef] [PubMed]
57. Turnbull, J.; Powell, A.; Guimond, S. Heparan sulfate: Decoding a dynamic multifunctional cell regulator. *Trends Cell Biol.* **2001**, *11*, 75–82. [CrossRef]

58. Lin, X. Functions of heparan sulfate proteoglycans in cell signaling during development. *Development* **2004**, *131*, 6009–6021. [CrossRef] [PubMed]
59. Eugster, C.; Panakova, D.; Mahmoud, A.; Eaton, S. Lipoprotein-heparan sulfate interactions in the hh pathway. *Dev. Cell* **2007**, *13*, 57–71. [CrossRef] [PubMed]
60. Panakova, D.; Sprong, H.; Marois, E.; Thiele, C.; Eaton, S. Lipoprotein particles are required for hedgehog and wingless signalling. *Nature* **2005**, *435*, 58–65. [CrossRef] [PubMed]
61. Vyas, N.; Goswami, D.; Manonmani, A.; Sharma, P.; Ranganath, H.A.; VijayRaghavan, K.; Shashidhara, L.S.; Sowdhamini, R.; Mayor, S. Nanoscale organization of hedgehog is essential for long-range signaling. *Cell* **2008**, *133*, 1214–1227. [CrossRef] [PubMed]
62. Chan, J.A.; Balasubramanian, S.; Witt, R.M.; Nazemi, K.J.; Choi, Y.; Pazyra-Murphy, M.F.; Walsh, C.O.; Thompson, M.; Segal, R.A. Proteoglycan interactions with sonic hedgehog specify mitogenic responses. *Nat. Neurosci.* **2009**, *12*, 409–417. [CrossRef] [PubMed]
63. Whalen, D.M.; Malinauskas, T.; Gilbert, R.J.; Siebold, C. Structural insights into proteoglycan-shaped hedgehog signaling. *Proc. Natl. Acad. Sci. USA* **2013**, *110*, 16420–16425. [CrossRef] [PubMed]
64. Pellegrini, L. Role of heparan sulfate in fibroblast growth factor signalling: A structural view. *Curr. Opin. Struct. Biol.* **2001**, *11*, 629–634. [CrossRef]
65. Filmus, J.; Capurro, M. The role of glypicans in hedgehog signaling. *Matrix Biol.* **2014**, *35*, 248–252. [CrossRef] [PubMed]
66. Capurro, M.I.; Xu, P.; Shi, W.; Li, F.; Jia, A.; Filmus, J. Glypican-3 inhibits hedgehog signaling during development by competing with patched for hedgehog binding. *Dev. Cell* **2008**, *14*, 700–711. [CrossRef] [PubMed]
67. Gallet, A.; Staccini-Lavenant, L.; Therond, P.P. Cellular trafficking of the glypican dally-like is required for full-strength hedgehog signaling and wingless transcytosis. *Dev. Cell* **2008**, *14*, 712–725. [CrossRef] [PubMed]
68. Witt, R.M.; Hecht, M.L.; Pazyra-Murphy, M.F.; Cohen, S.M.; Noti, C.; van Kuppevelt, T.H.; Fuller, M.; Chan, J.A.; Hopwood, J.J.; Seeberger, P.H.; et al. Heparan sulfate proteoglycans containing a glypican 5 core and 2-o-sulfo-iduronic acid function as sonic hedgehog co-receptors to promote proliferation. *J. Biol. Chem.* **2013**, *288*, 26275–26288. [CrossRef] [PubMed]
69. Touahri, Y.; Escalas, N.; Benazeraf, B.; Cochard, P.; Danesin, C.; Soula, C. Sulfatase 1 promotes the motor neuron-to-oligodendrocyte fate switch by activating shh signaling in olig2 progenitors of the embryonic ventral spinal cord. *J. Neurosci.* **2012**, *32*, 18018–18034. [CrossRef] [PubMed]
70. Danesin, C.; Agius, E.; Escalas, N.; Ai, X.; Emerson, C.; Cochard, P.; Soula, C. Ventral neural progenitors switch toward an oligodendroglial fate in response to increased sonic hedgehog (shh) activity: Involvement of sulfatase 1 in modulating shh signaling in the ventral spinal cord. *J. Neurosci.* **2006**, *26*, 5037–5048. [CrossRef] [PubMed]
71. Al Oustah, A.; Danesin, C.; Khouri-Farah, N.; Farreny, M.A.; Escalas, N.; Cochard, P.; Glise, B.; Soula, C. Dynamics of sonic hedgehog signaling in the ventral spinal cord are controlled by intrinsic changes in source cells requiring sulfatase 1. *Development* **2014**, *141*, 1392–1403. [CrossRef] [PubMed]
72. Ramsbottom, S.A.; Maguire, R.J.; Fellgett, S.W.; Pownall, M.E. Sulf1 influences the shh morphogen gradient during the dorsal ventral patterning of the neural tube in xenopus tropicalis. *Dev. Biol.* **2014**, *391*, 207–218. [CrossRef] [PubMed]
73. Ai, X.; Do, A.T.; Kusche-Gullberg, M.; Lindahl, U.; Lu, K.; Emerson, C.P., Jr. Substrate specificity and domain functions of extracellular heparan sulfate 6-o-endosulfatases, qsulf1 and qsulf2. *J. Biol. Chem.* **2006**, *281*, 4969–4976. [CrossRef] [PubMed]
74. Wojcinski, A.; Nakato, H.; Soula, C.; Glise, B. Dsulfatase-1 fine-tunes hedgehog patterning activity through a novel regulatory feedback loop. *Dev. Biol.* **2011**, *358*, 168–180. [CrossRef] [PubMed]
75. Bellaiche, Y.; The, I.; Perrimon, N. Tout-velu is a drosophila homologue of the putative tumour suppressor ext-1 and is needed for hh diffusion. *Nature* **1998**, *394*, 85–88. [PubMed]
76. Han, C.; Belenkaya, T.Y.; Khodoun, M.; Tauchi, M.; Lin, X.; Lin, X. Distinct and collaborative roles of drosophila ext family proteins in morphogen signalling and gradient formation. *Development* **2004**, *131*, 1563–1575. [CrossRef] [PubMed]
77. Ohlig, S.; Pickhinke, U.; Sirko, S.; Bandari, S.; Hoffmann, D.; Dreier, R.; Farshi, P.; Gotz, M.; Grobe, K. An emerging role of sonic hedgehog shedding as a modulator of heparan sulfate interactions. *J. Biol. Chem.* **2012**, *287*, 43708–43719. [CrossRef] [PubMed]

78. Ingham, P.W.; Taylor, A.M.; Nakano, Y. Role of the drosophila patched gene in positional signalling. *Nature* **1991**, *353*, 184–187. [CrossRef] [PubMed]
79. Carpenter, D.; Stone, D.M.; Brush, J.; Ryan, A.; Armanini, M.; Frantz, G.; Rosenthal, A.; de Sauvage, F.J. Characterization of two patched receptors for the vertebrate hedgehog protein family. *Proc. Natl. Acad. Sci. USA* **1998**, *95*, 13630–13634. [CrossRef] [PubMed]
80. Nieuwenhuis, E.; Motoyama, J.; Barnfield, P.C.; Yoshikawa, Y.; Zhang, X.; Mo, R.; Crackower, M.A.; Hui, C.C. Mice with a targeted mutation of patched2 are viable but develop alopecia and epidermal hyperplasia. *Mol. Cell. Biol.* **2006**, *26*, 6609–6622. [CrossRef] [PubMed]
81. Beachy, P.A.; Hymowitz, S.G.; Lazarus, R.A.; Leahy, D.J.; Siebold, C. Interactions between hedgehog proteins and their binding partners come into view. *Genes Dev.* **2010**, *24*, 2001–2012. [CrossRef] [PubMed]
82. Allen, B.L.; Song, J.Y.; Izzi, L.; Althaus, I.W.; Kang, J.S.; Charron, F.; Krauss, R.S.; McMahon, A.P. Overlapping roles and collective requirement for the coreceptors gas1, cdo, and boc in shh pathway function. *Dev. Cell* **2011**, *20*, 775–787. [CrossRef] [PubMed]
83. Chuang, P.T.; McMahon, A.P. Vertebrate hedgehog signalling modulated by induction of a hedgehog-binding protein. *Nature* **1999**, *397*, 617–621. [CrossRef] [PubMed]
84. Brazil, D.P.; Church, R.H.; Surae, S.; Godson, C.; Martin, F. Bmp signalling: Agony and antagony in the family. *Trends Cell Biol.* **2015**, *25*, 249–264. [CrossRef] [PubMed]
85. Holtz, A.M.; Griffiths, S.C.; Davis, S.J.; Bishop, B.; Siebold, C.; Allen, B.L. Secreted hhip1 interacts with heparan sulfate and regulates hedgehog ligand localization and function. *J. Cell Biol.* **2015**, *209*, 739–757. [CrossRef] [PubMed]
86. Kwong, L.; Bijlsma, M.F.; Roelink, H. Shh-mediated degradation of hhip allows cell autonomous and non-cell autonomous shh signalling. *Nat. Commun.* **2014**, *5*, 4849. [CrossRef] [PubMed]
87. Bourikas, D.; Pekarik, V.; Baeriswyl, T.; Grunditz, A.; Sadhu, R.; Nardo, M.; Stoeckli, E.T. Sonic hedgehog guides commissural axons along the longitudinal axis of the spinal cord. *Nat. Neurosci.* **2005**, *8*, 297–304. [CrossRef] [PubMed]
88. Wilson, N.H.; Stoeckli, E.T. Sonic hedgehog regulates its own receptor on postcrossing commissural axons in a glypican1-dependent manner. *Neuron* **2013**, *79*, 478–491. [CrossRef] [PubMed]
89. Alcedo, J.; Ayzenzon, M.; von Ohlen, T.; Noll, M.; Hooper, J.E. The drosophila smoothened gene encodes a seven-pass membrane protein, a putative receptor for the hedgehog signal. *Cell* **1996**, *86*, 221–232. [CrossRef]
90. Van den Heuvel, M.; Ingham, P.W. Smoothened encodes a receptor-like serpentine protein required for hedgehog signalling. *Nature* **1996**, *382*, 547–551. [CrossRef] [PubMed]
91. Chen, W.; Burgess, S.; Hopkins, N. Analysis of the zebrafish smoothened mutant reveals conserved and divergent functions of hedgehog activity. *Development* **2001**, *128*, 2385–2396. [PubMed]
92. Goodrich, L.V.; Milenkovic, L.; Higgins, K.M.; Scott, M.P. Altered neural cell fates and medulloblastoma in mouse patched mutants. *Science* **1997**, *277*, 1109–1113. [CrossRef] [PubMed]
93. Denef, N.; Neubuser, D.; Perez, L.; Cohen, S.M. Hedgehog induces opposite changes in turnover and subcellular localization of patched and smoothened. *Cell* **2000**, *102*, 521–531. [CrossRef]
94. Incardona, J.P.; Gruenberg, J.; Roelink, H. Sonic hedgehog induces the segregation of patched and smoothened in endosomes. *Curr. Biol.* **2002**, *12*, 983–995. [CrossRef]
95. Taipale, J.; Cooper, M.K.; Maiti, T.; Beachy, P.A. Patched acts catalytically to suppress the activity of smoothened. *Nature* **2002**, *418*, 892–897. [CrossRef] [PubMed]
96. Chen, J.K.; Taipale, J.; Cooper, M.K.; Beachy, P.A. Inhibition of hedgehog signaling by direct binding of cyclopamine to smoothened. *Genes Dev.* **2002**, *16*, 2743–2748. [CrossRef] [PubMed]
97. Chen, J.K.; Taipale, J.; Young, K.E.; Maiti, T.; Beachy, P.A. Small molecule modulation of smoothened activity. *Proc. Natl. Acad. Sci. USA* **2002**, *99*, 14071–14076. [CrossRef] [PubMed]
98. Rohatgi, R.; Scott, M.P. Patching the gaps in hedgehog signalling. *Nat. Cell. Biol.* **2007**, *9*, 1005–1009. [CrossRef] [PubMed]
99. Bijlsma, M.F.; Peppelenbosch, M.P.; Spek, C.A. A dual role for 7-dehydrocholesterol reductase in regulating hedgehog signalling? *Development* **2006**, *133*, 3951; author reply 3952–3953. [CrossRef] [PubMed]
100. Nedelcu, D.; Liu, J.; Xu, Y.; Jao, C.; Salic, A. Oxysterol binding to the extracellular domain of smoothened in hedgehog signaling. *Nat. Chem. Biol.* **2013**, *9*, 557–564. [CrossRef] [PubMed]

101. Dwyer, J.R.; Sever, N.; Carlson, M.; Nelson, S.F.; Beachy, P.A.; Parhami, F. Oxysterols are novel activators of the hedgehog signaling pathway in pluripotent mesenchymal cells. *J. Biol. Chem.* **2007**, *282*, 8959–8968. [CrossRef] [PubMed]
102. Khaliullina, H.; Panakova, D.; Eugster, C.; Riedel, F.; Carvalho, M.; Eaton, S. Patched regulates smoothened trafficking using lipoprotein-derived lipids. *Development* **2009**, *136*, 4111–4121. [CrossRef] [PubMed]
103. Yavari, A.; Nagaraj, R.; Owusu-Ansah, E.; Folick, A.; Ngo, K.; Hillman, T.; Call, G.; Rohatgi, R.; Scott, M.P.; Banerjee, U. Role of lipid metabolism in smoothened derepression in hedgehog signaling. *Dev. Cell* **2010**, *19*, 54–65. [CrossRef] [PubMed]
104. Khaliullina, H.; Bilgin, M.; Sampaio, J.L.; Shevchenko, A.; Eaton, S. Endocannabinoids are conserved inhibitors of the hedgehog pathway. *Proc. Natl. Acad. Sci. USA* **2015**, *112*, 3415–3420. [CrossRef] [PubMed]
105. Jia, J.; Tong, C.; Wang, B.; Luo, L.; Jiang, J. Hedgehog signalling activity of smoothened requires phosphorylation by protein kinase a and casein kinase i. *Nature* **2004**, *432*, 1045–1050. [CrossRef] [PubMed]
106. Zhao, Y.; Tong, C.; Jiang, J. Hedgehog regulates smoothened activity by inducing a conformational switch. *Nature* **2007**, *450*, 252–258. [CrossRef] [PubMed]
107. Chen, Y.; Sasai, N.; Ma, G.; Yue, T.; Jia, J.; Briscoe, J.; Jiang, J. Sonic hedgehog dependent phosphorylation by ck1alpha and grk2 is required for ciliary accumulation and activation of smoothened. *PLoS Biol.* **2011**, *9*, e1001083. [CrossRef] [PubMed]
108. Apionishev, S.; Katanayeva, N.M.; Marks, S.A.; Kalderon, D.; Tomlinson, A. Drosophila smoothened phosphorylation sites essential for hedgehog signal transduction. *Nat. Cell Biol.* **2005**, *7*, 86–92. [CrossRef] [PubMed]
109. Su, Y.; Ospina, J.K.; Zhang, J.; Michelson, A.P.; Schoen, A.M.; Zhu, A.J. Sequential phosphorylation of smoothened transduces graded hedgehog signaling. *Sci. Signal.* **2011**, *4*, ra43. [CrossRef] [PubMed]
110. Li, S.; Ma, G.; Wang, B.; Jiang, J. Hedgehog induces formation of pka-smoothened complexes to promote smoothened phosphorylation and pathway activation. *Sci. Signal.* **2014**, *7*, ra62. [CrossRef] [PubMed]
111. Chen, Y.; Li, S.; Tong, C.; Zhao, Y.; Wang, B.; Liu, Y.; Jia, J.; Jiang, J. G protein-coupled receptor kinase 2 promotes high-level hedgehog signaling by regulating the active state of smo through kinase-dependent and kinase-independent mechanisms in drosophila. *Genes Dev.* **2010**, *24*, 2054–2067. [CrossRef] [PubMed]
112. Jia, H.; Liu, Y.; Xia, R.; Tong, C.; Yue, T.; Jiang, J.; Jia, J. Casein kinase 2 promotes hedgehog signaling by regulating both smoothened and cubitus interruptus. *J. Biol. Chem.* **2010**, *285*, 37218–37226. [CrossRef] [PubMed]
113. Maier, D.; Cheng, S.; Faubert, D.; Hipfner, D.R. A broadly conserved g-protein-coupled receptor kinase phosphorylation mechanism controls drosophila smoothened activity. *PLoS Genet.* **2014**, *10*, e1004399. [CrossRef] [PubMed]
114. Zhao, Z.; Lee, R.T.; Pusapati, G.V.; Iyu, A.; Rohatgi, R.; Ingham, P.W. An essential role for grk2 in hedgehog signalling downstream of smoothened. *EMBO Rep.* **2016**, *17*, 739–752. [CrossRef] [PubMed]
115. Li, S.; Li, S.; Han, Y.; Tong, C.; Wang, B.; Chen, Y.; Jiang, J. Regulation of smoothened phosphorylation and high-level hedgehog signaling activity by a plasma membrane associated kinase. *PLoS Biol.* **2016**, *14*, e1002481. [CrossRef] [PubMed]
116. Forbes, A.J.; Nakano, Y.; Taylor, A.M.; Ingham, P.W. Genetic analysis of hedgehog signalling in the drosophila embryo. *Dev. Suppl.* **1993**, 115–124.
117. Aza-Blanc, P.; Ramirez-Weber, F.A.; Laget, M.P.; Schwartz, C.; Kornberg, T.B. Proteolysis that is inhibited by hedgehog targets cubitus interruptus protein to the nucleus and converts it to a repressor. *Cell* **1997**, *89*, 1043–1053. [CrossRef]
118. Hooper, J.E.; Scott, M.P. Communicating with hedgehogs. *Nat. Rev. Mol. Cell. Biol.* **2005**, *6*, 306–317. [CrossRef] [PubMed]
119. Wang, G.; Amanai, K.; Wang, B.; Jiang, J. Interactions with costal2 and suppressor of fused regulate nuclear translocation and activity of cubitus interruptus. *Genes Dev.* **2000**, *14*, 2893–2905. [CrossRef] [PubMed]
120. Wang, Q.T.; Holmgren, R.A. The subcellular localization and activity of drosophila cubitus interruptus are regulated at multiple levels. *Development* **1999**, *126*, 5097–5106. [PubMed]
121. Wang, G.; Jiang, J. Multiple cos2/ci interactions regulate ci subcellular localization through microtubule dependent and independent mechanisms. *Dev. Biol.* **2004**, *268*, 493–505. [CrossRef] [PubMed]
122. Ogden, S.K.; Ascano, M., Jr.; Stegman, M.A.; Suber, L.M.; Hooper, J.E.; Robbins, D.J. Identification of a functional interaction between the transmembrane protein smoothened and the kinesin-related protein costal2. *Curr. Biol.* **2003**, *13*, 1998–2003. [CrossRef] [PubMed]

123. Sisson, J.C.; Ho, K.S.; Suyama, K.; Scott, M.P. Costal2, a novel kinesin-related protein in the hedgehog signaling pathway. *Cell* **1997**, *90*, 235–245. [CrossRef]
124. Robbins, D.J.; Nybakken, K.E.; Kobayashi, R.; Sisson, J.C.; Bishop, J.M.; Therond, P.P. Hedgehog elicits signal transduction by means of a large complex containing the kinesin-related protein costal2. *Cell* **1997**, *90*, 225–234. [CrossRef]
125. Zhang, W.; Zhao, Y.; Tong, C.; Wang, G.; Wang, B.; Jia, J.; Jiang, J. Hedgehog-regulated costal2-kinase complexes control phosphorylation and proteolytic processing of cubitus interruptus. *Dev. Cell* **2005**, *8*, 267–278. [CrossRef] [PubMed]
126. Chen, Y.; Gallaher, N.; Goodman, R.H.; Smolik, S.M. Protein kinase a directly regulates the activity and proteolysis of cubitus interruptus. *Proc. Natl. Acad. Sci. USA* **1998**, *95*, 2349–2354. [CrossRef] [PubMed]
127. Wang, G.; Wang, B.; Jiang, J. Protein kinase a antagonizes hedgehog signaling by regulating both the activator and repressor forms of cubitus interruptus. *Genes Dev.* **1999**, *13*, 2828–2837. [CrossRef] [PubMed]
128. Jiang, J.; Struhl, G. Protein kinase A and hedgehog signaling in drosophila limb development. *Cell* **1995**, *80*, 563–572. [CrossRef]
129. Jiang, J.; Struhl, G. Regulation of the hedgehog and wingless signalling pathways by the f-box/wd40-repeat protein slimb. *Nature* **1998**, *391*, 493–496. [CrossRef] [PubMed]
130. Methot, N.; Basler, K. Suppressor of fused opposes hedgehog signal transduction by impeding nuclear accumulation of the activator form of cubitus interruptus. *Development* **2000**, *127*, 4001–4010. [PubMed]
131. Han, Y.; Shi, Q.; Jiang, J. Multisite interaction with sufu regulates Ci/Gli activity through distinct mechanisms in hh signal transduction. *Proc. Natl. Acad. Sci. USA* **2015**, *112*, 6383–6388. [CrossRef] [PubMed]
132. Zhou, Q.; Kalderon, D. Hedgehog activates fused through phosphorylation to elicit a full spectrum of pathway responses. *Dev. Cell* **2011**, *20*, 802–814. [CrossRef] [PubMed]
133. Akimaru, H.; Chen, Y.; Dai, P.; Hou, D.X.; Nonaka, M.; Smolik, S.M.; Armstrong, S.; Goodman, R.H.; Ishii, S. Drosophila CBP is a co-activator of cubitus interruptus in hedgehog signalling. *Nature* **1997**, *386*, 735–738. [CrossRef] [PubMed]
134. Ruiz i Altaba, A. Combinatorial gli gene function in floor plate and neuronal inductions by sonic hedgehog. *Development* **1998**, *125*, 2203–2212. [PubMed]
135. Pan, Y.; Bai, C.B.; Joyner, A.L.; Wang, B. Sonic hedgehog signaling regulates gli2 transcriptional activity by suppressing its processing and degradation. *Mol. Cell. Biol.* **2006**, *26*, 3365–3377. [CrossRef] [PubMed]
136. Dai, P.; Akimaru, H.; Tanaka, Y.; Maekawa, T.; Nakafuku, M.; Ishii, S. Sonic hedgehog-induced activation of the gli1 promoter is mediated by gli3. *J. Biol. Chem.* **1999**, *274*, 8143–8152. [CrossRef] [PubMed]
137. Sasaki, H.; Nishizaki, Y.; Hui, C.; Nakafuku, M.; Kondoh, H. Regulation of gli2 and gli3 activities by an amino-terminal repression domain: Implication of gli2 and gli3 as primary mediators of shh signaling. *Development* **1999**, *126*, 3915–3924. [PubMed]
138. Wang, B.; Li, Y. Evidence for the direct involvement of {beta}trcp in gli3 protein processing. *Proc. Natl. Acad. Sci. USA* **2006**, *103*, 33–38. [CrossRef] [PubMed]
139. Bhatia, N.; Thiyagarajan, S.; Elcheva, I.; Saleem, M.; Dlugosz, A.; Mukhtar, H.; Spiegelman, V.S. Gli2 is targeted for ubiquitination and degradation by beta-trcp ubiquitin ligase. *J. Biol. Chem.* **2006**, *281*, 19320–19326. [CrossRef] [PubMed]
140. Tempé, D.; Casas, M.; Karaz, S.; Blanchet-Tournier, M.F.; Concordet, J.P. Multisite protein kinase a and glycogen synthase kinase 3beta phosphorylation leads to gli3 ubiquitination by scfbetatrcp. *Mol. Cell. Biol.* **2006**, *26*, 4316–4326. [CrossRef] [PubMed]
141. Pan, Y.; Wang, B. A novel protein-processing domain in gli2 and gli3 differentially blocks complete protein degradation by the proteasome. *J. Biol. Chem.* **2007**, *282*, 10846–10852. [CrossRef] [PubMed]
142. Schrader, E.K.; Harstad, K.G.; Holmgren, R.A.; Matouschek, A. A three-part signal governs differential processing of gli1 and gli3 proteins by the proteasome. *J. Biol. Chem.* **2011**, *286*, 39051–39058. [CrossRef] [PubMed]
143. Kogerman, P.; Grimm, T.; Kogerman, L.; Krause, D.; Unden, A.B.; Sandstedt, B.; Toftgard, R.; Zaphiropoulos, P.G. Mammalian suppressor-of-fused modulates nuclear-cytoplasmic shuttling of gli-1. *Nat. Cell Biol.* **1999**, *1*, 312–319. [PubMed]
144. Liu, X.; Wang, X.; Du, W.; Chen, L.; Wang, G.; Cui, Y.; Liu, Y.; Dou, Z.; Wang, H.; Zhang, P.; et al. Suppressor of fused (sufu) represses gli1 transcription and nuclear accumulation, inhibits glioma cell proliferation, invasion and vasculogenic mimicry, improving glioma chemo-sensitivity and prognosis. *Oncotarget* **2014**, *5*, 11681–11694. [CrossRef] [PubMed]

145. Ding, Q.; Fukami, S.; Meng, X.; Nishizaki, Y.; Zhang, X.; Sasaki, H.; Dlugosz, A.; Nakafuku, M.; Hui, C. Mouse suppressor of fused is a negative regulator of sonic hedgehog signaling and alters the subcellular distribution of gli1. *Curr. Biol.* **1999**, *9*, 1119–1122. [CrossRef]
146. Tukachinsky, H.; Lopez, L.V.; Salic, A. A mechanism for vertebrate hedgehog signaling: Recruitment to cilia and dissociation of sufu-gli protein complexes. *J. Cell Biol.* **2010**, *191*, 415–428. [CrossRef] [PubMed]
147. Shimokawa, T.; Tostar, U.; Lauth, M.; Palaniswamy, R.; Kasper, M.; Toftgard, R.; Zaphiropoulos, P.G. Novel human glioma-associated oncogene 1 (gli1) splice variants reveal distinct mechanisms in the terminal transduction of the hedgehog signal. *J. Biol. Chem.* **2008**, *283*, 14345–14354. [CrossRef] [PubMed]
148. Lo, H.W.; Zhu, H.; Cao, X.; Aldrich, A.; Ali-Osman, F. A novel splice variant of gli1 that promotes glioblastoma cell migration and invasion. *Cancer Res.* **2009**, *69*, 6790–6798. [CrossRef] [PubMed]
149. Cao, X.; Geradts, J.; Dewhirst, M.W.; Lo, H.W. Upregulation of VEGF-A and CD24 gene expression by the tgli1 transcription factor contributes to the aggressive behavior of breast cancer cells. *Oncogene* **2012**, *31*, 104–115. [CrossRef] [PubMed]
150. Lee, J.; Platt, K.A.; Censullo, P.; Ruiz i Altaba, A. Gli1 is a target of sonic hedgehog that induces ventral neural tube development. *Development* **1997**, *124*, 2537–2552. [PubMed]
151. Carpenter, R.L.; Lo, H.W. Hedgehog pathway and gli1 isoforms in human cancer. *Discov. Med.* **2012**, *13*, 105–113. [PubMed]
152. Chen, M.H.; Gao, N.; Kawakami, T.; Chuang, P.T. Mice deficient in the fused homolog do not exhibit phenotypes indicative of perturbed hedgehog signaling during embryonic development. *Mol. Cell. Biol.* **2005**, *25*, 7042–7053. [CrossRef] [PubMed]
153. Preat, T. Characterization of suppressor of fused, a complete suppressor of the fused segment polarity gene of drosophila melanogaster. *Genetics* **1992**, *132*, 725–736. [PubMed]
154. Cooper, A.F.; Yu, K.P.; Brueckner, M.; Brailey, L.L.; Johnson, L.; McGrath, J.M.; Bale, A.E. Cardiac and cns defects in a mouse with targeted disruption of suppressor of fused. *Development* **2005**, *132*, 4407–4417. [CrossRef] [PubMed]
155. Svard, J.; Heby-Henricson, K.; Persson-Lek, M.; Rozell, B.; Lauth, M.; Bergstrom, A.; Ericson, J.; Toftgard, R.; Teglund, S. Genetic elimination of suppressor of fused reveals an essential repressor function in the mammalian hedgehog signaling pathway. *Dev. Cell* **2006**, *10*, 187–197. [CrossRef] [PubMed]
156. Varjosalo, M.; Li, S.P.; Taipale, J. Divergence of hedgehog signal transduction mechanism between drosophila and mammals. *Dev. Cell* **2006**, *10*, 177–186. [CrossRef] [PubMed]
157. Goetz, S.C.; Anderson, K.V. The primary cilium: A signalling centre during vertebrate development. *Nat. Rev. Genet.* **2010**, *11*, 331–344. [CrossRef] [PubMed]
158. Szymanska, K.; Johnson, C.A. The transition zone: An essential functional compartment of cilia. *Cilia* **2012**, *1*, 10. [CrossRef] [PubMed]
159. Chih, B.; Liu, P.; Chinn, Y.; Chalouni, C.; Komuves, L.G.; Hass, P.E.; Sandoval, W.; Peterson, A.S. A ciliopathy complex at the transition zone protects the cilia as a privileged membrane domain. *Nat. Cell Biol.* **2012**, *14*, 61–72. [CrossRef] [PubMed]
160. Coppieters, F.; Lefever, S.; Leroy, B.P.; de Baere, E. Cep290, a gene with many faces: Mutation overview and presentation of cep290base. *Hum. Mutat.* **2010**, *31*, 1097–1108. [CrossRef] [PubMed]
161. Ramsbottom, S.; Miles, C.; Sayer, J. Murine cep290 phenotypes are modified by genetic backgrounds and provide an impetus for investigating disease modifier alleles. *F1000Res* **2015**, *4*, 590. [CrossRef] [PubMed]
162. Wen, X.; Lai, C.K.; Evangelista, M.; Hongo, J.A.; de Sauvage, F.J.; Scales, S.J. Kinetics of hedgehog-dependent full-length gli3 accumulation in primary cilia and subsequent degradation. *Mol. Cell. Biol.* **2010**, *30*, 1910–1922. [CrossRef] [PubMed]
163. Kim, J.; Kato, M.; Beachy, P.A. Gli2 trafficking links hedgehog-dependent activation of smoothened in the primary cilium to transcriptional activation in the nucleus. *Proc. Natl. Acad. Sci. USA* **2009**, *106*, 21666–21671. [CrossRef] [PubMed]
164. Humke, E.W.; Dorn, K.V.; Milenkovic, L.; Scott, M.P.; Rohatgi, R. The output of hedgehog signaling is controlled by the dynamic association between suppressor of fused and the gli proteins. *Genes Dev.* **2010**, *24*, 670–682. [CrossRef] [PubMed]
165. Hynes, A.M.; Giles, R.H.; Srivastava, S.; Eley, L.; Whitehead, J.; Danilenko, M.; Raman, S.; Slaats, G.G.; Colville, J.G.; Ajzenberg, H.; et al. Murine joubert syndrome reveals hedgehog signaling defects as a potential therapeutic target for nephronophthisis. *Proc. Natl. Acad. Sci. USA* **2014**, *111*, 9893–9898. [CrossRef] [PubMed]

166. Barzi, M.; Berenguer, J.; Menendez, A.; Alvarez-Rodriguez, R.; Pons, S. Sonic-hedgehog-mediated proliferation requires the localization of pka to the cilium base. *J. Cell Sci.* **2010**, *123*, 62–69. [CrossRef] [PubMed]
167. Fumoto, K.; Hoogenraad, C.C.; Kikuchi, A. Gsk-3beta-regulated interaction of bicd with dynein is involved in microtubule anchorage at centrosome. *EMBO J.* **2006**, *25*, 5670–5682. [CrossRef] [PubMed]
168. Sillibourne, J.E.; Milne, D.M.; Takahashi, M.; Ono, Y.; Meek, D.W. Centrosomal anchoring of the protein kinase ck1delta mediated by attachment to the large, coiled-coil scaffolding protein cg-nap/akap450. *J. Mol. Biol.* **2002**, *322*, 785–797. [CrossRef]
169. Tuson, M.; He, M.; Anderson, K.V. Protein kinase a acts at the basal body of the primary cilium to prevent gli2 activation and ventralization of the mouse neural tube. *Development* **2011**, *138*, 4921–4930. [CrossRef] [PubMed]
170. Rohatgi, R.; Milenkovic, L.; Scott, M.P. Patched1 regulates hedgehog signaling at the primary cilium. *Science* **2007**, *317*, 372–376. [CrossRef] [PubMed]
171. Corbit, K.C.; Aanstad, P.; Singla, V.; Norman, A.R.; Stainier, D.Y.; Reiter, J.F. Vertebrate smoothened functions at the primary cilium. *Nature* **2005**, *437*, 1018–1021. [CrossRef] [PubMed]
172. Kovacs, J.J.; Whalen, E.J.; Liu, R.; Xiao, K.; Kim, J.; Chen, M.; Wang, J.; Chen, W.; Lefkowitz, R.J. Beta-arrestin-mediated localization of smoothened to the primary cilium. *Science* **2008**, *320*, 1777–1781. [CrossRef] [PubMed]
173. Rohatgi, R.; Milenkovic, L.; Corcoran, R.B.; Scott, M.P. Hedgehog signal transduction by smoothened: Pharmacologic evidence for a 2-step activation process. *Proc. Natl. Acad. Sci. USA* **2009**, *106*, 3196–3201. [CrossRef] [PubMed]
174. Pal, K.; Hwang, S.H.; Somatilaka, B.; Badgandi, H.; Jackson, P.K.; DeFea, K.; Mukhopadhyay, S. Smoothened determines beta-arrestin-mediated removal of the G protein-coupled receptor gpr161 from the primary cilium. *J. Cell Biol.* **2016**, *212*, 861–875. [CrossRef] [PubMed]
175. Mukhopadhyay, S.; Wen, X.; Ratti, N.; Loktev, A.; Rangell, L.; Scales, S.J.; Jackson, P.K. The ciliary g-protein-coupled receptor gpr161 negatively regulates the sonic hedgehog pathway via camp signaling. *Cell* **2013**, *152*, 210–223. [CrossRef] [PubMed]
176. Dorn, K.V.; Hughes, C.E.; Rohatgi, R. A smoothened-evc2 complex transduces the hedgehog signal at primary cilia. *Dev. Cell* **2012**, *23*, 823–835. [CrossRef] [PubMed]
177. Yang, C.; Chen, W.; Chen, Y.; Jiang, J. Smoothened transduces hedgehog signal by forming a complex with evc/evc2. *Cell Res.* **2012**, *22*, 1593–1604. [CrossRef] [PubMed]
178. Haycraft, C.J.; Banizs, B.; Aydin-Son, Y.; Zhang, Q.; Michaud, E.J.; Yoder, B.K. Gli2 and gli3 localize to cilia and require the intraflagellar transport protein polaris for processing and function. *PLoS Genet.* **2005**, *1*, e53. [CrossRef] [PubMed]
179. He, M.; Subramanian, R.; Bangs, F.; Omelchenko, T.; Liem, K.F., Jr.; Kapoor, T.M.; Anderson, K.V. The kinesin-4 protein kif7 regulates mammalian hedgehog signalling by organizing the cilium tip compartment. *Nat. Cell Biol.* **2014**, *16*, 663–672. [CrossRef] [PubMed]
180. Liu, A.; Wang, B.; Niswander, L.A. Mouse intraflagellar transport proteins regulate both the activator and repressor functions of gli transcription factors. *Development* **2005**, *132*, 3103–3111. [CrossRef] [PubMed]
181. Stasiulewicz, M.; Gray, S.D.; Mastromina, I.; Silva, J.C.; Bjorklund, M.; Seymour, P.A.; Booth, D.; Thompson, C.; Green, R.J.; Hall, E.A.; et al. A conserved role for Notch signaling in priming the cellular response to Shh through ciliary localisation of the key Shh transducer Smo. *Development* **2015**, *142*, 2291–2303. [CrossRef] [PubMed]
182. Kong, J.H.; Yang, L.; Dessaud, E.; Chuang, K.; Moore, D.M.; Rohatgi, R.; Briscoe, J.; Novitch, B.G. Notch activity modulates the responsiveness of neural progenitors to sonic hedgehog signaling. *Dev. Cell* **2015**, *33*, 373–387. [CrossRef] [PubMed]
183. Sasai, N.; Kutejova, E.; Briscoe, J. Integration of signals along orthogonal axes of the vertebrate neural tube controls progenitor competence and increases cell diversity. *PLoS Biol.* **2014**, *12*, e1001907. [CrossRef] [PubMed]

10

Functional and Comparative Genomics of *Hoxa2* Gene *cis*-Regulatory Elements: Evidence for Evolutionary Modification of Ancestral Core Element Activity

Adam Davis [1,*], Michael C. Reubens [2] and Edmund J. Stellwag [3]

1 Department of Biology and Physical Sciences, Gordon State College, Barnesville, GA 30204, USA
2 The Scripps Research Institute, 10550 N, Torrey Pines Road, MB3, La Jolla, CA 92037, USA; michaelreubens82@gmail.com
3 Department of Biology, Howell Science Complex, East Carolina University, Greenville, NC 27858, USA; stellwage@ecu.edu
* Correspondence: davisa0314@gmail.com

Academic Editor: Vincenzo Zappavigna

Abstract: *Hoxa2* is an evolutionarily conserved developmental regulatory gene that functions to specify rhombomere (r) and pharyngeal arch (PA) identities throughout the Osteichthyes. Japanese medaka (*Oryzias latipes*) *hoxa2a*, like orthologous *Hoxa2* genes from other osteichthyans, is expressed during embryogenesis in r2–7 and PA2-7, whereas the paralogous medaka pseudogene, *ψhoxa2b*, is expressed in noncanonical *Hoxa2* domains, including the pectoral fin buds. To understand the evolution of *cis*-regulatory element (CRE) control of gene expression, we conducted eGFP reporter gene expression studies with extensive functional mapping of several conserved CREs upstream of medaka *hoxa2a* and *ψhoxa2b* in transient and stable-line transgenic medaka embryos. The CREs tested were previously shown to contribute to directing mouse *Hoxa2* gene expression in r3, r5, and PA2-4. Our results reveal the presence of sequence elements embedded in the medaka *hoxa2a* and *ψhoxa2b* upstream enhancer regions (UERs) that mediate expression in r4 and the PAs (*hoxa2a* r4/CNCC element) or in r3–7 and the PAs *ψhoxa2b* r3–7/CNCC element), respectively. Further, these elements were shown to be highly conserved among osteichthyans, which suggests that the r4 specifying element embedded in the UER of *Hoxa2* is a deeply rooted rhombomere specifying element and the activity of this element has been modified by the evolution of flanking sequences that redirect its activity to alternative developmental compartments.

Keywords: Japanese medaka; *Hox* PG2 gene expression; *hoxa2a*; *ψhoxa2b*; embryonic development; *cis*-regulatory elements; vertebrate evolution; gene regulation

---

1. Introduction

Clustered *Hox* genes are a family of evolutionarily-related developmental regulatory genes that function to pattern regional tissue identities along the anterior-posterior (A-P) axis of animal species [1]. *Hox* clusters comprise up to 14 genes, which are expressed along the A-P axis during embryonic development collinear with their physical location within a cluster [2–4]. Multiple genome level duplications have expanded the total number of *Hox* clusters from one in chordates to four in tetrapods and at least seven or eight in most teleost fishes [5–11]. Post-genome duplication independent gene loss has generated clustered paralog groups (PGs) that differ in gene number depending on the historical timing of gene losses relative to genome duplications [6,12–15].

In addition to variation in gene number within and among paralog groups, results from expression and functional genetic studies have shown that duplicate *Hox* genes can exhibit either similar or heterogenous expression patterns. In the cases in which the expression patterns among paralog group members differ, they often appear to be the result of sequence divergence within *cis*-regulatory elements [6,12–21]. A particular case of heterogeneous expression patterns exhibited by duplicate genes is that of *Hoxa2* and *b2* of tetrapods. Mouse *Hoxa2* is expressed in rhombomeres (r) 2–8 of the developing hindbrain and in the cranial neural crest cells (CNCCs) that delaminate from r4 and r6/7 and populate the second pharyngeal arch (PA2) and the posterior arches, respectively [22,23]. *Cis*-regulatory elements (CREs) that direct mouse *Hoxa2* expression in r2, r4 and r3, r5 and the CNCCs are located in exon 2 of *Hoxa2*, the intron and exon 1 of *Hoxa2*, and the *Hoxa3-a2* intergenic region, respectively [21,24–34]. There has been extensive evolutionary conservation of the regulatory circuitry that directs the hindbrain and PA patterning activity of *Hoxa2*, as similar gene regulation was observed for *Hox2* of lamprey (*Petromzon marinus*) [35]. By contrast, mouse *Hoxb2* is expressed in r3–8 of the hindbrain but not in the CNCCs [36]. The CREs that direct mouse *Hoxb2* in r3, r4 and r5 are located in the mouse *Hoxb3-b2* intergenic region [36–38]. The variation in mouse *Hoxa2* and *b2* regulation and expression are mirrored by their functional divergence. Knockout experiments have shown that *Hoxa2* controls the segmentation of the anterior hindbrain and the axonal guidance of the trigeminal (Vth) and facial (VIIth) cranial motor nerve axons out of r2 and r4, respectively, while *Hoxb2* controls the specification of the somatic motor component of the VIIth cranial nerve exiting r4 [39–44]. Further, *Hoxa2*, but not *Hoxb2*, is involved in patterning the craniofacial elements arising from PA2 and the posterior arches in tetrapods [22,23,45–49].

Phylogenetic reconstructions that include a whole genome duplication event at the incipient stage of teleost evolution, which occurred 350–220 million years ago [7,50–54], support a post-genome duplication ancestral *Hox* PG2 gene complement consisting of two *Hoxa2* genes, *hoxa2a* and *a2b* (Figure 1A). The absence of a *hoxa2a* gene in zebrafish, but not in any members of the superorder Acanthopterygii, suggests that the loss of *hoxa2a* was restricted to a clade including zebrafish but not the acanthopterygians [5]. Results based on cloning and expression analyses of several acanthopterygians, including striped bass (*Morone saxatilis*), Nile tilapia (*Oreochromis niloticus*), fugu (*Takifugu rubripes*) and Japanese medaka (*Oryzias latipes*), have shown that while striped bass, Nile tilapia and fugu have two functional *Hoxa2* genes, *hoxa2a* and *a2b*, medaka has just one, *hoxa2a* [12,14,19]. The paralog of medaka *hoxa2a* was shown to be a transcribed pseudogene (*ψhoxa2b*), wherein several premature stop codons were discovered in the region corresponding to exon 2 upstream of the putative homeodomain [12]. However, the amino terminal 138 amino acid sequence was shown to be intact and contained a conserved hexapeptide motif, indicative that transcripts arising from this sequence could be translated into a product with activities like hexapeptide motif-related Pbx binding [55]. Further, while the *hoxa2a* and *a2b* genes of most teleosts are expressed in a conserved manner in much of the hindbrain (r2–5), PA2 and the posterior arches similar to orthologous genes in tetrapods [6,12,14,19,28,30,45,49,56,57], the medaka *ψhoxa2b* was shown to be expressed in noncanonical *Hox* PG2 domains, including the ventral-most aspect of the neural tube, the distal mesenchyme of the pectoral fin buds and the caudal-most region of the embryonic trunk [12]. These noncanonical expression patterns may be a reflection of sequence changes resulting from relaxed selection within the medaka *ψhoxa2b* CREs following the mutational inactivation of the homeodomain but not the hexapeptide region of the medaka *hoxa2b* coding sequence [12].

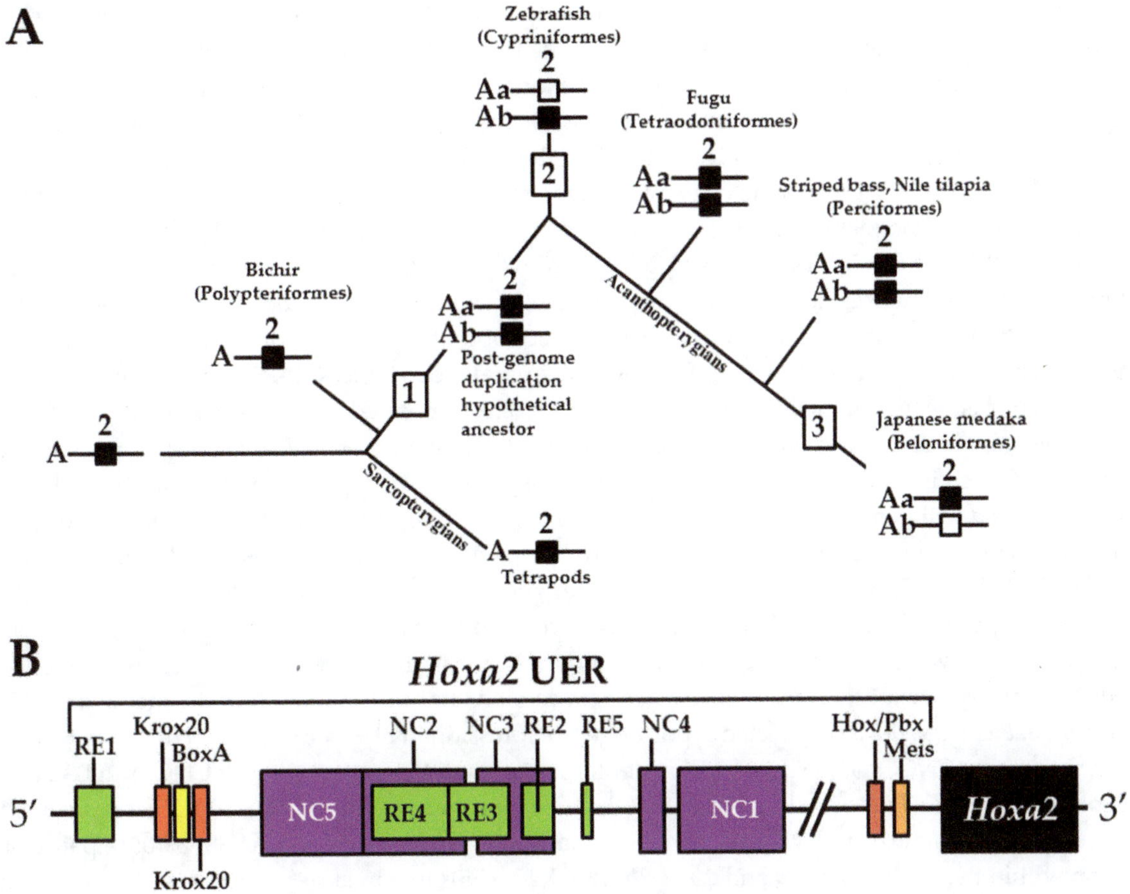

Figure 1. Evolution of *Hoxa2* gene complement and *Hoxa2* UER composition. (**A**) Phylogeny based on Steinke *et al.* (2006) [58]: (1) Genome duplication; (2) *hoxa2a* gene loss; and (3) *hoxa2b* gene loss. Filled in boxes represent functional *Hoxa2* genes. Open boxes represent loss of *Hoxa2* gene function. (**B**) Genomic map of the *Hoxa2* UER characterized in mouse. The genomic map was not drawn to scale. RE, rhombomeric element; NC, neural crest.

The *Hoxa3/Hoxa2* intergenic region is the most extensively studied functional genomic region that directs *Hoxa2* gene expression. Results from comparative and functional genomic analyses in mouse and chick embryos have identified a highly conserved upstream enhancer region (UER) that is responsible for directing *Hoxa2* in r3 and r5 of the hindbrain and the CNCCs in the PA2 and the posterior arches (Figure 1B) [21,24,25,27–32]. Identified rhombomeric CREs within this region include binding sites for Krox20, a protein shown to be integral for hindbrain development [21,30,31,34,59]. Other rhombomeric elements, termed RE1-5 and BoxA, have been shown to function in conjunction with Krox20 in potentiating *Hoxa2* expression in r3 and r5 across diverse taxa [21,25,28,30–32], which is indicative that these elements may represent a regulatory kernel subcircuit, *sensu* Davidson and Erwin (2006) [60]. Five functional CREs that direct *Hoxa2* expression in the CNCCs populating the PA2 and posterior arches that are referred to as neural crest (NC) 1–5 have been identified in the mouse [27,29]. One of these elements, NC4, encodes a binding site for AP-2, a protein that is crucial for CNCC survival and chondrocyte differentiation within the PAs [27,61]. A recent analysis provides evidence for Hox/Pbx and Meis protein binding elements within the proximal promoter region of mouse *Hoxa2* [24]. The proteins that bind these elements have been shown to be crucial for craniofacial skeleton morphogenesis, [22–24,27,45–49,56,62–66].

The present study was undertaken to examine the activity of the medaka *hoxa2a* and *ψhoxa2b* UERs in their natural background, and to understand the relationship between sequence differences of the two medaka UERs relative to the striking differences in the expression of their cognate genes, one of which, *ψhoxa2b*, is expressed in development well outside the compartments in which expression is observed for any other osteichthyan *Hox* PG2 gene [12]. Most of the CREs mentioned above have recently been shown using comparative and functional genomic analyses to be present within the *hoxa3a-a2a* and *hoxa9b-a2b* intergenic regions of teleost fishes, except for the NC4 and NC1 elements [29]. At the time that the present study was conducted for the CREs upstream of medaka *hoxa2a* and *ψhoxa2b*, only a subset of *Hoxa2* UER-specific CREs were identified within the teleost *hoxa3a-a2a* and *hoxa9b-a2b* intergenic regions, and included, 5′ to 3′, Krox20, BoxA, RE4, RE3, RE2 and RE5 [21], hereafter referred to as the UER(K20-RE5). This sequence also included NC5, NC2, and NC3 sequences, which overlap many of the RE sequences mentioned above. Further, when the medaka *hoxa2a* UER(K20-RE5) was tested for its function in chick embryos through electroporation, no reporter gene expression was observed in the hindbrain or PAs [21]. However, the paralogous UER(K20-RE5) of medaka *ψhoxa2b* yielded reporter gene staining in r3 and r5 when it was electroporated into chick embryos [21]. These results were inconsistent with *in situ* hybridization results for medaka *hoxa2a*, which showed that this gene is expressed in r2–7 of the hindbrain, PA2 and the posterior PAs, and *ψhoxa2b*, which was shown to be expressed within noncanonical *Hox* PG2 expression domains, including the pectoral fin buds [12]. Therefore, consistent with the results of *in situ* hybridization, it was hypothesized that the medaka *hoxa2a* UER(K20-RE5) region would direct reporter gene expression in r3, r5, PA2 and the posterior PAs when expressed in medaka. Similarly, it was hypothesized that the paralogous region cloned from medaka *ψhoxa2b* would direct expression of the reporter gene in the noncanonical domains mentioned above. Contrary to expectations, reporter gene expression analyses showed that the medaka *hoxa2a* and *ψhoxa2b* UER(K20-RE5)s directed reporter gene expression in r4 and r3–7, respectively, as well as in the migratory CNCCs of the hyoid and post-otic streams and the post-migratory CNCCs in PA2 and the posterior arches. These results suggest that the functional nature of the UER(K20-RE5) has diverged between medaka *hoxa2a* and *ψhoxa2b*. Deletion mapping also revealed a short 88/89 base pair (bp) region within the medaka *hoxa2a* and *ψhoxa2b* UER(K20-RE5)s that includes consensus Hox/Pbx and Prep/Meis sites that are known to direct expression in r4 for mouse *Hoxb1*, *a2*, and *b2* and the CNCCs in mouse *Hoxa2* [21,24,26,27,33,37,67,68] and likely represents a deeply rooted core CRE element. This core CRE element is present in the regulatory sequences common to gnathostome *Hox* A clusters and its effects on expression of cognate *Hoxa2* genes appears to be influenced by interactions with cis-elements outside and sequence variation within the core element itself.

2. Materials and Methods

2.1. Tol2 Plasmid Construction

Transient and stable-line transgenic analyses employed the pT2AL200L150G plasmid vector for transmitting transposon insertions into medaka embryos (generous gift from Koichi Kawakami) [69]. This plasmid vector contains the *Xenopus laevis EFI-αS* promoter upstream of *enhanced green fluorescent protein* (*eGFP*), and was used for positive controls to determine if the *Tol2* transposon system functions similarly in medaka relative to other osteichthyans. Positive controls were performed by co-microinjection of pT2AL200R150G and transposon mRNA. Negative controls were performed by co-microinjection of constructs containing the *Xenopus laevis EFI-αS* promoter without transposon mRNA. Medaka contains roughly 20–30 copies of the *Tol2* element in its genome, and these controls were performed to determine whether constructs could be integrated into the medaka genome independently of exogenously translated *Tol2* mRNA [70]. The microinjection procedure used for medaka zygotes is described below.

In order to analyze the function of the CREs in control of the medaka *hoxa2a* and *ψhoxa2b* UER(K20-RE5)s in RNA expression, the full length *EFIα-S* promoter of the pT2AL200L150G was truncated to include only the region encompassing the TATA box so as to diminish the amount of transcription occurring from the *Xenopus* promoter and maximize transcription resulting from cloned sequences of putative *cis*-regulatory elements. The TATA box from the full length *EF1α-S* promoter was amplified from pT2AL200R150G using the forward primer 5′-GATAGGGATCCAGTTCTCAGGATCGGTCG-3′ (containing the underlined BamHI sequence) and the reverse primer 5′-GGGGGCTCGAGTATATAAAGGGTGGTTAAGG-3′ (containing the underlined XhoI sequence) using standard PCR procedures. To generate pTolTATA-βMCS the *EF1α-S* promoter and *β-globin* intron were removed from pT2AL200R150G using the flanking XhoI and BamHI sites, and the amplified TATA box containing sequence was then ligated into the pT2AL200R150G digested backbone using the compatible BamHI and XhoI termini. To facilitate the ligation of potential enhancer sequences upstream of the resulting *EF1α-S* TATA box, a polylinker containing restriction sites for SmaI, EcoRI, PmeI, and EcoRV was cloned into the 5′ XhoI site. The pTolTATA-βMCS vector was also co-microinjected with and without transposon mRNA to test the efficacy of this vector system in the medaka model.

2.2. Medaka Genomic DNA Extraction

Adult medaka were anesthetized with MS-222 (0.04 *w*/*v*) prior to genomic DNA extraction. Tissues from the trunk of the fish behind the anus were homogenized in 2 mL of DNA extraction buffer (0.5% SDS, 50 mM· Tris, 100 mM· NaCl, 20 mM disodium EDTA, pH 8.0) and the resulting homogenate was poured into sterile 15 mL Falcon tubes (Becton Dickinson Labware, Franklin Lakes, NJ, USA). The tissue homogenizer was washed with an additional 2.0 mL of DNA extraction solution, which was combined with the previous homogenate in the sterile 15 mL polypropylene tube. Five microliters of RNase A (20 mg/mL) (Sigma-Aldrich, St. Louis, MO, USA) was added to the cell homogenate and the tube was incubated for 30 min at 37 °C. The homogenate was then treated with 100 μL of protease K (10 mg/mL) (Life Technologies, Carlsbad, CA, USA) and incubated at 55 °C for 6 h. The RNase A and protease K-treated homogenate was transferred to 1.5 mL centrifuge tubes (Fisher Scientific, Pittsburgh, PA, USA) and centrifuged for 30 s at 14,000 rpm using an Eppendorf 5514 C centrifuge (Hauppauge, NY, USA). The resulting supernatant was then divided into two 500 μL aliquots in sterile 1.5 mL centrifuge tubes (Fisher Scientific), and DNA was purified by two Phenol Chloroform Isoamyl alcohol (PCI) extractions as follows: 500 μL of PCI (Life Technologies, Carlsbad, CA, USA) was added to each 1.5 mL tube and mixed until a homogeneous emulsion formed. Microcentrifuge tubes containing the emulsion were centrifuged for 30 s at 14,000 rpm, and the aqueous layer was removed from the organic layer and then placed in a sterile 1.5 mL microcentrifuge tube. After a second PCI extraction, the aqueous phase was treated with 5 M· NaCl, which was added to the aqueous extract, such that the final concentration of NaCl in the tube would be 0.3 M after the addition of 2 volumes of 100% ethanol (EtOH, Pharmco-Aaper, Brookfield, CT, USA). After addition of the ethanol, the tubes were mixed gently to allow the DNA to precipitate, after which the precipitated DNA was collected by centrifugation at 10,000× *g* for 1 min. The supernatant was removed by pipetting and the sedimented DNA was washed twice with 70% EtOH and incubated at 37 °C until dry. Once dry, the DNA was suspended in 100 μL of Nuclease Free Water (Life Technologies, Carlsbad, CA, USA). The quality of the genomic DNA was assayed by agarose gel electrophoresis on a 0.5% gel. The migration of the genomic DNA was compared to that of the DNA fragments in the High Molecular Weight standard (Life Technologies, Carlsbad, CA, USA).

2.3. Amplification of Medaka hoxa2a and ψhoxa2b UER(K20-RE5)s

Amplification of genomic DNA corresponding to *cis*-regulatory elements in the upstream DNA sequences, exons 1 and 2 and the intronic DNA of medaka *hoxa2a* and *ψhoxa2b* was performed using long-range PCR. Primers and their hybridization coordinates to medaka genomic DNA with respect to ATG start site of medaka *hoxa2a* and *ψhoxa2b* are listed in Table 1. All primers for all analyses were developed using published genomic sequences of the Japanese medaka [71]. The PCR products generated from long-range PCR were cloned into pCR II vectors (Life Technologies, Carlsbad, CA, USA) according to the manufacturer's instructions. Confirmation and orientation of PCR products corresponding to inserts from plasmid genomic DNA clones were determined by restriction endonuclease digestion. These clones were then used to amplify the UER(K20-RE5)s of medaka *hoxa2a* and *ψhoxa2b*, which were tested for enhancer activity in the pTolTATA-βMCS plasmid vector.

Table 1. Primers used for the amplification of the medaka *hoxa2a* and *ψhoxa2b* UER(K20-RE5)s and the exclusion of sequence elements for the functional testing of these regulatory regions. The 5′-start site for primers pertains to the nucleotide base position with respect to the ATG start site of the downstream gene (*hoxa2a*) or pseudogene (*ψhoxa2b*).

Primer	Sequence 5′ to 3′	5′ Start Site
Medaka *hoxa2a* Genomic Primers		
A2a For	TTATTCCCACAACCCTTTCATTTCG	−2691
A2a Rev	CACACTCAGCCACAATCTCTTCTTC	1846
Medaka *ψhoxa2b* Genomic Primers		
A2b For	ACACAGCAGGGGTCAACAATAGGTC	−3093
A2b Rev	ATAGGCAGAGCACGAAAACAAAATG	3193
Medaka *hoxa2a* UER(K20-RE5) Forward Primers		
AF1	GATCGATATCGAACAGGCTGAAATCCACTGAATGC	−1778
AF2	GATCGATATCGCTTCTAATCTGAGAAGCCAGTGTTTC	−1468
AF3	GATCGATATCATGTGTTGCGAGGGCACCGAGCTGTC	−1392
AF4	GATCGATATCGAGTAAGATTGATCGCGCACAGGCTTC	−1354
Medaka *hoxa2a* UER(K20-RE5) Reverse Primers		
AR1	GATCGAATTCGTTTGCTGTGGAACAGAGGAAAGAAG	−1247
AR2	GATCGAATTCTTATATACCAAACAAAGAGTCCTGG	−1303
AR3	GATCGAATTCTTACTCGCCAAAAGGTCTGACAGCTC	−1348
Medaka *ψhoxa2b* UER(K20-RE5) Forward Primers		
BF1	GATCGATATCATGTGCCAACACCCACTCACCCCAG	−1068
BF2	GATCGATATCCTTCGCTCCGCACCGAGGGCATCCTC	−868
BF3	GATCGATATCATGTTCTCTAAGGGCAAAGAGCTGTC	−803
BF4	GATCGATATCTGGAAAGATTGATCACACAGAATACC	−765
Medaka *ψhoxa2b* UER(K20-RE5) Reverse Primers		
BR1	GATCGAATTCAAAAAGCTGCAGGAAAAGGAGGGGATC	−671
BR2	GATCGAATTCCCGGGCTCTGAACAAAAGATTCCTG	−715
BR3	GATCGAATTCTTTCCAGCCAAGAGCTCTGACAGCTC	−759

2.4. PCR-Mediated Deletion Mutagenesis of the Medaka hoxa2a and ψhoxa2b UER(K20-RE5)s

DNA sequences that contained the UER(K20-RE5)s of medaka *hoxa2a* and *ψhoxa2b* were amplified using PCR primers listed in Table 1. These primers were used to generate a product referred to as the UER(K20-RE5), which encompassed sequence elements corresponding to Krox20, BoxA, RE4, RE3, RE2, RE5, NC2, NC3 and NC5, and was similar to the region tested in chicken embryos [21], and a series of nested deletions that were derived from the UER(K20-RE5). Specific primer pairs and their amplified genomic DNA products are listed in Table 2. All 5′-located primers started with the sequence 5′-GATCGATATC-3′ and 3′-located primers started with the sequence 5′-GATCGAATTC-3 to ensure that the PCR products could be digested with EcoRI and EcoRV restriction digestion enzymes and oriented 5′ to 3′ with respect to the TATA box and green fluorescent protein (GFP) sequences located downstream in the pTolTATA-βMCS vector. PCR products were purified using the QIAquick PCR purification kit (Qiagen, Valencia, CA, USA) according to the manufacturer's instructions, restriction endonuclease digested with EcoRI and EcoRV and then ligated into the pTolTATA-βMCS plasmid vector using T4-DNA ligase (New England Biolabs, Ipswich, MA, USA). The pTolTATA-βMCS plasmid vector was pre-digested with EcoRI and EcoRV, purified using the QIAquick PCR sequence kit (Qiagen, Valencia, CA, USA) and treated with shrimp alkaline phosphatase (New England Biolabs, Ipswich, MA, USA) to remove phosphate groups at the digested sites of the vector to minimize vector religation. Plasmid vectors containing PCR-amplified genomic DNAs were transformed into *Escherichia coli (E. coli)* JM109 cells. *E. coli* cells were screened for recombinants by digestion of plasmid isolated from transformants with EcoRI and EcoRV after purification of plasmids using the GenElute HP Plasmid Miniprep kit (Sigma-Aldrich, St. Louis, MO, USA) according to the manufacturer's instructions. Further confirmation of cloned PCR products in the pTolTATA-βMCS plasmid vector was performed using DNA sequencing of the inserts by dideoxyterminator sequencing chemistry (Big Dye v. 3.0, Applied Biosystems, Foster City, CA, USA) on a 3130xl Genetic Analyzer (Applied Biosystems, Foster City, CA, USA).

Table 2. Primer pairs used for the amplification of sequences of the medaka *hoxa2a* and *ψhoxa2b* UER(K20-RE5)s used in functional genomic analyses. Schematics of constructs, frequencies of transient transgenic embryos (F0s) showing reporter gene expression in the hindbrain and CNCCs, and numbers of stable-line transgenic adults (F1s) are shown. K, Krox20, B, BoxA, R, Rhombomeric element, *, embryos showing nearly undetectable *eGFP* expression in the hindbrain and pharyngeal arches.

Construct	Primer Pairs	Amplicon Length	Construct Schematic	Hindbrain Expression (F0)	CNCC Expression (F0))	F1s
Medaka *hoxa2a* UER(K20-RE5) construct design and transgenic analysis						
1	AF1/AR1	531 bp	5′ K B R4 R3 R2 R5 3′	42/48 (87.5%)	42/48 (87.5%)	3
2	AF2/AR1	221 bp	5′ R4 R3 R2 R5 3′	64/84 (76%)	56/84 (67%)	3
3	AF3/AR1	145 bp	5′ R3 R2 R5 3′	39/49 (80%)	41/49 (84%)	0
4	AF4/AR1	107 bp	5′ R2 R5 3′	7/47 (15%) *	23/49 (49%) *	0
5	AF1/AR2	475 bp	5′ K B R4 R3 R2 3′	42/50 (84%)	42/50 (84%)	4
6	AF1/AR3	430 bp	5′ K B R4 R3 3′	0/52 (0%)	0/52 (0%)	0
7	AF3/AR2	89 bp	5′ R3 R2 3′	52/62 (84%)	52/62 (84%)	4
Medaka *ψhoxa2b* UER(K20-RE5) construct design and transgenic analysis						
8	BF1/BR1	397 bp	5′ K B R4 R3 R2 R5 3′	46/51 (90%)	46/51 (90%)	2
9	BF2/BR1	197 bp	5′ R4 R3 R2 R5 3′	47/52 (90%)	47/52 (90%)	4
10	BF3/BR1	132 bp	5′ R3 R2 R5 3′	33/38 (87%)	33/38 (87%)	3
11	BF4/BR1	94 bp	5′ R2 R5 3′	27/52 (52%) *	27/52 (52%)*	0
12	BF1/BR2	353 bp	5′ K B R4 R3 R2 3′	33/41 (80%)	33/41 (80%)	3
13	BF1/BR3	309 bp	5′ K B R4 R3 3′	9/64 (14%) *	15/64 (23%) *	0

2.5. Microinjection of Medaka Embryos

Cultivation of medaka was performed as previously described [12]. Zygotes (stage 1) were collected from several females to ensure genetic variation per each injection experiment [72]. Zygotes were transferred to ice-cold 1× medaka embryo rearing medium (ERM) (17 mM· NaCl, 0.4 mM· KCl, 0.66 mM· $MgSO_4 \cdot 7H_2O$, 0.27 mM· $CaCl_2 \cdot 2H_2O$) between 30 min to 2 h to arrest development [73]. Zygotes were then physically transferred to specially fabricated agarose "corrals" that were pre-chilled at 4 °C and that hold the zygotes in an orientation appropriate for microinjection. The agarose corrals were 1 mm wide × 1 mm deep × 4 mm long and were made using 60 mL of 1.5% agarose (Sigma-Aldrich, St. Louis, MO, USA). To reduce high internal egg pressures when microinjecting, the agarose corrals were overlaid with 10 mL of 15% Ficoll 400 (Sigma-Aldrich, St. Louis, MO, USA) pre-chilled at 4 °C. Zygotes were placed in their corrals and were allowed to equilibrate for at least

40 min in 15% Ficoll 400 on ice prior to microinjection. Zygotes were microinjected using needles derived from filamented borosilicate glass capillaries that had an outer diameter of 1 mm and an inner diameter of 0.58 mm (World Precision Instruments, Sarasota, FL, USA). Glass needles were pulled on a Sutter P-97 needle puller (Novato, CA, USA) housing a 1.5 mm trough filament using the following parameters: heat of 250, pull of 200, velocity of 100, time of 200 and pressure of 400. To avoid clogging from chorions, the tips of the needles were beveled at a 45° angle using a Narishige Micro-grinder (Tokyo, Japan). After beveling the tips, glass needles were stored by embedding their mid-sections in rounded stripes of Play-Doh (Hasbro, Pawtucket, RI, USA) housed in 100 mm petri dishes with dampened Kimwipes (Kimberly Clark®, Irving, TX, USA) at 4 °C. The Play-Doh prevented glass needles from movement and breakage in the Petri plates. The dampened Kimwipe maintained a moist environment inside the Petri dishes and kept the stored needles from clogging at the beveled ends. Microinjection of solutions into medaka zygotes with beveled needles was performed using a PV 820 pneumatic picopump (World Precision Instruments, Sarasota, FL, USA). Twenty pounds per square inch (psi) of eject pressure and 3 psi of hold pressure were used for the medaka zygote microinjection process. Zygotes were microinjected with the following solutions for analyzing transient transgenic embryos: 25 ng/μL plasmid DNA, 25 ng/μL *Tol2* transposase RNA, 1× Yamamoto Buffer (128 mM· NaCl, 2.7 mM· KCl, 1.8 mM· $CaCl_2$, 0.24 mM· $NaHCO_3$, pH 7.3). After microinjection, embryos were carefully removed from their individual agarose corrals and transferred in approximately 1 mL of 15% Ficoll to a Petri dish containing 20 mL 1× ERM. After transfer they were allowed to equilibrate to less than 1% Ficoll for roughly 2 h at room temperature (RT) without shaking. The medium housing the embryos was then replaced with fresh 1× ERM and the embryos were incubated at 28.5 °C to continue development.

2.6. Generation and Visualization of Transient and Stable-Line Transgenic Medaka Embryos

Microinjected embryos were reared in 1× ERM at 28.5 °C. After 24 h, embryos were visualized under a Leica dissecting microscope. Dead eggs and embryos that showed extremely defective morphologies (*i.e.*, gastrulation defects) were discarded. All other embryos were transferred in 1× ERM containing 0.1 mM phenylthiourea (Sigma-Aldrich, St. Louis, MO, USA) to reduce pigmentation [74]. Embryos were visualized for *eGFP* expression using a Zeiss inverted compound microscope (Thornwood, NY, USA) at selected times during development, focusing primarily on developmental stage 29/30 (74–82 hpf [72]). At stage 29/30, the hindbrain and pharyngeal arches are easily distinguished morphologically in medaka embryos when they are in the chorion. Since embryos at stage 29/30 have guanophores that auto-fluoresce under GFP filters, illumination of specimens using both GFP and rhodamine filters was performed to differentiate positive *eGFP* signal from auto-fluorescing pigmentation. Images of *eGFP* expression were processed using Axiovision AC 4.4 software. Depending on the strength of the *eGFP* signal, camera exposure times ranged from 1 to 2 s for transient transgenic embryos. Brightfield images were also taken of medaka embryos to determine the relative origin of the *eGFP* signal within the hindbrain and pharyngeal arches in transient transgenic embryos. The developing otic vesicle was used as a morphological landmark to determine whether the *eGFP* signal was occurring anteriorly or posteriorly in the hindbrain. Rhombomeres 3 and 4 develop dorsal to the anterior region of the otic vesicle while r5, r6 and r7 develop above the mid- to posterior region of the otic vesicle.

Transient transgenic medaka embryos that were observed to be positive for *eGFP* expression in the hindbrain and pharyngeal arches showed mosaic expression in comparisons among embryos. For this reason, transient transgenic embryos that showed strong *eGFP* signal in the hindbrain and/or pharyngeal arches were reared to adulthood and mated with wild-type fish. We observed that more robust *eGFP* expression patterns were detected in the hindbrain and pharyngeal arches of stable-line transgenic medaka embryos compared to transient transgenics. Transient transgenic embryos were reared until hatching in 1× ERM at 28.5 °C. Hatched embryos were reared in 1× ERM in breeding nets in 5 gallon tanks until they were large and strong enough to swim against filter-generated currents. F1 medaka embryos were assayed for *eGFP* expression in the hindbrain and pharyngeal arches according to the procedure outlined above for transient transgenic embryos.

2.7. Whole-Mount in Situ Hybridization

To corroborate the results of *eGFP* expression using a system independent of fluorescence, the expression of *eGFP* in embryos was visualized using *in situ* hybridization analysis with an anti-*eGFP* riboprobe. This method allowed for the visualization of *eGFP* expression in the absence of fluorescence originating from auto-fluorescing pigment cells, which was often the case when embryos were visualized under a GFP filter. We used whole-mount *in situ* hybridization to visualize *eGFP* transcripts at developmental stages 22 (nine somites), 29/30 and 34 (121 hpf) [72]. We assayed embryos at stage 22 because we wanted to determine if the CREs of the medaka *hoxa2a* and *ψhoxa2b* UER(K20-RE5)s were directing reporter gene expression in the migratory CNCCs of the hyoid and post-otic streams, at stage 29/30 to visualize the exact rhombomeres and pharyngeal arches that were expressing *eGFP* and at stage 34 to determine if the CREs of the medaka *hoxa2a* and *ψhoxa2b* UER(K20-RE5)s were directing expression in post-migratory CNCCs at the chondrogenic stages of pharyngeal arch development.

Embryos from stable-line transgenic medaka fish were collected, reared, anesthetized, fixed in 4% paraformaldehyde (PFA) and dehydrated as previously described [12]. Medaka embryos were developmentally staged according to Iwamatsu (2004) [72]. Whole-mount *in situ* hybridization was performed according to Davis *et al.* (2008) [12]. All experiments used digoxigenin (DIG)-labeled sense and antisense riboprobes that were produced and purified according to Scemama *et al.* (2006) [19]. Sense riboprobes were used in control experiments to assess nonspecific binding. Development of DIG-labeled probe signal, examination of embryos and digital photography of embryos was performed as described in Scemama *et al.* (2006) [19]. Morphological landmarks, including the midbrain/hindbrain boundary, rhombomeres (r), otic vesicles (OV), pectoral fins (PF), pharyngeal arches (PA) and somites (s) within developing embryos were used to define the location of *eGFP* expression. *eGFP* signal was also determined using double whole-mount *in situ* hybridization assays with DIG-labeled antisense-*eGFP* riboprobes and DIG-labeled antisense-*hoxd3a* riboprobes or fluorescein-labeled *hoxb1a* riboprobes. Medaka *hoxd3a* is expressed in r6–8 of the hindbrain and *hoxb1a* is expressed in r4 [13,16]. Production of fluorescein riboprobes and double whole-mount *in situ* hybridization using DIG and fluorescein-labeled riboprobes was performed as documented in Scemama *et al.* (2006) [19].

2.8. Comparative Genomic Sequence Analysis

Genomic DNA sequences corresponding to the UER(K20-RE5)s of medaka *hoxa2a* and *ψhoxa2b* that were shown to be required for directing reporter gene expression in the rhombomeric and PA embryonic domains using functional genomic analyses outlined above were examined for putative transcription factor binding sites using JASPAR [75]. We also performed a comparative genomic sequence analysis of the UER(K20-RE5)s between medaka *hoxa2a* and *ψhoxa2b* and the UER(K20-RE5)s of other vertebrate *Hoxa2* genes using the software programs Dialign-TX and CLUSTALX in order to determine if the putative transcription factor binding sites were conserved in other vertebrates [76,77]. The *Hoxa2* UER(K20-RE5)s from genomic sequences of medaka (accession numbers AB232918 and AB232919), fugu (accession numbers DQ481663 and DQ481664), zebrafish (accession number AL645795, direct submission), Nile tilapia (AF533976 and GCA_000188235.1), bichir, mouse (accession number: NC000072) , human (accession number: NG012078, direct submission), chicken (accession number: AC163712, direct submission), ceolacanth (accession number: FJ497005), horn shark (accession number: AF224262), dogfish (accession number: FQ032658), and western clawed frog (accession number: NW004668239) were used for sequence comparison [71,78–85]. All default and recommended parameters were used in the Dialign-TX and CLUSTALX programs for conducting genomic DNA sequence comparisons.

3. Results

3.1. *Validation of the Tol2 Transposon System for Medaka Embryos*

In order to determine if the *Tol2* transposon system functions similarly in medaka compared to zebrafish, we performed experiments using the pT2AL200L150G vector. This plasmid vector contains the *Xenopus laevis* eFI-αS promoter upstream of *eGFP*. We expected that positive control embryos, which were co-microinjected with pT2AL200L150G vector and *Tol2* transposon mRNA, would show green fluorescence during development. However, we were unsure if negative control embryos, which were microinjected solely with pT2AL200L150G vector, would not exhibit green fluorescence, since the *Tol2* element is present within the medaka genome [70]. All positive control embryos showed strong reporter gene expression throughout the body of the medaka embryos (68/68; 100%) (Figure 2A,C) while negative control embryos failed to show any reporter gene expression (0/52; 0%) (Figure 2B,D), which suggested that *eGFP* expression was not generated from microinjected vectors in the absence of exogenous *Tol2* transposon mRNA. Further, medaka zygotes microinjected with the pTolTATA-βMCS vector alone (0/49, 0%) or co-microinjected with vector and transposon mRNA (0/64, 0%) lacked detectable fluorescence (0/64, 0%) (Figure 2B,D). These results showed that the *Tol2* transposon system and the pTolTATA-βMCS vector can be used in medaka embryos for studying *cis*-regulatory element control of gene expression.

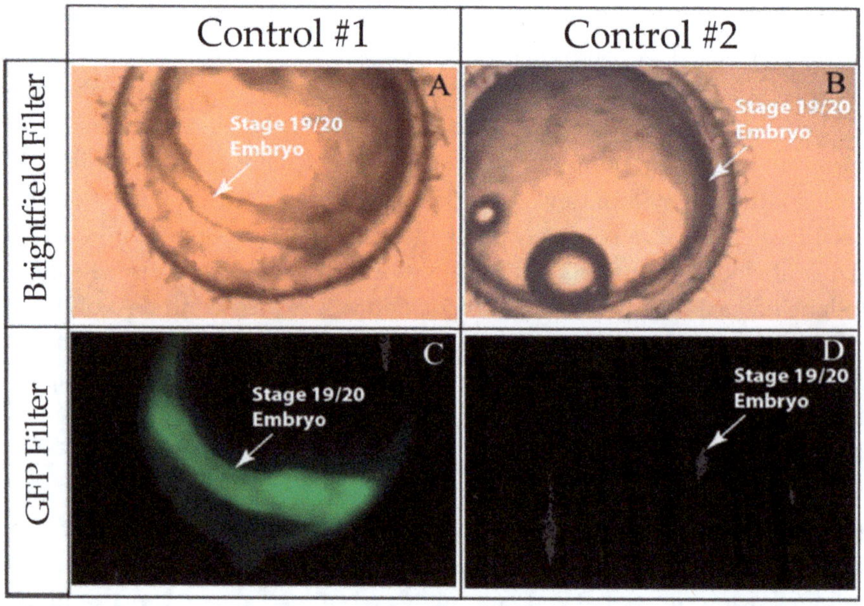

Control Experiment #	Constructs	% Positives
1	pT2AL200L150G + Transposon mRNA	68/68 (100%)
2	pT2AL200L150G - Transposon mRNA	0/52 (0%)
3	pTolTATA-βMCS + Transposon mRNA	0/64 (0%)
4	pTolTATA-βMCS - Transposon mRNA	0/49 (0%)

Figure 2. Control experiments used for the validation of the *Tol2* transposon system for medaka embryos. (**A**) Brightfield image of representative positive control embryo (pT2AL200L150G + Transposon mRNA); (**C**) Green fluorescent protein (GFP) filter image of the same embryo shown in A; (**B**) brightfield image of representative negative control embryo (pT2AL200L150G − Transposon mRNA); and (**D**) GFP filter image of the same embryo shown in B. No embryos in any of the other negative control experiments showed eGFP expression (pTolTATA-βMCS + Transposon mRNA or pTolTATA-βMCS-Transposon mRNA).

3.2. *Functional Genomic Analysis of the Medaka Hoxa2a UER(K20-RE5)*

Medaka *hoxa2a* is expressed in a relatively conserved manner in the hindbrain and pharyngeal arches similar to *hoxa2a* and *a2b* genes from teleosts, including zebrafish, tilapia, striped bass and fugu and *Hoxa2* genes of tetrapods, including the mouse, chicken and frog [6,12,14,19,23,24,31,45,49,56,57]. Based on this extensively conserved pattern of expression, we hypothesized that the medaka *hoxa2a* UER(K20-RE5) would direct reporter gene expression in r3 and r5 of the hindbrain and in the CNCCs entering PA2 and the posterior arches.

Contrary to our hypothesis, transient and stable-line transgenic medaka embryos showed that the construct containing the medaka *hoxa2a* UER(K20-RE5) (Construct #1) directed reporter gene expression in r4 of the hindbrain and in PA2 and the posterior PAs (Figure 3A–G). Even though we observed *eGFP* expression in the hindbrain (87.5%) and pharyngeal arches (87.5%) in a high percentage of transgenic embryos (Table 2), both direct fluorescence and whole-mount *in situ* hybridization with riboprobes directed against eGFP expressed in stable-line transgenic embryos showed that the *eGFP* expression in the hindbrain was restricted to r4 (Figure 3D–G). This was unexpected given that the UERs of chicken and mouse *Hoxa2* or zebrafish *hoxa2b* directed reporter gene expression in r3 and r5 [21,25,27–32]. This restricted and unusual pattern of expression prompted us to examine the validity of rhombomere assignments. To authenticate the rhombomere assignments, additional whole-mount *in situ* hybridization experiments were conducted using medaka *hoxb1a* and *hoxd3a*, which are expressed in r4, and r6–8 of the hindbrain, respectively, but not in the pharyngeal arches. [13,16]. Medaka *hoxb1a* antisense riboprobes were labeled with fluorescein. Both fluorescein-labeled *hoxb1a* and DIG-labeled *eGFP* riboprobes were observed to hybridize to their mRNA targets in r4 and to overlap the region that hybridizes to the *hoxa2a* UER(K20-RE5) reporter construct #1. (Figure 3H). When DIG-labeled *eGFP* and *hoxd3a* antisense riboprobes were used in tandem for *in situ* hybridization experiments, a one-rhombomere gap without any DIG-stained cells between the *eGFP* and *hoxd3a* expressing rhombomeres was observed (Figure 3I). These results are consistent with expression of *eGFP* in r4 with a gap in expression corresponding to r5 and *hoxd3a* expression in r6–8 (Figure 3I), and show that the medaka *hoxa2a* UER(K20-RE5) reporter expression was restricted to just r4. Further, *eGFP* riboprobes were observed to hybridize to their mRNA targets in the migratory CNCCs of the hyoid and post-otic streams at developmental stage 22, the post-migratory CNCCs in PA2 and the posterior pharyngeal arches at stage 29/30 and the chondrogenic CNCCs in PA2 and the posterior arches at stage 34 (Figure 4A–C). These results confirm that the medaka *hoxa2a* UER(K20-RE5) construct #1 directs reporter gene expression in the CNCCs. Moreover, the presence of reporter gene expression within the CNCCs in the PAs of medaka is consistent with results from previous studies for orthologous UER sequences from mouse *Hoxa2*, fugu *hoxa2a* and zebrafish *hoxa2b* [24,27,29]. Interestingly, we observed more robust staining in *in situ* hybridization analyses using *eGFP* riboprobes in the posterior PAs than in PA2, which suggests that the sequence elements in the medaka *hoxa2a* UER(K20-RE5) construct #1 are expressed more strongly in the posterior PAs than for PA2, which is the converse of that observed for the cognate gene transcripts.

To pinpoint the CREs responsible for directing the r4 and PA-specific reporter gene expression in medaka embryos, a comparison of expression in transient and stable-line transgenic embryos generated with a series of nested deletion constructs extending from the 5′- and 3′-ends of the medaka *hoxa2a* UER(K20-RE5) was performed (Constructs #2-7) (Table 2). These deletions eliminated regions corresponding to previously mapped enhancer elements including Krox20, BoxA, RE4, RE3, RE2 and RE5 [21,25,28–32]. Similar to transient transgenic embryos generated with Construct #1, microinjection of constructs that included 5′-deletions of Krox20 and BoxA binding sites (Construct #2) or Krox20, BoxA and RE4 (Construct #3) yielded high percentages of transient transgenic embryos with *eGFP* expression in the hindbrain (76%, Construct #2; 80%, Construct #3) and PAs (67%, Construct #2; 84%, Construct #3) (Table 2). Further, stable-line transgenic embryos generated with Construct #2 showed similar *eGFP* expression patterns to transgenic embryos generated with constructs containing the entire medaka *hoxa2a* UER(K20-RE5) (Construct #1). Specifically, we observed *eGFP* expression in r4 of the hindbrain, the migratory CNCCs of the hyoid and post-otic streams and the post-migratory and chondrogenic CNCCs of PA2 and the posterior PAs (Figure 4D–F). Like the full-length UER(K20-RE5) (construct #1) we observed stronger *eGFP* expression in the posterior PAs than in PA2. Unfortunately, we did not have success in obtaining stable-line transgenic embryos using Construct #3 (data not shown). Further, the extreme mosaic nature of the expression of the transient transgenic embryos treated with Construct #3 precluded us from drawing conclusions about the expression patterns. A much lower percentage of transient transgenic embryos generated with Construct #4, which included deletions of Krox20, BoxA, RE4 and RE3 sequences and only retained RE2 and RE5 sequences (Table 2), showed *eGFP* expression in the hindbrain (15%) and PAs (15%) when compared to embryos generated with Constructs #1, 2 and 3. Further, these embryos showed *eGFP* levels of expression that were barely detectable in the hindbrain and PAs and did not allow us to draw any conclusions of the expression patterns (data not shown). Unfortunately, we did not obtain any stable-line transgenic embryos treated with Construct #4 (data not shown). These results suggested that sequence elements downstream of RE4 within the medaka *hoxa2a* UER(K20-RE5), specifically, RE3, RE2 and RE5, are necessary for directing medaka *hoxa2a* expression in r4 and the CNCCs of PA2 and the posterior arches.

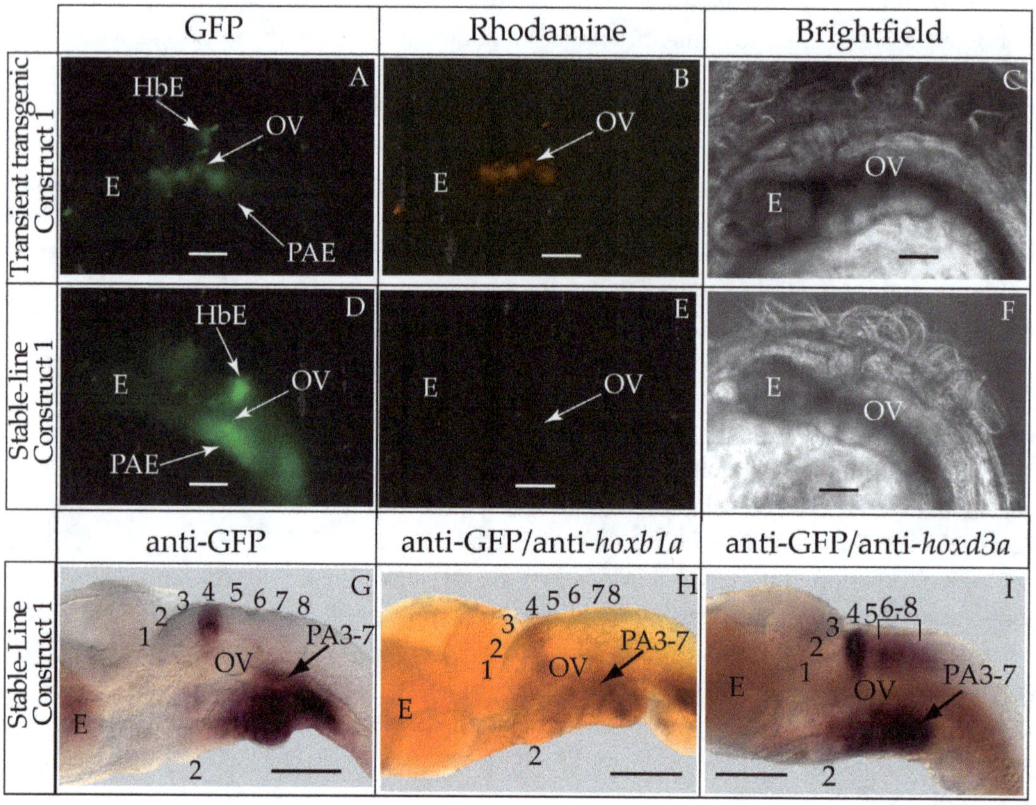

Figure 3. Transient (**A–C**) and stable-line (**D–I**) transgenic data from the medaka *hoxa2a* UER(K20-RE5) (Construct #1). (**A–F**) Images of transient (**A–C**) and stable-line (**D–F**) transgenic embryos at stage 29/30 (72–84 hpf) were taken using: GFP (**A**,**D**); rhodamine (**B**,**E**); and brightfield (**C**,**F**) filters. All embryos are still in their chorions and are positioned with their anterior sides to the left and their lateral sides to the reader. (**G–I**) Whole-mount *in situ* hybridization of medaka embryos using DIG-labeled anti-*eGFP* riboprobe (**G**). DIG-labeled anti-*eGFP* riboprobe with fluorescein-labeled anti-*hoxb1a* riboprobe (**H**) and DIG-labeled anti-*eGFP* and anti-*hoxd3a* riboprobes (**I**). Embryos were mounted with anterior sides facing left and lateral sides facing the reader. Rhombomere numbers are indicated by black numbers above the dorsal sides of the embryos. Pharyngeal arch 2 is indicated by the number 2 below the ventral sides of the embryos. Medaka *hoxd3a* expressing rhombomeres (r6–8) are indicated by the bracket in I. E, eye; HbE, hindbrain expression; OV, otic vesicle; PA, pharyngeal arch; PAE, pharyngeal arch expression. Scale bars equal 0.1 mm.

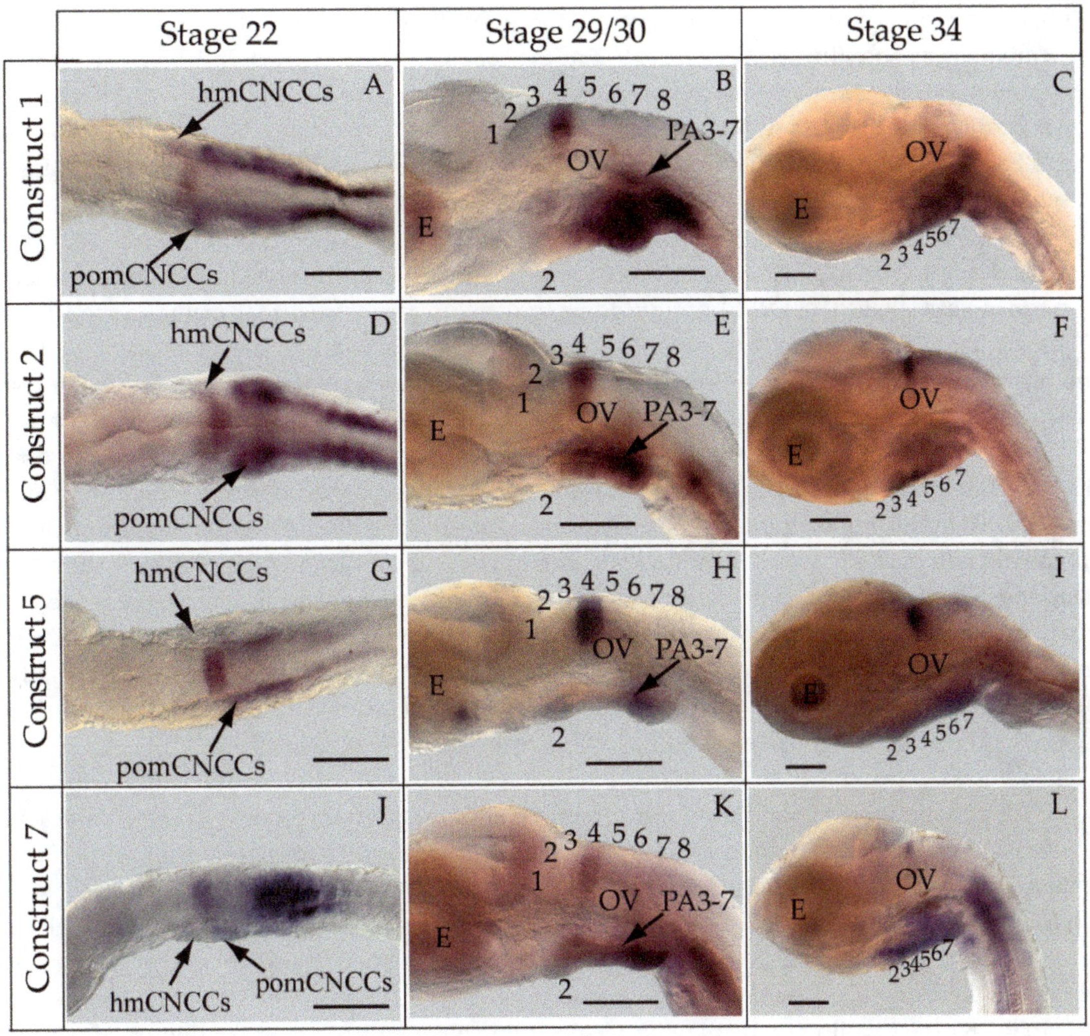

Figure 4. Whole-mount *in situ* hybridization of *eGFP* in stable-line *hoxa2a* UER(K20-RE5) transgenic medaka embryos generated with Construct #1 (**A–C**); Construct #2 (**D–F**); Construct #5 (**G–I**); and Construct #7 (**J–L**) at stages 22 (9 s) (**A,D,G,J**); 29/30 (72–84 hpf) (**B,E,H,K**); and 34 (121 hpf) (**C,F,I,L**). (**A,D,G,J**) Embryos were mounted with their anterior sides facing left and their dorsal sides facing the reader. (**B,C,E,F,H,I,K,L**) Embryos were mounted with their anterior sides facing left and lateral sides facing the reader. (**B,E,H,K**) Images are magnified to show rhombomere placement. Rhombomere numbers are indicated by black numbers above the dorsal sides of the embryos. Pharyngeal arches are indicated by black numbers below the ventral sides of the embryos. E, eye; hmCNCCs, hyoid migratory cranial neural crest cells; OV, otic vesicle; PA, pharyngeal arch; pomCNCCs, post-otic migratory cranial neural crest cells. Scale bars equal 0.1 mm.

The microinjection of constructs with the 3′-specific deletion of RE5 from the medaka *hoxa2a* UER(K20-RE5) but retention of Krox20, BoxA, RE4, RE3 and RE2 (Construct #5) yielded similar results to that of Construct #1, namely a high percentage of transient transgenic embryos with *eGFP* expression in the hindbrain (84%) and PAs (84%) (Table 2). Further, stable-line transgenic embryos generated with Construct #5 showed *eGFP* expression in r4 of the hindbrain, the migratory CNCCs of the hyoid and post-otic streams, and the post-migratory and chondrogenic CNCCs in PA2 and posterior arches (Figure 4G–I). However, embryos injected with Construct #5 showed reduced *eGFP* expression levels in the posterior PAs when compared to embryos treated with Constructs #1 and #2 (Figure 4B,E,H). Interestingly, embryos injected with Construct #6, in which both RE2 and RE5 were deleted but Krox20, BoxA, RE4, and RE3 were retained, did not show any *eGFP* expression in the hindbrain (0%) or in the pharyngeal arches (0%) (Table 2). Overall, these results showed that the sequence containing the elements corresponding to RE3 and RE2 of the medaka *hoxa2a* UER(K20-RE5) is required for r4 expression but the surrounding elements corresponding to Krox20, BoxA, RE4 and RE5 are not. Similar results were also observed for the reporter gene expression in the PAs, such that the loss of either RE3 or RE2 sequences yielded a loss in CNCC expression (Constructs #4 and 6) in transient transgenic embryos.

The aforementioned *eGFP* expression patterns obtained from the nested deletion constructs of the medaka *hoxa2a* UER(K20-RE5) prompted us to develop a construct that included the RE3 and RE2 elements only (Construct #7; Table 2). This construct was 89 bp in length and spanned from genomic bp positions −1392 to −1303 (Table 2). As expected based on the results from earlier tested constructs containing both RE3 and RE2 (Constructs #1, 2, 3 and 5), a high percentage of transient transgenic medaka embryos generated with Construct #7 showed *eGFP* expression in the hindbrain (84%) and PAs (84%) (Table 2). Stable-line transgenic medaka embryos confirmed that the sequence elements within this region directed *eGFP* expression in r4 of the hindbrain, the migratory CNCCs of the hyoid and post-otic streams and the post-migratory and chondrogenic CNCCs of PA2 and the posterior arches (Figure 4J–L). Further, in comparison to stable-line transgenic embryos generated with constructs containing RE3, RE2 and other surrounding elements (Constructs #1, 2 and 5), embryos generated with Construct #7 showed much more robust *eGFP* expression in PA2. Altogether, these results indicate that this 89 bp sequence contains elements necessary and sufficient to direct *Hoxa2* expression in r4, PA2 and the posterior PAs in the absence of other elements.

To identify putative transcription factor binding sites in the 89 bp region of the medaka *hoxa2a* UER(K20-RE5), hereafter referred to as the r4/CNCC-specifying element or core element, we performed a comparative genomic sequence analysis of this DNA sequence fragment with orthologous sequences from *Hoxa2* UERs of divergent gnathostomes , including horn shark, dogfish, coelacanth, frog, chicken, mouse, human, bichir, zebrafish, tilapia, fugu and medaka. Sequence alignments revealed several regions that were highly conserved among gnathostome *Hoxa2* UER genomic sequences but restricted to the region corresponding to the r4/CNCC-specifying element from medaka (Figure 5). Analysis of these sequences showed that they corresponded to a Prep/Meis binding site (5′-CTGTCA-3′) beginning at bp position −1371 of the medaka *hoxa2a* UER(K20-RE5) and a Hox/Pbx binding site (5′-AGATTGATCG-3′) beginning at bp postion −1349 (Figure 5). Interestingly, the Prep/Meis and Hox/Pbx sites of the medaka *hoxa2a* UER(K20-RE5) were observed to be identical in sequence to the functionally mapped Prep/Meis and Hox/Pbx binding sites in the mouse *Hoxb3-b2* intergenic region [37], which were shown to direct mouse *Hoxb2* expression in r4 and the migratory CNCCs [37]. To further support the concept of the r4/CNCC element as a core element, it is important to note that Hox/Pbx and Meis binding sites located upstream of mouse *Hoxa2*, but downstream of the sequence orthologous to the medaka r4/CNCC element, were shown to direct *Hoxa2* expression in the CNCCs of PA2 and the posterior arches [24]. Likewise, Hox/Pbx and Prep/Meis binding sites located in Exon 1 and the intron of *Hoxa2* of chicken and mouse were shown to direct expression in r4 [21,26,33]. Therefore, it is possible that flanking sequences not analyzed in this study that are orthologous to sequences of the *Hoxa2* UERs in other gnathostomes function to redirect the intrinsic r4 activity of the r4/CNCC-specifying element to r3 and r5.

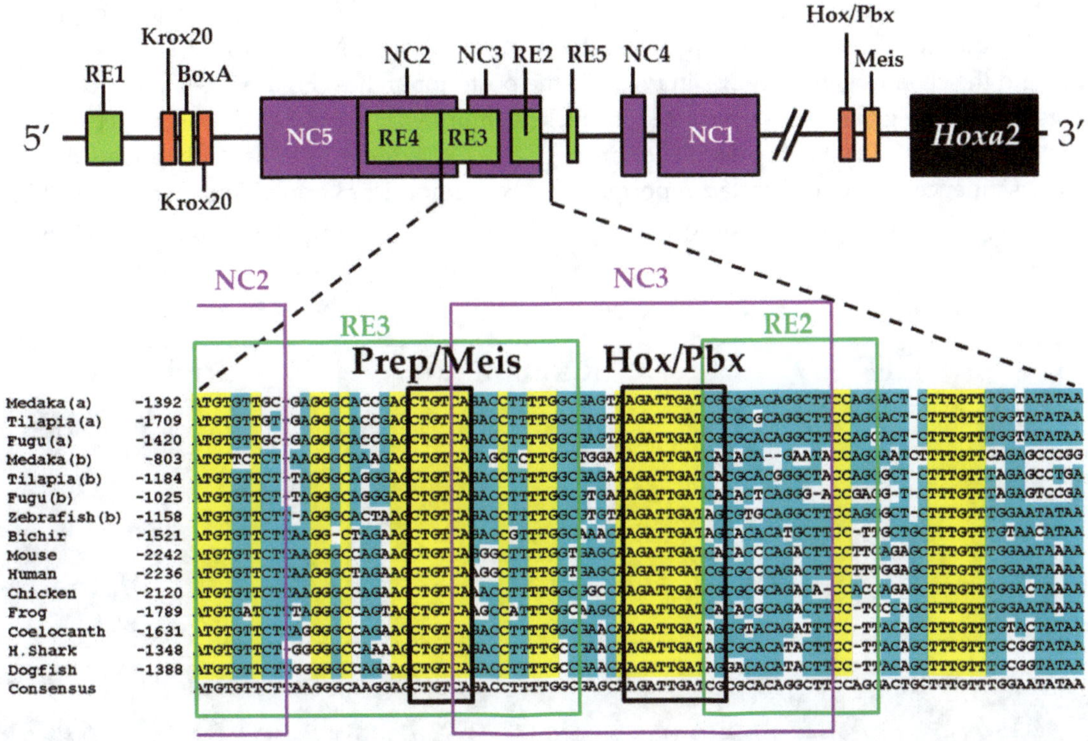

Figure 5. Comparative genomic sequence analysis of the 89 bp DNA fragment of the medaka *hoxa2a* UER(K20-RE5) (top sequence) that directs expression in r4 and the CNCCs. This sequence was compared to orthologous DNA sequences upstream of *hoxa2a* (denoted by a in parentheses after species name) and *hoxa2b* (denoted by b in parentheses after species name) of teleosts, *Hoxa2* of tetrapods and *Hoxa2* of coelacanth, horn shark and dogfish. Numbers correspond to genomic base pair positions relative to the ATG start site of the *Hoxa2* genes. The schematic diagram above the sequences corresponds to the mouse *Hoxa2* UER and the relative location of the 89-bp DNA fragment of the medaka *hoxa2a* r4/CNCC-specifying element. The schematic is not drawn to scale. Base pairs colored in yellow correspond to complete conservation at particular sites across all sequences examined. Base pairs colored in blue represent the majority of the sequences containing specific base pairs at specific sites. Purple boxed regions and green boxed elements correspond to neural crest and rhombomeric elements defined in the mouse *Hoxa2* UER [27,28]. Black boxed regions correspond to transcription factor binding sites identified in this study. A consensus sequence was derived from the aligned sequences. NC, neural crest; RE, rhombomeric element.

3.3. Functional Genomic Analysis of the Medaka ψHoxa2b UER(K20-RE5)

Unlike the conserved hindbrain and pharyngeal arch expression patterns common to many teleost *hoxa2a* and *a2b* and tetrapod *Hoxa2* genes, medaka *ψhoxa2b* is expressed in noncanonical *Hox* PG2 domains, which include the caudal-most region of the embryonic trunk, the ventral-most aspect of the neural tube and the distal regions of the pectoral fin buds [12]. Based on the divergence in expression observed in comparisons of medaka with other teleost and tetrapod *Hoxa2* genes, we hypothesized that the UER(K20-RE5) of medaka *ψhoxa2b* would direct expression in the noncanonical *Hox* PG2 domains instead of the hindbrain or CNCCs.

Contrary to our hypothesis, transient and stable-line transgenic medaka embryos showed that the 397 bp construct containing the entire medaka *ψhoxa2b* UER(K20-RE5) (Construct #8, Table 2) directed reporter gene expression in the hindbrain and PAs (Figure 6A–I, Table 2) instead of the expected noncanonical *Hox* PG2 expression domains, such as the pectoral fins. We observed a high percentage of transient transgenic embryos showing *eGFP* expression in the hindbrain (90%) and pharyngeal arches (90%) (Table 2). Whole-mount *in situ* hybridization analyses of stable-line transgenic embryos generated with construct #8 showed that *eGFP* was expressed in r3–7 of the hindbrain, the migratory CNCCs of the hyoid and post-otic streams, the post-migratory CNCCs in PA2 and the posterior arches and the chondrogenic CNCCs in PA2 and the posterior arches (Figure 7A–C). Interestingly, in transgenic reporter gene assays conducted in chicken embryos, a very similar construct containing a region corresponding to the the medaka *ψhoxa2b* UER(K20-RE5) used in this study, and also lacking the RE1 sequence element, directed reporter gene expression in r3 and r5 but not in the CNCCs, the pharyngeal arches or the noncanonical expression domains observed for *ψhoxa2b* expression in medaka [12,21].

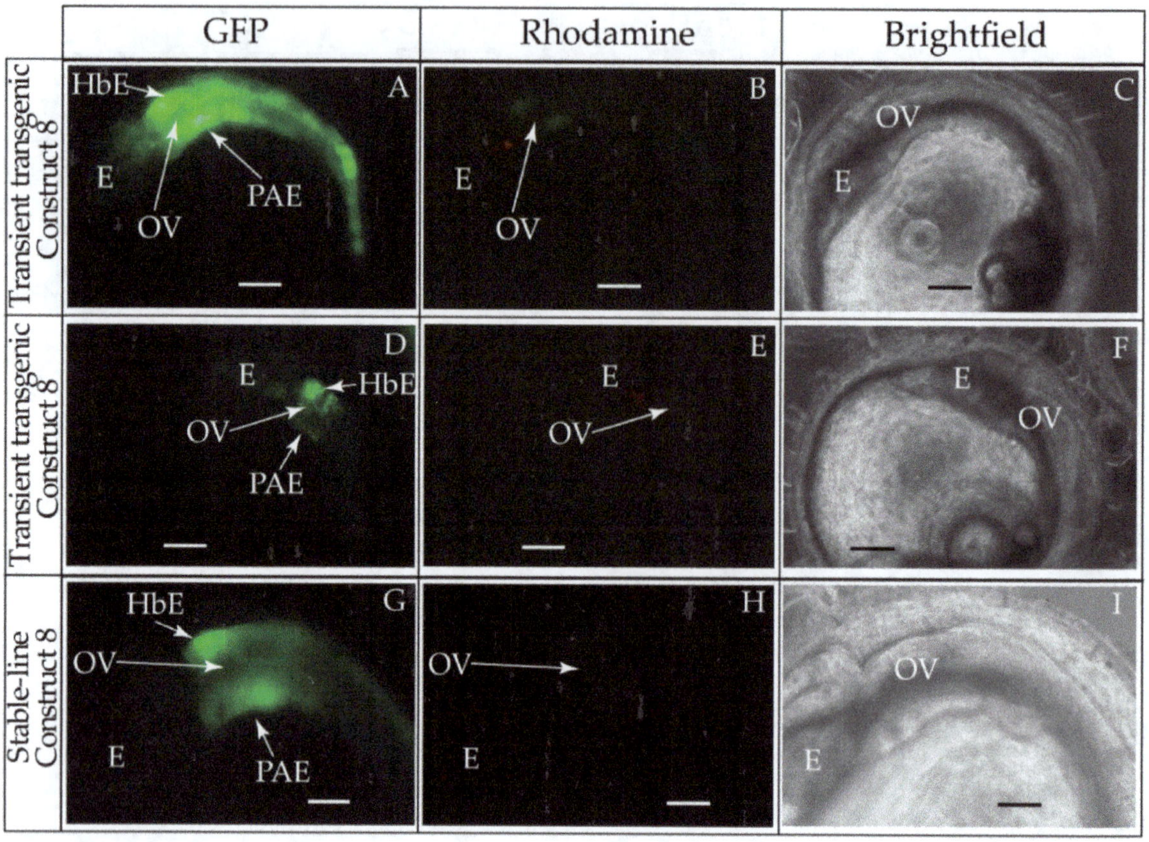

Figure 6. Transient (**A–F**) and stable-line (**G–I**) transgenic data from the medaka *ψhoxa2b* UER(K20-RE5)-(Construct #8). (**A–I**) Images of transient transgenic embryos at stage 29/30 (72–84 hpf) were taken using: GFP (**A**,**D**,**G**); rhodamine (**B**,**E**,**H**); and brightfield (**C**,**F**,**I**) filters. Transient transgenic analyses show varying degrees of *eGFP* signal between embryos (compare **A** and **D**). All embryos are still in their chorions and are positioned with their anterior sides to the left and their lateral sides to the reader. E, eye; HbE, hindbrain expression; OV, otic vesicle; PAE, pharyngeal arch expression. Scale bars equal 0.1 mm.

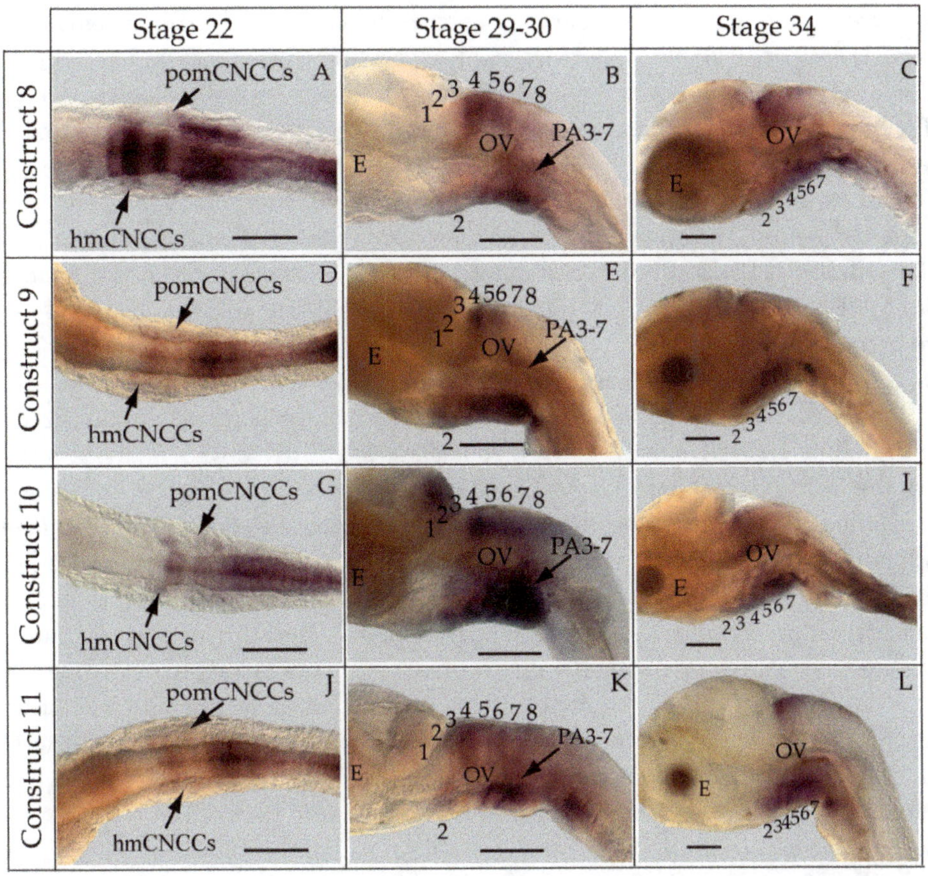

Figure 7. Whole-mount *in situ* hybridization of *eGFP* in stable-line *ψhoxa2b* UER(K20-RE5) transgenic medaka embryos generated with: Construct #8 (**A–C**); Construct #9 (**D–F**); Construct #10 (**G**, **H** and **I**); and construct #12 (**J–L**) at stages 22 (9 s) (**A**,**D**,**G**,**J**); 29/30 (72–84 hpf); (**B**,**E**,**H**,**K**); and 34 (121 hpf) (**C**,**F**,**I**,**L**). (**A**,**D**,**G**,**J**) Embryos were mounted with their anterior sides facing left and their dorsal sides facing the reader. (**B**,**C**,**E**,**F**,**H**,**I**,**K**,**L**) Embryos were mounted with their anterior sides facing left and lateral sides facing the reader. (**B**,**E**,**H**,**K**) Images are magnified to show rhombomere placement. Rhombomere numbers are indicated by black numbers above the dorsal sides of the embryos. Pharyngeal arches are indicated by black numbers below the ventral sides of the embryos. E, eye; hmCNCCs, hyoid migratory cranial neural crest cells; OV, otic vesicle; PA, pharyngeal arch; pomCNCCs, post-otic migratory cranial neural crest cells. Scale bars equal 0.1 mm.

To precisely map the functional activity of the medaka *ψhoxa2b* UER(K20-RE5) on control of expression, a series of nested deletion constructs from the medaka *ψhoxa2b* UER(K20-RE5) that selectively eliminated enhancer elements and transcription factor binding sites similar to those that were deleted in the *hoxa2a* UER(K20-RE5) series of nested deletions was performed (Constructs #9–13) (see Table 2). Similar to embryos microinjected with Construct #8, microinjection of constructs that included 5′-deletions of Krox20 and BoxA binding sites (Construct #9) or Krox20, BoxA and RE4 (Construct #10) yielded high percentages of transient transgenic embryos with *eGFP* expression in the hindbrain (90%, Construct #9; 97%, Construct #10) and PAs (90%, Construct #9; 87%, Construct #10) (Table 2). Stable-line transgenic embryos generated with Constructs #9 and #10 showed similar *eGFP* expression pattern to those treated with Construct #8, wherein reporter gene expression was observed in r3–7 of the hindbrain, the migratory CNCCs of the hyoid and post-otic streams, the post-migratory CNCCs in PA2 and the posterior arches and the chondrogenic CNCCs of PA2 and the posterior arches (Figure 7D–I). Similar to the medaka *hoxa2a* UER(K20-RE5) analysis with Construct #4 in which the Krox20, BoxA, RE4 and RE3 sequences were eliminated, a lower percentage of transient transgenic medaka embryos generated with Construct #11, which included the same deletions, showed *eGFP* expression in the hindbrain (52%) and pharyngeal arches (52%) when compared to embryos generated with Constructs #8, 9 or 10 (Table 2). Further, *eGFP* expression was almost undetectable in embryos generated with Construct #11 (Data not shown). Unfortunately, no stable-line transgenic embryos generated with Construct #11 showed any detectable *eGFP* expression in the hindbrain or pharyngeal arches (Data not shown). These results, like those for the medaka *hoxa2a* UER(K20-RE5), suggest that sequence elements downstream of the RE4 element function to direct medaka *ψhoxa2b* in r3–7, PA2 and the posterior arches. The microinjection of constructs with the 3′-specific deletion of RE5 from the medaka *ψhoxa2b* UER(K20-RE5) but retention of all sequence elements 5′ to this site, including Krox20, BoxA, RE4, RE3 and RE2 (Construct #12), yielded similar results to Construct #8 with a high percentage of transient transgenic embryos showing *eGFP* expression in the hindbrain (80%) and the PAs (80%) (Table 2) and stable-line transgenic embryos generated with Construct #12, showed a similar *eGFP* expression pattern relative to stable-line transgenic embryos generated with Constructs #8, 9 and 10 (Figure 7J–L). Interestingly, similar to the medaka *hoxa2a* UER(K20-RE5) analysis with Construct #6 in which the RE2 and RE5 sequences were eliminated a low percentage of transient transgenic embryos generated with Construct #13, which contained the same deletions, showed *eGFP* expression in the hindbrain (14%) and pharyngeal arches (23%) (Table 2). Further, in comparison to embryos generated with Constructs #8, 9, 10 and 12, *eGFP* expression in embryos generated with Construct #13 was significantly reduced and barely visible (data not shown). Unfortunately, we did not obtain any stable-line transgenic embryos generated with Construct #13 (Data not shown). Overall, these results showed that the sequence containing the elements corresponding to RE3 and RE2 of the medaka *ψhoxa2b* UER(K20-RE5) is required for r3–7 and CNCC expression in PA2 and the posterior PAs but the surrounding elements corresponding to Krox20, BoxA, RE4 and RE5 are not.

A comparative genomic sequence analysis revealed that the 88 bp DNA sequence fragment of the medaka *ψhoxa2b* UER(K20-RE5), hereafter referred to as the r3–7/CNCC-specifying element, like the paralogous *hoxa2a* r4/CNCC-specifying element and orthologous *Hoxa2* UER(K20-RE5) genomic sequences of other gnathostomes contains conserved Hox/Pbx, and Prep/Meis transcription factor binding sites (see Figure 5). While these sites may be involved in directing *eGFP* expression in r4 and the CNCCs, their activity may also be redirected to a much broader domain of expression, including r3, r5, r6 and r7. We observed 33 bp differences between the paralogous core sequences upstream of medaka *hoxa2a* and *ψhoxa2b* (Figure 8), and it is possible that the sequence differences between them have allowed the *ψhoxa2b* sequence, but not the *hoxa2a* sequence, to be responsive to transcription factors that mediate expression within r3, r5, r6 and r7 of the hindbrain. An analysis of these sequences in the JASPAR software program showed the presence of several Sox binding elements within the *ψhoxa2b* fragment but not for *hoxa2a*. Sox proteins have been shown to be integral factors in directing *Hox* gene expression among different rhombomeres of the hindbrain [59], indicative that the sequence differences leading to the evolution of additional Sox binding elements in the UER(K20-RE5) of *ψhoxa2b* fragment relative to the *hoxa2a* UER(K20-RE5) may be responsible for the broadened rhombomeric expression observed for the *ψhoxa2b* UER(K20-RE5) relative to the *hoxa2a* UER(K20-RE5).

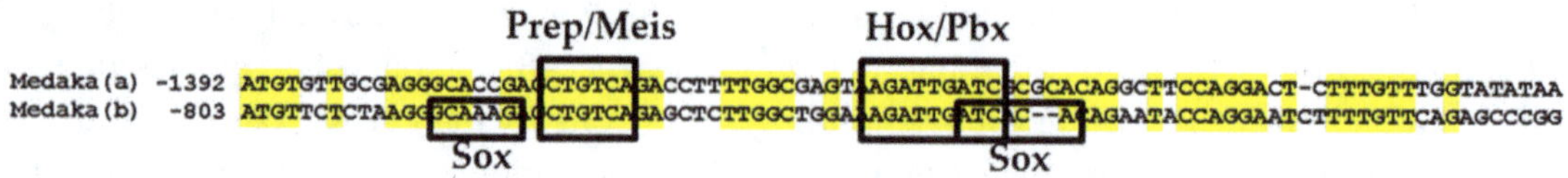

Figure 8. Sequence alignment of the medaka *hoxa2a* r4/CNCC and the *ψhoxa2b* r3–7/CNCC specifying elements. The sequence corresponding to the medaka *hoxa2a* r4/CNCC specifying element is denoted by "a" in parentheses. The sequence corresponding to the medaka *ψhoxa2b* r3–7/CNCC specifying element is denoted by "b" in parentheses. Numbers correspond to genomic base pair positions relative to the ATG start site of *hoxa2a* and *ψhoxa2b*. Base pairs colored in yellow correspond to complete conservation at particular sites across both sequences examined. Black boxed regions correspond to transcription factor binding sites identified in this study.

4. Discussion

4.1. The Use of Medaka in Reporter Gene Expression Analyses

The *Tol2* transposon system has been used for the generation of transgenic vertebrate lines for several osteichthyans, including zebrafish, *Xenopus*, chicken and mouse [70]. In this study, we showed that the *Tol2* transposon system can be used for both transient and stable-line transgenic analyses of genomic enhancer regions in the Japanese medaka. Although the *Tol2* element is found within the genome of medaka, our control experiments using constructs containing the *Xenopus* eFI-αS promoter upstream of *eGFP* showed that this system works similarly in medaka to that of other osteichthyans. Specifically, medaka zygotes that were co-microinjected with constructs containing the *Xenopus* eFI-αS promoter and transposon mRNA showed strong reporter gene expression throughout the body whereas embryos that were microinjected solely with constructs containing the *Xenopus* eFI-αS promoter were not able to direct reporter gene expression.

Our reporter gene expression results showed that the UER(K20-RE5)s from medaka *hoxa2a* and *ψhoxa2b* are functional, but that they have diverged from one another in their capacity to direct gene expression in the medaka hindbrain and pharyngeal arches. The medaka *hoxa2a* UER(K20-RE5)-directed reporter gene expression in r4 of the hindbrain and the CNCCs of PA2-7 while the medaka *ψhoxa2b* UER(K20-RE5) directed reporter gene expression in r3–7 of the hindbrain and PA2-7. These reporter gene expression results underscore the importance of using a homologous model system for analyzing *cis*-regulatory element control of gene expression during embryonic development. In support, neither the medaka *hoxa2a* nor *ψhoxa2b* UER(K20-RE5)s yielded similar reporter gene expression results when tested in chicken embryos [21]. Specifically, while the *hoxa2a* UER(K20-RE5) did not direct any reporter gene expression in the chicken embryonic head, the *ψhoxa2b* UER(K20-RE5) only directed expression in r3 and r5 but not the CNCCs [21]. The inconsistent results for the medaka *hoxa2a* and *ψhoxa2b* UER(K20-RE5)s within the medaka and chicken embryos suggest that these enhancer regions were not efficient in utilizing the *trans*-acting factors that were present in the heterologous chick model system. In support of this hypothesis, previous reporter gene experiments showed that the mouse *Hoxa2* UER was able to generate reporter gene expression in r3, r5 and the CNCCs in both mouse and zebrafish embryos but unable to direct any reporter gene expression in the hindbrain or CNCCs of chick embryos [29,32]. Similar results were also observed for the fugu *hoxa2a* UER, such that it was shown to be functional when it was tested in both mouse and zebrafish embryos, but not in the chicken [29,32]. Conversely, the UER of chick *Hoxa2* was unable to direct reporter gene expression in either mouse or zebrafish embryos [29]. These results along with our reporter gene expression results of the medaka *hoxa2a* and *ψhoxa2b* UER(K20-RE5)s in medaka embryos suggest that either the functional nature of the sequence elements of the UER, the transcription factors of the genetic regulatory networks that interact with these sequences or a combination of both have diverged greatly among chicken, and possibly other archosaurs, from evolutionarily divergent osteichthyans. Future analyses of the medaka *hoxa2a* and *ψhoxa2b* UER(K20-RE5)s should incorporate the use of zebrafish and/or mouse embryos to assay for species-specific differences in regulatory requirements among these species. Further, orthologous sequences from the mouse *Hoxa2* and zebrafish *hoxa2b* UERs should be tested within medaka embryos.

4.2. Medaka Hoxa2a-Directed Gene Expression in the Hindbrain

In this study we showed, using reporter gene assays and whole-mount *in situ* hybridization, that the medaka *hoxa2a* UER(K20-RE5) does not direct expression in r3 and r5 of the hindbrain as expected based on previous whole-mount *in situ* hybridization results of medaka *hoxa2a* [12], but instead directs expression in r4. We found these results interesting since this region of genomic DNA was shown to be conserved structurally with orthologous genomic regions in evolutionarily divergent osteichthyans [21,29,32,59,86]. Further, the *Hoxa2* UERs of zebrafish, mouse and chicken have been shown to direct reporter gene expression in r3 and r5 when tested in their respective homologous host systems [21,25,28–32,59]. An explanation for the lack of reporter gene expression in r3 and r5 in this study is that the genomic DNA sequence tested requires the cooperation of other surrounding sequences within the medaka *hoxa3a-a2a* intergenic region. Our reporter gene analysis of the medaka *hoxa2a* UER(K20-RE5) utilized a 531 bp genomic DNA sequence and contained all of the *cis*-regulatory elements that were tested in the chick embryonic model system [21], including Krox20 and BoxA binding sequences and sequence elements that pertain to RE4, RE3, RE2 and RE5. Functional genomic analyses of the mouse *Hoxa2* UER included regulatory elements orthologous to those mentioned above as well as a RE1 sequence element located upstream of Krox20 [28]. The deletion of the sequences upstream of the Krox20 elements, including the RE1 element, from the mouse *Hoxa2* UER resulted in the loss of reporter gene expression in r3 and r5 of stable-line transgenic mouse embryos [28]. Recent comparative and functional genomic sequence analyses utilizing the UERs of fugu *hoxa2a* and *a2b* and zebrafish *hoxa2b* have shown the presence of a RE1 sequence that is upstream of the Krox20 site in teleosts [29]. Further, these sequences were shown to be much farther upstream from Krox20 in teleosts in comparison to tetrapods [29]. For instance, in comparison to the mouse *Hoxa2* UER, in which the RE1 sequence begins 88 bp upstream of the Krox20 binding site, the orthologous RE1 sequence of medaka *hoxa2a* begins 290 bp upstream of Krox20 [29]. Our analysis of the medaka *hoxa2a* UER(K20-RE5) utilized a sequence that began at genomic bp position −1778, and it excluded the recently identified teleost RE1 sequence that begins at genomic bp postion −2041 [29]. Future reporter gene analyses in medaka embryos should be performed that utilize constructs containing medaka *hoxa3a-a2a* sequences that are orthologous to the "full-length" UERs tested in other gnathostomes to determine if the flanking sequences, including the RE1 element, function in conjunction with the *hoxa2a* UER(Krox20-RE5) sequence to direct reporter gene expression in r3 and r5.

Our reporter gene expression and whole-mount *in situ* hybridization analyses are the first to show the presence of sequences located in the vertebrate *Hoxa3-a2* intergenic regions that are involved in directing reporter gene expression in r4. The *eGFP* expression in r4 of medaka was authenticated using antisense *eGFP* riboprobes and antisense ribpoprobes of specific rhombomeric molecular markers. These molecular markers included medaka *hoxb1a*, which is expressed exclusively in r4 of the hindbrain [16], and medaka *hoxd3a*, which is expressed in r6, r7 and r8 [13]. Microinjection of a series of nested deletion constructs derived from the 531 bp fragment encompassing the medaka *hoxa2a* UER(K20-RE5) showed that the r4-specifying sequence lies within a 89 bp fragment that spans from genomic bp positions −1392 to −1303. Comparative genomic sequence analyses showed the presence of putative transcription factor binding sites, Prep/Meis and Hox/Pbx, located within this fragment. Further, these sites were shown to be highly conserved in *Hoxa2* genomic sequences across several evolutionarily divergent gnathostome vertebrates, including shark, bichir, coelacanth, teleosts and tetrapods (see Figure 5). Interestingly, Hox/Pbx and Prep/Meis binding sites have been shown in multiple studies to be crucial sequences in directing *Hox* expression in r4 [21,33,37]. Future analyses using gel-shift, chromatin immunoprecipitation (ChIP) and site-directed mutagenesis assays must be performed to determine if the Hox/Pbx and Prep/Meis sites within this fragment of the medaka *hoxa2a* UER(K20-RE5) are functional and responsible for directing medaka *hoxa2a* in r4. Further, reporter gene analyses must be performed to determine if the orthologous regions across evolutionarily divergent vertebrate *Hoxa3-a2* intergenic sequences direct reporter gene expression in r4.

A potential explanation for the presence of the r4-directed expression is that the surrounding genomic sequences that were not included in this analysis, such as the RE1 sequence, may function to restrict the r4/CNCC-specifying element from potentiating r4-specific expression or redirect it to direct expression in r3 and r5. In support, recent analyses of CREs responsible for directing fugu *pax2* paralogs, *pax2.1* and *pax2.2*, have shown that the functional activity of the paralogous CREs cognate to these two genes are influenced by interaction among clustered *cis*-regulatory elements in a manner very similar to those from the medaka *hoxa2a* UER(K20-RE5) and their immediately flanking sequences [87]. Specifically, the conserved noncoding element (CNE) pair 1, along with 50 bp extending both 5′ and 3′ of this element for *pax2.1* was shown to direct reporter gene expression in the notochord, muscle and fins [87]. However, when the flanking sequences were removed, the deletion construct containing the more tightly defined CNE pair 1 of fugu *pax2.1* directed an expanded expression pattern that included the otic vesicle, blood, heart, skin and pronephric region [87]. By contrast, the removal of flanking sequences from the paralogous element of *pax2.2* resulted in the gain of directed reporter gene expression in the fins but loss of expression from the eye and otic vesicle [87]. By comparison, we argue that the inclusion of flanking sequences to the medaka *hoxa2a* UER(K20-RE5), including the RE1 sequence, may change the functional activity of the medaka UER(K20-RE5) region to direct reporter gene expression in r3 and r5 but negate expression in r4.

4.3. *Hoxa2a-Directed Gene Expression in the Cranial Neural Crest Cells*

Beyond the hindbrain expression directed activity of the medaka *hoxa2a* UER(K20-RE5), this genomic DNA region was also shown to possess functional activity that directs *hoxa2a* expression in the CNCCs. Our whole-mount *in situ* hybridization results from stable-line transgenic medaka embryos have shown that the medaka *hoxa2a* UER(K20-RE5) directs reporter gene expression in the migratory CNCCs and is involved in maintaining *hoxa2a* expression in post-migratory CNCCs in PA2 and the posterior arches up until the chondrogenic stages of craniofacial skeleton development. Several functional genetic studies have shown that *Hox* PG2 gene expression that persists late into post-migratory CNCC stages of development is necessary for the proper patterning of the craniofacial skeleton [45,56,62,64,88]. Our analyses suggest that the genomic sequences of the medaka UER(K20-RE5) are utilized *in vivo* to maintain medaka *hoxa2a* expression up until chondrogenesis in PA2 and the posterior arches to allow for proper craniofacial skeleton development.

Although we observed *eGFP* expression patterns in the post-migratory CNCCs that were directed by the medaka *hoxa2a* UER(K20-RE5), they did not fully phenocopy the post-migratory CNCC expression patterns shown by medaka *hoxa2a* whole-mount *in situ* hybridization experiments [12]. For instance, we observed very low levels of *eGFP* expression within PA2 but strong *eGFP* expression in the posterior arches in stable-line transgenic medaka embryos generated with Constructs #1, 2 and 5 (see Figure 4). However, we observed strong *eGFP* expression in PA2 and the posterior arches in stable-line transgenic embryos generated with Construct #7 (see Figure 4). Thus, our reporter gene expression results suggest that the genomic sequences analyzed in this study, as well as other surrounding sequences, are necessary for medaka *hoxa2a* to be expressed and ultimately function in patterning the jaw support elements that are derived from PA2 and the pharyngeal jaw apparatus from the posterior pharyngeal arches. Future functional genomic studies using the medaka *hoxa2a* UER(K20-RE5) and flanking genomic sequences will be required to fully phenocopy the medaka *hoxa2a* expression pattern in the post-migratory CNCCs.

Our functional and comparative genomic analyses showed that the 89-bp region (Construct #7) responsible for directing reporter gene expression in the CNCCs of PA2 and the posterior arches contains Hox/Pbx and Prep/Meis binding sites (see Figure 5). These results are suggestive of auto- and/or cross-regulation of medaka *Hox* PG2 genes in the PAs. Recent functional genomic analyses in mouse showed that Hox/Pbx and Meis binding sequences located within the proximal promoter region upstream of mouse *Hoxa2* are necessary for *Hoxa2* expression and function in PA2 and posterior arches [24]. Further, this study showed that mouse *Hoxa2*, *Pbx* and *Meis* utilize these binding sites to function in a positive feedback loop that maintains and amplifies *Hoxa2* expression in the PAs [24]. Specifically, *Hoxa2* was shown to positively regulate *Meis* within the PAs, and *Meis* along with *Hoxa2*, were shown to work in concert in amplifying *Hoxa2* expression within these domains [24]. Unlike tetrapods, many species of ray-finned fishes contain multiple *Hox* PG2 genes that are expressed in PA2 and the posterior arches up until chondrogenesis, including *hoxa2a*, *a2b* and *b2a* [12,14,19,56,89]. Further, functional and comparative genomic sequence analyses have shown that Hox/Pbx and Prep/Meis sites are located within the *hoxb3a-b2a* intergenic region of ray-finned fishes and *Hoxb3-b2* of lobe-finned fishes [37,89]. These results suggest that the *Hox/Pbx* and *Prep/Meis* sites are used by teleost *Hox* PG2 genes in auto- and cross-regulatory mechanisms of gene expression in the PAs. In support, the independent knockdown of *hoxa2a* and *b2a* in tilapia resulted in altered expression levels of tilapia *Hox* PG2 genes in the PAs [64]. Specifically, the knockdown of tilapia *hoxa2a* resulted in a reduction of *hoxa2a* expression from PA2 and the posterior arches and *hoxb2a* expression from PA2, and the knockdown of *hoxb2a* resulted in the reduction of *hoxa2a* and *a2b* expression from PA2 [64]. These altered expression levels corresponded to the varying severity of loss-of-function phenotypes in tilapia, wherein the loss of *hoxa2a* resulted in a full homeotic transformation the PA2-derived skeletal elements but the loss of *hoxb2a* resulted in altered bony morphology of PA2 elements without a full homeotic transformation [64]. By contrast, in zebrafish the knockdown of both *hoxa2b* and *b2a* were required to obtain a full homeotic transformation of PA2-derived skeletal elements [56], which suggests that both zebrafish *Hox* PG2 genes function redundantly in utilizing the Hox/Pbx and Prep/Meis binding sites to auto- and cross-regulate each other's expression in PA2. Future functional genetic analyses must be performed to understand if medaka *hoxa2a* and *b2a* utilize the Hox/Pbx and Prep/Meis binding sites similar to *hoxa2b* and *b2a* of zebrafish or *hoxa2a* and *b2a* of tilapia.

4.4. Functional Nature of the Medaka ψHoxa2b r3/5 Enhancer Region

Our results from transient and stable-line transgenesis and whole-mount *in situ* hybridization analyses using anti-*eGFP* riboprobes have shown that the medaka *ψhoxa2b* UER(K20-RE5) is functional and is able to direct robust reporter gene expression in r3–7 of the hindbrain, PA2 and the posterior PAs. These results were contrary to our expectations based on whole-mount *in situ* hybridization analyses of medaka *ψhoxa2b*, which showed that this pseudogene is expressed strongly in noncanonical *Hox* PG2 expression domains but not at all in the characteristic hindbrain and CNCC compartments [12]. These contradictory results suggest that regions of genomic sequences surrounding the medaka *ψhoxa2b* UER(K20-RE5) function to redirect the activity of this region to mediate expression in the noncanonical *Hox* PG2 embryonic compartments. In addition to the unexpected functional nature of the medaka *ψhoxa2b* UER(K20-RE5), this region was observed to be divergent in function from the UER(K20-RE5) of medaka *hoxa2a*, which directs gene expression in r4, PA2 and the posterior arches. Further, deletion mutagenesis of the medaka *ψhoxa2b* UER(K20-RE5) showed that a relatively short sequence that is paralagous to the r4/CNCC-specifying element of the medaka *hoxa2a* UER(K20-RE5) was responsible for directing reporter gene expression in r3–7, PA2 and the posterior PAs. Comparative genomic analysis of the medaka *ψhoxa2b* r3–7/CNCC-specifying element revealed the presence of conserved Hox/Pbx and Prep/Meis binding sites (see Figure 5). However, this sequence was shown to be divergent from the medaka *hoxa2a* r4/CNCC-specifying element by 33 bp (see Figure 8). It is possible that these substitutions between the medaka *hoxa2a* and *ψhoxa2b* genomic sequences have allowed the *ψhoxa2b*-specific sequence, but not the *hoxa2a*, to be receptive to transcription factors expressed within r3, r5, r6 and r7 of the hindbrain. In support, analyses of the CREs that direct fugu *pax2.1* and *pax2.2* gene expression showed that while there is high sequence similarity between several paralogous CREs, these elements direct divergent expression patterns [87]. For instance, although the pair 1 CNEs of *pax2.1* and *pax2.2* show roughly 90% similarity in sequence, the pair 1 CNE of fugu *pax2.1* directs gene expression in the spinal cord, muscle, notochord and fins while the paralogous sequence of *pax2.2* directs expression in the forebrain, midbrain, hindbrain, spinal cord, eye, otic vesicle, blood, heart, skin, thyroid region and pronephric region [87]. Interestingly, our analyses of the medaka *Hox cis*-regulatory sequences using the JASPAR software program revealed several Sox binding elements within the *ψhoxa2b* r3–7/CNCC-specifying element but not in the *hoxa2a* r4/CNCC-specifying element (see Figure 8). Sox proteins have been shown to be involved in directing *Hox* gene expression in several rhombomeres of the hindbrain [59], so it is possible that these elements were involved in directing reporter gene expression in r3, r5, r6 and r7. The base pair mutations between the r3–7/CNCC-specifying element of medaka *ψhoxa2b* and the paralogous r4/CNCC-specifying element may have occurred due to relaxation of selective pressures on the *ψhoxa2b* genomic sequence after the inactivation of the *hoxa2b* gene in the lineage leading to medaka [12].

Our reporter gene expression results of the medaka *ψhoxa2b* UER(K20-RE5) coupled with previous expression pattern results of medaka *ψhoxa2b* during embryonic development may provide an excellent example of how evolutionary changes in noncoding sequences contribute to morphological evolution. An increasing body of evidence in the field of evolutionary and developmental biology has shown that sequence changes in conserved noncoding sequences can affect expression patterns of developmentally important genes, and the co-option of such genes in divergent embryonic domains can lead to morphological novelties (see [90]). Assuming that the medaka *ψhoxa2b* transcript gives rise to a functional product that can affect developmental specification, the heterotopic shift in expression that lead to medaka *ψhoxa2b* co-option into noncanonical *Hox* PG2 developmental compartments may have contributed to unique morphological and functional traits. Based on *in silico* translation of the likely mRNA product derived from *ψhoxa2b* cDNA clones, it is clear that this pseudogene cannot generate a fully functional Hox protein product [12]. However, it can generate a truncated protein that possesses a hexapeptide that is conserved with orthologous hexapeptides of *hoxa2b* genes from evolutionarily divergent teleosts [12]. Since the hexapeptide of *Hox* genes are known to mediate interactions with other transcription factors, especially *Pbx*, there is the tantalizing possibility that the truncated product of *ψhoxa2b* translation could interact with *Pbx* and influence transcriptional activity in developmental compartments in which it is expressed. In the case of medaka, which is a member of the beloniform fishes (an order that includes the flying fishes), these mechanisms may be responsible for generating divergent morphological characters specific to beloniforms. Of particular interest in this regard is the fact that we have documented expression of *ψhoxa2b* in the developing pectoral fin buds [12]. Several functional genetic studies have shown that *Meis*, *Pbx*, and posterior *Hox* genes within the A and D clusters function to pattern the proximal-distal axis of the developing limbs [91–96]. Likewise, *Pbx* has been shown to be involved in the upregulation of posterior *HoxA* and *D* genes, and that the loss of function of either *Pbx* or posterior *HoxA* and *D* cluster genes results in truncated limbs [92,93]. Given the location of expression of the pseudogene and its potential to interact with *Hox* gene products in the fin bud progress zone, it is possible that the co-option of *ψhoxa2b* expression into this compartment may have influenced the evolution of elongated pectoral fins in the lineage leading to flying fish, a remarkable adaptation that is restricted to this taxonomic clade. In order to test these hypotheses, functional genomic analyses of the *hoxa2b* UER(K20-RE5)) of other beloniform fishes must be tested within the medaka. Further, phylogenetic analyses must be performed in order to understand if the *hoxa2b* inactivation is specific to the lineage leading to medaka or if it occurred earlier in the beloniform radiation to include other members of this taxonomic clade.

5. Conclusions

In conclusion, our reporter gene analyses showed that the medaka *hoxa2a* and *ψhoxa2b* UER(K20-RE5)s are functionally divergent from one another. While both UER(K20-RE5)s were shown to direct reporter gene expression in the CNCCs of PA2 and the posterior PAs, the medaka *hoxa2a* UER(K20-RE5) directed expression in r4 of the hindbrain while the medaka *ψhoxa2b* directed expression in r3–7. These results are different from heterologous reporter gene expression results of the medaka *hoxa2a* and *ψhoxa2b* UER(K20-RE5)s when they were tested in chicken embryos. In chicken, the medaka *hoxa2a* UER(K20-RE5) did not direct expression in the hindbrain or pharyngeal arches while the *ψhoxa2b* UER(K20-RE5) directed reporter gene expression in just r3 and r5 of the hindbrain. Overall, our analyses show that when tested independent of their natural genomic surroundings, the UER(K20-RE5)s of medaka *hoxa2a* and *ψhoxa2b* are able to direct expression within r4 and *Hox* PG2 canonical domains, respectively.

Acknowledgments: We acknowledge Koichi Kawakami for the generous gift of the pT2AL200L150G plasmid vector. We acknowledge Jean-Luc Scemama for providing guidance with this project. We acknowledge Micheal Odom for aiding in the development of the medaka microinjection method.

Author Contributions: A.D. designed and developed all constructs, collected all transgenic medaka data for his Ph.D. dissertation, and wrote much of this manuscript under the guidance of E.J.S. E.J.S. advised with the design of this project and aided in the writing of this manuscript. M.C.R. designed the pTolTATA-βMCS vector system for this M.S. thesis under the guidance of E.J.S and aided in the writing of this manuscript.

Conflicts of Interest: The authors declare no conflict of interest.

Abbreviations

The following abbreviations are used in this manuscript:

A-P	anterior-posterior
bp	base pair
CNCC	cranial neural crest cell
CNE	conserved noncoding element
CRE	*cis*-regulatory element
eGFP	enhanced green fluorescent protein
ERM	embryo rearing medium
E. coli	*Escherichia coli*
GFP	green fluorescent protein
NC	neural crest
PA	pharyngeal arch
PG	paralog group
r	rhombomere
RE	rhombomeric element
UER	upstream enhancer region
UER(K20-RE5)	upstream enhancer region spanning 5′ from Krox20 binding element to the 3′ RE5 element

References

1. McGinnis, W.; Krumlauf, R. Homeobox genes and axial patterning. *Cell* **1992**, *68*, 283–302. [CrossRef]
2. Ferrier, D.E.; Minguillon, C.; Holland, P.W.; Garcia-Fernandez, J. The amphioxus Hox cluster: Deuterostome posterior flexibility and Hox14. *Evolut. Dev.* **2000**, *2*, 284–293. [CrossRef]
3. Holland, P.W.; Garcia-Fernandez, J. Hox genes and chordate evolution. *Dev. Biol.* **1996**, *173*, 382–395. [CrossRef] [PubMed]
4. Powers, T.P.; Amemiya, C.T. Evidence for a Hox14 paralog group in vertebrates. *Curr. Biol.* **2004**, *14*, R183–R184. [CrossRef] [PubMed]
5. Amores, A.; Force, A.; Yan, Y.L.; Joly, L.; Amemiya, C.; Fritz, A.; Ho, R.K.; Langeland, J.; Prince, V.; Wang, Y.L.; *et al.* Zebrafish *hox* clusters and vertebrate genome evolution. *Science* **1998**, *282*, 1711–1714. [CrossRef] [PubMed]
6. Amores, A.; Suzuki, T.; Yan, Y.L.; Pomeroy, J.; Singer, A.; Amemiya, C.; Postlethwait, J.H. Developmental roles of pufferfish *Hox* clusters and genome evolution in ray-fin fish. *Genome Res.* **2004**, *14*, 1–10. [CrossRef] [PubMed]
7. Hoegg, S.; Boore, J.L.; Kuehl, J.V.; Meyer, A. Comparative phylogenomic analyses of teleost fish Hox gene clusters: Lessons from the cichlid fish *Astatotilapia burtoni*. *BMC Genom.* **2007**, *8*, 317. [CrossRef] [PubMed]
8. Moghadam, H.; Ferguson, M.; Danzmann, R. Evolution of Hox Clusters in Salmonidae: A Comparative Analysis Between Atlantic Salmon (*Salmo salar*) and Rainbow Trout (*Oncorhynchus mykiss*). *J. Mol. Evol.* **2005**, *61*, 636–649. [CrossRef] [PubMed]
9. Mungpakdee, S.; Seo, H.; Angotzi, A.R.; Dong, X.; Akalin, A.; Chourrout, D. Differential evolution of the 13 Atlantic salmon hox clusters. *Mol. Biol. Evol.* **2008**, *25*, 1333–1343. [CrossRef] [PubMed]
10. Prince, V. The *Hox* paradox: More complex(es) than imagined. *Dev. Biol.* **2002**, *249*, 1–15. [CrossRef] [PubMed]
11. Stellwag, E.J. Hox gene duplication in fish. *Semin. Cell Dev. Biol.* **1999**, *10*, 531–540. [CrossRef] [PubMed]

12. Davis, A.; Scemama, J.L.; Stellwag, E.J. Japanese medaka *Hox* paralog group 2: Insights into the evolution of *Hox* PG2 gene composition and expression in the Osteichthyes. *J. Exp. Zool. B Mol. Dev. Evol.* **2008**, *310*, 623–641. [CrossRef] [PubMed]
13. Davis, A.; Stellwag, E.J. Spatio-temporal patterns of *Hox* paralog group 3–6 gene expression during Japanese medaka (*Oryzias latipes*) embryonic development. *Gene Expr. Patterns* **2010**, *10*, 244–250. [CrossRef] [PubMed]
14. Le Pabic, P.; Stellwag, E.J.; Brothers, S.N.; Scemama, J.L. Comparative analysis of Hox paralog group 2 gene expression during Nile tilapia (Oreochromis niloticus) embryonic development. *Dev. Genes Evol.* **2007**, *217*, 749–758. [CrossRef] [PubMed]
15. Lyon, R.S.; Davis, A.; Scemama, J.L. Spatio-temporal expression patterns of anterior *Hox* genes during Nile tilapia (*Oreochromis niloticus*) embryonic development. *Gene Expr. Patterns* **2013**, *13*, 104–108. [CrossRef] [PubMed]
16. Hurley, I.A.; Scemama, J.; Prince, V.E. Consequences of Hoxb1 duplication in teleost fish. *Evol. Dev.* **2007**, *9*, 540–554. [CrossRef] [PubMed]
17. Le Pabic, P.; Stellwag, E.J.; Scemama, J. Embryonic development and skeletogenesis of the pharyngeal jaw apparatus in the cichlid Nile tilapia (*Oreochromis niloticus*). *Anat. Rec.* **2009**, *292*, 1780–1800. [CrossRef] [PubMed]
18. Mungpakdee, S.; Seo, H.; Chourrout, D. Spatio-temporal expression patterns of anterior *Hox* genes in Atlantic salmon (*Salmo salar*). *Gene Expr. Patterns* **2008**, *8*, 508–514. [CrossRef] [PubMed]
19. Scemama, J.L.; Vernon, J.L.; Stellwag, E. Differential expression of Hoxa2a and Hoxa2b genes during striped bass embryonic development. *Gene Expr. Patterns* **2006**, *6*, 843–848. [CrossRef] [PubMed]
20. Soshnikova, N.; Dewaele, R.; Janvier, P.; Krumlauf, R.; Duboule, D. Duplications of hox gene clusters and the emergence of vertebrates. *Dev. Biol.* **2013**, *378*, 194–199. [CrossRef] [PubMed]
21. Tümpel, S.; Cambronero, F.; Wiedemann, L.M.; Krumlauf, R. Evolution of cis elements in the differential expression of two Hoxa2 coparalogous genes in pufferfish (Takifugu rubripes). *Proc. Natl. Acad. Sci. USA* **2006**, *103*, 5419–5424. [CrossRef] [PubMed]
22. Gendron-Maguire, M.; Mallo, M.; Zhang, M.; Gridley, T. *Hoxa-2* mutant mice exhibit homeotic transformation of skeletal elements derived from cranial neural crest. *Cell* **1993**, *75*, 1317–1331. [CrossRef]
23. Rijli, F.M.; Mark, M.; Lakkaraju, S.; Dierich, A.; Dolle, P.; Chambon, P. A homeotic transformation is generated in the rostral branchial region of the head by disruption of *Hoxa-2*, which acts as a selector gene. *Cell* **1993**, *75*, 1333–1349. [CrossRef]
24. Amin, S.; Donaldson, I.J.; Zannino, D.A.; Hensman, J.; Rattray, M.; Losa, M.; Spitz, F.; Ladam, F.; Sagestrom, C.; Bobola, N. Hoxa2 selectively enhances Meis binding to change a branchial arch ground state. *Dev. Cell* **2015**, *32*, 265–277. [CrossRef] [PubMed]
25. Frasch, M.; Chen, X.; Lufkin, T. Evolutionary-conserved enhancers direct region-specific expression of the murine *Hoxa-1* and *Hoxa-2* loci in both mice and Drosophila. *Development* **1995**, *121*, 957–974. [PubMed]
26. Lampe, X.; Samad, O.A.; Guiguen, A.; Matis, C.; Remacle, S.; Picard, J.J.; Rijli, F.M.; Rezsohazy, R. An ultraconserved Hox-Pbx responsive element resides in the coding sequence of *Hoxa2* and is active in rhombomere 4. *Nucleic Acids Res.* **2008**, *36*, 3214–3225. [CrossRef] [PubMed]
27. Maconochie, M.K.; Krishnamurthy, R.; Nonchev, S.; Meier, P.; Manzanares, M.; Mitchell, P.J.; Krumlauf, R. Regulation of *Hoxa2* in cranial neural crest cells involves members of the *AP-2* family. *Development* **1999**, *126*, 1483–1494. [PubMed]
28. Maconochie, M.K.; Nonchev, S.; Manzanares, M.; Marshall, H.; Krumlauf, R. Differences in Krox20-dependent regulation of Hoxa2 and Hoxb2 during hindbrain development. *Dev. Biol.* **2001**, *233*, 468–481. [CrossRef] [PubMed]
29. McEllin, J.A.; Alexander, T.B.; Tümpel, S.; Wiedemann, L.M.; Krumlauf, R. Analyses of fugu *hoxa2* genes provide evidence for subfunctionalization of neural crest cell and rhombomere *cis*-regulatory modules during vertebrate evolution. *Dev. Biol.* **2016**, *409*, 530–542. [CrossRef] [PubMed]
30. Nonchev, S.; Maconochie, M.; Vesque, C.; Aparicio, S.; Ariza-McNaughton, L.; Manzanares, M.; Maruthainar, K.; Kuroiwa, A.; Brenner, S.; Charnay, P.; *et al.* The conserved role of Krox-20 in directing Hox gene expression during vertebrate hindbrain segmentation. *Proc. Natl. Acad. Sci. USA* **1996**, *93*, 9339–9345. [CrossRef] [PubMed]

31. Nonchev, S.; Vesque, C.; Maconochie, M.; Seitanidou, T.; Ariza-McNaughton, L.; Frain, M.; Marshall, H.; Sham, M.H.; Krumlauf, R.; Charnay, P. Segmental expression of Hoxa-2 in the hindbrain is directly regulated by Krox-20. *Development* **1996**, *122*, 543–554. [PubMed]
32. Tümpel, S.; Maconochie, M.; Wiedemann, L.M.; Krumlauf, R. Conservation and diversity in the cis-regulatory networks that integrate information controlling expression of Hoxa2 in hindbrain and cranial neural crest cells in vertebrates. *Dev. Biol.* **2002**, *246*, 45–56. [CrossRef] [PubMed]
33. Tümpel, S.; Cambronero, F.; Ferretti, E.; Blasi, F.; Wiedemann, L.M.; Krumlauf, R. Expression of Hoxa2 in rhombomere 4 is regulated by a conserved cross-regulatory mechanism dependent upon Hoxb1. *Dev. Biol.* **2007**, *302*, 646–660. [CrossRef] [PubMed]
34. Tümpel, S.; Cambronero, F.; Sims, C.; Krumlauf, R.; Wiedemann, L.M. A regulatory module embedded in the coding region of *Hoxa2* controls expression in rhombomere 2. *Proc. Natl. Acad. Sci. USA* **2008**, *105*, 20077–20082. [CrossRef] [PubMed]
35. Parker, H.A.; Bronner, M.E.; Krumlauf, R. A *Hox* regulatory network of hindbrain segmentation is conserved to the base of vertebrates. *Nature* **2014**, *514*, 490–493. [CrossRef] [PubMed]
36. Sham, M.H.; Vesque, C.; Nonchev, S.; Marshall, H.; Frain, M.; Gupta, R.D.; Whiting, J.; Wilkinson, D.; Charnay, P.; Krumlauf, R. The zinc finger gene *Krox20* regulates *HoxB2* (*Hox2.8*) during hindbrain segmentation. *Cell* **1993**, *72*, 183–196. [CrossRef]
37. Ferretti, E.; Marshall, H.; Popperl, H.; Maconochie, M.; Krumlauf, R.; Blasi, F. Segmental expression of Hoxb2 in r4 requires two separate sites that integrate cooperative interactions between Prep1, Pbx and Hox proteins. *Development* **2000**, *127*, 155–166. [PubMed]
38. Vesque, C.; Maconochie, M.; Nonchev, S.; Ariza-McNaughton, L.; Kuroiwa, A.; Charnay, P.; Krumlauf, R. *Hoxb-2* transcriptional activation in rhombomeres 3 and 5 requires an evolutionarily conserved cis-acting element in addition to the Krox-20 binding site. *EMBO J.* **1996**, *15*, 5383–5396. [PubMed]
39. Barrow, J.R.; Capecchi, M.R. Targeted disruption of the Hoxb-2 locus in mice interferes with expression of Hoxb-1 and Hoxb-4. *Development* **1996**, *122*, 3817–3828. [PubMed]
40. Barrow, J.R.; Stadler, H.S.; Capecchi, M.R. Roles of Hoxa1 and Hoxa2 in patterning the early hindbrain of the mouse. *Development* **2000**, *127*, 933–944. [PubMed]
41. Davenne, M.; Maconochie, M.K.; Neun, R.; Pattyn, A.; Chambon, P.; Krumlauf, R.; Rijli, F.M. Hoxa2 and Hoxb2 control dorsoventral patterns of neuronal development in the rostral hindbrain. *Neuron* **1999**, *22*, 677–691. [CrossRef]
42. Gavalas, A.; Davenne, M.; Lumsden, A.; Chambon, P.; Rijli, F.M. Role of *Hoxa-2* in axon pathfinding and rostral hindbrain patterning. *Development* **1997**, *124*, 3693–3702. [PubMed]
43. Gavalas, A.; Ruhrberg, C.; Livet, J.; Henderson, C.E.; Krumlauf, R. Neuronal defects in the hindbrain of *Hoxa1*, *Hoxb1* and *Hoxb2* mutants reflect regulatory interactions among these *Hox* genes. *Development* **2003**, *130*, 5663–5679. [CrossRef] [PubMed]
44. Oury, F.; Murakami, Y.; Renaud, J.S.; Pasqualetti, M.; Charnay, P.; Ren, S.Y.; Rijli, F.M. Hoxa2- and rhombomere-dependent development of the mouse facial somatosensory map. *Science* **2006**, *313*, 1408–1413. [CrossRef] [PubMed]
45. Baltzinger, M.; Ori, M.; Pasqualetti, M.; Nardi, I.; Rijli, F.M. *Hoxa2* knockdown in *Xenopus* results in hyoid to mandibular homeosis. *Dev. Dyn.* **2005**, *234*, 858–867. [CrossRef] [PubMed]
46. Grammatopoulos, G.A.; Bell, E.; Toole, L.; Lumsden, A.; Tucker, A.S. Homeotic transformation of branchial arch identity after Hoxa2 overexpression. *Development* **2000**, *127*, 2355–5365.
47. Kitazawa, T.; Fujisawa, K.; Narboux-Nême, N.; Arima, Y.; Kawamura, Y.; Inoue, T.; Wada, Y.; Kohro, T.; Aburatani, H.; Kodama, T.; *et al.* Distinct effects of Hoxa2 overexpression in cranial neural crest populations reveal that the mammalian hyomandibular-ceratohyal boundary maps within the styloid process. *Dev. Biol.* **2015**, *402*, 162–174. [CrossRef] [PubMed]
48. Minoux, M.; Antonarakis, G.S.; Kmita, M.; Duboule, D.; Rijli, F.M. Rostral and caudal pharyngeal arches share a common neural crest ground pattern. *Development* **2009**, *136*, 637–645. [CrossRef] [PubMed]
49. Pasqualetti, M.; Ori, M.; Nardi, I.; Rijli, F.M. Ectopic *Hoxa2* induction after neural crest migration results in homeosis of jaw elements in *Xenopus*. *Development* **2000**, *127*, 5367–5378. [PubMed]
50. Hurley, I.A.; Mueller, R.L.; Dunn, K.A.; Schmidt, K.J.; Friedman, M.; Ho, R.K.; Prince, V.E.; Yang, Z.; Thomas, M.G.; Coates, M.I. A new time-scale for ray-finned fish evolution. *Proc. R. Soc. B* **2007**, *274*, 489–498. [CrossRef] [PubMed]

51. Meyer, A.; van de Peer, Y. From 2R to 3R: Evidence for a fish-specific genome duplication (FSGD). *Bioessays* **2005**, *27*, 937–945. [CrossRef] [PubMed]
52. Postlethwait, J.; Amores, A.; Cresko, W.; Singer, A.; Yan, Y.L. Subfunction partitioning, the teleost radiation and the annotation of the human genome. *Trends Genet.* **2004**, *20*, 481–490. [CrossRef] [PubMed]
53. Santini, F.; Harmon, L.J.; Carnevale, G.; Alfaro, M.E. Did genome duplication drive the origin of teleosts? A comparative study of diversification in ray-finned fishes. *BMC Evol. Biol.* **2009**, *9*, 194. [CrossRef] [PubMed]
54. Taylor, J.S.; van de Peer, Y.; Meyer, A. Revisiting recent challenges to the ancient fish-specific genome duplication hypothesis. *Curr. Biol.* **2001**, *11*, R1005–R1008. [CrossRef]
55. Moens, C.B.; Selleri, L. Hox cofactors in vertebrate development. *Dev. Biol.* **2006**, *291*, 193–206. [CrossRef] [PubMed]
56. Hunter, M.; Prince, V.E. Zebrafish Hox paralogue group 2 genes function redundantly as selector genes to pattern the second pharyngeal arch. *Dev. Biol.* **2002**, *247*, 367–389. [CrossRef] [PubMed]
57. Prince, V.; Lumsden, A. *Hoxa-2* expression in normal and transposed rhombomeres: Independent regulation in the neural tube and neural crest. *Development* **1994**, *120*, 911–923. [PubMed]
58. Steinke, D.; Salzburger, W.; Meyer, A. Novel relationships among ten fish model species based on a phylogenomic analysis using ESTs. *J. Mol. Evol.* **2006**, *62*, 772–784. [CrossRef] [PubMed]
59. Tümpel, S.; Wiedemann, L.M.; Krumlauf, R. Hox genes and segmentation of the vertebrate hindbrain. *Curr. Top. Dev. Biol.* **2009**, *88*, 103–137. [PubMed]
60. Davidson, E.H.; Erwin, D.H. Gene regulatory networks and the evolution of animal body plans. *Science* **2006**, *311*, 796–800. [CrossRef] [PubMed]
61. Barrallo-Gimeno, A.; Holzschuh, J.; Driever, W.; Knapik, E.W. Neural crest survival and differentiation in zebrafish depends on mont blanc/tfap2a gene function. *Development* **2004**, *131*, 1463–1477. [CrossRef] [PubMed]
62. Crump, J.G.; Swartz, M.E.; Eberhart, J.K.; Kimmel, C.B. Moz-dependent Hox expression controls segment-specific fate maps of skeletal precursors in the face. *Development* **2006**, *133*, 2661–2669. [CrossRef] [PubMed]
63. Deflorian, G.; Tiso, N.; Ferretti, E.; Meyer, D.; Blasi, F.; Bortolussi, M.; Argenton, F. Prep1.1 has essential genetic functions in hindbrain development and cranial neural crest cell differentiation. *Development* **2004**, *131*, 613–627. [CrossRef] [PubMed]
64. Le Pabic, P.; Scemama, J.; Stellwag, E.J. Role of *Hox* PG2 genes in Nile tilapia pharyngeal arch specification: Implications for gnathostome pharyngeal arch evolution. *Evol. Dev.* **2010**, *12*, 45–60. [CrossRef] [PubMed]
65. Melvin, V.S.; Feng, W.; Hernandez-Lagunas, L.; Artinger, K.B.; Williams, T. A morpholino-based screen to identify novel genes involved in craniofacial morphogenesis. *Dev. Dyn.* **2013**, *242*, 817–831. [CrossRef] [PubMed]
66. Selleri, L.; Depew, M.J.; Jacobs, Y.; Chanda, S.K.; Tsang, K.Y.; Cheah, K.S.; Rubenstein, J.L.; O'Gorman, S.; Cleary, M.L. Requirement for Pbx1 in skeletal patterning and programming chondrocyte proliferation and differentiation. *Development* **2001**, *128*, 3543–3557. [PubMed]
67. Ferretti, E.; Cambronero, F.; Tümpel, S.; Longobardi, E.; Wiedemann, L.M.; Blasi, F.; Krumlauf, R. The *Hoxb1* enhancer and control of rhombomere 4 expression: Complex interplay between PREP1-PBX1-HOXB1 binding sites. *Mol. Cell. Biol.* **2005**, *25*, 8541–8552. [CrossRef] [PubMed]
68. Pöpperl, H.; Bienz, M.; Studer, M.; Chan, S.; Aparicio, S.; Brenner, S.; Mann, R.; Krumlauf, R. Segmental expression of *Hoxb1* is controlled by a highly conserved autoregulatory loop dependent on *exd/Pbx*. *Cell* **1995**, *81*, 1031–1042. [CrossRef]
69. Urasaki, A.; Morvan, G.; Kawakami, K. Functional dissection of the *Tol2* transposable element identified the minimal *cis*-sequences and a highly repetitive sequence in the subterminal region essential for transposition. *Genetics* **2006**, *174*, 639–649. [CrossRef] [PubMed]
70. Kawakami, K. *Tol2*: A versatile gene transfer vector in vertebrates. *Genome Biol.* **2007**, *8*, S7. [CrossRef] [PubMed]
71. Kurosawa, G.; Takamatsu, N.; Takahashi, M.; Sumitomo, M.; Sanaka, E.; Yamada, K.; Nishii, K.; Matsuda, M.; Asakawa, S.; Ishiguro, H.; *et al.* Organization and structure of hox gene loci in medaka genome and comparison with those of pufferfish and zebrafish genomes. *Gene* **2006**, *370*, 75–82. [CrossRef] [PubMed]
72. Iwamatsu, T. Stages of normal development in the medaka *Oryzias latipes*. *Mech. Dev.* **2004**, *12*, 605–618. [CrossRef] [PubMed]

73. Oxendine, S.L.; Cowden, J.; Hinton, D.E.; Padilla, S. Adapting the medaka embryo assay to a high-throughput approach for developmental toxicity testing. *Neuro Toxicol.* **2006**, *27*, 840–845. [CrossRef] [PubMed]
74. Karlsson, J.; Hofsten, J.V.; Olsson, P.E. Generating transparent zebrafish: A refined method to improve detection of gene expression during embryonic development. *Mar. Biotechnol.* **2001**, *3*, 522–527. [CrossRef] [PubMed]
75. Sandelin, A.; Alkema, W.; Engstrom, P.; Wasserman, W.W.; Lenhard, B. JASPAR: An open-access database for eukaryotic transcription factor binding profiles. *Nucleic Acids Res.* **2004**, *32*, D91–D94. [CrossRef] [PubMed]
76. Larkin, M.A.; Blackshields, G.; Brown, N.P.; Chenna, R.; McGettigan, P.A.; McWilliam, H.; Valentin, F.; Wallace, I.M.; Wilm, A.; Lopez, R.; *et al.* Clustal W and Clustal X version 2.0. *Bioinformatics* **2007**, *23*, 2947–2948. [CrossRef] [PubMed]
77. Subramanian, A.; Kaufmann, M.; Morgenstern, B. DIALIGN-TX: Greedy and Progressive Approaches for Segment-Based Multiple Sequence Alignment. *Algorithms Mol Biol.* **2008**, *3*, 6. [CrossRef] [PubMed]
78. Amemiya, C.T.; Powers, T.P.; Prohaska, S.J.; Grimwood, J.; Schmutz, J.; Dickson, M.; Miyake, T.; Schoenborn, M.A.; Myers, R.M.; Ruddle, F.H.; *et al.* Complete *Hox* cluster characterization of the coelacanth provides further evidence for slow evolution of its genome. *Proc. Natl. Acad. Sci. USA* **2010**, *107*, 3622–3627. [CrossRef] [PubMed]
79. Chiu, C.; Dewar, K.; Wagner, G.P.; Takahashi, K.; Ruddle, F.; Ledje, C.; Bartsch, P.; Scemama, J.; Stellwag, E.J.; Fried, C.; *et al.* Bichir *HoxA* cluster sequence reveals surpising trends in ray-finned fish genomic evolution. *Genome Res.* **2004**, *14*, 11–17. [CrossRef] [PubMed]
80. Church, D.M.; Goodstadt, L.; Hillier, L.W.; Zody, M.C.; Goldstein, S.; She, X.; Bult, C.J.; Agarwala, R.; Cherry, J.L.; DiCuccio, M.; *et al.* Lineage-specific biology revealed by a finished genome assembly of the mouse. *PLoS Biol.* **2009**, *7*, e1000112. [CrossRef] [PubMed]
81. Hellsten, U.; Harland, R.M.; Gilchrist, M.J.; Hendrix, D.; Jurka, J.; Kapitonov, V.; Ovcharenko, I.; Putnam, N.H.; Shu, S.; Taher, L.; *et al.* The genome of the western clawed frog *Xenopus tropicalis*. *Science* **2010**, *328*, 633–636. [CrossRef] [PubMed]
82. Kim, C.B.; Amemiya, C.; Bailey, W.; Kawasaki, K.; Mezey, J.; Miller, W.; Minoshima, S.; Shimizu, N.; Wagner, G.; Ruddle, F. *Hox* cluster genomics in the horn shark, *Heterodontus francisci*. *Proc. Natl. Acad. Sci. USA* **2000**, *97*, 1655–1660. [CrossRef] [PubMed]
83. Lee, A.P.; Koh, E.G.L.; Tay, A.; Brenner, S.; Venkatesh, B. Highly conserved syntenic blocks at the vertebrate *Hox* loci and conserved regulatory elements within and outside *Hox* gene clusters. *Proc. Natl. Acad. Sci. USA* **2006**, *103*, 6994–6999. [CrossRef] [PubMed]
84. Malaga-Trillo, E.; Meyer, A. Genome duplications and accelerated evolution of *Hox* genes and cluster architecture in teleost fishes. *Am. Zool.* **2001**, *41*, 676–686.
85. Oulion, S.; Debiais-Thibaud, M.; D'Aubenton-Carafa, Y.; Thermes, C.; da Silva, C.; Bernard-Samain, S.; Gavory, F.; Wincker, P.; Mazan, S.; Casane, D. Evolution of *Hox* gene clusters in gnathostomes: Insights from a survey of a shark (*Scyliorhinus canicula*) transcriptome. *Mol. Biol. Evol.* **2010**, *27*, 2829–2838. [CrossRef] [PubMed]
86. Raincrow, J.D. *Hox* Cluster Intergenic Sequence Evolution. Ph.D. Thesis, Rutgers University, NJ, USA, 2010.
87. Goode, D.K.; Callaway, H.A.; Cerda, G.A.; Lewis, K.E.; Elgar, G. Minor change, major divergence: Divergent functions of highly conserved *cis*-regulatory elements subsequent to whole genome duplication events. *Development* **2011**, *138*, 879–884. [CrossRef] [PubMed]
88. Santagati, F.; Minoux, M.; Ren, S.Y.; Rijli, F.M. Temporal requirement of *Hoxa2* in cranial neural crest skeletal morphogenesis. *Development* **2005**, *132*, 4927–4936. [CrossRef] [PubMed]
89. Scemama, J.L.; Hunter, M.; McCallum, J.; Prince, V.; Stellwag, E. Evolutionary divergence of vertebrate *Hoxb2* expression patterns and transcriptional regulatory loci. *J. Exp. Zool. B Mol. Evol. Dev.* **2002**, *294*, 285–299. [CrossRef] [PubMed]
90. Caroll, S.B. Evo-Devo and an expanding evolutionary synthesis: A genetic theory of morphological evolution. *Cell* **2008**, *134*, 25–36. [CrossRef] [PubMed]
91. Capellini, T.D.; di Giacomo, G.; Salsi, V.; Brendolan, A.; Ferretti, E.; Srivastava, D.; Zappavigna, V.; Selleri, L. Pbx1/Pbx2 requirement for distal limb patterning is mediated by the hierarchical control of Hox gene spatial distribution and Shh expression. *Development* **2006**, *133*, 2263–2273. [CrossRef] [PubMed]
92. Kmita, M.; Tarchini, B.; Zakany, J.; Logan, M.; Tabin, C.J.; Duboule, D. Early developmental arrest of mammalian limbs lacking HoxA/HoxD gene function. *Nature* **2005**, *435*, 1113–1116. [CrossRef] [PubMed]

93. Mercader, N.; Leonardo, E.; Azpiazu, N.; Serrano, A.; Morata, G.; Martinez, C.; Torres, M. Conserved regulation of proximodistal limb axis development by Meis1/Hth. *Nature* **1999**, *402*, 425–429. [PubMed]
94. Tamura, K.; Yokouchi, Y.; Kuroiwa, A.; Ide, H. Retinoic acid changes the proximodistal developmental competence and affinity of distal cells in the developing chick limb bud. *Dev. Biol.* **1997**, *188*, 224–234. [CrossRef] [PubMed]
95. Tamura, K.; Yonei-Tamura, S.; Yano, T.; Yokoyama, H.; Ide, H. The autopod: Its formation during limb development. *Dev. Growth Differ.* **2008**, *50*, S177–S187. [CrossRef] [PubMed]
96. Zeller, R. The temporal dynamics of vertebrate limb development, teratogenesis and evolution. *Curr. Opin. Genet. Dev.* **2010**, *20*, 384–390. [CrossRef] [PubMed]

11

Hoxa5: A Key Player in Development and Disease

Lucie Jeannotte [1,2,*], Florian Gotti [1,2] and Kim Landry-Truchon [1,2]

1 Centre de recherche sur le cancer de l'Université Laval; CRCHU de Québec, L'Hôtel-Dieu de Québec, QC G1R 3S3, Canada; florian.gotti.1@ulaval.ca (F.G.); kim.landry-truchon.1@ulaval.ca (K.L.-T.)
2 Department of Molecular Biology, Medical Biochemistry and Pathology, Université Laval, QC G1V 0A6, Canada
* Correspondence: lucie.jeannotte@crhdq.ulaval.ca

Academic Editor: Vincenzo Zappavigna

Abstract: A critical position in the developmental hierarchy is occupied by the *Hox* genes, which encode transcription factors. *Hox* genes are crucial in specifying regional identity along the embryonic axes and in regulating morphogenesis. In mouse, targeted mutations of *Hox* genes cause skeletal transformations and organ defects that can impair viability. Here, we present the current knowledge about the *Hoxa5* gene, a paradigm for the function and the regulation of *Hox* genes. The phenotypic survey of $Hoxa5^{-/-}$ mice has unveiled its critical role in the regional specification of the skeleton and in organogenesis. Most $Hoxa5^{-/-}$ mice die at birth from respiratory distress due to tracheal and lung dysmorphogenesis and impaired diaphragm innervation. The severity of the phenotype establishes that *Hoxa5* plays a predominant role in lung organogenesis *versus* other *Hox* genes. *Hoxa5* also governs digestive tract morphogenesis, thyroid and mammary glands development, and ovary homeostasis. Deregulated *Hoxa5* expression is reported in cancers, indicating *Hoxa5* involvement in tumor predisposition and progression. The dynamic *Hoxa5* expression profile is under the transcriptional control of multiple *cis*-acting sequences and *trans*-acting regulators. It is also modulated by epigenetic mechanisms, implicating chromatin modifications and microRNAs. Finally, lncRNAs originating from alternative splicing and distal promoters encompass the *Hoxa5* locus.

Keywords: *Hox* genes; organogenesis; tumorigenesis; gene regulation; long non-coding RNA

---

1. Introduction

Embryonic development can be viewed as a hierarchy of molecular events controlled by the concerted action of regulatory networks. *Hox* genes occupy a central position in the control of the formation of body segment-specific structures by regulating the transcription of downstream effectors that, in turn, direct the morphogenetic events leading to complex body forms [1]. Consequently, mutations in *Hox* genes alter segmental identity and cause morphological defects [2]. In human and mouse, 39 *Hox* genes are distributed over four complexes located on different chromosomes. They are expressed sequentially in time and space according to their position in the complexes. Accordingly, the 3′ most genes are expressed earlier and in more anterior domains along the embryonic axes than the 5′ located ones [3]. This clustered organization is fundamental for the precise spatio-temporal regulation and the function of each gene and hence for the correct patterning of the embryo [4,5]. Different *Hox* genes are expressed in discrete but overlapping domains along the developing anterior-posterior axis. A specific combination of HOX proteins at a particular anterior-posterior level thus provides a unique genetic address that determines the structures at this level. Analyses of *Hox* mutant mice phenotypes endorse the collinear relationship existing between the position of individual genes within the *Hox* clusters and the structural defects observed along the anterior-posterior axis.

Based on sequence homology and their relative position within the cluster, *Hox* genes are subdivided into 13 paralog groups. The similarities in protein structure and expression pattern among

Hox paralogs indicate that they can perform overlapping functions. Indeed, compound mutant mice for *Hox* paralogs exhibit new or more severe phenotypes than single *Hox* mutant mice [6]. Furthermore, knock-in substitutions of *Hox* genes by their paralogs confirm that they can fulfill similar roles [7,8].

Hox genes encode transcription factors that bind via their homeodomain DNA motifs in HOX-responsive elements. Despite similar DNA-binding specificity *in vitro*, individual HOX proteins confer different regulatory actions *in vivo*. This differential selectivity is largely achieved by their interaction with cofactors, mainly members of the PBC and MEIS homeodomain protein families [9,10]. While data on the developmental role of *Hox* genes are rapidly expanding, understanding how *Hox* genes regulate regional identity and organogenesis remains an issue and still awaits the identification of target genes. So far, HOX proteins have been shown to control the expression of other "high-level executive" genes encoding transcription factors, including *Hox* genes, morphogen signals, as well as effectors mediating cell behavior [11]. Not surprisingly, mutations or inappropriate expression of *Hox* genes can disrupt normal programs of growth, leading to various genetic disorders and diseases, including cancer [12,13].

The *Hoxa5* (initially named *Hox1.3*) gene belongs to this large gene family. It is positioned near the middle of the *HOXA* cluster located on mouse chromosome 6 and human chromosome 7. *Hoxa5* encodes a 270 amino acid ANTP-class homeodomain protein [14].

2. *Hoxa5*: An Imperative in Morphogenesis

Analysis of *Hox* mutant mice has revealed a plethora of phenotypes including skeletal homeotic transformations, organ defects, and postnatal phenotypes that are indicative of the broad range of actions of *Hox* genes throughout life. The same situation prevails for the *Hoxa5* gene as revealed by the phenotypic survey of *Hoxa5* mutant mice, which unveiled the importance of *Hoxa5* in the development of several tissues and organs (Table 1). Moreover, the loss of *Hoxa5* function is among the few single *Hox* gene mutations that cause death at birth.

2.1. *Axial Skeleton*

As most *Hox* mutants, *Hoxa5* null mutants present homeotic transformations of the axial skeleton. They are essentially localized in the cervical and upper thoracic regions, between the third cervical (C3) and second thoracic (T2) vertebrae [15,16]. The anterior transformation of C6 into the likeness of C5 with the loss of the tuberculum anterior and the posterior transformation of C7, which adopts the identity of a thoracic vertebra with the presence of ectopic ribs, are the most penetrant skeletal anomalies observed. They are present in all *Hoxa5*$^{-/-}$ mutants from the different genetic backgrounds tested. Other skeletal malformations are also detected at a lesser prevalence. They include the presence of an ectopic dorsal process on T1, which then resembles T2, and the loss or the reduction of the dorsal process normally found on T2 [16]. *Hoxa4* and *Hoxa6* single mutants also present homeotic transformations in the cervicothoracic transition region but to a smaller extent [17,18]. However, these shared phenotypes appear to be due to the loss of *Hoxa5* expression in the C3–T2 region of the prevertebral column in *Hoxa4* and *Hoxa6* mutant embryos resulting from deleterious long-range *cis* effects of the *neo* cassettes inserted into the *Hoxa4* and *Hoxa6* alleles that hinder the transcription from the *Hoxa5* promoter [16,19]. Taken together, these data establish the importance of *Hoxa5* in the patterning of the skeletal cervicothoracic region.

Table 1. *Hoxa5* mutant phenotypes.

Organs/Structures		Mutant Phenotypes	References
Axial skeleton		*Homeotic transformations:* *C6 → C5, loss of tuberculum anterior* *C7 → T1, ectopic ribs* *T1 → T2, gain of dorsal process on T1* *T2 → T1, loss of dorsal process on T2* *Sternum malformations:* *Fusion of cervical ribs to the manubrium sterni* *Malformed xiphoid process* *Disorganized sternebrae*	[15,16]
Appendicular skeleton		*Reduced/interrupted/missing acromion*	[16,20]
Respiratory system	Tracheal mesenchyme	*Reduced luminal surface to complete occlusion* *Thickening of the lamina propria* *Abnormal patterning of cartilage rings*	[21,22]
	Tracheal epithelium	*Epithelium disorganization:* *Pseudostratified → stratified epithelium*	[21]
	Lung mesenchyme	*Lung hypoplasia:* *Reduced cell proliferation* *Reduced branching morphogenesis* *Disorganization of the lung mesenchyme:* *Abnormal compact appearance/Thicker mesenchyme* *Narrower airspaces in embryo* *Impaired motility of alveolar myofibroblast precursors:* *Mispositioning of the alveolar myofibroblasts* *Elastic fiber disorganization and altered alveogenesis* *Lung airspace enlargement in adult*	[21–23]
	Lung epithelium	*Decreased surfactant protein expression* *Reduced number of secretory club cells* *Reduced number of alveolar type I pneumocytes* *Goblet cell metaplasia and mucus hypersecretion*	[21–24]
	Lung endothelium	*Undeveloped lung microvasculature:* *Less and trapped endothelial cells within the dense mesenchyme*	[22]
	Phrenic motor column	*Impaired diaphragm innervation*	[22,25]
	Respiratory function	*Impaired breathing:* *Increased upper airway resistance* *Increased breathing frequency and overall minute ventilation in resting conditions* *Adaptation of the tidal volume and breathing frequency to maintain a higher minute ventilation when facing hypoxia*	[26]
Digestive system	Stomach	*Perturbed cell specification of the gastric epithelium* *Goblet cell hyperplasia* *Reduced number of zymogenic and enteroendocrine cells:* *Reduced pepsin enzymatic activity*	[27]
	Intestine	*No obvious morphological defects* *Delay in the functional enzymatic activity of intestinal epithelial cells*	[28]
	Colon	*Abnormal distribution of goblet cells*	[27]
Thyroid gland		*Embryo:* *Smaller and disorganized follicles* *Increased proportion of thyroglobulin-depleted follicles* *Adult:* *Hypothyroidism symptoms:* *Delayed eye opening and ear elevation* *Growth retardation*	[29]
Mammary gland		*Abnormal development:* *Mutant females cannot feed properly their pups* *Increased proliferation and accelerated differentiation of mammary epithelium*	[30]
Ovary		*Precocious puberty* *Early onset of estrous acyclicity that worsens with age* *Ovarian epithelial cyst formation in older females*	[31]

2.2. Appendicular Skeleton

Besides transformations of the axial skeleton, the *Hoxa5* mutation affects the appendicular skeleton, more specifically the acromion, a digit-like projection emerging from the spina scapula of the pectoral girdle to articulate with the clavicula [16]. Depending on the genetic environment, *Hoxa5* mutants present a reduced, interrupted or missing acromion. This defect is also seen in *undulated* (*un*) mice, which carry a point mutation in the *Pax1* gene [32–34]. *Hoxa5* and *Pax1* genes are coexpressed over a large domain along the developing vertebral column but they show a genetic cooperation at the cervicothoracic transition level where the pectoral girdle is aligned. The expressivity and the penetrance of the skeletal anomalies affecting C6, T1 and T2 are augmented when one mutant allele is introduced into the other mutant background [20]. In acromion formation, *Hoxa5* and *Pax1* act in a complementary way: *Hoxa5* provides regional cues for the correct onset of *Pax1* expression in the developing pectoral girdle, while *Pax1* promotes the recruitment of acromion precursor cells. Whether HOXA5 acts directly on *Pax1* expression remains to be elucidated. *Hoxa5* is also involved in the control of chondrogenesis by negatively regulating the expression of *Sox9*, a master regulator of cartilage development [35]. The negative action of *Hoxa5* on *Sox9* expression was reported in chick somites suggesting that it may constitute an evolutionary-conserved pathway used by *Hoxa5* to regulate cartilage development and morphology [36].

Mutant mice for the *Hoxb5* paralog gene also present anomalies in the axial and appendicular skeletons, distinct from those observed in *Hoxa5* mutants but covering a similar territory [37]. These observations add the concept of regional functional complementarity of *Hox* genes to the frequently evoked notion of functional redundancy [38].

2.3. Respiratory System

The *Hoxa5* null mutation distinguishes itself by a high mortality rate at birth of the *Hoxa5* homozygous mutant pups (*Hoxa5*$^{-/-}$), whereas heterozygous mutants (*Hoxa5*$^{+/-}$) survive with a normal lifespan and no obvious phenotype [15]. *Hoxa5*$^{-/-}$ newborns die from respiratory distress due to severe alterations of the respiratory tract, a phenotype that is coherent with the strong *Hoxa5* mesenchymal expression along the entire respiratory system [21,22].

Laryngotracheal malformations are fully penetrant in *Hoxa5*$^{-/-}$ mice. The phenotype includes an important reduction of the luminal surface of the trachea, a profound disorganization of the epithelium and a thickening of the lamina propria with the formation of polyp-like structures. In *Hoxa5*$^{-/-}$ mice, the C-shaped cartilage rings, which normally encircle the ventrolateral surfaces of the trachea and primary bronchi to prevent collapse during respiration, present ventral gaps along the trachea, a reduced number of rings and an abnormal banding pattern. Moreover, the cartilage does not extend as dorsally as in controls. In the worst cases, a near-complete tracheal occlusion occurs contributing to the death of *Hoxa5*$^{-/-}$ pups at birth. This phenotype is reminiscent of human tracheal stenosis [21,22,39]. The tracheal malformations are similar to those observed in mutant mice for the *Fstl1* gene, which encodes a bone morphogenetic protein (BMP) antagonist [40]. *Fstl1* expression is reduced in the upper airways of *Hoxa5* mutants and ChIP experiments pinpointed *Fstl1* as a HOXA5 transcriptional target [22].

Profound anomalies affect the developing lungs of *Hoxa5* mutant embryos. They include the disorganization of the lung mesenchyme during early lung development and reduced branching morphogenesis that affects the subsequent formation of the saccula and results in an abnormal compact lung appearance with narrower airspaces and thicker mesenchyme prior to birth [21]. Decreased lung branching together with reduced cell proliferation also contribute to lung hypoplasia in *Hoxa5* mutant embryos, a phenotype that is rescued in surviving mutants by the time they are weaned [22,23]. The expression of the *Nkx2-1* and *Foxa2* genes is diminished in the lung epithelium of *Hoxa5*$^{-/-}$ embryos [21]. The NKX2-1 and FOXA2 transcriptional factors are known to govern lung epithelial cell differentiation and to regulate the expression of surfactant proteins [41,42]. Accordingly, *Hoxa5* mutant collapsed lungs present decreased surfactant protein expression. As well, lung epithelial cell fate is perturbed in *Hoxa5* mutants. A reduced number of secretory club (Clara) cells is displayed in upper airways and less alveolar type I pneumocytes, which are normally in close association with vascular endothelial cells for gas exchange, are detected in the distal lung epithelium [22,24]. As *Hoxa5* expression is restricted to the lung mesenchyme, the epithelial alterations seen in *Hoxa5*$^{-/-}$ lungs imply that HOXA5 acts indirectly on the respiratory epithelium.

Hoxa5 has also cell-autonomous functions. HOXA5 protein expression is detected in alveolar myofibroblast progenitors as shown by its colocalization with GFP from the *GFP* knock-in allele at the *Pdgfr* locus [23]. Alveolar myofibroblasts are interstitial contractile cells responsible for elastin deposition and alveolar formation, two processes affected in *Hoxa5*$^{-/-}$ mutants. Indeed, the motility of alveolar myofibroblast precursors is impaired in *Hoxa5* mutants. This leads to the mispositioning of alveolar fibroblasts that become confined to the alveoli parenchyma, which causes elastic fiber disorganization and altered alveogenesis [23,43]. Moreover, *Hoxa5* mutants show an undeveloped lung microvasculature characterized by fewer endothelial cells trapped within the dense mesenchyme, a phenotype that is consistent with *Hoxa5* expression in the lung endothelium [22]. Thus, *Hoxa5* can act directly on alveolar myofibroblast progenitors and endothelial cells, but necessitates mesenchymal-epithelial communication to control lung epithelial processes like branching and cell fate specification.

A small proportion of *Hoxa5*$^{-/-}$ mice survives and reaches adulthood. Surviving *Hoxa5* mutants develop an emphysema-like phenotype characterized by lung airspace enlargement, as mentioned above. The decreased lung surface area available for gas exchange and the increased upper airway resistance due to tracheal obstruction perturb the respiration of *Hoxa5* mutants. To compensate for these morphological defects, *Hoxa5*$^{-/-}$ mice present a higher breathing frequency and overall minute ventilation in resting conditions. When facing hypoxia, *Hoxa5*$^{-/-}$ mice adapt their tidal volume and breathing frequency to maintain a higher minute ventilation [26]. Thus, *Hoxa5* mutants develop breathing strategies to counteract their deficit in gas exchange capacity.

Hoxa5$^{-/-}$ mice also present mucus hypersecretion caused by goblet cell metaplasia [23]. The latter results from the transdifferentiation of club cells into goblet cells, an event that initiates prior to birth [24]. Goblet cell metaplasia in *Hoxa5* mutants is accompanied by increased expression of the SPDEF and FOXA3 transcription factors that govern goblet cell specification [44,45]. Goblet cell metaplasia is often associated with the loss of FOXA2 expression in airway epithelium [46,47]. However in *Hoxa5* mutants, the process is FOXA2-independent, a result analogous to that observed when Notch signaling is activated in the lung epithelium [48]. Similarly, an increased Notch signaling activity occurs in the lung airway epithelium from *Hoxa5*$^{-/-}$ mice and in areas of goblet cell metaplasia in patients suffering from chronic obstructive pulmonary disease (COPD) [24]. Treatment of *Hoxa5*$^{-/-}$ mice with a γ-secretase inhibitor blocking Notch signaling attenuates the goblet cell metaplasia phenotype, pinpointing a potential therapy to inhibit mucus overproduction in human airway diseases. Thus, the loss of *Hoxa5* function in the lung mesenchyme impacts on epithelial cell fate by modulating Notch signaling in the lung epithelium. Studies of a mouse line carrying mutations in all three *Hox5* paralog genes show that the Wnt canonical signaling pathway also mediates *Hox5* mesenchyme-to-epithelium action in the lung [49]. The strong *Hoxa5* expression in the developing lung and the severity of the *Hoxa5* lung phenotype *versus* other single *Hox* mutant mice suggest that *Hoxa5* plays a predominant role in the combinatorial control of signaling pathways during lung formation [22,50].

Hoxa5 is expressed in the phrenic motor column, a group of motor neurons that provides the exclusive source of diaphragm innervation [25]. In agreement, *Hoxa5*$^{-/-}$ mutants present innervation defects of the diaphragm, which likely participate to the lung hypoplasia phenotype as lung morphogenesis is influenced by physical forces such as fetal breathing movements [22,51,52].

In summary, *Hoxa5* contributes extensively to lung development, maturation and function as revealed by the severe anomalies that affect the respiratory system of *Hoxa5*$^{-/-}$ mice. These phenotypes are in agreement with the strong expression of the *Hoxa5* gene along the entire respiratory track mesenchyme and in the phrenic motor neurons responsible for diaphragm innervation.

Interestingly, several *Hoxa5* phenotypes share characteristics with human pathologies. As well, changes in *HOXA5* expression are associated with pulmonary diseases. For instance, *HOXA5* expression is altered in patients suffering from developmental diseases such as bronchopulmonary dysplasia, and in adult diseases like lung cancer, primary lung hypertension and COPD, one of the leading mortality causes worldwide [53–55]. Accumulated evidences on the mechanisms contributing to adult lung pathogenesis support the notion that genetic alterations affecting lung developmental processes in early life may later function as a trigger for an apparently adult onset of lung diseases [56]. The results accumulated so far indicate that deciphering the developmental role of the *Hoxa5* gene may contribute resolving the molecular mechanisms underlying lung pathologies.

2.4. Digestive System

The loss of *Hoxa5* function also affects gut morphogenesis. In mice, the gut derives from two endodermal folds, the anterior and posterior intestinal portals that fuse ventrally and join at the yolk stalk level. The oesophagus and the stomach both originate from the foregut, while the midgut develops into the gastrointestinal tract and the hindgut forms the colon [57,58]. *Hoxa5* is expressed in a dynamic spatio-temporal fashion during gut formation [15,27,28,59,60]. *Hoxa5* transcripts are first detected at embryonic (E) day 9 in the primitive gut mesenchyme. In the prospective stomach, *Hoxa5* expression presents a rostro-caudal gradient from E12.5 to E17.5. Then, *Hoxa5* transcripts become mainly confined to the submucosal and muscular layers of the hindstomach. No *Hoxa5* expression is detected in the stomach after weaning age. In the midgut, *Hoxa5* expression becomes restricted to the myenteric plexus of the enteric nervous system around E14.5, a pattern maintained up to adulthood. A similar situation prevails in the hindgut. Coherent with the *Hoxa5* expression along the gut, the loss of *Hoxa5* function affects the entire gastrointestinal tract. In absence of *Hoxa5* function, goblet cells are abnormally distributed along the colon [27]. In the intestine, no obvious morphological or cellular alterations accompany the delay in the functional enzymatic activity of epithelial cells in *Hoxa5*$^{-/-}$ mutants [28]. In contrast, cell specification is perturbed in the gastric epithelium [27]. As for the lung and colon, goblet cell hyperplasia occurs in the stomach, strengthening the importance of *Hoxa5* for the correct specification of this secretory cell lineage. Concomitantly, less zymogenic cells, which release pepsinogen, and fewer enteroendocrine cells, a source of secretagogues stimulating pepsinogen production, are observed in the glandular epithelium, explaining the reduced pepsin enzymatic activity detected in *Hoxa5*$^{-/-}$ mice. These changes in cell fate are compatible with a homeotic transformation of the hindstomach toward an intestine identity. They also concur with shifts in *Shh* and *Ihh* complementary expression domains along the gastric mucosa as well as with the altered expression of *Fgf10* and *Tgfβs*, leading to the model that *Hoxa5* provides regional cues that ensure the establishment of signaling networks warranting proper gut patterning [27].

2.5. Thyroid Gland

Surviving *Hoxa5* mutants present hypothyroidism symptoms, including growth retardation and delays in eye opening and ear elevation [29]. However, growth retardation cannot be attributed exclusively to hypothyroidism as the delayed acquisition of an adult mode of digestion that accompanies the impaired maturation of the digestive tract in *Hoxa5* mutants can also contribute to the growth deficit [28]. The lack of *Hoxa5* function perturbs the development and the structural organization of the thyroid gland at late gestation, as shown by the smaller and disorganized follicles and the large proportion of thyroglobulin-depleted follicles. *Hoxa5* expression is restricted to the mesenchyme adjoining the thyroid gland, the latter originating in part from the foregut as the lung and the stomach [29]. In *Hoxa5* mutants, the expression of *Nkx2-1*, *Pax8* and *Titf2* genes, key regulators of thyroid ontogeny and function, is altered suggesting that *Hoxa5* acts on thyroid development in a non-cell autonomous manner that requires mesenchyme-epithelium signaling.

2.6. Mammary Gland

Despite their transient hypothyroidism phenotype, surviving *Hoxa5* homozygous mice are fertile. However, *Hoxa5*$^{-/-}$ females cannot feed properly their pups, due to abnormal post-natal mammary gland development [30]. Mammary glands develop under the control of a complex interplay involving systemic hormones, local growth factors and mesenchyme-epithelium communications [61]. The *Hoxa5* mutation causes inappropriate precocious mammary epithelium development with no major change in hormonal levels. Proliferation is augmented and accelerated differentiation occurs in nulliparous and pregnant females preceding the abnormal secretory activity at parturition that underlies the incapacity of *Hoxa5*$^{-/-}$ dams to correctly nourish their progeny [30]. The accelerated lobuloalveolar epithelium development can be rescued upon grafting of mutant mammary epithelium into wild-type fat pads. Conversely, reciprocal grafting experiments demonstrate that *Hoxa5*$^{-/-}$ stroma cannot support normal proliferation of wild-type epithelium, a result that is in accordance with the restricted expression of *Hoxa5* transcripts to the mammary stroma in mice. This unveils the importance of *Hoxa5* in the delicate balance between cell growth and differentiation and establishes the essential contribution of *Hoxa5* to mammary epithelium instruction via a mesenchyme-epithelium crosstalk.

2.7. Ovary

Hoxa5$^{-/-}$ female mice also exhibit a precocious puberty and an early onset of estrous acyclicity that worsens with age, a phenotype coherent with an implication of *Hoxa5* in ovarian biology [31]. *Hoxa5* is mainly expressed in the stroma of the ovary and oviduct throughout the estrous cycle and expression levels increase with age. *Hoxa5* transcripts are also detected during gestation with a progressive raise due to the increasing expression in the corpus lutea. This dynamic expression profile of *Hoxa5* suggests that it may be subjected to regulation by sexual hormones. It also proposes specific roles for *Hoxa5* according to the cell type and the hormonal status.

The loss of *Hoxa5* function leads to ovarian epithelial cyst formation in older females, a phenotype reminiscent of human endosalpingiosis, a pelvic condition characterized by the presence of epithelium-lined cystic structures [62]. The ovarian epithelial cysts detected in *Hoxa5*$^{-/-}$ females likely derive from the ovarian surface epithelium (OSE). OSE is a simple mesothelium surrounding the outer surface of the ovary that has a very plastic phenotype [63]. Following release of the oocyte during ovulation, OSE cells proliferate and cover the wound [64]. Invaginations of the epithelium result in crypts that can be pinched off to form inclusion cysts within the stroma [65]. In normal conditions, OSE cells trapped in the stroma undergo an epithelial-mesenchymal transition to become integrated into the ovarian stroma. Failure of OSE cells to switch to a fibroblast-like identity favors cyst formation [66]. The presence of ovarian epithelial cysts in *Hoxa5*$^{-/-}$ females indicates that the loss of *Hoxa5* function in ovarian stroma may affect OSE cell behavior. This is supported by the decreased expression of proteins involved in EGFR signaling, a key pathway for ovarian homeostasis [67]. Thus, *Hoxa5* is necessary to warrant proper EGFR signaling that is essential for the harmonious postovulatory epithelium repair process and ovarian physiology.

2.8. Hematopoiesis

Finally, *Hoxa5* is associated with hematopoiesis [68,69]. *HOXA5* is expressed in human bone marrow progenitor cells. Removal of *HOXA5* expression inhibits the differentiation towards a granulocytic/monocytic cell fate and favors the proliferation of a population of multipotential cells of an erythroid-committed subtype. Conversely, sustained *HOXA5* expression promotes myelopoiesis and prevents erythropoiesis.

In summary, perturbed epithelium represents a common theme in the different organs affected by the *Hoxa5* mutation in mice. As *Hoxa5* expression is mainly restricted to the mesenchyme in lung, gut, thyroid, mammary gland and ovary, the control of mesenchymal-epithelial interactions appears as the *modus operandi* of *Hoxa5* action. These observations support the concept that *Hoxa5* may govern similar pathways in organs undergoing related developmental processes. They also establish that *Hoxa5* is a key developmental regulator in organogenesis.

3. *Hoxa5*: Deregulated Expression and Tumorigenesis

The ability of *Hox* genes to control morphogenesis implies their role in multiple cellular processes. Since aberrant proliferation, survival, differentiation, adhesion, and migration are hallmarks of cancer cells, it is not surprising that deregulated *Hox* expression is associated with oncogenesis. Several studies have revealed the potential role of *Hox* genes in tumor development, invasion and metastasis. In numerous types of cancer, expression of specific *Hox* genes is either increased or decreased, suggesting that they may be involved in tumor promotion or suppression [70,71]. Changes in *Hox* gene expression in various cancers have been associated with altered proliferation, angiogenesis, apoptosis, DNA repair and metastatic behavior [72–75]. *Hox* misregulation can perturb the expression of downstream effectors, causing improper activation of embryonic developmental cascade(s), thereby disrupting normal programs of growth and differentiation and leading to neoplasia. Increased incidence of malignancies also correlates with ectopic cervical ribs in humans, a frequent skeletal transformation found in *Hox* mutant mice. Thus, *Hox* genes may be the molecular link between congenital anomalies and cancer [76].

Hoxa5$^{-/-}$ mice are not prone to spontaneous tumorigenesis, indicating that the loss of *Hoxa5* function is not a genetic lesion sufficient to initiate oncogenesis. In normal human breast tissue, the HOXA5 protein is detected in ductal epithelial, myoepithelial and stromal cells at the parous, nulliparous and post-menopausal stages. Expression levels are low during lactation [77]. Nearly 70% of human breast carcinomas have decreased HOXA5 protein levels compared to normal tissue. Moreover, the loss of *HOXA5* gene expression in human breast cancer correlates with progression to higher-grade lesions, suggesting that it may act as a tumor suppressor gene [77,78]. In breast cancer cell lines and patient tumors, *HOXA5* silencing was proposed to result from the methylation of CpG islands in the *HOXA5* promoter region [78]. Reduced *HOXA5* expression is also associated with decreased *p53* expression. Transactivation and electrophoretic mobility shift assays showing a direct binding of the HOXA5 protein to a putative HOX-binding motif in the *p53* promoter support the idea that HOXA5 may possess growth-suppressive properties through activation of *p53* expression and apoptosis [78]. HOXA5 was also shown to interact with the transcription factor TWIST in breast cancer cell lines [79]. When overexpressed, TWIST demonstrates oncogenic potential by compromising the p53 response and cell cycle progression. HOXA5 can reverse the p53-suppressive effects of TWIST, acting as a safeguard of the p53 response via its ability to augment *p53* expression and by its capacity to bind TWIST thus limiting its negative action on p53. Alternatively, HOXA5 can induce cell death through a p53-independent program involving caspases 2 and 8 [80]. HOXA5 may also have an indirect effect on the integrity of the genome by regulating the expression of the mismatch repair gene *hMLH1* [81].

Although *Hoxa5*$^{-/-}$ mice do not develop mammary tumors, the epithelial mispecification and the hyperplasia seen in the mammary glands of *Hoxa5*$^{-/-}$ females suggest a protective role for *Hoxa5* toward cancer predisposition and reinforce the notion that *Hoxa5* may possess tumor-suppressive properties [30]. In *Hoxa5;p53* compound mutant mice, the presence of *Hoxa5* null alleles increases the susceptibility of *p53*$^{-/-}$ mice to develop tumors with a higher prevalence for thymic lymphomas [82]. Grafting experiments of whole mammary glands from *Hoxa5;p53* compound mutants into wild-type recipients reveal that the loss of both *Hoxa5* functional alleles in *p53*$^{+/-}$ mammary grafts triggers mammary tumor development, establishing the cooperative action of *Hoxa5* and *p53* in mammary tumorigenesis. However, this collaboration does not imply *p53* misregulation, which may reflect differences between species [82].

As mentioned, the ovarian epithelial inclusion cysts detected in *Hoxa5*$^{-/-}$ females likely originate from the OSE. Although usually benign, OSE-derived inclusion cysts are thought to be a potential source of ovarian cancer and their more frequent occurrence in women with hereditary risk of ovarian cancer strengthens this hypothesis [83]. Human epithelial inclusion cysts express proteins detected in ovarian cancers, such as E-cadherin, p53, c-KIT, HER-2/neu, PAX8 and WT1, consistent with the hypothesis that they are preneoplastic lesions [84–89]. The ovarian epithelial inclusion cysts in *Hoxa5*$^{-/-}$ mice also express the ovarian cancer markers PAX8 and WT1 suggesting that they constitute preneoplastic lesions and that the loss of *Hoxa5* function confers ovarian cancer predisposition [31].

Reduced *HOXA5* expression is a biomarker for poor prognostic in human non-small cell lung cancer (NSCLC). Indeed, HOXA5 controls NSCLC cell proliferation by positively regulating the expression of *Cdkn1a*, encoding the cyclin-dependent kinase inhibitor p21 [90]. Downregulation of *HOXA5* gene in NSCLC can occur due to aberrant promoter methylation or following *HOXA5* suppression by the microRNA-196a, which directly binds the 3′ untranslated region (UTR) of the *HOXA5* transcript [55,91].

A progressive downregulation of HOXA5 expression is observed from normal colon tissue to adenoma to carcinoma [92]. The decreased HOXA5 expression in colorectal tumors correlates with high levels of nuclear β-catenin, a hallmark of Wnt signaling activity. Moreover, *HOXA5* overexpression in colorectal tumors following a retinoid treatment suppresses the self-renewal capacity of cancer stem cells by inhibiting Wnt signaling. It also induces epithelial differentiation and reduces tumor size and metastatic incidence [92]. Interestingly, the negative action of *Hoxa5* on the Wnt canonical pathway was also reported in the developing lungs, reinforcing the notion that *Hoxa5* is a critical regulator of the Wnt signaling cascade [24].

HOXA5 is expressed in quiescent endothelial cells, but absent in angiogenic endothelial cells. Forced expression of HOXA5 in activated angiogenic endothelial cells was reported to prevent angiogenesis by inhibiting the expression of pro-angiogenic molecules, like the VEGF receptor 2, while promoting the expression of the anti-angiogenic factor Thrombospondin-2 [93]. *HOXA5* downregulation in endothelial cells is mediated by the microRNA-130a that specifically targets a consensus sequence in the 3′-UTR of the *HOXA5* transcript [94]. Altogether, these data raise the possibility that HOXA5 may become a potential novel therapeutic agent limiting tumor progression.

In all previous cases, decreased *HOXA5* expression is associated with tumorigenesis. Involvement of *Hoxa5* in cancer was also revealed in acute leukemias induced by the translocation t(10;11)(p13;q14) that produces the fusion of the *AF10* gene, encoding a transcription factor, with the *CALM* (Clathrin Assembly Lymphoid Myeloid) leukemia gene [95]. The leukemic transformation was shown to require *Hoxa5* upregulation through the methylation of lysine 79 on histone H3 (H3K79) at the *Hoxa5* locus by hDOT1L, a methyltransferase that interacts with the CALM-AF10 fusion protein via its AF10 moiety [96]. In bone marrow cells from *Hoxa5*$^{-/-}$ mice, the CALM-AF10 fusion protein cannot cause transformation. Thus, *Hoxa5* overexpression is critical for CALM-AF10-mediated leukemic transformation. Moreover, this does not appear to be limited to leukemia involving CALM-AF10 as *Hoxa5* upregulation was observed in other leukemic cell lines [97].

In summary, in all organs affected in *Hoxa5* mutant mice, perturbed cell proliferation and impaired differentiation are common denominators, indicating that *Hoxa5* resides at a key position in the cell growth and differentiation hierarchies. The same situation prevails in tumors. Thus, either a gain or a loss of *Hoxa5* gene expression may disrupt normal growth and differentiation programs causing neoplasia. It now remains to be determined how *Hoxa5* action is mediated by identifying its direct and indirect downstream effectors. In addition to the HOXA5 transcriptional targets mentioned above (the tumor suppressor gene *p53*, the mismatch repair gene *MutL homolog 1* and the follistatin-like 1 gene *Fstl1*), reported HOXA5 targets include the genes encoding the progesterone receptor and the cytokine pleiotrophin [98,99]. All these data also raise the possibility that acting on *Hoxa5* expression levels could provide a potent therapeutic strategy.

4. *Hoxa5*: Transcriptional Complexity and Mechanistic Integration

Hox function is intimately linked to the correct developmental expression of *Hox* genes as illustrated by the homeotic transformations observed when *Hox cis*-acting regulatory elements are mutated [100–102]. The critical role of *Hox* genes in development is clearly recognized but many gaps remain regarding the mechanisms that tightly control the establishment and maintenance of *Hox* gene expression in a precise spatio-temporal fashion in the embryo. A complex array of modes of regulation governs *Hox* gene expression [103,104]. Regulation primarily occurs at the transcriptional level via the combinatorial interplay of several signaling pathways and transcriptional factors that interact with positive and negative *cis*-acting sequences to differentially control *Hox* expression in a spatio-temporal and tissue-specific manner. The proximity of *Hox* genes in the clusters implies the integrated regulation of adjacent *Hox* promoters through the sharing, the competition and/or the selective use of defined enhancers [105]. Global regulatory elements located outside the *Hox* clusters are able of long-distance action to coordinate the expression of several genes along the *Hox* complexes [106]. Large-scale chromatin remodeling events also participate to the regulation of the collinear expression of *Hox* genes [107]. Finally, the discovery of non-coding RNAs throughout *Hox* clusters has unveiled additional levels of regulation of *Hox* expression implicating epigenetic control [108–110].

A complex organization of overlapping transcriptional units encompassing the *Hoxa5* locus exists in the mouse embryo, which results from alternative splicing and the use of three promoters: one proximal producing the 1.8-kb transcript and two distal ones (D1 and D2) giving rise to four long RNAs (5.0- (2×), 9.5- and 11.0-kb transcripts; Figure 1). The distal promoter D1 corresponds to the putative *Hoxa6* promoter, while the most distal one (D2) is located in the *Hoxa6–Hoxa7* intergenic region downstream the *Hoxa7* gene [19]. The D1 and D2 putative promoters possess *Hox*-like transcriptional activity as shown by transgenesis. Sequence comparison of the D2 promoter region reveals a highly conserved DNA sequence of 160-bp, including a putative TATA box and a transcription start site (TSS), thus arguing for important evolutionary conserved regulatory DNA elements involved in the production of the larger transcripts. The 5.0-, 9.5- and 11.0-kb transcripts are expressed later and in more posterior embryonic structures than the 1.8-kb transcript, which recapitulates the *Hoxa7* expression domain [19,111]. All the *Hoxa5*-associated transcripts include the ORF encoding the HOXA5 protein but only the 1.8-kb form produces the protein, which is restricted to the cervico-thoracic region, where solely the 1.8-kb transcript is expressed. However, the 5-kb *Hoxa6/a5* transcript can produce the HOXA6 protein as shown by transfection assays in HEK293 cells. Thus, the 5-kb *Hoxa5*, the 9.5-kb and the 11.0-kb transcripts, all transcribed from the D2 promoter, can be considered as long non-coding (lnc) RNAs. The characterization of *Hoxa6*$^{-/-}$ mutant embryos shows that the *Hoxa6* null mutation does not preclude the transcription of the 5.0-, 9.5- and 11.0-kb *Hoxa5* lncRNAs but it causes the production of 1-kb larger mutant transcripts due to the presence of the *neo* cassette [18]. This modification of the lncRNAs does not result in phenotypic consequences in the posterior domain of the embryo where they are expressed indicating that an alteration of the *Hoxa5* lncRNA sequences does not trigger major physiological effects [19]. The functional significance of the *Hoxa5*-associated lncRNAs remains to be determined. LncRNAs are present along the *Hox* clusters, some participating to the epigenetic regulation of *Hox* expression. LncRNAs can control negatively or positively gene expression acting either in *cis* or in *trans* [109,112,113]. All these data refute the initial view that lncRNAs correspond to transcriptional noise.

The presence of complex and overlapping transcriptional units at the *Hoxa5* locus implies dispersed and shared regulatory regions in the cluster. Even though the *Hoxa5* null mutation perturbs all *Hoxa5* transcripts [15], it only affects regions that express the 1.8-kb transcript, indicating that the 1.8-kb transcript is the functional form of the *Hoxa5* gene. Hence, studies combining transgenesis and molecular approaches were done to identify the regulatory elements of the *Hoxa5* proximal promoter (Figure 1). A 14.5-kb genomic fragment, starting within the neighboring *Hoxa6* gene and containing sequences up to the *Hoxa4* gene, largely reproduces the *Hoxa5* spatio-temporal expression driven by the proximal promoter in mouse embryos [114]. It encompasses *cis*-acting DNA control elements, such as a 604-bp brachial spinal cord (BSC) enhancer and a 650-bp temporal control region, both contained in the *Hoxa5* 5′-flanking sequences [115–118]. A 2.1-kb mesodermal enhancer sequence (MES) situated downstream the *Hoxa5* coding sequences was shown to be essential for *Hoxa5* paraxial and lateral plate mesoderm expression in the cervical and upper thoracic region [118]. The MES includes DNA elements that limit the *Hoxa5* regional specific expression domain along the anterior-posterior axis, including a 164-bp DNA sequence that binds CDX proteins to correctly position the *Hoxa5* expression domain [119]. A 1.5-kb DNA region that targets *Hoxa5* lung and gut developmental expression was also identified in the *Hoxa4-Hoxa5* intergenic sequences [120]. The 1.5-kb sequence includes a retinoic acid response element (RARE). *Hox* genes, mainly from paralog groups 1 to 5, are directly responsive to retinoic acid (RA), which activates retinoic acid receptors that then interact with RARE identified near *Hox* genes [101,121]. The identified RARE sequence is important for the establishment of the correct *Hoxa5* expression domain in the neural tube. However, despite the fact that it was reported to be necessary to drive *Hoxa4* expression in embryonic lung and gut, this RARE site does not play a key role in *Hoxa5* lung and gut expression during development [114]. Thus, the action of the RARE in organogenesis appears to be *Hox* promoter-specific [122]. Another RARE located at the 3′ end of the human *HOXA5* gene and conserved in the mouse genome was shown to mediate RA responsiveness of *HOXA5* in breast cancer cells [123].

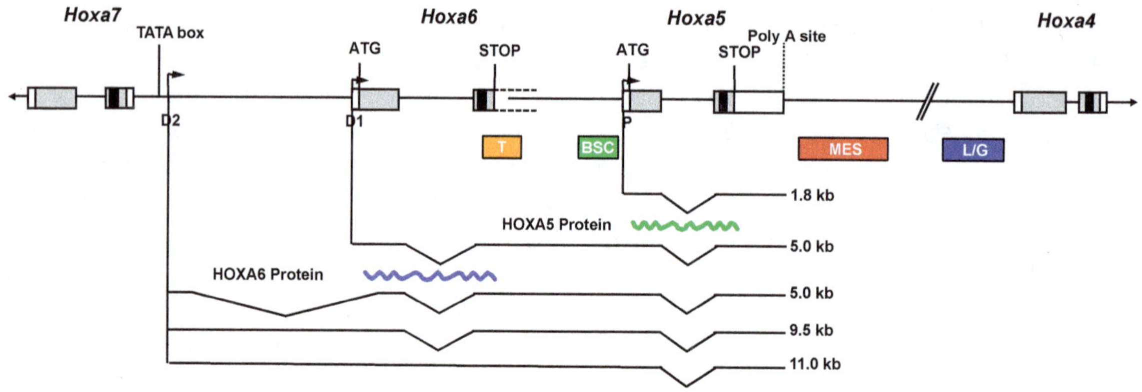

Figure 1. Schematic representation of the transcripts encompassing *Hoxa5* sequences and of the regulatory regions of the *Hoxa5* proximal promoter in the mouse embryo. Genomic organization of the *Hoxa4*, *Hoxa5*, *Hoxa6* and *Hoxa7* genes along the *HoxA* cluster. Black, grey and open boxes indicate homeobox, translated and transcribed sequences, respectively. The two known exons of *Hoxa5* and the start codon ATG of the HOXA5 protein are represented. The 3′ non-coding sequences of *Hoxa6* exon 2 extend further downstream into the *Hoxa6-Hoxa5* intergenic region and they are indicated by dotted lines. The start codon ATG of the putative HOXA6 protein is indicated. The promoters driving expression of the different transcripts are shown: proximal promoter, P; distal promoters D1 and D2. The transcripts are represented underneath. The waved lines correspond to the translated HOXA5 and HOXA6 proteins. Colored boxes define the DNA control sequences regulating *Hoxa5* expression driven by the P promoter. T, temporal; BSC, brachial spinal cord; MES, mesodermal enhancer sequence; L/G, Lung and gut. (Adapted from [19]).

The 1.5-kb *Hoxa5* lung/gut-specific *cis*-acting enhancer also interacts with the ubiquitous zinc-finger-containing transcription factor YY1. YY1 contains diverse domains enabling a plethora of protein-protein interactions. It can recruit coactivators or corepressors, which determine whether YY1 will execute inhibitory or activating functions on targets [124]. Even though the role of YY1 in *Hox* regulation was mainly associated with gene repression, YY1 acts as a positive regulator of *Hoxa5* gene expression in the developing respiratory and digestive tracts [114]. Moreover, the conditional deletion of *Yy1* function in lung mesenchyme results in mutant mice presenting a *Hoxa5*-like lung phenotype with decreased *Hoxa5* and *Hoxa4* gene expression. Therefore, the regulation driven by the YY1 binding sites located in the lung/gut enhancer in the intergenic *Hoxa5-Hoxa4* region is shared between the *Hoxa5* and *Hoxa4* genes. Interestingly, the specific ablation of *Yy1* in the lung epithelium impairs lung branching and causes airway dilation similar to that seen in pleuropulmonary blastoma, a rare pediatric lung cancer [125]. This reveals how critical YY1 is for lung morphogenesis. YY1 binding sites are also found in *Hoxa5* upstream sequences. Upon binding to these sites, YY1 mediates Polycomb Group (PcG) repression of *Hoxa5* expression in the anterior domain of the trunk, participating in the establishment of the correct *Hoxa5* expression domain in the prevertebral column [126]. Thus, depending on the developmental context, YY1 can mediate repression or activation of *Hoxa5* gene expression.

DNA methylation contributes to *Hoxa5* gene expression. In the mouse, *Hoxa5* exhibits a tissue-specific pattern of methylation of CpG islands in the promoter region that is established postnatally. The extent of *Hoxa5* methylation negatively correlates with the expression levels in adult tissues, suggesting that DNA methylation may participate in the temporal control of *Hoxa5* activity after birth [127]. Several microRNAs have also been shown to mediate *Hoxa5* downregulation [55,94,128–131]. miR-130a is frequently cited to contribute to *Hoxa5* post-transcriptional regulation. For example, RA treatment decreases miR-130a levels, which derepress HOXA5 translation. The induction of HOXA5 upon RA addition is also mediated by the RNA binding protein Human antigen R (HuR), which binds the 3′-UTR of *HOXA5* transcript to increase its stability [129]. Thus, RA modulates *Hoxa5* expression at the transcriptional and post-transcriptional levels. Not surprisingly, changes in the epigenetic control of *Hoxa5* have profound effects on its expression that are often linked to cancer. In some instances, *HOXA5* hypermethylation correlates with *HOXA5* downregulation and tumor progression [91,132,133]. As well, overexpression of miRNAs that target *Hoxa5* inhibits its expression and favors tumorigenesis [55,94,130,131].

5. Conclusion

The data accumulated over the years on the *Hoxa5* gene function clearly demonstrate that *Hoxa5* occupies a critical position in the developmental hierarchy that governs embryo patterning and organogenesis. Despite that, several questions remain. We still have to elucidate which signals transduce the information produced during early embryogenesis to correctly trigger *Hoxa5* expression in time and space and the regulatory mechanisms that maintain it throughout development. It is also essential to decode how *Hoxa5* provides regional cues to the developing organs and tissues by identifying the effectors and gene regulatory networks that mediate *Hoxa5* functions. Deciphering *Hoxa5* modes of action during development will provide insights into the molecular etiology of developmental and adult pathologies that may uncover novel therapeutic approaches.

Acknowledgments: We thank all past and present members of the laboratory as well as colleagues for their valuable contribution to our understanding of *Hoxa5* gene function and regulation. We are also grateful to Jean Charron, Amélie-Fradet-Turcotte and Nicolas Bisson for their critical comments on the manuscript. This work received the financial support of the National Science and Engineering Research Council of Canada (to L.J.).

Author Contributions: All authors contribute to the redaction of the manuscript.

Conflicts of Interest: The authors declare no conflict of interest.

References

1. Akam, M. The molecular basis for metameric pattern in the Drosophila embryo. *Development* **1987**, *101*, 1–22. [PubMed]
2. Lewis, E.B. A gene complex controlling segmentation in Drosophila. *Nature* **1978**, *276*, 565–570. [CrossRef] [PubMed]
3. Krumlauf, R. Hox genes in vertebrate development. *Cell* **1994**, *78*, 191–201. [CrossRef]
4. Duboule, D. The rise and fall of Hox gene clusters. *Development* **2007**, *134*, 2549–2560. [CrossRef] [PubMed]
5. Tschopp, P.; Tarchini, B.; Spitz, F.; Zakany, J.; Duboule, D. Uncoupling time and space in the collinear regulation of Hox genes. *PLoS Genet.* **2009**, *5*, e1000398. [CrossRef] [PubMed]
6. Mallo, M.; Wellik, D.M.; Deschamps, J. Hox genes and regional patterning of the vertebrate body plan. *Dev. Biol.* **2010**, *344*, 7–15. [CrossRef] [PubMed]
7. Greer, J.M.; Puetz, J.; Thomas, K.R.; Capecchi, M.R. Maintenance of functional equivalence during paralogous Hox gene evolution. *Nature* **2000**, *403*, 661–665. [PubMed]

8. Tvrdik, P.; Capecchi, M.R. Reversal of Hox1 gene subfunctionalization in the mouse. *Dev. Cell* **2006**, *11*, 239–250. [CrossRef] [PubMed]
9. Mann, R.S.; Lelli, K.M.; Joshi, R. Hox specificity: Unique roles for cofactors and collaborators. *Curr. Top. Dev. Biol.* **2009**, *88*, 63–101. [PubMed]
10. Rezsohazy, R.; Saurin, A.J.; Maurel-Zaffran, C.; Graba, Y. Cellular and molecular insights into Hox protein action. *Development* **2015**, *142*, 1212–1227. [CrossRef] [PubMed]
11. Pearson, J.C.; Lemons, D.; McGinnis, W. Modulating Hox gene functions during animal body patterning. *Nat. Rev. Genet.* **2005**, *6*, 893–904. [CrossRef] [PubMed]
12. Quinonez, S.C.; Innis, J.W. Human HOX gene disorders. *Mol. Genet. Metab.* **2014**, *111*, 4–15. [CrossRef] [PubMed]
13. Bhatlekar, S.; Fields, J.Z.; Boman, B.M. HOX genes and their role in the development of human cancers. *J. Mol. Med.* **2014**, *92*, 811–823. [CrossRef] [PubMed]
14. Odenwald, W.F.; Taylor, C.F.; Palmer-Hill, F.J.; Friedrich, V.; Tani, M.; Lazzarini, R.A. Expression of a homeo domain protein in noncontact-inhibited cultured cells and postmitotic neurons. *Genes Dev.* **1987**, *1*, 482–496. [CrossRef] [PubMed]
15. Jeannotte, L.; Lemieux, M.; Charron, J.; Poirier, F.; Robertson, E.J. Specification of axial identity in the mouse: Role of the Hoxa-5 (Hox1. 3) gene. *Genes Dev.* **1993**, *7*, 2085–2096. [CrossRef] [PubMed]
16. Aubin, J.; Lemieux, M.; Tremblay, M.; Behringer, R.R.; Jeannotte, L. Transcriptional interferences at the Hoxa4/Hoxa5 locus: Importance of correct Hoxa5 expression for the proper specification of the axial skeleton. *Dev. Dyn.* **1998**, *212*, 141–156. [CrossRef]
17. Horan, G.S.; Wu, K.; Wolgemuth, D.J.; Behringer, R.R. Homeotic transformation of cervical vertebrae in Hoxa-4 mutant mice. *Proc. Natl. Acad. Sci. USA* **1994**, *91*, 12644–12648. [CrossRef] [PubMed]
18. Kostic, D.; Capecchi, M.R. Targeted disruptions of the murine Hoxa-4 and Hoxa-6 genes result in homeotic transformations of components of the vertebral column. *Mech. Dev.* **1994**, *46*, 231–247. [CrossRef]
19. Coulombe, Y.; Lemieux, M.; Moreau, J.; Aubin, J.; Joksimovic, M.; Bérubé-Simard, F.A.; Tabariès, S.; Boucherat, O.; Guillou, F.; Larochelle, C.; *et al.* Multiple promoters and alternative splicing: Hoxa5 transcriptional complexity in the mouse embryo. *PLoS ONE* **2010**, *12*, e10600. [CrossRef] [PubMed]
20. Aubin, J.; Lemieux, M.; Moreau, J.; Lapointe, J.; Jeannotte, L. Cooperation of Hoxa5 and Pax1 genes during formation of the pectoral girdle. *Dev. Biol.* **2002**, *244*, 96–113. [CrossRef] [PubMed]
21. Aubin, J.; Lemieux, M.; Tremblay, M.; Bérard, J.; Jeannotte, L. Early postnatal lethality in Hoxa-5 mutant mice is attributable to respiratory tract defects. *Dev. Biol.* **1997**, *192*, 432–445. [CrossRef] [PubMed]
22. Boucherat, O.; Montaron, S.; Bérubé-Simard, F.A.; Aubin, J.; Philippidou, P.; Wellik, D.M.; Dasen, J.S.; Jeannotte, L. Partial functional redundancy between Hoxa5 and Hoxb5 paralog genes during lung morphogenesis. *Am. J. Physiol. Lung Cell Mol. Physiol.* **2013**, *304*, 817–830. [CrossRef] [PubMed]
23. Mandeville, I.; Aubin, J.; LeBlanc, M.; Lalancette-Hébert, M.; Janelle, M.F.; Tremblay, G.M.; Jeannotte, L. Impact of the loss of Hoxa5 function on lung alveogenesis. *Am. J. Pathol.* **2006**, *169*, 1312–1327. [CrossRef] [PubMed]
24. Boucherat, O.; Chakir, J.; Jeannotte, L. The loss of Hoxa5 function promotes Notch-dependent goblet cell metaplasia in lung airways. *Biol. Open* **2012**, *1*, 677–691. [CrossRef] [PubMed]
25. Philippidou, P.; Walsh, C.M.; Aubin, J.; Jeannotte, L.; Dasen, J.S. Sustained Hox5 gene activity is required for respiratory motor neuron development. *Nat. Neurosci.* **2012**, *15*, 1636–1644. [CrossRef] [PubMed]
26. Kinkead, R.; Leblanc, M.; Gulemetova, R.; Lalancette-Hébert, M.; Lemieux, M.; Mandeville, I.; Jeannotte, L. Respiratory adaptations to lung morphological defects in adult mice lacking Hoxa5 gene function. *Pediatr. Res.* **2004**, *56*, 553–562. [CrossRef] [PubMed]
27. Aubin, J.; Déry, U.; Lemieux, M.; Chailler, P.; Jeannotte, L. Stomach regional specification requires Hoxa5-driven mesenchymal-epithelial signaling. *Development* **2002**, *129*, 4075–4087. [PubMed]
28. Aubin, J.; Chailler, P.; Ménard, D.; Jeannotte, L. Loss of Hoxa5 gene function in mice perturbs intestinal maturation. *Am. J. Physiol. Cell Physiol.* **1999**, *277*, 965–973.
29. Meunier, D.; Aubin, J.; Jeannotte, L. Perturbed thyroid morphology and transient hypothyroidism symptoms in Hoxa5 mutant mice. *Dev. Dyn.* **2003**, *227*, 367–378. [CrossRef] [PubMed]
30. Garin, É.; Lemieux, M.; Coulombe, Y.; Robinson, G.W.; Jeannotte, L. Stromal Hoxa5 function controls the growth and differentiation of mammary alveolar epithelium. *Dev. Dyn.* **2006**, *235*, 1858–1871. [CrossRef] [PubMed]

31. Gendronneau, G.; Boucherat, O.; Aubin, J.; Lemieux, M.; Jeannotte, L. The loss of Hoxa5 function causes estrous acyclicity and ovarian epithelial inclusion cysts. *Endocrinology* **2012**, *153*, 1484–1497. [CrossRef] [PubMed]
32. Grüneberg, H. Genetical studies on the skeleton of the mouse. *J. Genet.* **1954**, *52*, 441–454. [CrossRef]
33. Balling, R.; Deutsch, U.; Gruss, P. Undulated, a mutation affecting the development of the mouse skeleton, has a point mutation in the paired box of Pax 1. *Cell* **1988**, *55*, 531–535. [CrossRef]
34. Wallin, J.; Wilting, J.; Koseki, H.; Fritsch, R.; Christ, B.; Balling, R. The role of Pax-1 in axial skeleton development. *Development* **1994**, *120*, 1109–1121. [PubMed]
35. Bi, W.; Deng, J.M.; Zhang, Z.; Behringer, R.R.; de Crombrugghe, B. Sox9 is required for cartilage formation. *Nat. Genet.* **1999**, *22*, 85–89. [PubMed]
36. Chen, J.W.; Zahid, S.; Shilts, M.H.; Weaver, S.J.; Leskowitz, R.M.; Habbsa, S.; Aronowitz, K.P.R.; Chang, Y.; Pinnella, Z.; Holloway, L.; *et al.* Hoxa-5 acts in segmented somites to regulate cervical vertebral morphology. *Mech. Dev.* **2013**, *130*, 226–240. [CrossRef] [PubMed]
37. Rancourt, D.E.; Tsuzuki, T.; Capecchi, M.R. Genetic interaction between hoxb-5 and hoxb-6 is revealed by nonallelic noncomplementation. *Genes Dev.* **1995**, *9*, 108–122. [CrossRef] [PubMed]
38. McIntyre, D.C.; Rakshit, S.; Yallowitz, A.R.; Loken, L.; Jeannotte, L.; Capecchi, M.R.; Wellik, D.M. Hox patterning of the vertebrate rib cage. *Development* **2007**, *134*, 2981–2989. [CrossRef] [PubMed]
39. Desir, A.; Ghaye, B. Congenital abnormalities of intrathoracic airways. *Radiol. Clin. North Am.* **2009**, *47*, 203–225. [CrossRef] [PubMed]
40. Geng, Y.; Dong, Y.; Yu, M.; Zhang, L.; Yan, X.; Sun, J.; Qiao, L.; Geng, H.; Nakajima, M.; Furuichi, T.; *et al.* Follistatin-like 1 (Fstl1) is a bone morphogenetic protein (BMP) 4 signaling antagonist in controlling mouse lung development. *Proc. Natl. Acad. Sci. USA* **2011**, *108*, 7058–7063. [CrossRef] [PubMed]
41. Bohinski, R.J.; di Lauro, R.; Whitsett, J.A. The lung-specific surfactant protein B gene promoter is a target for thyroid transcription factor 1 and hepatocyte nuclear factor 3, indicating common factors for organ-specific gene expression along the foregut axis. *Mol. Cell. Biol.* **1994**, *14*, 5671–5681. [CrossRef] [PubMed]
42. Cardoso, W.V. Transcription factors and pattern formation in the developing lung. *Am. J. Physiol. Lung Cell Mol. Physiol.* **1995**, *269*, 429–442.
43. Kim, N.; Vu, T.H. Parabronchial smooth muscle cells and alveolar myofibroblasts in lung development. *Birth Defects Res. C. Embryo Today* **2006**, *78*, 80–89. [CrossRef] [PubMed]
44. Park, K.S.; Korfhagen, T.R.; Bruno, M.D.; Kitzmiller, J.A.; Wan, H.; Wert, S.E.; Khurana Hershey, G.K.; Chen, G.; Whitsett, J.A. SPDEF regulates goblet cell hyperplasia in the airway epithelium. *J. Clin. Invest.* **2007**, *117*, 978–988. [CrossRef] [PubMed]
45. Park, S.W.; Verhaeghe, C.; Nguyenvu, L.T.; Barbeau, R.; Eisley, C.J.; Nakagami, Y.; Huang, X.; Woodruff, P.C.; Fahy, J.V.; Erle, D.J. Distinct roles of FOXA2 and FOXA3 in allergic airway disease and asthma. *Am. J. Respir. Crit. Care Med.* **2009**, *180*, 603–610. [CrossRef] [PubMed]
46. Wan, H.; Kaestner, K.H.; Ang, S.L.; Ikegami, M.; Finkelman, F.D.; Stahlman, M.T.; Fulkerson, P.C.; Rothenberg, M.E.; Whitsett, J.A. Foxa2 regulates alveolarization and goblet cell hyperplasia. *Development* **2004**, *131*, 953–964. [CrossRef] [PubMed]
47. Chen, G.; Korfhagen, T.R.; Xu, Y.; Kitzmiller, J.; Wert, S.E.; Maeda, Y.; Gregorieff, A.; Clevers, H.; Whitsett, J.A. SPDEF is required for mouse pulmonary goblet cell differentiation and regulates a network of genes associated with mucus production. *J. Clin. Invest.* **2009**, *119*, 2914–2924. [CrossRef] [PubMed]
48. Guseh, J.S.; Bores, S.A.; Stanger, B.Z.; Zhou, Q.; Anderson, W.J.; Melton, D.A.; Rajagopal, J. Notch signaling promotes airway mucous metaplasia and inhibits alveolar development. *Development* **2009**, *136*, 1751–1759. [CrossRef] [PubMed]
49. Hrycaj, S.M.; Dye, B.R.; Baker, N.C.; Larsen, B.M.; Burke, A.C.; Spence, J.R.; Wellik, D.M. Hox5 genes regulate the Wnt2/2b-Bmp4-signaling axis during lung development. *Cell Rep.* **2015**, *12*, 903–912. [CrossRef] [PubMed]
50. Gaunt, S.J.; Coletta, D.; Pravtcheva, D.; Sharpe, P.T. Mouse Hox-3.4: Homeobox sequence and embryonic expression patterns with other members of the Hox gene gen network. *Development* **1990**, *109*, 329–339. [PubMed]
51. Liggins, G.C.; Vilos, G.A.; Campos, G.A.; Kitterman, J.A.; Lee, C.H. The effect of spinal cord transection on lung development in fetal sheep. *J. Dev. Physiol.* **1981**, *3*, 267–274. [PubMed]

52. Inanlou, M.R.; Baguma-Nibasheka, M.; Kablar, B. The role of fetal breathing-like movements in lung organogenesis. *Histol. Histopathol.* **2005**, *20*, 1261–1266. [PubMed]
53. Golpon, H.A.; Geraci, M.W.; Moore, M.D.; Miller, H.L.; Miller, G.J.; Tuder, R.M.; Voelkel, N.F. HOX genes in human lung: Altered expression in primary pulmonary hypertension and emphysema. *Am. J. Pathol.* **2001**, *158*, 955–966. [CrossRef]
54. Boucherat, O.; Franco-Montoya, M.L.; Thibault, C.; Incitti, R.; Chailley-Heu, B.; Delacourt, C.; Bourbon, J.R. Gene expression profiling in lung fibroblasts reveals new players in alveolarization. *Physiol. Genomics* **2007**, *32*, 128–141. [CrossRef] [PubMed]
55. Liu, X.H.; Lu, K.H.; Wang, K.M.; Sun, M.; Zhang, E.B.; Yang, J.S.; Yin, D.D.; Liu, X.L.; Zhou, J.; Liu, Z.J.; *et al.* MicroRNA-196a promotes non-small cell lung cancer cell proliferation and invasion through targeting HOXA5. *BMC Cancer* **2012**, *12*, 348. [CrossRef] [PubMed]
56. Shi, W.; Warburton, D. Is COPD in adulthood really so far removed from early development? *Eur. Respir. J.* **2010**, *35*, 12–13. [CrossRef] [PubMed]
57. Grapin-Botton, A.; Melton, D.A. Endoderm development: From patterning to organogenesis. *Trends Genet.* **2000**, *16*, 124–130. [CrossRef]
58. Roberts, D.J. Molecular mechanisms of development of the gastrointestinal tract. *Dev. Dynam.* **2000**, *219*, 109–120. [CrossRef]
59. Dony, C.; Gruss, P. Specific expression of the Hox 1.3 homeo box gene in murine embryonic structures originating from or induced by the mesoderm. *EMBO J.* **1987**, *6*, 2965–2975. [PubMed]
60. James, R.; Kazenwadel, J. Homeobox gene expression in the intestinal epithelium of adult mice. *J. Biol. Chem.* **1991**, *266*, 3246–3251. [PubMed]
61. Hennighausen, L.; Robinson, G.W. Signaling pathways in mammary gland development. *Dev. Cell* **2001**, *1*, 467–475. [CrossRef]
62. Ryuko, K.; Miura, H.; Abu-Musa, A.; Iwanari, O.; Kitao, M. Endosalpingiosis in association with ovarian surface papillary tumor of borderline malignancy. *Gynecol. Oncol.* **1992**, *46*, 107–110. [CrossRef]
63. Auersperg, N.; Wong, A.S.; Choi, K.C.; Kang, S.K.; Leung, P.C. Ovarian surface epithelium: Biology, endocrinology, and pathology. *Endocr. Rev.* **2001**, *22*, 255–288. [CrossRef] [PubMed]
64. Murdoch, W.J.; McDonnel, A.C. Roles of the ovarian surface epithelium in ovulation and carcinogenesis. *Reproduction* **2002**, *123*, 743–750. [CrossRef] [PubMed]
65. Nicosia, S.V. The aging ovary. *Med. Clin. North Am.* **1987**, *71*, 1–9. [PubMed]
66. Auersperg, N. The origin of ovarian carcinomas: A unifying hypothesis. *Int. J. Gynecol. Pathol.* **2011**, *30*, 12–21. [CrossRef] [PubMed]
67. Ahmed, N.; Maines-Bandiera, S.; Quinn, M.A.; Unger, W.G.; Dedhar, S.; Auersperg, N. Molecular pathways regulating EGF-induced epithelio-mesenchymal transition in human ovarian surface epithelium. *Am. J. Physiol. Cell Physiol.* **2006**, *290*, 1532–1542. [CrossRef] [PubMed]
68. Fuller, J.F.; McAdara, J.; Yaron, Y.; Sakaguchi, M.; Fraser, J.K.; Gasson, J.C. Characterization of HOX gene expression during myelopoiesis: Role of HOXA5 in lineage commitment and maturation. *Blood* **1999**, *93*, 3391–3400. [PubMed]
69. Crooks, G.M.; Fuller, J.; Petersen, D.; Izadi, P.; Malik, P.; Pattengale, P.K.; Kohn, D.B.; Gasson, J.C. Constitutive HOXA5 expression inhibits erythropoiesis and increases myelopoiesis from human hematopoietic progenitors. *Blood* **1999**, *94*, 519–528. [PubMed]
70. Abate-Shen, C. Deregulated homeobox gene expression in cancer: cause or consequence? *Nat. Rev. Cancer* **2002**, *2*, 777–785. [CrossRef] [PubMed]
71. Shah, N.; Sukumar, S. The Hox genes and their roles in oncogenesis. *Nat. Rev. Cancer* **2010**, *10*, 361–371. [CrossRef] [PubMed]
72. Carrio, M.; Arderiu, G.; Myers, C.; Boudreau, N.J. Homeobox D10 induces phenotypic reversion of breast tumor cells in a three-dimensional culture model. *Cancer Res.* **2005**, *65*, 7177–7185. [CrossRef] [PubMed]
73. Rhoads, K.; Arderiu, G.; Charboneau, A.; Hansen, S.L.; Hoffman, W.; Boudreau, N. A role for HoxA5 in regulating angiogenesis and vascular patterning. *Lymphat. Res. Biol.* **2005**, *3*, 240–252. [CrossRef] [PubMed]
74. Rubin, E.; Wu, X.; Zhu, T.; Cheung, J.C.; Chen, H.; Lorincz, A.; Pandita, R.K.; Sharma, G.G.; Ha, H.C.; Gasson, J.; *et al.* A role for the HOXB7 homeodomain protein in DNA repair. *Cancer Res.* **2007**, *67*, 1527–1535. [CrossRef] [PubMed]

75. Chen, H.; Lee, J.S.; Liang, X.; Zhang, H.; Zhu, T.; Zhang, Z.; Taylor, M.E.; Zahnow, C.; Feigenbaum, L.; Rein, A.; *et al.* Hoxb7 inhibits transgenic HER-2/neu–induced mouse mammary tumor onset but promotes progression and lung metastasis. *Cancer Res.* **2008**, *68*, 3637–3644. [CrossRef] [PubMed]
76. Anbazhagan, R.; Raman, V. Homeobox genes: Molecular link between congenital anomalies and cancer. *Eur. J. Cancer* **1997**, *33*, 635–637. [CrossRef]
77. Henderson, G.S.; van Diest, P.J.; Burger, H.; Russo, J.; Raman, V. Expression pattern of a homeotic gene, HOXA5, in normal breast and in breast tumors. *Cell. Oncol.* **2006**, *28*, 305–313. [PubMed]
78. Raman, V.; Martensen, S.A.; Reisman, D.; Evron, E.; Odenwald, W.F.; Jaffee, E.; Marks, J.; Sukumar, S. Compromised HOXA5 function can limit p53 expression in human breast tumours. *Nature* **2000**, *405*, 974–978. [PubMed]
79. Stasinopoulos, I.A.; Mironchik, Y.; Raman, A.; Wildes, F.; Winnard, P.; Raman, V. HOXA5-twist interaction alters p53 homeostasis in breast cancer cells. *J. Biol. Chem.* **2005**, *280*, 2294–2299. [CrossRef] [PubMed]
80. Chen, H.; Chung, S.; Sukumar, S. HOXA5-induced apoptosis in breast cancer cells is mediated by caspases 2 and 8. *Mol. Cell. Biol.* **2004**, *24*, 924–935. [CrossRef] [PubMed]
81. Duriseti, S.; Winnard, P.T.; Mironchik, Y.; Vesuna, F.; Raman, A.; Raman, V. HOXA5 regulates hMLH1 expression in breast cancer cells. *Neoplasia* **2006**, *8*, 250–258. [CrossRef] [PubMed]
82. Gendronneau, G.; Lemieux, M.; Morneau, M.; Paradis, J.; Têtu, B.; Frenette, N.; Aubin, J.; Jeannotte, L. Influence of Hoxa5 on p53 tumorigenic outcome in mice. *Am. J. Pathol.* **2010**, *176*, 995–1005. [CrossRef] [PubMed]
83. Salazar, H.; Godwin, A.K.; Daly, M.B.; Laub, P.B.; Hogan, W.M.; Rosenblum, N.; Boente, M.P.; Lynch, H.T.; Hamilton, T.C. Microscopic benign and invasive malignant neoplasms and a cancer-prone phenotype in prophylactic oophorectomies. *J. Natl. Cancer Inst.* **1996**, *88*, 1810–1820. [CrossRef] [PubMed]
84. Wang, D.P.; Konishi, I.; Koshiyama, M.; Nanbu, Y.; Iwai, T.; Nonogaki, H.; Mori, T.; Fujii, S. Immunohistochemical localization of c-erbB-2 protein and epidermal growth factor receptor in normal surface epithelium, surface inclusion cysts, and common epithelial tumours of the ovary. *Virchows Arch. A Pathol. Anat. Histopathol.* **1992**, *421*, 393–400. [CrossRef] [PubMed]
85. Hutson, R.; Ramsdale, J.; Wells, M. p53 protein expression in putative precursor lesions of epithelial ovarian cancer. *Histopathology* **1995**, *27*, 367–371. [CrossRef] [PubMed]
86. Sundfeldt, K.; Piontkewitz, Y.; Ivarsson, K.; Nilsson, O.; Hellberg, P.; Brännström, M.; Janson, P.O.; Enerback, S.; Hedin, L. E-cadherin expression in human epithelial ovarian cancer and normal ovary. *Int. J. Cancer* **1997**, *74*, 275–280. [CrossRef]
87. Tonary, A.M.; Macdonald, E.A.; Faught, W.; Senterman, M.K.; Vanderhyden, B.C. Lack of expression of c-KIT in ovarian cancers is associated with poor prognosis. *Int. J. Cancer* **2000**, *89*, 242–250. [CrossRef]
88. Bowen, N.J.; Logani, S.; Dickerson, E.B.; Kapa, L.B.; Akhtar, M.; Benigno, B.B.; McDonald, J.F. Emerging roles for PAX8 in ovarian cancer and endosalpingeal development. *Gynecol. Oncol.* **2007**, *104*, 331–337. [CrossRef] [PubMed]
89. Madore, J.; Ren, F.; Filali-Mouhim, A.; Sanchez, L.; Köbel, M.; Tonin, P.N.; Huntsman, D.; Provencher, D.M.; Mes-Masson, A.M. Characterization of the molecular differences between ovarian endometrioid carcinoma and ovarian serous carcinoma. *J. Pathol.* **2010**, *220*, 392–400. [CrossRef] [PubMed]
90. Zhang, M.L.; Nie, F.Q.; Sun, M.; Xia, R.; Xie, M.; Lu, K.H.; Li, W. HOXA5 indicates poor prognosis and suppresses cell proliferation by regulating p21 expression in non-small cell lung cancer. *Tumor Biol.* **2015**, *36*, 3521–3531. [CrossRef] [PubMed]
91. Kim, D.S.; Kim, M.J.; Lee, J.Y.; Lee, S.M.; Choi, J.Y.; Yoon, G.S.; Na, Y.K.; Hong, H.S.; Kim, S.G.; Choi, J.E.; *et al.* Epigenetic inactivation of Homeobox A5 gene in non-small cell lung cancer and its relationship with clinicopathological features. *Mol. Carcinog.* **2009**, *48*, 1109–1115. [CrossRef] [PubMed]
92. Ordóñez-Morán, P.; Dafflon, C.; Imajo, M.; Nishida, E.; Huelsken, J. HOXA5 counteracts stem cell traits by inhibiting Wnt signaling in colorectal cancer. *Cancer Cell* **2015**, *28*, 815–829. [CrossRef] [PubMed]
93. Cuevas, I.; Layman, H.; Coussens, L.; Boudreau, N. Sustained endothelial expression of HoxA5 *in vivo* impairs pathological angiogenesis and tumor progression. *PloS ONE* **2015**, *10*, e0121720. [CrossRef] [PubMed]
94. Chen, Y.; Gorski, D.H. Regulation of angiogenesis through a microRNA (miR-130a) that down-regulates antiangiogenic homeobox genes GAX and HOXA5. *Blood* **2008**, *111*, 1217–1226. [CrossRef] [PubMed]

95. Dreyling, M.H.; Martinez-Climent, J.A.; Zheng, M.; Mao, J.; Rowley, J.D.; Bohlander, S.K. The t (10; 11)(p13; q14) in the U937 cell line results in the fusion of the AF10 gene and CALM, encoding a new member of the AP-3 clathrin assembly protein family. *Proc. Natl. Acad. Sci. USA* **1996**, *93*, 4804–4809. [CrossRef] [PubMed]
96. Okada, Y.; Jiang, Q.; Lemieux, M.; Jeannotte, L.; Su, L.; Zhang, Y. Leukaemic transformation by CALM–AF10 involves upregulation of Hoxa5 by hDOT1L. *Nat. Cell. Biol.* **2006**, *8*, 1017–1024. [CrossRef] [PubMed]
97. Quentmeier, H.; Dirks, W.G.; Macleod, R.A.; Reinhardt, J.; Zaborski, M.; Drexler, H.G. Expression of HOX genes in acute leukemia cell lines with and without MLL translocations. *Leuk. Lymphoma* **2004**, *45*, 567–574. [CrossRef] [PubMed]
98. Raman, V.; Tamori, A.; Vali, M.; Zeller, K.; Korz, D.; Sukumar, S. HOXA5 regulates expression of the progesterone receptor. *J. Biol. Chem.* **2000**, *275*, 26551–26555. [CrossRef] [PubMed]
99. Chen, H.; Rubin, E.; Zhang, H.; Chung, S.; Jie, C.C.; Garrett, E.; Biswal, S.; Sukumar, S. Identification of transcriptional targets of HOXA5. *J. Biol. Chem.* **2005**, *280*, 19373–19380. [CrossRef] [PubMed]
100. Gérard, M.; Chen, J.Y.; Gronemeyer, H.; Chambon, P.; Duboule, D.; Zakany, J. *In vivo* targeted mutagenesis of a regulatory element required for positioning the Hoxd-11 and Hoxd-10 expression boundaries. *Genes Dev.* **1996**, *10*, 2326–2334. [CrossRef] [PubMed]
101. Dupé, V.; Davenne, M.; Brocard, J.; Dollé, P.; Mark, M.; Dierich, A.; Chambon, P.; Rijli, F.M. *In vivo* functional analysis of the Hoxa-1 3′ retinoic acid response element (3′ RARE). *Development* **1997**, *124*, 399–410. [PubMed]
102. Zákány, J.; Gérard, M.; Favier, B.; Duboule, D. Deletion of a HoxD enhancer induces transcriptional heterochrony leading to transposition of the sacrum. *EMBO J.* **1997**, *16*, 4393–4402. [CrossRef] [PubMed]
103. Mallo, M.; Alonso, C.R. The regulation of Hox gene expression during animal development. *Development* **2013**, *140*, 3951–3963. [CrossRef] [PubMed]
104. Montavon, T.; Soshnikova, N. Hox gene regulation and timing in embryogenesis. *Semin. Cell. Dev. Biol.* **2014**, *34*, 76–84. [CrossRef] [PubMed]
105. Sharpe, J.; Nonchev, S.; Gould, A.; Whiting, J.; Krumlauf, R. Selectivity, sharing and competitive interactions in the regulation of Hoxb genes. *EMBO J.* **1998**, *17*, 1788–1798. [CrossRef] [PubMed]
106. Spitz, F.; Duboule, D. Global control regions and regulatory landscapes in vertebrate development and evolution. *Adv. Genet.* **2008**, *61*, 175–205. [PubMed]
107. Chambeyron, S.; Bickmore, W.A. Chromatin decondensation and nuclear reorganization of the HoxB locus upon induction of transcription. *Genes Dev.* **2004**, *18*, 1119–1130. [CrossRef] [PubMed]
108. Ponting, C.P.; Oliver, P.L.; Reik, W. Evolution and functions of long noncoding RNAs. *Cell* **2009**, *136*, 629–641. [CrossRef] [PubMed]
109. Rinn, J.L.; Kertesz, M.; Wang, J.K.; Squazzo, S.L.; Xu, X.; Brugmann, S.A.; Goodnough, L.H.; Helms, J.A.; Farnham, P.J.; Segal, E.; *et al.* Functional demarcation of active and silent chromatin domains in human HOX loci by noncoding RNAs. *Cell* **2007**, *129*, 1311–1323. [CrossRef] [PubMed]
110. Sessa, L.; Breiling, A.; Lavorgna, G.; Silvestri, L.; Casari, G.; Orlando, V. Noncoding RNA synthesis and loss of Polycomb group repression accompanies the colinear activation of the human HOXA cluster. *RNA* **2007**, *13*, 223–239. [CrossRef] [PubMed]
111. Püschel, A.W.; Balling, R.; Gruss, P. Position-specific activity of the Hox1.1 promoter in transgenic mice. *Development* **1990**, *108*, 435–442. [PubMed]
112. Wang, K.C.; Yang, Y.W.; Liu, B.; Sanyal, A.; Corces-Zimmerman, R.; Chen, Y.; Lajoie, B.R.; Protacio, A.; Flynn, R.A.; Gupta, R.A.; *et al.* A long noncoding RNA maintains active chromatin to coordinate homeotic gene expression. *Nature* **2011**, *472*, 120–124. [CrossRef] [PubMed]
113. Li, L.; Liu, B.; Wapinski, O.L.; Tsai, M.-C.; Qu, K.; Zhang, J.; Carlson, J.C.; Lin, M.; fang, F.; Gupta, R.A.; *et al.* Targeted disruption of Hotair leads to homeotic transformation and gene derepression. *Cell Rep.* **2013**, *5*, 3–12. [CrossRef] [PubMed]
114. Bérubé-Simard, F.A.; Prudhomme, C.; Jeannotte, L. YY1 Acts as a Transcriptional Activator of Hoxa5. *PLoS ONE* **2014**, *9*, e93989.
115. Zakany, J.; Tuggle, C.K.; Patel, M.D.; Nguyen-Huu, M.C. Spatial regulation of homeobox gene fusions in the embryonic central nervous system of transgenic mice. *Neuron* **1988**, *1*, 679–691. [CrossRef]
116. Tuggle, C.K.; Zakany, J.; Cianetti, L.; Peschle, C.; Nguyen-Huu, M.C. Region-specific enhancers near two mammalian homeo box genes define adjacent rostrocaudal domains in the central nervous system. *Genes Dev.* **1990**, *4*, 180–189. [CrossRef] [PubMed]

117. Nowling, T.; Zhou, W.; Krieger, K.E.; Larochelle, C.; Nguyen-Huu, M.C.; Jeannotte, L.; Tuggle, C.K. Hoxa5 gene regulation: A gradient of binding activity to a brachial spinal cord element. *Dev. Biol.* **1999**, *208*, 134–146. [CrossRef] [PubMed]
118. Larochelle, C.; Tremblay, M.; Bernier, D.; Aubin, J.; Jeannotte, L. Multiple cis-acting regulatory regions are required for restricted spatio-temporal Hoxa5 gene expression. *Dev. Dyn.* **1999**, *214*, 127–140. [CrossRef]
119. Tabariès, S.; Lapointe, J.; Besch, T.; Carter, M.; Woollard, J.; Tuggle, C.K.; Jeannotte, L. Cdx protein interaction with Hoxa5 regulatory sequences contributes to Hoxa5 regional expression along the axial skeleton. *Mol. Cell. Biol.* **2005**, *25*, 1389–1401. [CrossRef] [PubMed]
120. Moreau, J.; Jeannotte, L. Sequence analysis of a Hoxa4-Hoxa5 intergenic region including shared regulatory elements. *DNA Seq.* **2002**, *13*, 203–209. [CrossRef] [PubMed]
121. Oosterveen, T.; van Vliet, P.; Deschamps, J.; Meijlink, F. The direct context of a Hox retinoic acid response element is crucial for its activity. *J. Biol. Chem.* **2003**, *278*, 24103–24107. [CrossRef] [PubMed]
122. Packer, A.I.; Crotty, D.A.; Elwell, V.A.; Wolgemuth, D.J. Expression of the murine Hoxa4 gene requires both autoregulation and a conserved retinoic acid response element. *Development* **1998**, *125*, 1991–1998. [PubMed]
123. Chen, H.; Zhang, H.; Lee, J.; Liang, X.; Wu, X.; Zhu, T.; Lo, P.K.; Zhang, X.; Sukumar, S. HOXA5 acts directly downstream of retinoic acid receptor β and contributes to retinoic acid–induced apoptosis and growth inhibition. *Cancer Res.* **2007**, *67*, 8007–8013. [CrossRef] [PubMed]
124. Deng, Z.; Cao, P.; Wan, M.M.; Sui, G. Yin Yang 1: A multifaceted protein beyond a transcription factor. *Transcription* **2010**, *1*, 81–84. [CrossRef] [PubMed]
125. Boucherat, O.; Landry-Truchon, K.; Bérubé-Simard, F.A.; Houde, N.; Beuret, L.; Lezmi, G.; Foulkes, W.D.; Delacourt, C.; Charron, J.; Jeannotte, L. Epithelial inactivation of Yy1 abrogates lung branching morphogenesis. *Development* **2015**, *142*, 2981–2995. [CrossRef] [PubMed]
126. Kim, S.Y.; Paylor, S.W.; Magnuson, T.; Schumacher, A. Juxtaposed Polycomb complexes co-regulate vertebral identity. *Development* **2006**, *133*, 4957–4968. [CrossRef] [PubMed]
127. Hershko, A.Y.; Kafri, T.; Fainsod, A.; Razin, A. Methylation of HoxA5 and HoxB5 and its relevance to expression during mouse development. *Gene* **2003**, *302*, 65–72. [CrossRef] [PubMed]
128. Mujahid, S.; Nielsen, H.C.; Volpe, M.V. MiR-221 and miR-130a regulate lung airway and vascular development. *PLoS ONE* **2013**, *8*, e55911.
129. Yang, F.; Miao, L.; Mei, Y.; Wu, M. Retinoic acid-induced HOXA5 expression is co-regulated by HuR and miR-130a. *Cell. Signal.* **2013**, *25*, 1476–1485. [PubMed]
130. Lee, D.H.; Forscher, C.; di Vizio, D.; Koeffler, H.P. Induction of p53-independent apoptosis by ectopic expression of HOXA5 in human liposarcomas. *Sci. Rep.* **2015**, *5*, 12580. [CrossRef] [PubMed]
131. Wang, Y.; Xu, L.; Jiang, L. miR-1271 promotes non-small-cell lung cancer cell proliferation and invasion via targeting HOXA5. *Biochem. Biophys. Res. Commun.* **2015**, *458*, 714–719. [CrossRef] [PubMed]
132. Watson, R.E.; Curtin, G.M.; Hellmann, G.M.; Doolittle, D.J.; Goodman, J.I. Increased DNA methylation in the HoxA5 promoter region correlates with decreased expression of the gene during tumor promotion. *Mol. Carcinog.* **2004**, *41*, 54–66. [PubMed]
133. Strathdee, G.; Sim, A.; Soutar, R.; Holyoake, T.L.; Brown, R. HOXA5 is targeted by cell-type-specific CpG island methylation in normal cells and during the development of acute myeloid leukaemia. *Carcinogenesis* **2006**, *28*, 299–309. [PubMed]

12

The Roles of Aquaporins in Plant Stress Responses

Zunaira Afzal [1], T. C. Howton [1], Yali Sun [1] and M. Shahid Mukhtar [1,2,*]

1 Department of Biology, University of Alabama at Birmingham, Birmingham, AL 35294, USA; zunaira1@uab.edu (Z.A.); tchowton@uab.edu (T.C.H.); yalisun@uab.edu (Y.S.)
2 Nutrition Obesity Research Center, University of Alabama at Birmingham, Birmingham, AL 35294, USA
* Correspondence: smukhtar@uab.edu

Academic Editor: Andy Wessels

Abstract: Aquaporins are membrane channel proteins ubiquitously present in all kingdoms of life. Although aquaporins were originally discovered as water channels, their roles in the transport of small neutral solutes, gasses, and metal ions are now well established. Plants contain the largest number and greatest diversity of aquaporin homologs with diverse subcellular localization patterns, gating properties, and solute specificity. The roles of aquaporins in physiological functions throughout plant growth and development are well known. As an integral regulator of plant–water relations, they are presumed to play an important role in plant defense responses against biotic and abiotic stressors. This review highlights involvement of various aquaporin homologs in plant stress responses against a variety of environmental stresses that disturb plant cell osmotic balance and nutrient homeostasis.

Keywords: aquaporins; abiotic stress; biotic stress; drought; salinity; cold stress; osmotic stress; nutrient homeostasis

---

1. Background and Discovery of Aquaporins

Water, quite distinctly, is the universal solvent paramount for all living cells. The transportation of water within an organism or between the organism and its environment is crucial to accomplishing all fundamental life processes. Since the discovery of the lipid bilayer in the 1920s, the flow of water across cells and subcellular compartments was assumed to be carried out by simple diffusion across biological membranes. However, the diffusion of water across membranes occurs too slowly to account for particular physiological processes such as secretory pathways in mammals and stomatal aperture regulation in plants. These processes require a rapid, reversible flow of large volumes of water across membranes. Moreover, subcellular membranes within the same cell exhibit remarkably different levels of water permeability that cannot be justified solely by simple diffusion [1].

To account for these processes, the transport of water across biological membranes through specialized pores rather than by simple diffusion was first proposed by Koefoed-Johnsen *et al.* in 1953 [2] and then confirmed by Macey *et al.* in 1970 [3]. The proteinaceous nature of this pore came to light in the late 1980s with the advent of CHIP28, a highly abundant, channel-forming, integral membrane protein of 28-kDa isolated from human erythrocytes [4]. The molecular identity of the first water channel protein, aquaporin 1 (AQP1) was established by Agre and coworkers in 1992 based upon its ability to dramatically increase the water permeability of *Xenopus oocytes* expressing the *CHIP28* gene [5]. Although the first aquaporin member, Nodulin-26, GmNOD26 was identified in soybeans as early as 1987 [6], Wayne *et al.* in 1990 suggested the presence of pertinacious pores in plant membranes even before its discovery in humans. Quite uniquely, the existence of water channels in plants was not clearly hypothesized until 1993 when Maurel *et al.* demonstrated the functional expression of the first plant aquaporin, Arabidopsis tonoplast intrinsic protein homolog (AtTIP1;1), in *Xenopus. laevis* oocytes [7] (Figure 1).

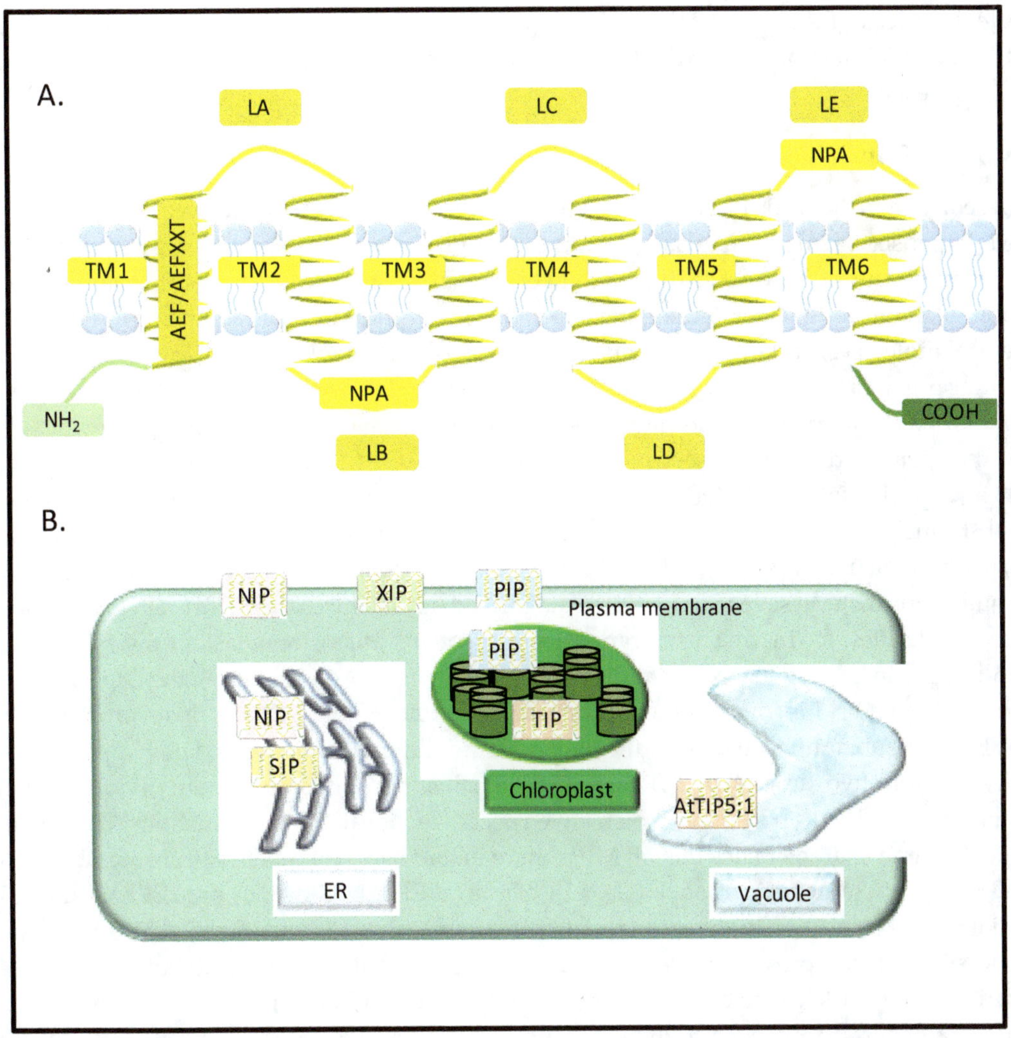

Figure 1. General structure and localization of aquaporins. (**A**) Major intrinsic protein (MIP)monomer includes six alpha helical transmembrane helices (TM1–TM6), five inter-helical loops (LA-LE), an AEF(Ala-Glu-Phe) or AEFXXT motif in the N terminal domain and two highly conserved NPA (Asp-Pro-Ala) motifs; (**B**) General localization of aquaporins. Plasma membrane intrinsic proteins (PIPs), nodulin-26 like intrinsic proteins (NIPs), and uncategorized X intrinsic proteins (XIPs) are generally localized to the plasma membrane, and expressed on the entire cell surface. Small basic intrinsic proteins (SIPs) and some NIPs have been found localizing to endoplasmic reticulum (ER). Tonoplast intrinsic proteins (TIPs) are localized to tonoplast, the membrane of vacuole. Some PIPs and TIPs have been predicted to localized to the inner envelop and thylakoids, while AtTIP5;1 has been found to be located to tonoplast.

Aquaporins belong to a highly conserved super family of membrane proteins known as major intrinsic protein (MIP). MIPs are ubiquitously present in all living organisms except thermophilic Archaea and intracellular bacteria [8] (Figure 1). Currently, aquaporins are recognized as the most abundant transmembrane transporters of substrates like glycerol, urea, CO_2, NH_3, metalloids, and reactive oxygen species (ROS) in addition to water. [9]. Due to their sedentary nature, the absence of a specialized circulatory system, and a large number of intracellular compartments, plants exhibit a great necessity for fine-tuned water regulation to adapt to environmental fluctuations. Presently, 35 aquaporin encoding genes in *Arabidopsis thaliana* (thereafter Arabidopsis) [10], 31 in maize [11], 33 in rice [12], 34 in sweet orange [13], 47 in tomato [14], 55 in *Populus trichocarpa* [15], 66 in soybean [16], 50 in banana [17], 71 in cotton [18], and 41 in sorghum [19] have been identified. This review will explicate the roles of the seven subfamilies of aquaporins in drought and salinity responses, stomatal regulation, chilling responses, nutrient uptake and transport during growth and development, and biotic stress responses.

2. Diversity of Aquaporins in Plants

Presently, more than 800 MIPs [20,21] have been identified in bacteria [22], yeast [23], protozoa [24], archaea [25], insects [26] mammals [5,27] and plants [6]. Based on substrate specificity and protein sequence similarities, these 800 MIPs have been classified into three main subfamilies: (1) AQPs (aquaporins) involved in water and ion transport; (2) GLPs (glycerol-facilitators) permeable to glycerol and neutral molecules; and (3) GLAs (aquaglyceroporins) permeable to both water and glycerol. [20]. Unicellular organisms such as bacteria or yeast present the least MIP diversity, typically possessing only one or two aquaporin encoding genes. Plants do not have GLP orthologs [28], but they do exhibit the greatest diversity of ubiquitously localized AQPs [29,30] in comparison to mammals with 13 aquaporin isoforms (AQP0-12). Mammalian orthologs are primarily restricted to secretory glands [31] and fluid-conducting organs. Owing to the abundance and diversity of aquaporins in plants, MIPs are usually assimilated in a broader sense as aquaporins in literature (Figure 1).

Aquaporins include seven subfamilies categorized according to their intracellular locations and sequence similarities: the plasma membrane intrinsic proteins (PIPs), tonoplast intrinsic proteins (TIPs), NOD26-like intrinsic proteins (NIPs) and small, basic intrinsic proteins (SIPs), the GlpF-like intrinsic proteins (GIPs), hybrid intrinsic protein (HIP), and the uncategorized X intrinsic protein (XIP) [32] (Figure 1). Each subfamily could be further divided into different subgroups. For example, PIPs have been divided into two subgroups, PIP1 and PIP2. Each subgroup is then again divided to different isoforms, such as PIP1;1 and PIP1;2, which have specialized locations and functions (Figure 1).

The common characteristics that all MIPs share include six alpha-helical transmembrane helices (TM-1-TM-6), five inter-helical loops (LA-LE) [33] an AEF(Ala-Glu-Phe) or AEFXXT motif in the *N*-termial domain, and two highly conserved NPA (Asp-Pro-Ala) motifs (the "NPA box") (Figure 1). Transport substrates, expression pattern, different levels of modification, regulation, and intracellular localizations contribute to the differences between the subfamilies of plant aquaporins. PIPs, NIPs, and XIPs are generally localized to the plasma membrane and expressed on the entire cell surface, while TIPs are localized to the tonoplast, the membrane of the vacuole. For most plant aquaporins, localization on endoplasmic reticulum (ER) can be observed during the processes of post-transcription, translation, and modification (Figure 1). However, SIPs and some NIPs have been found to localize to the ER [9], although the mechanism of targeting and their cellular functions are still not clear. SIPs tagged with green fluorescent protein were transiently expressed in Arabidopsis cells, showing the ER subcellular localization [34]. SIP1;1 and SIP1;2 may function as water channels in the ER, while SIP2;1 might act as an ER channel for other small molecules or ions [34]. Based on proteomic analyses in Arabidopsis, some PIPs and TIPs have been predicted to localized to the inner envelop and thylakoids [9], while it has also been found that AtTIP5;1 is located to tonoplast [35] (Figure 1).

Another difference between various MIP subfamilies is their substrate selectivity. Besides their function as water transporters, MIPs have been reported to transport atypical substrates, including ammonia, antimony, arsenite, boron, carbon dioxide, formamide, glycerol, hydrogen peroxide (H_2O_2), lactic acid, silicon, and urea. Two factors that contribute to their substrate selectivity are the conserved NPA motifs and amino acid residues including the ar/R (aromatic/arginine) region. The conserved NPA motifs contribute to the selectivity for water molecules [32], which are highly conserved in plant PIPs and TIPs, while alternative motifs have been found only in the NIP or SIP groups [36,37]. PIPs function as the transporters of water, glycerol, H_2O_2, carbon dioxide, and urea. PIP aquaporins are also involved in abiotic stress responses. However, several studies demonstrated altered responses of PIPs and TIPs family members upon salt, drought, or cold stresses [32]. The main role of TIPs has been described in the permeability of water. The water permeability function of plant aquaporins was first demonstrated in a TIP from Arabidopsis. Further studies have shown that TIPs may control water exchange between cytosolic and vacuolar compartments, which implies that they have a role in regulating cell turgor. Besides its water permeability function, TIPs also play roles in glycerol, urea, and ammonia transport and abiotic stress response. Specific TIP isoforms of rice and maize or Arabidopsis also show differential responses to water stress, salt, and cold stress. NIPs are able to transport water, glycerol, ammonia, silicic acid, and other solutes between plant and bacterial symbionts. Compared to PIPs and TIPs, NIPs have less water transport activity, but higher permeability to small organic molecules and mineral nutrients. The functions of NIPs include transporting beneficial and toxic metal molecules [38,39]. For example, OsNIP2;1 not only functions as a silicon influx transporter but is also involved in selenite uptake in rice [40]. SIPs have moderate water transport activity and may also function in original pore conformation. XIPs work as multifunctional channels permeable to water, metalloids and ROS. For example, *PtXIP2;1* exhibited a differential expression in leaves and stems under the stress of drought, salicylic acid, or wounding [41].

3. Roles of Aquaporins in Plant–Water Relations

Due to their autotrophic nature, plants require light and CO_2 from their aerial environment culminated with water and nutrients from soil to carry out photosynthesis. A constant water flow starting from the absorption of water from soil to its distribution throughout the plant body and evaporation in the atmosphere is crucial for carrying out all the physiological activities of the plant and managing stress imposed by fluctuating environment. Plant–water relations result in establishing a soil-plant-atmosphere continuum [42]. Although water is a key element of all physiological processes, the plant, for its growth and metabolism, uses only a small fraction. The remaining 99.5% is lost during transpiration [43]. In order to fix one kilogram of carbon during photosynthesis, plants transport several hundred kilograms of water [44]. This bulk flow of water through plants can take three different routes: the apoplastic route along cell wall structure, the symplastic route from cell to cell through the plasmodesmata, and the transcellular path across the cellular membranes [45] (Figure 2).

Long-distance bulk flow of water during transpiration and sugar transport takes the apoplastic route through vascular bundles and is generally not limited by membrane barriers [46]. Water ascends through the xylem and phloem by capillary action driven by the gradient of water potential ($\Delta\Psi$) in which water moves from a region higher Ψ to a region of lower Ψ. In the light phase, stomatal opening leads to evaporation that causes a decreases in leaf Ψ that pulls water by creating tension in the xylem vessels resulting in the movement of water from the soil through the roots [42,47] (Figure 2).

Short distance non-vascular water transport across cellular membranes is crucial to maintaining intracellular hydrostatic pressure (turgor) and cell water homeostasis at each level of organization. This includes (1) at the cell level: cellular/subcellular metabolic reactions, cell division, differentiation and elongation; (2) at the tissue level: stomatal movement; and (3) at the organ level: leaf petiole movement and maintenance of upright status of whole plant all processes depend upon cell water relations. Aquaporins as transmembrane water and solute transporter channels could be speculated as potential regulators of plant cell water relations that reflect their key roles in plant cell osmoregulation [48], root hydraulic conductivity (Lp_r), leaf hydraulic conductivity [49], transpiration [50], and cell elongation [51] (Figure 2).

Maintenance of the cell's osmotic potential under stress conditions such as pathogen infection, drought, flooding, salinity, high or low temperatures, and biotic stresses is a major challenge for plant growth and development. The majority of abiotic stress conditions directly impact plant water relations and stimulate an array of complicated cellular and physiological responses that lead to turning on plant water-saving strategies such as stomatal closure to cut off water loss during transpiration. As a tradeoff, water-saving photosynthetic activity is decreased due to the unavailability of CO_2, ultimately leading to decreased plant biomass production. Consequently, a dire need exists to understand the underlying mechanisms which control plant–water relations in relation to photosynthesis and their response to biotic and abiotic stresses [52]. Aquaporins, as vital regulators of plant–water relations, are potential targets in developing stress resistant crop plants. Their significance in all facets of plant growth and development is well-established, but the mechanistic pathways behind their roles under plant defense responses remains to be elucidated [9,32]. Numerous comparative transcriptome studies under various abiotic stress conditions have shown a differential response of different aquaporin homologs in diverse plant tissues.

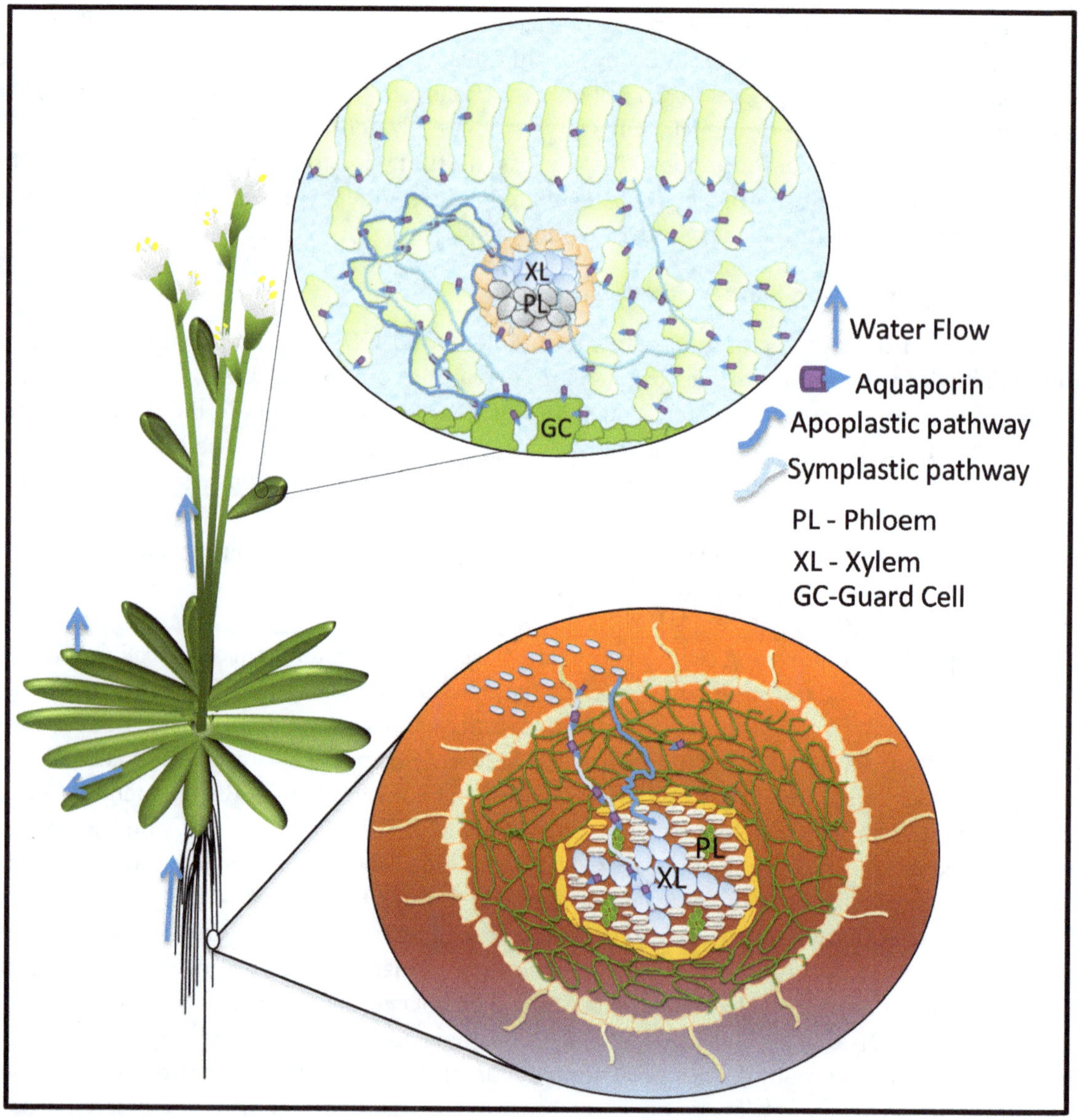

Figure 2. Water flow in plant from roots to aerial parts. Leaf cross-section shows flow of water through apoplastic and symplastic water pathways. In apoplastic pathway, water flows across the cells through cell wall. In symplastic pathway, water moves through cytoplasm and cellular membranes. Aquaporins are involved in symplastic path facilitating transmembrane water flow across cells and subcellular compartments.

Reverse genetic approaches have also been extensively used to fine-tune the role of different aquaporin encoding genes in response to diverse environmental stresses. However, the presence of a large number of diverse aquaporins, integrated complex expression patterns, and technique limitations for measuring accurate solute and water movement across transmembrane aquaporin channels are major hurdles in establishing their conclusive roles in plant growth and survival under abiotic and biotic stresses [46].

4. Plant Aquaporins in Water Stress

Based on conclusive evidence, it is widely accepted that, in most plant species, water uptake and transcellular water flow in roots are largely mediated by PIPs and TIPs. These are the most abundant aquaporins in the plasma membrane and tonoplast of the plant cells, respectively [53]. Comparative transcriptome studies revealed differential expression of multiple aquaporin homologs in response to drought stress suggesting definite roles in stress responses. Induced expression of *AtPIP2;3* under drought stress conditions is one of the earliest evidence of a drought responsive aquaporin [54]. In 2005, Alexandersson *et al.* monitored the expression of all 35 aquaporin homologs in Arabidopsis in response to drought stress alone and found that most *PIP* and some *TIP* genes have high levels of expression, while *NIP* genes have very low expression [55]. The authors also showed that all PIP genes are down-regulated in drought stress response in leaves except *AtPIP1;4* and *AtPIP2;5*, which are up-regulated. Moreover, *AtPIP2;6* and *AtSIP1;1* are constitutively expressed and are not significantly affected by the drought stress [55]. Expression of *AtPIP2;5* was also significantly up-regulated in leaves in response to a combination of drought and heat stresses [56]. Consistent with these results, several other studies in Arabidopsis have shown that among all subfamilies of aquaporins PIPs are most responsive to drought stress and most of them undergo a transcriptional down-regulation. Only a few genes were found to be up-regulated [57,58]. All of these *PIP* genes that are down-regulated in response to drought are highly expressed in the roots. Strong down-regulation of *PIP genes* transcription under drought stress was also observed in the roots and twigs of olive plants [59], in the roots of tobacco [60], and in the fruits of peach [61]. Moreover, most of these drought responsive transcriptional changes are conserved among different Arabidopsis accessions [57]. Distinctly, in many other plant species, differential responses by the same aquaporin homologs have been seen among different cultivars of the same plant species. For example, in the grapevines the expression of *VvPIP1;1* in the root was up-regulated by drought stress in an anisohydric cultivar but not in an isohydric cultivar [62]. In another experiment, expression of the *VvPIP2;1* gene was down-regulated under drought conditions [63]. Therefore, contrasting results have been reported between similar aquaporin homologs within the same species under drought stress. One transcriptome comparison of a field-grown cotton (*Gossypium. hirsutum*) under normal and drought stress conditions revealed that two highly homologous genes, *PIP1;3* and *PIP1;1* have contrasting expression patterns in leaves and roots [18] indicating that their roles in drought stress response differs despite their structural similarity. In addition to water deficiency dehydration caused by other environmental stimuli exhibited differential responses in some species. For example, osmotic stress induced by 10% polyethylene glycol (PEG) in rice revealed no effect on *OsPIP1;3*, but expression of *OsPIP1;1* and *OsPIP1;2* was up-regulated [64]. Contrastingly, expression of *OsPIP1;1* was down-regulated in osmotic stress administrated by manitol [65] and also by drought stress [66]. Similar alterations were seen in reddish aquaporins in response to salt, PEG, and mannitol induced osmotic stress [67]. Recently, characterization of *GoPIP1* from a legume forage *Galega orientalis* showed its association with drought tolerance. In contrast to rice and reddish, the transcripts levels of *GoPIP1* increased significantly in roots upon exposure to the osmotic stress imposed by both high NaCl concentration and PEG. Overexpression of this gene in transgenic Arabidopsis made the plants more vulnerable to drought stress but not to salinity stress [68]. Stress levels also lead to differential responses, for example, in grapevine leaves, moderate drought stress led to a significant down regulation of the five PIP genes investigated, but prolonged stress increased their expression levels [69]. Water transport by aquaporins is also coupled with diurnal rhythms. Diurnal expression of PIPs in response to different intensities of drought was investigated in *Fragaria vesca*. Researchers found that most of the PIPs are down-regulated in roots and the expression of *FvPIP1;1* and *FvPIP2;1* was strongly correlated to the decrease in substrate moisture contents. In leaves, the amplitude of diurnal aquaporin expression was down-regulated in response to drought [70]. Responses of aquaporins to drought stress also vary based on plant symbiosis. In a study of *Phaseolus vulgaris*, it was observed that PIPs' responses to drought stress vary depending on whether the plants is inoculated with arbuscular mycorrhizal fungi or not [71]. Arbuscular mycorrhizal symbiosis with maize plants has been shown to regulate aquaporin expression differentially in short term and prolonged drought stress conditions. Furthermore, various ZmPIPs, specifically the expression of *ZmTIP1;2*, *ZmTIP2;3*, and *ZmNIP2;1* also varies in response to drought and symbiosis [72].

In transcriptome studies, the roles of PIPs in drought response is well pronounced in most of the cases but some reports specifically point out the role of TIPs in water deficient conditions. In rice, the expression of *OsTIP1;1* was up-regulated in roots and shoot in response to water stress [73]. Root transcriptome analysis of chickpea plants under drought stressed showed both up and down-regulation of different *PIPs*, *TIPs*, and *NIPs* homologs suggesting their complex integrated roles in regulation of water balance under water deficit [74]. Reports about involvement of NIPs, SIPs, and XIPs in drought stress responses are very few. Detailed transcriptome studies of all aquaporin homologs in citrus plants under drought stress conditions revealed that all *CsTIPs* (11) and *CsXIPS* (3) were up-regulated in leaves but down-regulated in roots, whereas in leaves only two *CsPIPs* (*CsPIP1;1* and *CsPIP2;4*), one *CsNIP* (*CsNIP1;1*), and one *CsSIP* (*CsSIP1;2*) were differentially up-regulated [13].

On the basis of transcriptome data, it is difficult to identify a concerted pattern of aquaporin expression in response to drought stress. Reverse genetics approaches based on one or a few genes provide a better snapshot than the whole transcriptome studies. Crucial involvement of PIPs in modulating Lp_r in response to various environmental stresses has been widely observed [11,49,75]. Increased susceptibility of plants to water stress due to reduction in Lp_r by silencing PIPs has been reported extensively in various plant species. For example, in Arabidopsis knockout mutants of *AtPIP1;2* and *AtPIP2;2* genes, there was a noteworthy reduction in water permeability of protoplasts [76] and a 14% decrease in Lp_r, respectively making these mutants more vulnerable to drought stress [77]. Double antisense lines with reduced expression of *AtPIP1* and *AtPIP2* in Arabidopsis showed up to a 30-fold decrease in Lp_r. In tobacco a 55% reduction in Lp_r and increased sensitivity to drought was observed by targeting *NtPIP1* gene using antisense technology. In addition to decreased Lp_r, reduced expression of *NtPIP1* also showed a significant decrease in transpiration rate [49]. In moss, *Physcomitrella patens*, knockout mutants of *PpPIP2;1* and *PpPIP2;2* exhibited severe stress phenotypes when grown under water-limited conditions. Based on these observations, it is plausible that these targeted aquaporins play a cumulative role as water transporters, and their decreased expressions make plants vulnerable to water stress due to a decreased Lp_r. Further, it is also possible that decreased transpiration rates cause a reduction in photosynthesis, in turn affecting the overall survivability of the plant [78].

Consistently, transgenic plants overexpressing aquaporins showed enhanced drought tolerance. For example, the overexpression of a BnPIP1 from *Brassica napus* in transgenic tobacco plants resulted in increased tolerance to drought stress [79]. Similarly, transgenic tobacco plants overexpressing the wheat aquaporin gene *TaAQP7* (*PIP2*) were more drought tolerant in comparison to non-transgenic tobacco plants due to enhanced water retention capabilities of transgenic plants [80]. Transgenic Arabidopsis plants expressing a *Vicia faba PIP1* (*VfPIP1*) showed improved drought resistance by preventing water loss through transpiration due to the induction of stomatal closure [81]. In Arabidopsis, the overexpression of a banana *PIP* gene *MaPIP1;1* showed increased root growth and enhanced survival rates of transgenic plants under drought stress, when compared to wild type plants [82]. Similarly, transgenic banana plants expressing banana *PIP1;2* driven through two diverse promoters conferred enhanced tolerance to drought stress [17]. Overexpression of a tomato *SlTIP2;2* gene in transgenic tomato plants resulted in increased drought-tolerance due to ability of plant to regulate its transpiration rate under drought stress conditions [83]. The above experimental evidence suggests that the overexpression of aquaporins make plants more resistant to drought stress. However, some contrasting results have also been observed because rapid water loss due to increased leaf and root hydraulic conductivity makes some plants even more vulnerable to drought stress conditions. For example, the overexpression of *AtPIP1;4* or *AtPIP2;5* in transgenic Arabidopsis plants showed no advantage under normal conditions but presented a rapid water loss under drought stress conditions.

Similarly, overexpression of an Arabidopsis aquaporin gene *AtPIP1b*, in tobacco showed drastic water loss under drought stress [84].

In order to understand the role of aquaporins in drought tolerance many studies have been conducted in naturally drought tolerant plant species. In a drought tolerant plant *Eragrostis nindensis* (resurrection grass), *TIP3;1* was found in small vacuoles of desiccation tolerant leaves, suggesting its importance in enhanced mobilization of water and solutes from these vacuoles upon rehydration [85]. Functional characterization of PIPs in two differential drought responsive poplar clones *Populus balsamifera* and *P. simonii* × *P. balsamifera* revealed a clear correlation between differential expressions of PIP genes in drought avoiding and drought tolerating strategies adopted by these two clones. The rapid reduction of stomatal conductance by reduction in PIPs activities was observed in leaves of drought avoiding *P. simonii* × *P. balsamifera* [86]. Water deficit exposure to a drought tolerant rice cultivar revealed up-regulation of the *OsPIP1;3* gene suggesting its role in drought avoidance [87]. Strong up-regulation of some PIPs in a dehydration tolerant succulent plant, *Craterostigma plantagineum*, was observed under water stress [88]. Characterization of a *FaPIP2;1* gene isolated from a drought-tolerant perennial grass species *Festuca arundinacea* showed its involvement in leaf dehydration status during water stress by overexpressing this gene in Arabidopsis. Transgenic plants showed enhanced resistance and better growth and development under drought stress [89]. In a drought tolerant Vitis hybrid, Richter-110 (*Vitis berlandieri* × *Vitis rupestris*) expression of five PIPs and two TIPs was monitored at different levels of water stress, and it was found that the aquaporin genes in leaves show differential regulation in response to moderate and drastic water stress. A moderate decrease in water availability results in down-regulation of the aquaporins. However, in roots, aquaporin expression showed complex patterns, with no generality among different aquaporins [69].

It can be concluded that drought stress response of aquaporins is highly variable depending on stress levels, aquaporin isoform, tissue, species, presence of symbionts, and the nature of stimuli causing dehydration similar to drought stress. However, a general down-regulation of most of the *PIP* genes is thought to reduce water loss and to help prevent backflow of water to drying soil. Although *TIPs* are found to play a key role in controlling cell water homeostasis by rapid water transport between the vacuole and cytoplasm of plant cells, experimental evidence on their roles in response to drought stress are limited in comparison to *PIPs*.

5. Aquaporins in CO_2 Homeostasis in Water Deficit Conditions

CO_2 conductivity and transpiration are intrinsically associated with stomatal movement. In order to cut off the loss of water through transpiration, stomatal closure is the paramount water-saving strategy of plants under water-limiting conditions, which result in a reduced rate of photosynthesis due to the unavailability of CO_2. In addition to their pivotal role in the transmembrane, water transport, and regulation of stomatal aperture, reports have also shown the involvement of *PIPs* in CO_2 conductivity relevant to the photosynthetic capacity of plant under water-limiting conditions. In tobacco, the involvement of *NtAQP1* has been shown in leaf CO_2 transport [90]. An *AtPIP1;2* T-DNA insertion line facilitates the diffusion of CO_2 suggesting a relevance of CO_2 transport through aquaporins [91,92]. The *Mesembryanthemum crystallinum* aquaporin McMIPB, which is classified as a PIP1 that interacts with PIP2 by forming heterotetramers, showed enhanced mesophyll and stomatal conductance, CO_2 diffusion, photosynthesis, and overall plant growth. This suggests that the co-expression of *PIP2* and *PIP1* may increase the activity of PIP1 not only for water flow but also for CO_2 diffusion [93]. In Eucalyptus trees, the expression of *EgPIP1* and *EgPIP2* was co-suppressed by using *RsPIP1;1* (*Raphanus sativus* aquaporin) and a low rate of CO_2 assimilation was observed [94]. In rice, when *HvPIP2;1* is overexpressed, enhanced CO_2 conductance and CO_2 assimilation is observed, but there is also greater sensitivity to salt stress [95]. In tomato (*Solanum lycopersicum*), overexpression of *SlTIP2;2* regulates transpiration rates under stress conditions resulting in improved CO_2 uptake and a more balanced water and nutrient supply [83].

Subsequently, under water stress conditions aquaporins are not only responsible for minimizing water loss from plant tissues, but they also serve a prized function in facilitating CO_2 homeostasis.

6. Roles of Aquaporins in Salt Stress

Soil salinity is the combination of water and ionic stress that exerts noxious effects on plant growth and development by disturbing the cell's osmotic balance. The primary responses of plants to salt stress are the inhibition of root-water uptake and a resultant decrease in Lp_r [96], which is also manifested in cases of drought stress responses. Thus, most of the experimental evidence of aquaporin responses in salinity is in consistent with drought stress. For example, in Arabidopsis a significant reduction in Lp_r, coupled with a 60% to 75% decrease in *PIP* and *TIP* aquaporin transcripts abundance, was observed after exposure to salt stress. Transgenic banana plants overexpressing *MusaPIP1;2* and *MusaPIP1;2* displayed enhanced tolerance to both salt and drought stresses [17]. Transcript and protein levels of the barley *HvPIP2;1* gene were found to be down-regulated in roots but up-regulated in the shoots of plants under salt stress [97]. Additionally, the overexpression in transgenic barely revealed increased salt sensitivity, furthering suggesting a relevant role in salt tolerance [98]. A salinity-tolerant ice plant (*M. crystallinum*) also showed down-regulation of many PIP genes [99] and a *TIP* gene (*MIPF*) in roots and leaves, respectively under salt stress [100]. In Maize plants, an ABA mediated, salt-induced down regulation of most of the members of *ZmPIP1* and *ZmPIP2* were observed, but a transiently enhanced expression of *ZmPIP1;1*, *ZmPIP1;5*, and *ZmPIP2;4* was also seen preferentially in the outer parts of the roots. No change in expression of *ZmTIPs* was observed in these experiments [101]. Subsequent studies showed involvement of some TIPs, therefore it can be anticipated that in addition to their role in increased water transport across the tonoplast, TIPs are also involved in the accumulation of ions in vacuoles in response to salt stress. For example, the overexpression of a*AtTIP5;1* in Arabidopsis resulted in the tolerance of transgenic plants to high levels of borate, suggesting its involvement in vacuolar compartmentation of borate [102]. In rice, the expression of *OsTIP1;1* was down-regulated in response to cold stress [12] but up-regulated during response to water and salinity stress [73]. The overexpression of the *Panax ginseng* aquaporin, *PgTIP1, in* Arabidopsis showed increased plant growth under optimal conditions and also enhanced tolerance to salt and drought stress [103]. Several studies involving the localization of aquaporins during salt stress have revealed relocalization or redistribution of aquaporins in response to high salt concentrations. For example, salt stress induced re-localization of AtTIP1;1 into intravacuolar invaginations is also shown in Arabidopsis [96]. In addition to redistribution, an alternative mechanism regulating PIP abundance in the plasma membrane is endocytosis of PIPs either through the clathrin-dependent pathway [104] or a salt- stimulated membrane raft–associated pathway [105]. Relocalization of a TIP1;1-GFP into intracellular spherical structures and internalization of PIP2;1-GFP in response to salt stress was also observed in Arabidopsis [96]. Both endocytosis and exocytosis caused by the cycling of AtPIP2;1 to and from the plasma membrane is also observed under salt stress [106].

Similar trends of *PIPs* and *TIPs* regulation in response to salt stress have been observed with only a few exceptions. For example, in drought-tolerant, salinity-sensitive grapevine the expression of the *PIP2;1* gene is up-regulated under salt stress but down-regulated under drought [63]. In addition to PIPs and TIPs, being small solute and ion transporters, NIPs have also been found to be involved in salinity stress. The overexpression of wheat *NIP*, *TaNIP*, in Arabidopsis plants showed increased tolerance to salt stress. Expression of this *TaNIP* gene is up-regulated after ABA treatment, suggesting its association with other ABA regulated pathways [107]. Transcriptome analyses of citrus roots and leaves under salt stress revealed that in addition to *CsPIPs* and *CsTIPs*, most of the *CsNIPs*, *CsXIPs*, and *CsSIPs* are also up-regulated in roots, whereas in leaves both up and down regulation patterns were observed for some homologs of each aquaporin family except CsSIPs [13]. Similar to drought stress, most of the experimental evidence shows involvement of PIPs and TIPs, while only a few experiments point towards the involvement NIPs. Interestingly, only one account exists of XIPs' involvement in salt stress.

7. Cold Stress and Aquaporins

Similar to water and salt stresses, cold stress is an important abiotic stress factor that significantly limits plant growth and development. The early response of Aquaporins to cold stress has been frequently reported in literature. For example, transgenic banana plants over expressing *MusaPIP1;2* and *MusaPIP1;2* displayed enhanced tolerance to both cold and drought stress [17]. Similarly, transgenic tobacco plants overexpressing a wheat aquaporin *TaAQP7* (PIP2) gene exhibited enhanced cold-tolerance as well as drought tolerant in comparison to non-transgenic tobacco plants [80,108]. Distinctly, other experiments reveal opposite patterns. For example, transgenic Arabidopsis plants overexpressing *AtPIP1;4* or *AtPIP2;5* showed enhanced tolerance to cold stress but are more susceptible to drought due to rapid water loss. In cold sensitive plants like rice, prolonged exposure to cold stress cause an increase in Lp_r which should be regulated through root aquaporins, as most, particularly *OsPIP2;5*, are found to be up-regulated [109]. In another experiment in rice, the mRNA levels of ten genes including *OsTIP1;1* and *OsTIP2;2* were significantly down-regulated but the expression of *OsPIP1;3* increased up to 60% in roots on exposure to chilling treatment [12]. *OsPIP1;3* is also significantly up-regulated in response to drought stress in drought-tolerant rice cultivar [87]. In contrast, *OsTIP1;1* was up-regulated in response to water and salinity stress [73]. While some homologs present similarities in all dehydration inducing stresses like drought, salinity, and cold, others like *OsTIP1;1*, in the following example, are counter-regulated in cold in comparison to drought and salt. *OsPIP1;1* presents different patterns of expression upon exposure to diverse osmotic stresses. Moreover, its involvement in all salt, drought and chilling tolerance has been suggested [64,65].

In Arabidopsis, only *PIP2;5* and *PIP2;6*, which are normally among the low- expressed PIPs, are significantly up regulated by cold stress in both the roots and aerial parts of the plant. All other PIP genes were found to be down regulated by cold stress and their pattern of expression vary with the application of salt or drought stress. Most abiotic stresses, including chilling, induce the production of ABA. Plant responses to stress are generally ABA mediated. However, in an experiment, responses of each aquaporin to ABA treatment were different, suggesting that the regulation of aquaporins might follow both ABA-dependent and ABA-independent signaling pathways [58,67]. Aroca *et al.* in 2005 investigated the effect of long and short term chilling stress on the expression of aquaporins in cold-tolerant and cold-susceptible maize cultivars. They reported that long and short term chilling differentially induce aquaporin responses in similar patterns in both genotypes. These responses lead to recovery in tolerant genotypes, but not in sensitive genotypes including those that are unable to cope with the oxidative stress caused by chilling. Down regulation of *PIPs* in both cultivars is consistent with previous findings in rice and sugarcane [65,110,111].

The response of aquaporins to cold stress is linked with an increase in root hydraulic conductivity that may be regulated through increased aquaporin expression levels in plant roots upon exposure to low temperatures. Differential responses have been observed between short term and long term exposure to cold stress. Moreover, recovery from cold stress has been linked with aquaporin expression, particularly in PIPs.

8. Aquaporins in Micronutrient Homeostasis and Heavy Metal Stress

Although the water transport capacity of NIPs is very limited, they play a fundamental role in transporting other substrates involved in various cellular processes. Most *NIP* homologs appear to transport nutrients in plants. Arabidopsis knockout mutant of the *NIP5;1* gene displayed striking growth retardation in response to low supply of boron (B). B is an essential element for plant growth, development, and reproduction, and its deficiency in arable areas has drastically effected crop production worldwide [112]. In the event of high B supply, a feedback inhibition of the *AtNIP5;1* gene is observed, thus providing a strong evidence of the involvement of *NIP5;1* in B homeostasis the adaptation of plants to B toxicity in soil. These results were further confirmed in maize, where a loss of function mutation in *ZmNIP3;1*, a maize ortholog of *AtNIP5;1*, was shown to be responsible for an abnormal phenotype caused by B deficiency. In addition, *AtNIP6;1* and *AtNIP7;1* homologs serve in

B transport in Arabidopsis, facilitating its distribution in shoots and anthers, respectively [113,114]. Tolerance to high B was observed in barley by a reduced expression of the *HvNIP2;1* gene. [115]. In addition to *NIPs*, an overexpression of *TIP5;1* in Arabidopsis suggested its involvement in vacuolar compartmentation of borate [102]. Silicon (Si) plays an important role in plant defense against biotic and abiotic stresses. Si-mediated alleviation of plant stress is a well-known phenomenon [116]. The Si transporter *OsNIP2;1* (Lsi1) was identified by quantitative trait loci (QTL) mapping in a rice mutant defective in Si uptake [117]. Furthermore, the Si transporter *NIP2;2* (Lsi6) was also identified from rice [116]. Therefore, OsNIPs seems to play a significant role in Si transport in rice plants.

In addition to nutrient transport, NIPs are also found to be involved in both the transport and responses to heavy metal stress. A T-DNA knockout mutant of Arabidopsis *NIP1;1* showed arsenite (As) tolerance suggesting its role as an As transporter [118]. Recently, *AtNIP3;1* has shown to be involved in As transport. Double knockout mutants for both *AtNIP1;1* and *AtNIP3;1* show more pronounced tolerance against As stress. An additional four other isoforms, *NIP5;1*, *NIP6;1*, *NIP7;1*, and *NIP1;2*, are reported to be capable of As(III) transport on the basis of their expressions in yeast and oocytes [119]. The transport of antimony (Sb) by *AtNIP1;1* has also been established, and its loss of function mutant showed an increased tolerance to high Sb stress [120]. A lactic acid transporter *AtNIP2;1* was shown to be associated with water logging, leading to oxygen deprivation. Thus it can be anticipated that *AtNIP2;1* is an anaerobic-induced lactic acid transporter that may play a role in plant adaptation to lactic fermentation under anaerobic stress conditions [121].

9. Aquaporins in Plant Symbiotic and Pathogenic Relations

Plant symbiotic relationships are known to affect plant aquaporin responses under different abiotic stress conditions [71,72]. It is hypothesized that in plant symbiotic relations, the re-distribution of water and nutrients takes place between the host and symbionts further shedding light on a relationship with aquaporins. The roles of NIPs in plant interactions with symbionts and pathogens are obvious but experimental evidence is scarce. The first identified aquaporin, nodulin-26, was found in soybean root nodules and is speculated to have formed as a result of a symbiotic interaction between plant and nitrogen fixing rhizobium. While the roles of NIPs in various forms of nitrogen transport (urea, ammonia) are well known, the mechanistic involvement of NIPs in driving these nutrient exchanges between plants and symbionts or plants and pathogens needs further investigation. The roles of plant aquaporin-mediated solute transport during plant symbiosis with arbuscular mycorrhizae have been examined by gene profiling. Such results have indicated that the transport of glycerol from plant to the microbe, in addition toNH_4/NH_3 from microbe to plant, is mediated by aquaporin genes including *NIPs* [72].

Like symbionts, plant pathogens rely on plant nutrients for their growth and survival on their plant hosts. Dehydration, as a result of pathogen infection, is a common observation responsible for regulating plant water relations and nutrient homeostasis. The production of reactive oxygen species is an integral component of plant immune system. The transport of H_2O_2 by aquaporin homologs also highlights functions in plant defense, despite not having a full understanding of these pathways [122,123]. Expression profiling of soybean leaves upon *Pseudomonas syringae* infection revealed down-regulation in 24 of 32 soybean aquaporin genes [124]. In citrus plants, six CsMIPs (*CsPIP1;2*, *CsPIP2;2*, *CsNIP2;2*, *CsNIP5;2*, *CsNIP6;1*, and *CsSIP1;1*) were found to be differentially expressed exclusively under the biotic stress imposed by the citrus infecting proteobacterium, *Candidatus Liberibacter*. Comparisons of *CsPIP2;2*, *CsTIP1;2*, *CsTIP2;1*, *CsTIP2;2*, and *CsNIP5;1* expression patterns of susceptible sweet orange and tolerant rough lemon cultivars revealed that most *CsMIPs* are down regulated. Thus, they could be correlated with the disease development [13,125]. As a result, the roles of aquaporins in response to pathogen infection are highly anticipated. Evidence of NIPs' involvement in biotic stress responses in maize was reported [126]. The induction of a tobacco root specific *TIP1* gene expression upon nematode infection was observed [127]. In grasses (Festuca spp.), Si accumulation offers an effective defense against herbivore attacks [128]. Thus, the regulation of Si uptake through NIPs can be correlated with plant defense against herbivory. The precise modes of aquaporin regulation during infection and their response to pathogen signals are still unknown. Some interactions of aquaporins with bacterial and oomycete effectors were observed in yeast two hybrid systems [129]. TIP1 and TIP2 were found to interact with a cucumber mosaic virus (CMV) replication protein, CMV1a, in the SOS recruitment system suggesting that this TIPs-CMV1a interaction potentially affects CMV replication in the host plant's tonoplasts [130]. Interactome and transcriptome data provide important clues about the involvement of aquaporins in host-pathogen interactions. Extensive reverse genetic-based characterization of individual or multiple putative aquaporin homologs needs to be undertaken in relation to pathogen infection to reach a definite conclusion about role of aquaporins in plant diseases.

10. Complex Integrated Roles of Aquaporins in Plant Stresses

Despite enormous experimental evidence about the involvement of aquaporins genes in various stress responses (Figure 3), precise roles of individual genes or particular subfamilies are still hard to define because of their highly complex and integrated roles in the response to different environmental stimuli and involvement in other plant growth and developmental processes. These integrated contributions in multiple physiological facets can be more pronounced in transgenic plants containing foreign aquaporin genes. In transgenic plants expressing an exogenous aquaporin gene, tolerance or susceptibility is collectively effects by interactions with other plant genes. Transgenic Arabidopsis plants expressing a *Vicia faba PIP1* (*VfPIP1*) showed improved drought resistance due to induction of stomatal closure, thereby preventing water loss in water- deficient conditions [81]. Hence, the interaction of *VfPIP1* with other genes and gene products involved in stomatal movement contributes to overall drought tolerance. The silencing of endogenous genes by expressing exogenous aquaporins has also been observed when the *Raphanus sativus* aquaporin *RsPIP1;1* gene is expressed in eucalyptus causing the silencing of endogenous *EgPIP1* and *EgPIP2* PIP aquaporins [94]. The overexpression of *AtPIP1b* in tobacco results in an increased stomatal density and increased growth rates under favorable conditions but drastic water loss under drought stress [84]. Transgenic tobacco plants overexpressing a wheat aquaporin *TaAQP7* (*PIP2*) gene showed increased activity of detoxification enzymes resulting in an improved antioxidant defense and a reduction in H_2O_2 levels under osmotic stress [80]. Similarly, the overexpression of a wheat *PIP*, *TdPIP1;1*, in tobacco showed an enhanced stress tolerance coupled with increased leaf area and increased root size [131]. The overexpression of two foreign aquaporin isoforms, *CfPIP2;1* (*Cucurbita ficifolia*) and *CsPIP1;1* (*C. sativus*) in Arabidopsis, noticeably perturb the natural expression of other aquaporins, resulting in altered stress responses of the plants under various stress conditions [132]. In addition, the overexpression of a *BnPIP1* from *Brassica napus* in transgenic tobacco plants results in an increased tolerance to drought stress, increased plant growth, and higher seed germination rates, whereas antisense plants showed developmental deformities and an increased susceptibility to drought stress [79].

In addition to these outcomes of over-expressing aquaporins in different transgenic backgrounds, advances in interactomics has shown physical interactions among different aquaporin homologs [133–135] as well as aquaporin interactions with several other proteins involved in various diverse functions in plants, alluding to their complex, integrated role in different physiological and stress responses [130,136–141].

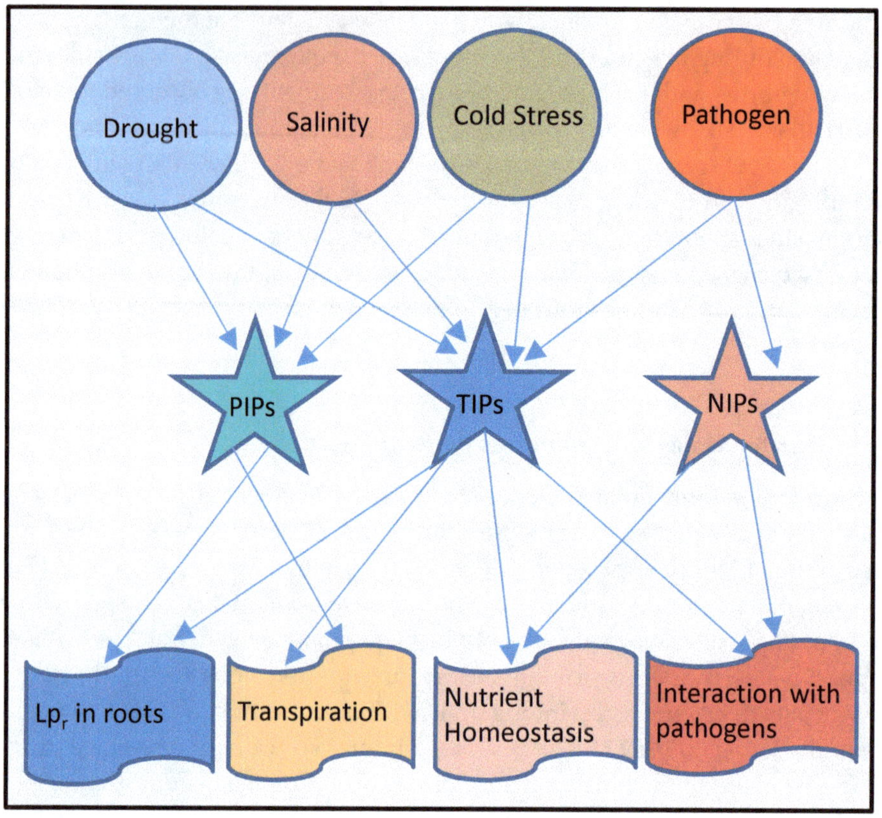

Figure 3. A generalized summary of plant aquaporins in response to environmental stimuli. Both PIPs and TIPs are more responsive to drought, salt and cold stress that disturb cell osmotic balance, they regulate root hydraulic conductivity (Lp_r) and transpiration rates. TIPs along with NIPs are involved in biotic stress responses and involved in regulating nutrient homeostasis between host and its pathogen. NIPs and TIPs are also found to interact with some pathogen proteins.

11. Conclusions

Current data suggest that drought and salt stress induces more dramatic changes in aquaporin expression than any other abiotic stress [19]. A profound role of PIPs in response to osmotic stress stimuli like drought and salinity can be established (Figure 3). Knockout mutants of *PIPs* and their overexpressing transgenic plants revealed that PIPs respond to osmotic stresses in an integrated, complex way by regulating root water uptake and increasing transpiration rates as well as affecting the overall plant growth and development processes. Like PIPs, TIPs are involved in regulating osmotic stress responses, but experimental evidence is far less in comparison to PIPs. TIPs are also found to be involved in the uptake of some micronutrients facilitating responses to higher micronutrient concentrations. NIPs' roles in nutrient homeostasis and heavy metal stress responses are more pronounced than any other aquaporin subfamily. Information on the involvement of aquaporins in biotic stress responses is still scarce. However, current analyses on the involvement of TIPs and NIPs in nutrient uptake points to a pronounced involvement of NIPs in plant pathogen interactions (Figure 3).

Since the discovery of aquaporins two decades ago, more information has been unveiled regarding structure, substrate specificity, gating properties, subcellular localization, and roles in various physiological processes in plant growth and development. Forward and reverse genetic approaches have also been used extensively to define their roles in plant defense mechanisms against environmental stresses. However, a clear cut mechanistic role of individual aquaporin homologs in response to a particular environmental stress has not yet been established because of the complex, integrated responses that vary among different plant species. Numerous studies of plant aquaporins under osmotic stress conditions have revealed their importance in regulating plant stress responses.

Rapid changes in the expression levels of aquaporins, in response to diverse stresses, have been documented often, but the molecular and cellular mechanisms underlying these responses are still unknown. The localization patterns in specific membranes and subcellular compartments combined with their redistribution with exposure to diverse osmotic stresses could be crucial for efficacy of aquaporins [142,143]. Moreover, the responses of aquaporins to various stress induced hormones indicate their involvement in stress induced hormonal pathways. The interactions of aquaporins with other plant proteins also provide an insight into the complex mechanisms involved in aquaporin regulations and diverse stress responses. In order to arrive at a definite conclusion, more detailed studies are needed to map different pathways of aquaporin activities in plant physiology and stress responses. Differential responses of aquaporin homologs to stress induced hormones need thorough investigation to uncover their mechanistic involvement in plant stress responses.

Acknowledgments: We acknowledge funding from Fulbright graduate fellowship to Z.A., U.A.B., College of Arts and Science graduate fellowship to Y.S. and N.S.F. (IOS-1557796) to M.S.M. We greatly appreciate Jessica Lopez for critical reading the manuscript.

Author Contributions: M.S.M. conceived the ideas and designed the layout the article. Z.A., T.C.H. and Y.S. wrote the paper and contributed the figures

Conflicts of Interest: The authors declare no conflict of interest.

Abbreviations

The following abbreviations are used in this manuscript:

Lp_r	Root hydraulic Conductivity
MIPs	Major intrinsic proteins
MIPs	Major intrinsic protein genes
PIPs	Plasma membrane intrinsic proteins
PIPs	Plasma membrane intrinsic protein genes
TIPs	Tonoplast intrinsic proteins
TIPs	Tonoplast intrinsic protein genes
NIPs	Nodulin-26 like intrinsic proteins
NIPs	Nodulin-26 like intrinsic protein genes
SIPs	Small basic intrinsic proteins
SIPs	Small basic intrinsic protein genes
GIPs	GlpF-like intrinsic proteins
HIPs	Hybrid intrinsic proteins
AQPs	Aquaporins
XIPs	Uncategorized X intrinsic proteins
XIPs	Uncategorized X intrinsic protein genes
ROS	Reactive oxygen species
GLPs	Glycerol-facilitators
ER	Endoplasmic reticulum
$\Delta\Psi$	Water potential
CMV	Cucumber mosaic virus

References

1. Maurel, C.; Tacnet, F.; Guclu, J.; Guern, J.; Ripoche, P. Purified vesicles of tobacco cell vacuolar and plasma membranes exhibit dramatically different water permeability and water channel activity. *Proc. Natl. Acad. Sci. USA* **1997**, *94*, 7103–7108. [CrossRef] [PubMed]

2. Koefoed-Johnsen, V. The contributions of diffusion and flow to the passage of D_2O through living membranes: Effect of neurohypophysenl hormone 011 isolated anuran skin. *Acta Physiol. Scand.* **1953**, *28*, 60–76. [CrossRef] [PubMed]
3. Macey, R.L.; Farmer, R.E. Inhibition of water and solute permeability in human red cells. *Biochim. Biophys. Acta Biomembr.* **1970**, *211*, 104–106. [CrossRef]
4. Denker, B.M.; Smith, B.L.; Kuhajda, F.P.; Agre, P. Identification, purification, and partial characterization of a novel MR 28,000 integral membrane protein from erythrocytes and renal tubules. *J. Biol. Chem.* **1988**, *263*, 15634–15642. [PubMed]
5. Preston, G.M.; Carroll, T.P.; Guggino, W.B.; Agre, P. Appearance of water channels in xenopus oocytes expressing red cell chip28 protein. *Science* **1992**, *256*, 385–387. [CrossRef] [PubMed]
6. Fortin, M.G.; Morrison, N.A.; Verma, D.P.S. Nodulin-26, a peribacteroid membrane nodulin is expressed independently of the development of the peribacteroid compartment. *Nucleic Acids Res.* **1987**, *15*, 813–824. [CrossRef] [PubMed]
7. Maurel, C.; Reizer, J.; Schroeder, J.I.; Chrispeels, M.J. The vacuolar membrane protein gamma-tip creates water specific channels in xenopus oocytes. *EMBO J.* **1993**, *12*, 2241–2247. [PubMed]
8. Abascal, F.; Irisarri, I.; Zardoya, R. Diversity and evolution of membrane intrinsic proteins. *Biochim. Biophys. Acta Gen. Subj.* **2014**, *1840*, 1468–1481. [CrossRef] [PubMed]
9. Maurel, C.; Boursiac, Y.; Luu, D.T.; Santoni, V.R.; Shahzad, Z.; Verdoucq, L. Aquaporins in plants. *Physiol. Rev.* **2015**, *95*, 1321–1358. [CrossRef] [PubMed]
10. Johanson, U.; Karlsson, M.; Johansson, I.; Gustavsson, S.; Sjovall, S.; Fraysse, L.; Weig, A.R.; Kjellbom, P. The complete set of genes encoding major intrinsic proteins in arabidopsis provides a framework for a new nomenclature for major intrinsic proteins in plants. *Plant Physiol.* **2001**, *126*, 1358–1369. [CrossRef] [PubMed]
11. Chaumont, F.O.; Barrieu, F.O.; Wojcik, E.; Chrispeels, M.J.; Jung, R. Aquaporins constitute a large and highly divergent protein family in maize. *Plant Physiol.* **2001**, *125*, 1206–1215. [CrossRef] [PubMed]
12. Sakurai, J.; Ishikawa, F.; Yamaguchi, T.; Uemura, M.; Maeshima, M. Identification of 33 rice aquaporin genes and analysis of their expression and function. *Plant Cell Physiol.* **2005**, *46*, 1568–1577. [CrossRef] [PubMed]
13. Martins, C.D.P.S.; Pedrosa, A.M.; Du, D.; Goncalves, L.P.; Yu, Q.; Gmitter, F.G., Jr.; Costa, M.G.C. Genome-wide characterization and expression analysis of major intrinsic proteins during abiotic and biotic stresses in sweet orange (*Citrus sinensis* L. Osb.). *PLoS ONE* **2015**, *10*, e0138786.
14. Reuscher, S.; Akiyama, M.; Mori, C.; Aoki, K.; Shibata, D.; Shiratake, K. Genome-wide identification and expression analysis of aquaporins in tomato. *PLoS ONE* **2013**, *8*, e79052. [CrossRef] [PubMed]
15. Gupta, A.B.; Sankararamakrishnan, R. Genome-wide analysis of major intrinsic proteins in the tree plant populus trichocarpa: Characterization of XIP subfamily of aquaporins from evolutionary perspective. *BMC Plant Biol.* **2009**, *9*. [CrossRef] [PubMed]
16. Ali, Z.; Wang, C.B.; Xu, L.; Yi, J.X.; Xu, Z.L.; Liu, X.Q.; He, X.L.; Huang, Y.H.; Khan, I.A.; Trethowan, R. Genome-wide sequence characterization and expression analysis of major intrinsic proteins in soybean (*Glycine max* L.). *PLoS ONE* **2013**, *8*, e56312.
17. Shekhawat, U.K.; Ganapathi, T.R. Overexpression of a native plasma membrane aquaporin for development of abiotic stress tolerance in banana. *Plant Biotechnol. J.* **2013**, *11*, 942–952.
18. Park, W.; Scheffler, B.E.; Bauer, P.J.; Campbell, B.T. Genome-wide identification of differentially expressed genes under water deficit stress in upland cotton (*Gossypium hirsutum* L.). *BMC Plant Biol.* **2012**, *12*. [CrossRef] [PubMed]
19. Reddy, P.S.; Rao, T.S.R.B.; Sharma, K.K.; Vadez, V. Genome-wide identification and characterization of the aquaporin gene family in *Sorghum bicolor* (L.). *Plant Gene* **2015**, *1*, 18–28. [CrossRef]
20. Karkouri, K.; Gueune, H.; Delamarche, C. MIPDB: A relational database dedicated to MIP family proteins. *Biol. Cell* **2005**, *97*, 535–543. [CrossRef] [PubMed]
21. Zardoya, R. Phylogeny and evolution of the major intrinsic protein family. *Biol. Cell* **2005**, *97*, 397–414. [CrossRef] [PubMed]
22. Calamita, G.; Bishai, W.R.; Preston, G.M.; Guggino, W.B.; Agre, P. Molecular cloning and characterization of AQPZ, a water channel from escherichia coli. *J. Biol. Chem.* **1995**, *270*, 29063–29066. [PubMed]
23. Carbrey, J.M.; Bonhivers, M.I.; Boeke, J.D.; Agre, P. Aquaporins in saccharomyces: Characterization of a second functional water channel protein. *Proc. Natl. Acad. Sci. USA* **2001**, *98*, 1000–1005. [CrossRef] [PubMed]

24. Mitra, B.N.; Yoshino, R.; Morio, T.; Yokoyama, M.; Maeda, M.; Urushihara, H.; Tanaka, Y. Loss of a member of the aquaporin gene family, AQPA affects spore dormancy in dictyostelium. *Gene* **2000**, *251*, 131–139. [CrossRef]
25. Kozono, D.; Ding, X.; Iwasaki, I.; Meng, X.; Kamagata, Y.; Agre, P.; Kitagawa, Y. Functional expression and characterization of an archaeal aquaporin aqpm from methanothermobacter marburgensis. *J. Biol. Chem.* **2003**, *278*, 10649–10656. [CrossRef] [PubMed]
26. Beuron, F.; le Caherec, F.; Guillam, M.T.; Cavalier, A.; Garret, A.; Tassan, J.P.; Delamarche, C.; Schultz, P.; Mallouh, V.; Rolland, J.P. Structural analysis of a mip family protein from the digestive tract of cicadella viridis. *J. Biol. Chem.* **1995**, *270*, 17414–17422. [CrossRef] [PubMed]
27. Agre, P.; Preston, G.M.; Smith, B.L.; Jung, J.; Raina, S.; Moon, C.; Guggino, W.B.; Nielsen, S. Aquaporin chip: The archetypal molecular water channel. *Am. J. Physiol. Renal Physiol.* **1993**, *265*, F463–F476.
28. Zardoya, R.; Ding, X.; Kitagawa, Y.; Chrispeels, M.J. Origin of plant glycerol transporters by horizontal gene transfer and functional recruitment. *Proc. Natl. Acad. Sci. USA* **2002**, *99*, 14893–14896. [CrossRef] [PubMed]
29. Johansson, I.; Karlsson, M.; Johanson, U.; Larsson, C.; Kjellbom, P. The role of aquaporins in cellular and whole plant water balance. *Biochim. Biophys. Acta Biomembr.* **2000**, *1465*, 324–342. [CrossRef]
30. Maurel, C.; Javot, H.; Lauvergeat, V.; Gerbeau, P.; Tournaire, C.; Santoni, V.; Heyes, J. Molecular physiology of aquaporins in plants. *Int. Rev. Cytol.* **2002**, *215*, 105–148. [PubMed]
31. Nielsen, S.R.; King, L.S.; Christensen, B.M.N.; Agre, P. Aquaporins in complex tissues. II. Subcellular distribution in respiratory and glandular tissues of rat. *Am. J. Physiol. Cell Physiol.* **1997**, *273*, C1549–C1561.
32. Forrest, K.L.; Bhave, M. Major intrinsic proteins (MIPs) in plants: A complex gene family with major impacts on plant phenotype. *Funct. Integr. Genomics* **2007**, *7*, 263–289. [CrossRef] [PubMed]
33. Hove, R.M.; Bhave, M. Plant aquaporins with non-aqua functions: Deciphering the signature sequences. *Plant Mol. Biol.* **2011**, *75*, 413–430. [CrossRef] [PubMed]
34. Ishikawa, F.; Suga, S.; Uemura, T.; Sato, M.H.; Maeshima, M. Novel type aquaporin SIPs are mainly localized to the ER membrane and show cell-specific expression in arabidopsis thaliana. *FEBS Lett.* **2005**, *579*, 5814–5820. [CrossRef] [PubMed]
35. Wudick, M.M.; Luu, D.T.; Tournaire-Roux, C.; Sakamoto, W.; Maurel, C. Vegetative and sperm cell-specific aquaporins of arabidopsis highlight the vacuolar equipment of pollen and contribute to plant reproduction. *Plant Physiol.* **2014**, *164*, 1697–1706. [CrossRef] [PubMed]
36. Ishibashi, K. Aquaporin subfamily with unusual NPA boxes. *Biochim. Biophys. Acta Biomembr.* **2006**, *1758*, 989–993. [CrossRef] [PubMed]
37. Ishibashi, K.; Kuwahara, M.; Kageyama, Y.; Sasaki, S.; Suzuki, M.; Imai, M. Molecular cloning of a new aquaporin superfamily in mammals. In *Molecular Biology and Physiology of Water and Solute Transport*; Springer: NewYork, NY, USA, 2000; pp. 123–126.
38. Ma, J.F.; Yamaji, N.; Mitani, N.; Xu, X.Y.; Su, Y.H.; McGrath, S.P.; Zhao, F.J. Transporters of arsenite in rice and their role in arsenic accumulation in rice grain. *Proc. Natl. Acad. Sci. USA* **2008**, *105*, 9931–9935. [CrossRef] [PubMed]
39. Zangi, R.; Filella, M. Transport routes of metalloids into and out of the cell: A review of the current knowledge. *Chemico Biol. Interact.* **2012**, *197*, 47–57. [CrossRef] [PubMed]
40. Zhao, X.Q.; Mitani, N.; Yamaji, N.; Shen, R.F.; Ma, J.F. Involvement of silicon influx transporter OsNIP2;1 in selenite uptake in rice. *Plant Physiol.* **2010**, *153*, 1871–1877. [CrossRef] [PubMed]
41. Lopez, D.; Bronner, G.L.; Brunel, N.; Auguin, D.; Bourgerie, S.; Brignolas, F.; Carpin, S.; Tournaire-Roux, C.; Maurel, C.; Fumanal, B. Insights into populus XIP aquaporins: Evolutionary expansion, protein functionality, and environmental regulation. *J. Exp. Bot.* **2012**, *63*, 2217–2230. [CrossRef] [PubMed]
42. Steudle, E. The cohesion-tension mechanism and the acquisition of water by plant roots. *Annu. Rev. Plant Biol.* **2001**, *52*, 847–875. [CrossRef] [PubMed]
43. Freeman, S. *Biological Science*, 3rd ed.; Benjamin Cummins: Upper Saddle River, NJ, USA, 2007.
44. Tyerman, S.; Niemietz, C.; Bramley, H. Plant aquaporins: Multifunctional water and solute channels with expanding roles. *Plant Cell Environ.* **2002**, *25*, 173–194. [CrossRef] [PubMed]
45. Steudle, E.; Peterson, C.A. How does water get through roots? *J. Exp. Bot.* **1998**, *49*, 775–788. [CrossRef]
46. Kaldenhoff, R.; Ribas-Carbo, M.; Sans, J.F.; Lovisolo, C.; Heckwolf, M.; Uehlein, N. Aquaporins and plant water balance. *Plant Cell Environ.* **2008**, *31*, 658–666. [CrossRef] [PubMed]

47. Lopez, D.; Venisse, J.S.P.; Fumanal, B.; Chaumont, F.O.; Guillot, E.; Daniels, M.J.; Cochard, H.; Julien, J.L.; Gousset-Dupont, A.L. Aquaporins and leaf hydraulics, poplar sheds new light. *Plant Cell Physiol.* **2013**, *54*, 1963–1975. [CrossRef] [PubMed]
48. Wallace, I.S.; Choi, W.G.; Roberts, D.M. The structure, function and regulation of the nodulin 26-like intrinsic protein family of plant aquaglyceroporins. *Biochim. Biophys. Acta Biomembr.* **2006**, *1758*, 1165–1175. [CrossRef] [PubMed]
49. Siefritz, F.; Tyree, M.T.; Lovisolo, C.; Schubert, A.; Kaldenhoff, R. PIP1 plasma membrane aquaporins in tobacco from cellular effects to function in plants. *Plant Cell* **2002**, *14*, 869–876. [CrossRef] [PubMed]
50. Sade, N.; Gebretsadik, M.; Seligmann, R.; Schwartz, A.; Wallach, R.; Moshelion, M. The role of tobacco aquaporin1 in improving water use efficiency, hydraulic conductivity, and yield production under salt stress. *Plant Physiol.* **2010**, *152*, 245–254. [CrossRef] [PubMed]
51. Hukin, D.; Doering-Saad, C.; Thomas, C.; Pritchard, J. Sensitivity of cell hydraulic conductivity to mercury is coincident with symplasmic isolation and expression of plasmalemma aquaporin genes in growing maize roots. *Planta* **2002**, *215*, 1047–1056. [PubMed]
52. Park, W.J.; Campbell, B.T. Aquaporins as targets for stress tolerance in plants: Genomic complexity and perspectives. *Turk. J. Bot.* **2015**, *39*, 879–886. [CrossRef]
53. Boursiac, Y.; Boudet, J.; Postaire, O.; Doan, T.L.; Colette, T.R.; Maurel, C. Stimulus induced downregulation of root water transport involves reactive oxygen species activated cell signalling and plasma membrane intrinsic protein internalization. *Plant J.* **2008**, *56*, 207–218. [CrossRef] [PubMed]
54. Yamaguchi-Shinozaki, K.; Koizumi, M.; Urao, S.; Shinozaki, K. Molecular cloning and characterization of 9 cDNAs for genes that are responsive to desiccation in *Arabidopsis thaliana*: Sequenceanalysis of one cDNA clone that encodes a putative transmembrane channel protein. *Plant Cell Physiol.* **1992**, *33*, 217–224.
55. Alexandersson, E.; Fraysse, L.; Sjovall-Larsen, S.; Gustavsson, S.; Fellert, M.; Karlsson, M.; Johanson, U.; Kjellbom, P. Whole gene family expression and drought stress regulation of aquaporins. *Plant Mol. Biol.* **2005**, *59*, 469–484. [CrossRef] [PubMed]
56. Rizhsky, L.; Liang, H.; Shuman, J.; Shulaev, V.; Davletova, S.; Mittler, R. When defense pathways collide. The response of arabidopsis to a combination of drought and heat stress. *Plant Physiol.* **2004**, *134*, 1683–1696. [CrossRef] [PubMed]
57. Alexandersson, E.; Danielson, J.A.; Rade, J.; Moparthi, V.K.; Fontes, M.; Kjellbom, P.; Johanson, U. Transcriptional regulation of aquaporins in accessions of arabidopsis in response to drought stress. *Plant J.* **2010**, *61*, 650–660. [CrossRef] [PubMed]
58. Jang, J.Y.; Kim, D.G.; Kim, Y.O.; Kim, J.S.; Kang, H. An expression analysis of a gene family encoding plasma membrane aquaporins in response to abiotic stresses in arabidopsis thaliana. *Plant Mol. Biol.* **2004**, *54*, 713–725. [CrossRef] [PubMed]
59. Secchi, F.; Lovisolo, C.; Schubert, A. Expression of OePIP2;1 aquaporin gene and water relations of OLEA europaea twigs during drought stress and recovery. *Ann. Appl. Biol.* **2007**, *150*, 163–167. [CrossRef]
60. Mahdieh, M.; Mostajeran, A.; Horie, T.; Katsuhara, M. Drought stress alters water relations and expression of PIP-type aquaporin genes in Nicotiana tabacum plants. *Plant Cell Physiol.* **2008**, *49*, 801–813. [CrossRef] [PubMed]
61. Sugaya, S.; Ohshima, I.; Gemma, H.; Iwahori, S. Expression analysis of genes encoding aquaporins during the development of peach fruit. In Proceedings of the XXVI International Horticultural Congress: Environmental Stress and Horticulture Crops 618, Toronto, AB, Canada, 11–17 August 2002; pp. 363–370.
62. Vandeleur, R.K.; Mayo, G.; Shelden, M.C.; Gilliham, M.; Kaiser, B.N.; Tyerman, S.D. The role of plasma membrane intrinsic protein aquaporins in water transport through roots: Diurnal and drought stress responses reveal different strategies between isohydric and anisohydric cultivars of grapevine. *Plant Physiol.* **2009**, *149*, 445–460. [CrossRef] [PubMed]
63. Cramer, G.R.; Ergul, A.; Grimplet, J.; Tillett, R.L.; Tattersall, E.A.; Bohlman, M.C.; Vincent, D.; Sonderegger, J.; Evans, J.; Osborne, C. Water and salinity stress in grapevines: Early and late changes in transcript and metabolite profiles. *Funct. Integr. Genomics* **2007**, *7*, 111–134. [CrossRef] [PubMed]
64. Guo, L.; Wang, Z.Y.; Lin, H.; Cui, W.E.; Chen, J.; Liu, M.; Chen, Z.L.; Qu, L.J.; Gu, H. Expression and functional analysis of the rice plasma-membrane intrinsic protein gene family. *Cell Res.* **2006**, *16*, 277–286. [CrossRef] [PubMed]

65. Li, L.G.; Li, S.F.; Tao, Y.; Kitagawa, Y. Molecular cloning of a novel water channel from rice: Its products expression in xenopus oocytes and involvement in chilling tolerance. *Plant Sci.* **2000**, *154*, 43–51. [CrossRef]
66. Malz, S.; Sauter, M. Expression of two pip genes in rapidly growing internodes of rice is not primarily controlled by meristem activity or cell expansion. *Plant Mol. Biol.* **1999**, *40*, 985–995. [CrossRef] [PubMed]
67. Suga, S.; Komatsu, S.; Maeshima, M. Aquaporin isoforms responsive to salt and water stresses and phytohormones in radish seedlings. *Plant Cell Physiol.* **2002**, *43*, 1229–1237. [CrossRef] [PubMed]
68. Li, J.; Ban, L.; Wen, H.; Wang, Z.; Dzyubenko, N.; Chapurin, V.; Gao, H.; Wang, X. An aquaporin protein is associated with drought stress tolerance. *Biochem. Biophys. Res. Commun.* **2015**, *459*, 208–213. [CrossRef] [PubMed]
69. Galmes, J.; Pou, A.; Alsina, M.M.; Tomas, M.; Medrano, H.l.; Flexas, J. Aquaporin expression in response to different water stress intensities and recovery in Richter-110 (*Vitis sp.*): Relationship with ecophysiological status. *Planta* **2007**, *226*, 671–681. [CrossRef] [PubMed]
70. Surbanovski, N.; Sargent, D.J.; Else, M.A.; Simpson, D.W.; Zhang, H.; Grant, O.M. Expression of fragaria vesca PIP aquaporins in response to drought stress: PIP down-regulation correlates with the decline in substrate moisture content. *PLoS ONE* **2013**, *8*, e74945.
71. Aroca, R.; Porcel, R.; Ruiz-Lozano, J.M. How does arbuscular mycorrhizal symbiosis regulate root hydraulic properties and plasma membrane aquaporins in phaseolus vulgaris under drought, cold or salinity stresses? *New Phytol.* **2007**, *173*, 808–816. [CrossRef] [PubMed]
72. Barzana, G.; Aroca, R.; Bienert, G.P.; Chaumont, F.O.; Ruiz-Lozano, J.M. New insights into the regulation of aquaporins by the arbuscular mycorrhizal symbiosis in maize plants under drought stress and possible implications for plant performance. *Mol. Plant Microb. Interact.* **2014**, *27*, 349–363. [CrossRef] [PubMed]
73. Liu, Q.; Umeda, M.; Uchimiya, H. Isolation and expression analysis of two rice genes encoding the major intrinsic protein. *Plant Mol. Biol.* **1994**, *26*, 2003–2007. [CrossRef] [PubMed]
74. Molina, C.; Rotter, B.R.; Horres, R.; Udupa, S.M.; Besser, B.; Bellarmino, L.; Baum, M.; Matsumura, H.; Terauchi, R.; Kahl, G.N. Supersage: The drought stress-responsive transcriptome of chickpea roots. *BMC Genomics* **2008**, *9*. [CrossRef] [PubMed]
75. Martre, P.; Morillon, R.L.; Barrieu, F.O.; North, G.B.; Nobel, P.S.; Chrispeels, M.J. Plasma membrane aquaporins play a significant role during recovery from water deficit. *Plant Physiol.* **2002**, *130*, 2101–2110. [CrossRef] [PubMed]
76. Kaldenhoff, R.; Grote, K.; Zhu, J.J.; Zimmermann, U. Significance of plasmalemma aquaporins for water transport in arabidopsis thaliana. *Plant J.* **1998**, *14*, 121–128. [CrossRef] [PubMed]
77. Javot, H.L.N.; Lauvergeat, V.; Santoni, V.R.; Martin-Laurent, F.; Guclu, J.; Vinh, J.L.; Heyes, J.; Franck, K.I.; Schaffner, A.R.; Bouchez, D. Role of a single aquaporin isoform in root water uptake. *Plant Cell* **2003**, *15*, 509–522. [CrossRef] [PubMed]
78. Lienard, D.; Durambur, G.; Kiefer-Meyer, M.C.; Nogue, F.; Menu-Bouaouiche, L.; Charlot, F.; Gomord, V.R.; Lassalles, J.P. Water transport by aquaporins in the extant plant physcomitrella patens. *Plant Physiol.* **2008**, *146*, 1207–1218. [CrossRef] [PubMed]
79. Yu, Q.; Hu, Y.; Li, J.; Wu, Q.; Lin, Z. Sense and antisense expression of plasma membrane aquaporin bnpip1 from brassica napus in tobacco and its effects on plant drought resistance. *Plant Sci.* **2005**, *169*, 647–656. [CrossRef]
80. Zhou, S.; Hu, W.; Deng, X.; Ma, Z.; Chen, L.; Huang, C.; Wang, C.; Wang, J.; He, Y.; Yang, G. Overexpression of the wheat aquaporin gene, TaAQP7, enhances drought tolerance in transgenic tobacco. *PLoS ONE* **2012**, *7*, e52439. [CrossRef] [PubMed]
81. Cui, X.H.; Hao, F.S.; Chen, H.; Chen, J.; Wang, X.C. Expression of the vicia faba VFPIP1 gene in arabidopsis thaliana plants improves their drought resistance. *J. Plant Res.* **2008**, *121*, 207–214. [CrossRef] [PubMed]
82. Xu, Y.; Hu, W.; Liu, J.; Zhang, J.; Jia, C.; Miao, H.; Xu, B.; Jin, Z. A banana aquaporin gene, MaPIP1;1, is involved in tolerance to drought and salt stresses. *BMC Plant Biol.* **2014**, *14*. [CrossRef] [PubMed]
83. Sade, N.; Vinocur, B.J.; Diber, A.; Shatil, A.; Ronen, G.; Nissan, H.; Wallach, R.; Karchi, H.; Moshelion, M. Improving plant stress tolerance and yield production: Is the tonoplast aquaporin SlTIP2;2 a key to isohydric to anisohydric conversion? *New Phytol.* **2009**, *181*, 651–661. [CrossRef] [PubMed]
84. Aharon, R.; Shahak, Y.; Wininger, S.; Bendov, R.; Kapulnik, Y.; Galili, G. Overexpression of a plasma membrane aquaporin in transgenic tobacco improves plant vigor under favorable growth conditions but not under drought or salt stress. *Plant Cell* **2003**, *15*, 439–447. [CrossRef] [PubMed]

85. Vander Willigen, C.; Pammenter, N.; Mundree, S.G.; Farrant, J.M. Mechanical stabilization of desiccated vegetative tissues of the resurrection grass eragrostis nindensis: Does a TIP3;1 and/or compartmentalization of subcellular components and metabolites play a role? *J. Exp. Bot.* **2004**, *55*, 651–661. [CrossRef] [PubMed]
86. Almeida-Rodriguez, A.M.; Cooke, J.E.; Yeh, F.; Zwiazek, J.J. Functional characterization of drought responsive aquaporins in populus balsamifera and populus simonii X balsamifera clones with different drought resistance strategies. *Physiol. Plant.* **2010**, *140*, 321–333. [CrossRef] [PubMed]
87. Lian, H.L.; Yu, X.; Ye, Q.; Ding, X.S.; Kitagawa, Y.; Kwak, S.S.; Su, W.A.; Tang, Z.C. The role of aquaporin RWC3 in drought avoidance in rice. *Plant Cell Physiol.* **2004**, *45*, 481–489. [CrossRef] [PubMed]
88. Smith-Espinoza, C.; Richter, A.; Salamini, F.; Bartels, D. Dissecting the response to dehydration and salt (NACL) in the resurrection plant craterostigma plantagineum. *Plant Cell Environ.* **2003**, *26*, 1307–1315. [CrossRef]
89. Zhuang, L.; Liu, M.; Yuan, X.; Yang, Z.; Huang, B. Physiological effects of aquaporin in regulating drought tolerance through overexpressing of festuca arundinacea aquaporin gene *FaPIP2;1*. *J. Am. Soc. Hortic. Sci.* **2015**, *140*, 404–412.
90. Flexas, J.; Miquel, R.C.; Hanson, D.T.; Bota, J.; Otto, B.; Cifre, J.; McDowell, N.; Medrano, H.; Kaldenhoff, R. Tobacco aquaporin NTAQP1 is involved in mesophyll conductance to CO_2 *in vivo*. *Plant J.* **2006**, *48*, 427–439. [CrossRef] [PubMed]
91. Boudichevskaia, A.; Heckwolf, M.; Kaldenhoff, R. T-DNA insertion in aquaporin gene *AtPIP1;2* generates transcription profiles reminiscent of a low CO_2 response. *Plant Cell Environ.* **2015**, *38*, 2286–2298. [CrossRef] [PubMed]
92. Uehlein, N.; Sperling, H.; Heckwolf, M.; Kaldenhoff, R. The arabidopsis aquaporin PIP1;2 rules cellular CO_2 uptake. *Plant Cell Environ.* **2012**, *35*, 1077–1083. [CrossRef] [PubMed]
93. Flexas, J.; Barbour, M.M.; Brendel, O.; Cabrera, H.N.M.; CarriquÃ, M.; DÃaz-Espejo, A.; Douthe, C.; Dreyer, E.; Ferrio, J.P.; Gago, J. Mesophyll diffusion conductance to CO_2: An unappreciated central player in photosynthesis. *Plant Sci.* **2012**, *193*, 70–84. [CrossRef] [PubMed]
94. Tsuchihira, A.; Hanba, Y.T.; Kato, N.; Doi, T.; Kawazu, T.; Maeshima, M. Effect of overexpression of radish plasma membrane aquaporins on water-use efficiency, photosynthesis and growth of eucalyptus trees. *Tree Physiol.* **2010**, *30*, 417–430. [CrossRef] [PubMed]
95. Hanba, Y.T.; Shibasaka, M.; Hayashi, Y.; Hayakawa, T.; Kasamo, K.; Terashima, I.; Katsuhara, M. Overexpression of the barley aquaporin HvPIP2;1 increases internal CO_2 conductance and CO_2 assimilation in the leaves of transgenic rice plants. *Plant Cell Physiol.* **2004**, *45*, 521–529. [CrossRef] [PubMed]
96. Boursiac, Y.; Chen, S.; Luu, D.T.; Sorieul, M.; van den Dries, N.; Maurel, C. Early effects of salinity on water transport in arabidopsis roots. Molecular and cellular features of aquaporin expression. *Plant Physiol.* **2005**, *139*, 790–805. [CrossRef] [PubMed]
97. Katsuhara, M.; Akiyama, Y.; Koshio, K.; Shibasaka, M.; Kasamo, K. Functional analysis of water channels in barley roots. *Plant Cell Physiol.* **2002**, *43*, 885–893. [CrossRef] [PubMed]
98. Katsuhara, M.; Koshio, K.; Shibasaka, M.; Hayashi, Y.; Hayakawa, T.; Kasamo, K. Over-expression of a barley aquaporin increased the shoot/root ratio and raised salt sensitivity in transgenic rice plants. *Plant Cell Physiol.* **2003**, *44*, 1378–1383. [CrossRef] [PubMed]
99. Yamada, S.; Katsuhara, M.; Kelly, W.B.; Michalowski, C.B.; Bohnert, H.J. A family of transcripts encoding water channel proteins: Tissue-specific expression in the common ice plant. *Plant Cell* **1995**, *7*, 1129–1142. [CrossRef] [PubMed]
100. Kirch, H.H.; Vera-Estrella, R.; Golldack, D.; Quigley, F.; Michalowski, C.B.; Barkla, B.J.; Bohnert, H.J. Expression of water channel proteins in mesembryanthemum crystallinum. *Plant Physiol.* **2000**, *123*, 111–124. [CrossRef] [PubMed]
101. Zhu, C.; Schraut, D.; Hartung, W.; SchÃffner, A.R. Differential responses of maize MIP genes to salt stress and ABA. *J. Exp. Bot.* **2005**, *56*, 2971–2981. [CrossRef] [PubMed]
102. Pang, Y.; Li, L.; Ren, F.; Lu, P.; Wei, P.; Cai, J.; Xin, L.; Zhang, J.; Chen, J.; Wang, X. Overexpression of the tonoplast aquaporin AtTIP5;1 conferred tolerance to boron toxicity in arabidopsis. *J. Genet. Genomics* **2010**, *37*, 389–397. [CrossRef]
103. Peng, Y.; Lin, W.; Cai, W.; Arora, R. Overexpression of a panax ginseng tonoplast aquaporin alters salt tolerance, drought tolerance and cold acclimation ability in transgenic arabidopsis plants. *Planta* **2007**, *226*, 729–740. [CrossRef] [PubMed]

104. Dhonukshe, P.; Aniento, F.; Hwang, I.; Robinson, D.G.; Mravec, J.; Stierhof, Y.D.; Friml, J.Ã. Clathrin-mediated constitutive endocytosis of PIN auxin efflux carriers in arabidopsis. *Curr. Biol.* **2007**, *17*, 520–527. [CrossRef] [PubMed]
105. Li, X.; Wang, X.; Yang, Y.; Li, R.; He, Q.; Fang, X.; Luu, D.T.; Maurel, C.; Lin, J. Single-molecule analysis of PIP2;1 dynamics and partitioning reveals multiple modes of arabidopsis plasma membrane aquaporin regulation. *Plant Cell* **2011**, *23*, 3780–3797. [CrossRef] [PubMed]
106. Luu, D.T.; Martiniere, A.; Sorieul, M.; Runions, J.; Maurel, C. Fluorescence recovery after photobleaching reveals high cycling dynamics of plasma membrane aquaporins in arabidopsis roots under salt stress. *Plant J.* **2012**, *69*, 894–905. [CrossRef] [PubMed]
107. Gao, Z.; He, X.; Zhao, B.; Zhou, C.; Liang, Y.; Ge, R.; Shen, Y.; Huang, Z. Overexpressing a putative aquaporin gene from wheat, TaNIP, enhances salt tolerance in transgenic arabidopsis. *Plant Cell Physiol.* **2010**, *51*, 767–775. [CrossRef] [PubMed]
108. Huang, C.; Zhou, S.; Hu, W.; Deng, X.; Wei, S.; Yang, G.; He, G. The wheat aquaporin gene TAAQP7 confers tolerance to cold stress in transgenic tobacco. *Z. Naturforschung. C* **2014**, *69*, 142–148. [CrossRef]
109. Ahamed, A.; Murai-Hatano, M.; Ishikawa-Sakurai, J.; Hayashi, H.; Kawamura, Y.; Uemura, M. Cold stress-induced acclimation in rice is mediated by root-specific aquaporins. *Plant Cell Physiol.* **2012**, *53*, 1445–1456. [CrossRef] [PubMed]
110. Aroca, R.; Amodeo, G.; Fernandez-Illescas, S.; Herman, E.M.; Chaumont, F.; Chrispeels, M.J. The role of aquaporins and membrane damage in chilling and hydrogen peroxide induced changes in the hydraulic conductance of maize roots. *Plant Physiol.* **2005**, *137*, 341–353. [CrossRef] [PubMed]
111. Nogueira, F.B.T.; de Rosa, V.E.; Menossi, M.; Ulian, E.N.C.; Arruda, P. RNA expression profiles and data mining of sugarcane response to low temperature. *Plant Physiol.* **2003**, *132*, 1811–1824. [CrossRef] [PubMed]
112. Takano, J.; Wada, M.; Ludewig, U.; Schaaf, G.; von Wiren, N.; Fujiwara, T. The arabidopsis major intrinsic protein NIP5;1 is essential for efficient boron uptake and plant development under boron limitation. *Plant Cell* **2006**, *18*, 1498–1509. [CrossRef] [PubMed]
113. Li, T.; Choi, W.G.; Wallace, I.S.; Baudry, J.; Roberts, D.M. Arabidopsis thaliana NIP7;1: An anther-specific boric acid transporter of the aquaporin superfamily regulated by an unusual tyrosine in HELIX 2 of the transport pore. *Biochemistry* **2011**, *50*, 6633–6641. [CrossRef] [PubMed]
114. Tanaka, M.; Wallace, I.S.; Takano, J.; Roberts, D.M.; Fujiwara, T. NIP6;1 is a boric acid channel for preferential transport of boron to growing shoot tissues in arabidopsis. *Plant Cell* **2008**, *20*, 2860–2875. [CrossRef] [PubMed]
115. Schnurbusch, T.; Hayes, J.; Hrmova, M.; Baumann, U.; Ramesh, S.A.; Tyerman, S.D.; Langridge, P.; Sutton, T. Boron toxicity tolerance in barley through reduced expression of the multifunctional aquaporin HvNIP2;1. *Plant Physiol.* **2010**, *153*, 1706–1715. [CrossRef] [PubMed]
116. Yamaji, N.; Mitatni, N.; Ma, J.F. A transporter regulating silicon distribution in rice shoots. *Plant Cell* **2008**, *20*, 1381–1389. [CrossRef] [PubMed]
117. Ma, J.F.; Tamai, K.; Yamaji, N.; Mitani, N.; Konishi, S.; Katsuhara, M.; Ishiguro, M.; Murata, Y.; Yano, M. A silicon transporter in rice. *Nature* **2006**, *440*, 688–691. [CrossRef] [PubMed]
118. Kamiya, T.; Tanaka, M.; Mitani, N.; Ma, J.F.; Maeshima, M.; Fujiwara, T. NIP1;1, an aquaporin homolog, determines the arsenite sensitivity of arabidopsis thaliana. *J. Biol. Chem.* **2009**, *284*, 2114–2120. [CrossRef] [PubMed]
119. Bienert, G.P.; Thorsen, M.; Schussler, M.D.; Nilsson, H.R.; Wagner, A.; Tamas, M.J.; Jahn, T.P. A subgroup of plant aquaporins facilitate the bi-directional diffusion of $As(OH)_3$ and $Sb(OH)_3$ across membranes. *BMC Biol.* **2008**, *6*. [CrossRef] [PubMed]
120. Kamiya, T.; Fujiwara, T. Arabidopsis NIP1;1 transports antimonite and determines antimonite sensitivity. *Plant Cell Physiol.* **2009**, *50*, 1977–1981. [CrossRef] [PubMed]
121. Choi, W.G.; Roberts, D.M. Arabidopsis NIP2;1, a major intrinsic protein transporter of lactic acid induced by anoxic stress. *J. Biol. Chem.* **2007**, *282*, 24209–24218. [CrossRef] [PubMed]
122. Dynowski, M.; Schaaf, G.; Loque, D.; Moran, O.; Ludewig, U. Plant plasma membrane water channels conduct the signalling molecule H_2O_2. *Biochem. J.* **2008**, *414*, 53–61. [CrossRef] [PubMed]
123. Hooijmaijers, C.; Rhee, J.Y.; Kwak, K.J.; Chung, G.C.; Horie, T.; Katsuhara, M.; Kang, H. Hydrogen peroxide permeability of plasma membrane aquaporins of arabidopsis thaliana. *J. Plant Res.* **2012**, *125*, 147–153. [CrossRef] [PubMed]

124. Zou, J.; Rodriguez-Zas, S.; Aldea, M.; Li, M.; Zhu, J.; Gonzalez, D.O.; Vodkin, L.O.; DeLucia, E.; Clough, S.J. Expression profiling soybean response to pseudomonas syringae reveals new defense-related genes and rapid HR-specific downregulation of photosynthesis. *Mol. Plant Microb. Interact.* **2005**, *18*, 1161–1174. [CrossRef] [PubMed]

125. Aritua, V.; Achor, D.; Gmitter, F.G.; Albrigo, G.; Wang, N. Transcriptional and microscopic analyses of citrus stem and root responses to *Candidatus* Liberibacter asiaticus infection. *PLoS ONE* **2013**, *8*, e73742.

126. Lawrence, S.; Novak, N.; Xu, H.; Cooke, J. Herbivory of maize by southern corn rootworm induces expression of the major intrinsic protein ZmNIP1;1 and leads to the discovery of a novel aquaporin ZmPIP2;8. *Plant Signal. Behav.* **2013**, *8*, e24937. [CrossRef] [PubMed]

127. Opperman, C.H.; Taylor, C.G.; Conkling, M.A. Root-knot nematode-directed expression of a plant root-specific gene. *Science* **1994**, *263*, 221–223. [CrossRef] [PubMed]

128. Hartley, S.E.; Fitt, R.N.; McLarnon, E.L.; Wade, R.N. Defending the leaf surface: Intra-and inter-specific differences in silicon deposition in grasses in response to damage and silicon supply. *Front. Plant Sci.* **2015**, *6*, 35. [CrossRef] [PubMed]

129. Mukhtar, M.S.; Carvunis, A.R.; Dreze, M.; Epple, P.; Steinbrenner, J.; Moore, J.; Tasan, M.; Galli, M.; Hao, T.; Nishimura, M.T. Independently evolved virulence effectors converge onto hubs in a plant immune system network. *Science* **2011**, *333*, 596–601. [CrossRef] [PubMed]

130. Kim, M.J.; Kim, H.R.; Paek, K.H. Arabidopsis tonoplast proteins Tip1 and Tip2 interact with the cucumber mosaic virus 1A replication protein. *J. Gen. Virol.* **2006**, *87*, 3425–3431. [CrossRef] [PubMed]

131. Ayadi, M.; Cavez, D.; Miled, N.; Chaumont, F.O.; Masmoudi, K. Identification and characterization of two plasma membrane aquaporins in durum wheat (*Triticum turgidum* L. Subsp. Durum) and their role in abiotic stress tolerance. *Plant Physiol. Biochem.* **2011**, *49*, 1029–1039. [CrossRef] [PubMed]

132. Jang, J.Y.; Lee, S.H.; Rhee, J.Y.; Chung, G.C.; Ahn, S.J.; Kang, H. Transgenic arabidopsis and tobacco plants overexpressing an aquaporin respond differently to various abiotic stresses. *Plant Mol. Biol.* **2007**, *64*, 621–632. [CrossRef] [PubMed]

133. Zelazny, E.; Borst, J.W.; Muylaert, M.L.; Batoko, H.; Hemminga, M.A.; Chaumont, F.O. FRET imaging in living maize cells reveals that plasma membrane aquaporins interact to regulate their subcellular localization. *Proc. Natl. Acad. Sci. USA* **2007**, *104*, 12359–12364. [CrossRef] [PubMed]

134. Chen, W.; Yin, X.; Wang, L.; Tian, J.; Yang, R.; Liu, D.; Yu, Z.; Ma, N.; Gao, J. Involvement of rose aquaporin RhPIP1;1 in ethylene-regulated petal expansion through interaction with RhPIP2;1. *Plant Mol. Biol.* **2013**, *83*, 219–233. [CrossRef] [PubMed]

135. Li, D.D.; Ruan, X.M.; Zhang, J.; Wu, Y.J.; Wang, X.L.; Li, X.B. Cotton plasma membrane intrinsic protein 2s (PIP2s) selectively interact to regulate their water channel activities and are required for fibre development. *New Phytol.* **2013**, *199*, 695–707. [CrossRef] [PubMed]

136. Besserer, A.; Burnotte, E.; Bienert, G.P.; Chevalier, A.S.; Errachid, A.; Grefen, C.; Blatt, M.R.; Chaumont, F.O. Selective regulation of maize plasma membrane aquaporin trafficking and activity by the snare SYP121. *Plant Cell* **2012**, *24*, 3463–3481. [CrossRef] [PubMed]

137. Hachez, C.; Laloux, T.E.; Reinhardt, H.; Cavez, D.; Degand, H.; Grefen, C.; de Rycke, R.; InzÃ, D.; Blatt, M.R.; Russinova, E. Arabidopsis snares syp61 and SYP121 coordinate the trafficking of plasma membrane aquaporin PIP2;7 to modulate the cell membrane water permeability. *Plant Cell* **2014**, *26*, 3132–3147. [CrossRef] [PubMed]

138. Hachez, C.; Veljanovski, V.; Reinhardt, H.; Guillaumot, D.; Vanhee, C.; Chaumont, F.O.; Batoko, H. The arabidopsis abiotic stress-induced tspo-related protein reduces cell-surface expression of the aquaporin PIP2;7 through protein-protein interactions and autophagic degradation. *Plant Cell* **2014**, *26*, 4974–4990. [CrossRef] [PubMed]

139. Lee, H.K.; Cho, S.K.; Son, O.; Xu, Z.; Hwang, I.; Kim, W.T. Drought stress-induced RMA1H1, a ring membrane-anchor E3 ubiquitin ligase homolog, regulates aquaporin levels via ubiquitination in transgenic arabidopsis plants. *Plant Cell* **2009**, *21*, 622–641. [CrossRef] [PubMed]

140. Masalkar, P.; Wallace, I.S.; Hwang, J.H.; Roberts, D.M. Interaction of cytosolic glutamine synthetase of soybean root nodules with the *C*-terminal domain of the symbiosome membrane nodulin 26 aquaglyceroporin. *J. Biol. Chem.* **2010**, *285*, 23880–23888. [CrossRef] [PubMed]

141. Wu, X.N.; Rodriguez, C.S.; Pertl-Obermeyer, H.; Obermeyer, G.; Schulze, W.X. Sucrose-induced receptor kinase SIRK1 regulates a plasma membrane aquaporin in arabidopsis. *Mol. Cell. Proteomics* **2013**, *12*, 2856–2873. [CrossRef] [PubMed]
142. Chevalier, A.S.; Chaumont, F.O. Trafficking of plant plasma membrane aquaporins: Multiple regulation levels and complex sorting signals. *Plant Cell Physiol.* **2015**, *56*, 819–829. [CrossRef] [PubMed]
143. Vera-Estrella, R.; Barkla, B.J.; Bohnert, H.J.; Pantoja, O. Novel regulation of aquaporins during osmotic stress. *Plant Physiol.* **2004**, *135*, 2318–2329. [CrossRef] [PubMed]

13

Cell Fate Decisions During Breast Cancer Development

Kayla Gross [1,2,3], Ania Wronski [1,2,3], Adam Skibinski [1,3], Sarah Phillips [1,3] and Charlotte Kuperwasser [1,2,3,*]

1 Department of Developmental, Molecular and Chemical Biology, Sackler School of Graduate Biomedical Sciences, Tufts University School of Medicine, 136 Harrison Ave., Boston, MA 02111, USA

2 Raymond and Beverly Sackler Convergence Laboratory, Tufts University School of Medicine, 145 Harrison Ave., Boston, MA 02111, USA

3 Molecular Oncology Research Institute, Tufts Medical Center, 800 Washington St., Boston, MA 02111, USA

* Correspondence: Charlotte.Kuperwasser@tufts.edu

Academic Editors: Robert W. Dettman and Andy Wessels

Abstract: During the formation of breast cancer, many genes become altered as cells evolve progressively from normal to a pre-malignant to a malignant state of growth. How mutations in genes lead to specific subtypes of human breast cancer is only partially understood. Here we review how initial genetic or epigenetic alterations within mammary epithelial cells (MECs) can alter cell fate decisions and put pre-malignant cells on a path towards cancer development with specific phenotypes. Understanding the early stages of breast cancer initiation and progression and how normal developmental processes are hijacked during transformation has significant implications for improving early detection and prevention of breast cancer. In addition, insights gleaned from this understanding may also be important for developing subtype-specific treatment options.

Keywords: breast cancer; cell fate; mammary epithelial cells; plasticity; cancer heterogeneity; mammary gland

1. Introduction

Over the past three decades, enormous strides have been made in understanding the genetic and biochemical defects in cancer cells that are responsible for deregulated growth and proliferation. Research has revealed critical mutations in many key cellular genes, specifically oncogenes and tumor suppressor genes, within the genomes of a wide variety of human breast cancer cells. During the formation of the majority of breast cancers, these genes become altered as cells evolve progressively from normal to a pre-malignant to a malignant state of growth.

Although these findings have been most illustrative, they fail to inform us about the details of how specific subtypes of human breast cancer begin. A significant number of genes have been catalogued in the genomes of highly advanced breast carcinomas that are detected in the clinic [1]. While this research has revealed the final steps leading up to the appearance of aggressive breast cancers, we have only recently begun to uncover the nature of the early steps of breast cancer development. Here we review how initial genetic or epigenetic alterations in mammary epithelial cells (MECs) can alter cell fate decisions during the earliest steps of neoplastic transformation. We focus on work that has identified the cell-of-origin for various breast cancer subtypes and how cell fate changes in different precursor cells leads to tumors with different phenotypes and behaviors. This new understanding has significant implications for early detection and prevention of breast cancer, as well as possible development of improved therapeutics unique for each tumor subtype. However, in order to prevent breast cancer formation, much remains to be uncovered regarding how the cell-of-origin responds to mutation, stress, DNA damage and aging.

2. Breast Cancer and Its Complex Heterogeneity

Breast cancer is a multifaceted disease, consisting of tumors with varying levels of heterogeneity, which includes differences in clinical characteristics, molecular profiles, metastatic behavior, and therapeutic responsiveness. Breast tumors are classified based on the tumor's morphology and structural organization, with the most common histological type of invasive breast tumor being invasive ductal carcinoma [2,3]. In addition to morphological profiling, immunohistochemical analyses for specific markers—including estrogen receptor (ERα), progesterone receptor (PR), and human epidermal growth factor receptor 2 (HER2/NEU/ERBB2)—have allowed breast tumors to be classified into different subtypes. These categories include: ERα^+ (ERα^+/HER2$^-$); HER2$^+$ (ER$^-$/HER2$^+$); triple negative (TN; ERα^-/PR$^-$/HER2$^-$), and triple positive (ERα^+/PR$^+$/HER2$^+$) [2]. These subtypes also have been correlated with trends in overall prognosis and helped identify tumors that may respond favorably to targeted therapies [2]. These general immunohistochemical subtypes also contain heterogeneity within their respective groups. For example, TN breast cancers demonstrate diverse histological patterning and few consistently recurring mutational trends aside from TP53, PTEN, and PICK3CA, the latter concern being a large obstacle for developing targeted therapies [4]. Recently, TN breast cancer has been subcategorized into six subtypes; the majority of those subtypes correlate with a basal-like profile, though some overlap with other molecular subtypes [4]. Though currently less used in clinical settings, molecular profiling, as determined by high-throughput gene expression and hierarchical clustering, is another way breast tumor heterogeneity is defined. This method classifies tumors into several "intrinsic" breast cancer subtypes: luminal-like (also further subdivided into luminal A and luminal B), basal-like, HER2-enriched, and claudin-low [5–7].

Two models have emerged that attempt to explain the diverse heterogeneity observed in breast cancer: (1) the "mutation-of-origin" model or (2) the "cell-of-origin" model. The first model posits that all tumor subtypes are derived from a common precursor cell—a bipotent mammary stem cell—but that specific genetic and epigenetic alterations acquired during the process of neoplastic transformation affects that cell's ability to commit to either the luminal or basal lineage (Figure 1a). Thus, mutations that lock the stem cell's progeny towards differentiating down the luminal lineage would result in luminal-type tumors, while those that trap stem cells into the basal fate would give rise to basal-like tumors. Based on this assumption, one would expect to observe common and recurring mutations in specific breast cancer subtypes. Indeed, large-scale sequencing efforts have shown that luminal- and basal-like breast cancers exhibit specific and generally non-overlapping mutations. For example, amplification of cyclin D1 and MDM2 or mutations in TBX3 and RUNX1 are preferentially observed in luminal-like breast cancers, while mutations in p53, pRb, BRCA1 and PTEN are commonly found among basal-like tumors [1].

In contrast, the "cell-of-origin" model hypothesizes that a cell's differentiation program is so tightly encoded it survives the neoplastic transformation process (Figure 1b). Thus, luminal-type tumors would be derived from luminal progenitor cells while basal-type tumors would be derived from basal/myoepithelial progenitor cells. This theory is supported by the fact that molecular profiles of luminal-type tumors share many similarities with that of normal luminal mammary epithelial cells; they express luminal cluster genes (cytokeratins (CKs) *8*, *18*, *19*, and *7*) as well as luminal differentiation-associated genes (*GATA3*, *EPCAM*, *MUC1*, *CD24*, and *ERα*) [3,5]. Likewise, basal-like tumors closely resemble gene expression programs of their normal basal mammary epithelial counterparts, with expression of basal cluster genes including CKs *5*, *6*, and *17*, as well as *CD49f*, *CD44*, *LAMININ*, and *p63* [3,5]. In addition, experimental support for this model comes from studies where transforming different cell types with the same set of oncogenes gives rise to tumors with different characteristics and phenotypes that are reminiscent of the cell-of-origin [8,9].

In reality however, a third hybrid model has emerged: breast cancer heterogeneity and subtype differences appear to arise from a combination of both the mutation-of-origin and the cell-of-origin (Figure 1c). How these two categorical drivers of breast tumor identity and diversity interact is still being investigated, but recent exciting discoveries have begun to tease apart the convergence of cellular and mutational origins of breast cancer.

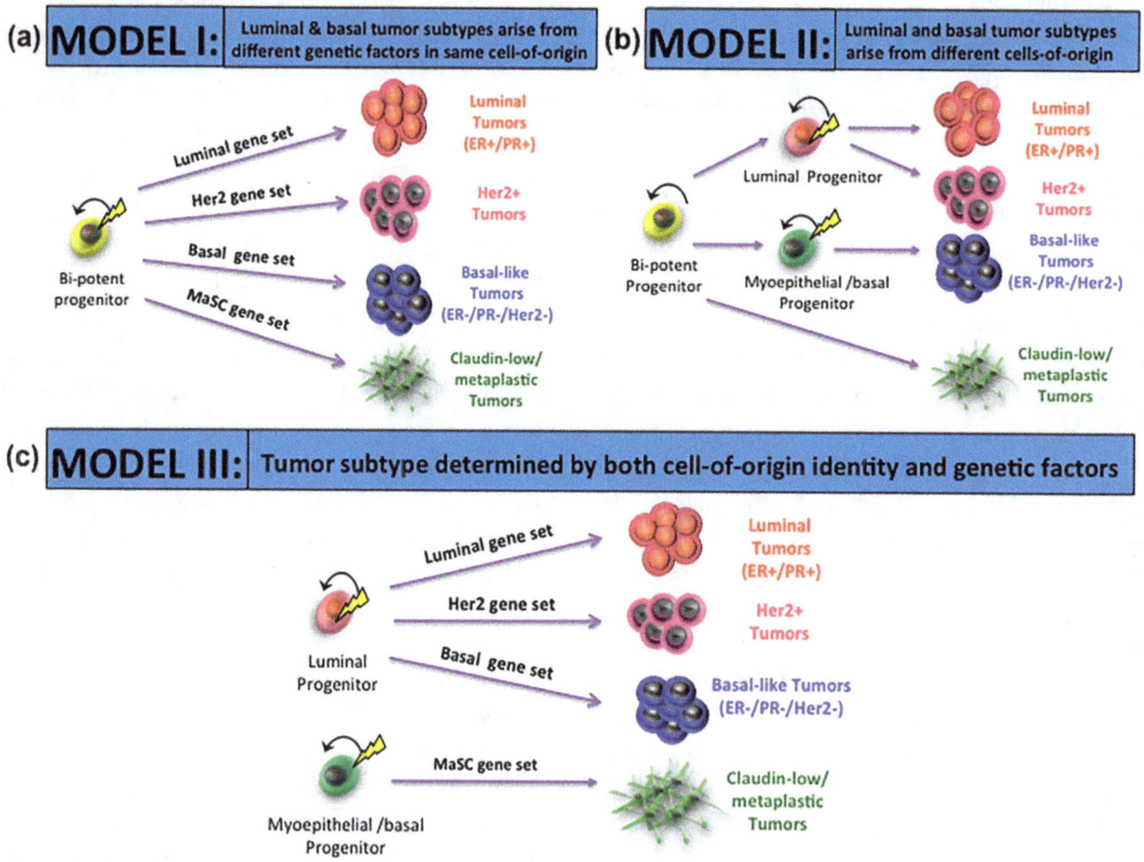

Figure 1. Schematic models of origins of breast cancer heterogeneity (**a**) Model I posits that mutations associated with neoplastic transformation determine subtype; (**b**) Model II posits that the cell-of-origin identity determines subtype during neoplastic transformation; (**c**) Model III posits that both mutation-of-origin and cell-of-origin contribute to the path of disease progression in regards to subtype development.

3. Cells-of-Origin and Impact on Breast Cancer Subtypes

Several complementary and important advances in the field have occurred to identify the cellular origin of breast cancer. First, gene signature sets for the various subsets of normal mammary epithelial cells—including mammary stem cell (MaSC)-enriched, luminal progenitor, mature luminal, and stromal populations—were derived [3,10,11]. These normal gene signatures were subsequently compared to gene expression datasets for the breast cancer molecular subtypes. As hypothesized, the luminal gene signature had the highest degree of overlap with the luminal-like tumor type. Similarly, the stromal gene signature correlated with claudin-low type tumors, a result consistent with the mesenchymal features characteristic of this subtype [10]. The claudin-low transcriptional profile is also similar to the metaplastic $CD10^+$ profiles and expresses markers of the epithelial-to-mesenchymal transition (EMT) and cancer stem cells [12]. Remarkably however, the expression signature of basal-like tumors showed a remarkable similarity to the luminal progenitor gene signature [10]. This finding was further supported by immunophenotypic profiling of human breast cancer tissues that revealed that basal-like tumors as well as Her2 tumors were comprised of luminal ($EpCAM^+/CD49f^+$) cells [13].

Therefore, not surprisingly, the gene signature of HER2 tumors does not overlap with any normal mammary epithelial cell type [3,10].

Second, experimental evidence functionally defining the cellular origins of breast cancer was reported [14,15]. Normal luminal and basal mammary epithelial cells were isolated and sorted from reduction mammoplasty tissue and infected using various combinations of transforming oncogenes; these infected cells were immediately implanted into immunocompromised humanized mice to create spontaneous tumors [14,15]. When luminal cells were infected and implanted into mice, they formed both ER^+ luminal-like tumors and ER^- basal-like tumors. In contrast, when basal cells were infected and implanted into mice, rare metaplastic tumors formed, which resembled the claudin-lo subtype. These studies were the first to show that in human breast tissue, cells from the luminal lineage contained precursors to basal-like breast cancer [14,15]. The finding that the cell of origin to luminal and basal tumors are within the luminal lineage has also been supported in mouse tumor models. MMTV-PyMT and MMTV-Neu mice develop tumors of the luminal-like subtype, while Etv6-NTRK3 mice form basal-like mammary tumors. In all three of these models, tumors were found to originate from $CD61^+$ alveolar progenitor-enriched luminal cells [16]. Using lineage-specific drivers of cancer, targeted loss of *BRCA1* in luminal cells, but not basal cells, produced basal-like tumors [17]. A more recent study using targeted deletion of Brca2, Pten and p53 in mice also showed that when these genes were lost in basal MECs, the same tumor phenotype always emerged—one that resembled claudin-low tumors. In contrast, depending on the initiating genetic lesion in luminal MECs, tumor-initiating cells from this lineage gave rise to basal-like, luminal-like, and normal-like tumors [18]. This important study not only demonstrated that multiple mammary tumor subtypes can arise from the same cell-of-origin pool, but also that molecular subtype cannot be used to infer tumor cell-of-origin identity. This study also illustrates how both the mutation-of-origin and the order in which the mutations occur influences the path of neoplastic transformation.

4. Mutations-of-Origin and Impact on Breast Cancer Subtypes

If luminal cells, and most likely luminal progenitor cells, are the precursors to the most common forms of breast cancer, then it stands to reason that genetic mutations contribute strongly to the fate of luminal cells during cancer formation. A classic example of this has been shown in women who inherit mutations in the Breast Cancer Associated 1 gene (*BRCA1*). Women with mutations in *BRCA1* preferentially develop basal-like tumors, and only recently has exploration into the underlying mechanism for this correlation begun to yield results. Gene expression analysis of *BRCA1*-associated breast cancer strongly correlates with the gene expression signature of luminal progenitor cells; *BRCA1*-mutation carriers in this study appeared to contain an expanded luminal progenitor population [10]. However, a different study also examining cells from *BRCA1*-mutation carriers failed to observe this expansion; rather they found an expansion of the basal progenitor population [14]. Despite the increased number of basal cells in *BRCA1*-mutation carriers, this study showed that the luminal cells, and not the basal cells, were the origins of basal-like breast cancer [14]. When probed more deeply, it was found that luminal progenitor cells from *BRCA1*-mutation carriers exhibited a defect in differentiation, one in which progenitor cells could not give rise to mature cells. Instead, progenitor cells gave rise to luminal cells that expressed basal markers [14]. This revealed that an early genetic mutation, even prior to cancer formation, could alter the fate of the cell-of-origin so that it was predisposed for forming tumors of a specific subtype.

Mouse tumor models have also supported the notion that the fate of luminal progenitor cells can be affected by transforming oncogenes. In mice harboring mutations in the PI3K-pathway effector *PIK3CA*, mammary tumors arise with various characteristics including those of basal-like breast cancer. Interestingly, luminal cells in these mice exhibit an altered fate and de-differentiate into multipotent stem-like cells [19–22]. Viral introduction of tagged oncogenes (ErbB2, PyMT) into luminal cells *in vivo* can also influence cell fate during neoplastic transformation [21]. In this model, a small percentage of luminal cells demonstrated basal marker ($K5^+$) expression following oncogene transduction indicating that luminal cells were adopting a basal fate [23]. These findings all support the concept that gene mutations do affect the fate of the cell-of-origin, directing tumor phenotype during neoplastic transformation.

5. Mechanisms Affecting Cell Fate Changes

If luminal-committed progenitor cells are the precursors for the most common forms of breast cancer, including basal-like, how is it that lineage-restricted cells can adopt phenotypes of a different lineage? Can other precursor sources also undergo similar cell fate switching mechanisms that contribute to subtype development as well? De-differentiation and trans-differentiation both have been established as viable options for cell fate switching in the mammary gland, though the exact mechanisms have only begun to be elucidated.

In the case of *BRCA1*-mutation carriers, aberrant expression of the transcription factor Slug was shown to be the culprit for cell fate switching. Mutation in *BRCA1* was reported to promote stability of the Slug protein [14]. Slug in turn could enhance the expression of basal and stem-related genes as well as associate with and recruit the lysine-specific histone demethylase LSD1 to the promoters of luminal genes to prevent their expression [24]. This role as a cell fate regulator was confirmed *in vivo* using a functional Slug knockout model, which revealed the expression of luminal genes in basal cells, solidifying the importance of Slug as a determinant of basal cell identity [24].

In other contexts, the mechanisms that induce cell state transitions or promote lineage switching are less clear. Whether the switch in cell fate occurs though trans-differentiation or de-differentiation into a more unrestricted primitive state is not known. Examples from other tissues suggest that both can occur and that the cell fate decision may be cell type- or mutation-specific. In the pancreas, trans-differentiation of acinar cells into duct-like cells after *Notch*- and *Kras*-drive reprogramming eventually leads to ductal intraepithelial neoplasia [25,26]. However, in basal cell carcinoma, interfollicular epidermal cells serve as the cells-of-origin, and are reprogrammed back into a more progenitor-like state before progressing towards carcinoma, possibly by transitioning through an EMT phase [27–29]. Likewise in the intestine, while $LGR5^+$ or BMI1-expressing stem cells do give rise to adenomas, villus cells with abnormal Wnt signaling are able to de-differentiate and exhibit tumor-forming capacity [28–32]. These studies show that identity changes within specific cell populations can occur through multiple routes during tumor formation and may be dependent on the cells-of-origin as well as the early oncogenic events during neoplastic transformation.

In the mammary gland, transplantation can promote lineage switching by inducing de-differentiation of restricted progenitors into bipotent stem cells [33–35]. Interestingly, Slug appears to control this multipotent fate potential of MECs during transplantation but also during tumorigenesis. MECs from mice expressing a mutant of Slug that lacks transcriptional activity are unable to unlock stem cell programs needed to regenerate mammary tissue following transplantation or to form tumors driven by Myc [24,36].

While it has been shown that lineage switching of MECs can occur through activating multipotency, there is also evidence demonstrating that luminal cells can directly trans-differentiate into basal cells without having to transition through a stem cell intermediate. Expression of the hippo transducer TAZ was sufficient to promote basal cell fate through regulating expression of basal lineage genes in concert with the SWI/SNF chromatin-remodeling complex [37]. Interestingly, elevated TAZ expression and gene amplification is associated with a basal-like tumor phenotype and correlates with poor survival, possibly suggesting that selection for a basal MEC lineage regulator during tumorigenesis may encourage a shift towards a more basal-like cell state [1,37].

Another regulator of basal cell identity is LBH (limb bud and heart development), which is a Wnt-controlled transcriptional co-factor expressed in a rare population of stem-like cells that reside within the mammary basal lineage [38]. It was shown to control basal cell identity by inducing ΔNp63 gene expression while also repressing ERα gene expression. Loss of LBH impeded normal mammary gland development and consequently its expression is crucial for mammary stem cell maintenance and function, as well as lineage specification [38]. As LBH is overexpressed in basal-like breast cancer, as suggested above for Taz, its selective activation during neoplastic transformation may influence the path of disease progression in relation to cell fate and thus ultimately affect breast tumor subtype determination [39,40].

Cell fate switching has also been shown not only to occur in the luminal cell compartment of the mammary gland, but in the basal cell compartment as well. Mice lacking the transcriptional repressor *Ovol2* exhibit disrupted mammary morphogenesis because loss of Ovol2 in basal cells causes them to become non-epithelial cell types, including fibroblast and muscle [41]. As Ovol2 is significantly downregulated in claudin-low-type breast cancers, this suggests that cell identity-related disruptions within the basal cell compartment may also contribute to subtype specification during breast cancer development [41]. In addition, this fortifies the notion that basal cells can serve as the precursors to claudin-low/metaplastic tumors that exhibit non-epithelial features.

While transcription factors are the major orchestrators of directing and maintaining cell identity programs, epigenetic mechanisms—such as DNA methylation and histone modifications—enable activation or repression of transcription factors and thus are also master controllers of such transcriptional programs. In the breast, the relationship between gene expression and DNA methylation was shown in sorted hMECs [42]. Sorted CD44$^+$ progenitor cells demonstrated hypomethylation on promoters of genes related to pluripotency and self-renewal functions. Many of these genes were targets of the polycomb repressive complex 2 (PCR2) component Suz12 and were also highly expressed in the CD44$^+$ progenitor-enriched cell population. In contrast, more differentiated CD24$^+$ cell populations exhibited increased DNA methylation at promoters of Suz12 target genes, suggesting that differentiation may induce epigenetic marks that block activation of stem programs and contribute to directing cell identity [42]. A more recent study also established a similar but more comprehensive epigenetic/transcriptional map by comparing the epigenetic landscape of cells with their transcriptome profiles. Using diverse RNA, DNA, and chromatin integrative analyses, Gascard *et al.* identified and defined various unmethylated regions (UMR) in regulatory elements of genes that direct cell type [43]. When comparing these UMRs in luminal *versus* basal cells, there were almost twice as many UMRs present in the proximal regulatory regions of various transcription factor genes in the luminal cells and that these cell-type specific UMRs correlated with increased gene expression of these transcription factors [43].

How DNA methylation fate is marked in these different cell populations is beginning to be elucidated. The DNA methyltransferase DNMT1 was shown to be necessary for MaSC maintenance; its loss *in vivo* resulted in decreased basal (CD24$^+$/CD49f^{hi}) and luminal progenitor (CD24$^+$/CD49f^{lo}CD61$^+$) populations, lowered their regenerative and repopulating ability, and hindered tumorigenesis [44]. Uncovering the ability of this epigenetic regulator to impact both normal mammary function and tumor progression is a first step in beginning to understand the mechanism behind DNA methylation influences on mammary cell fate and how that may impact tumorigenic potential.

Histone modifications are another class of epigenetic marks that are critical regulators of MEC identity and fate switching. Global histone methylation profiles of sorted mouse MEC populations were shown to correlate with cell type-specific gene expression profiles. For example, genes upregulated in luminal progenitor (CD24$^+$/CD49f^{lo}/CD61$^+$) *versus* basal/MaSC (CD24$^-$/CD49f^{hi}) cells had increased H3K4me3 and decreased H3K27me3 marks, and vice versa for downregulated genes. Similar correlative trends between gene expression and histone methylation patterns were observed when comparing the luminal progenitor and mature luminal (CD24$^+$/CD49f^{lo}/CD61$^-$) populations [45]. There was also an interesting association between H3K27me3 marks during hormone-dependent changes during pregnancy; H3K27me3 patterns correlated with gene expression patterns. This trend was linked with EZH2-dependent maintenance of H3K27me3 marks, as loss of EZH2 decreased the presence of this mark [45]. Another regulator of histone modifications, the histone methylation sensor Pygo2, was shown to be crucial for proper stem/progenitor maintenance in the mammary gland, as knockout of Pygo2 in a mouse model exhibited a decrease in basal/MaSC and luminal progenitor populations [46]. Additionally, in these mice, the basal/MaSC population showed a luminal differentiation bias, suggesting that loss of this epigenetic regulator was shifting basal cell identity towards a more luminal one. Mechanistically, the increase in luminal fate of the Pygo2-deficient basal/MaSC was reported to occur through Pygo2 normally maintaining a bivalent (H3K4me3/H3K27me3) mark at the Notch3 gene promoter and thus its absence allowed activation of Notch signaling and its related luminal differentiation programs [46]. These initial characterizations of cell type-specific and developmental stage-specific trends in histone methylation further emphasize the importance of epigenetics and the molecular regulators of those marks in regulating mammary cell identity and function.

One characteristic of histone modifications that is particularly crucial in cell fate decisions and how they may be important during malignant transformation are bivalent chromatin marks. Changes in bivalency may shift cell identity into one lineage or another and may therefore affect tumor phenotype. Putative bivalent domains are present in various gene subsets in both luminal and progenitor cell populations, demonstrating that MECs in either lineage may be poised for lineage switching, highlighting the innate plasticity of epithelial cells in the mammary gland [47]. In cancer, bivalent chromatin marks were shown to be important in determining cancer stem cell (CSC) fate. The EMT transcription factor Zeb1 has bivalent chromatin marks at its gene promoter in non-CSCs of basal breast cancer tissue yet upon appropriate signaling can shift to allow for active transcription of Zeb1, thereby enhancing cell plasticity and activating CSC programs [48]. Although the plasticity of bivalency was observed in cancer cells, it is plausible that normal progenitor cells might contain bivalent marks in key transcription factors making them poised for cell state changes. More work is needed to fully understand how the epigenetic landscape of normal cells can influence cell fate and contribute to cancer.

Taken together, regulators of MEC identity are crucial to cell fate determination and switching, and manipulation of MEC lineage regulators can affect the ability of a progenitor cell to make normal cell fate decisions. Additionally, abnormal versions or expression of these regulators during cancer initiation seem to impact tumor phenotype. Identifying additional epigenetic and transcriptional regulators of mammary epithelial cell fate and linage transitioning will provide deeper insight into the complex regulation of differentiation and also will define important molecular drivers of tumor phenotype.

6. Exploiting Cell Fate Changes During Development

As initial genetic mutations can alter the fate of tumor initiating/precursor cells, thereby dictating tumor phenotype, it is also plausible that naturally occurring cell fate changes during development might also be co-opted to drive tumor phenotype. For example, pregnancy is associated with acute expansion of the mammary epithelium, particularly the luminal population. Various stem/progenitor populations have been identified prior to and after pregnancy, and these primitive bipotent cells demonstrate varying longevities and contributions to gland restructuring [49–54]. Given that parity, while protective in the long term, increases short-term breast cancer risk, it is possible that the pregnancy-related changes to luminal MECs primes the gland to be more receptive to cell state transitioning and might be exploited and cooperate with early drivers of malignancy that are associated with pregnancy [3,55].

As another example, the actions of the steroid hormones progesterone and estrogen through the actions of their receptors might play a role in cell fate decision upon transformation. Progesterone exhibits different modes of action in luminal and basal compartments during development. In luminal cells, progesterone increases Rankl expression, which in turn promotes the proliferation of adjacent luminal epithelial cells [56]. Progesterone also can induce the expression of growth hormone, resulting in increased progenitor activity when measured *in vitro* [57]. In basal cells, progesterone stimulates Wnt signaling to promote expansion of the basal progenitor cells as well as induce basal cell proliferation [58–60]. Thus, it is possible that progesterone might increase the number of potential cancer precursor cells, which might be targets for mutation and subsequent transformation. Consistent with this, large epidemiological studies have revealed that women who were on hormone replacement therapy containing estrogen and progesterone, but not estrogen alone, had a significantly higher incicence of breast cancer [61–63]. The mechanism for this remains unclear, but it is plausible that it could be due in part, to the effects of progesterone on cell fate changes and progenitor cell biology.

Historically, ERα expression has been thought to be limited to mature luminal cells within the developing mammary gland as estrogen is hypothesized to function through an indirect proliferative response [64]. However, rare populations of ERα+ cells have been identified in populations with progenitor-like characteristics [65]. The independent function of ER signaling has been difficult to ascertain as ER and PR are commonly co-expressed and estrogen can increase the expression of PR [64,66]. Regardless, ER is a powerful mediator of a number of transcription factors, which in turn regular cell fate within the mammary gland. For example, both PR and ER regulate Stat5a, which is necessary for ductal branching and tumorigenesis [67].

Aging is another naturally occurring stage where cell fate changes may contribute to breast cancer subtype proclivity. Unlike in young women where basal-like breast cancers are found more frequently, luminal-like hormone-positive breast cancers are the major tumor subtype observed in older women [68]. The underlying mechanism behind this tumor subtype bias associated with aging is unclear; however, evidence has emerged that indicates that luminal cells in breast tissues of older women exhibit altered phenotypes and behaviors. Firstly, the number of MECs expressing ERα increase, as well as a concurrent trend towards increased co-expression with the proliferation marker Ki67, suggesting shifts in differentiation and proliferative potential of luminal cells [69–72]. Second, disease-free breast tissue of older women contains an abnormal luminal progenitor population with impaired differentiation capacity, as its progeny gives rise to luminal cells expressing basal markers [73,74]. Third, while luminal progenitor cells contain abnormally short telomeres independent of age compared to other populations of normal mammary epithelial cells, a decline in telomerase activity—often associated with activation of differentiation programs—was found with increasing age [75,76]. Taken together, these findings suggest that aging affects the ability of the luminal lineage to carry out normal differentiation and stem cell function, thus resulting in defective capacity of these cells to properly maintain and/or shift identity.

Whether or not these changes contribute to the luminal subtype bias observed during aging has yet to be established. However, there is some emerging evidence that might support this connection. Age-related gene expression signatures from disease-free pre- or perimenopausal breast tissue of young women (40 years) correlated with higher grade and aggressive breast tumors, while gene expression signatures from disease-free pre- or perimenopausal breast tissue of older women (>40 years) correlated with lower grade and less aggressive breast tumors [77]. This suggests that aging may alter the transcriptional landscape of the breast, potentially in a way that influences breast tumor development. Interestingly, this same study also tested for gene expression changes based on menopausal status alone and found no significant trends separating gene expression profiles of pre/peri and post-menopausal tissue. Despite this, the hormonal changes and their downstream effects that occur with menopause need to be better characterized to truly determine what effects this important age-related shift in endocrine signaling might have on the breast and its cell populations and how it interacts with other aging-related processes.

Epigenetic changes have also been reported to contribute to aging-related changes to the breast. In normal human breast tissue, older women exhibited increased DNA methylation of CpG islands and polycomb group protein target gene (PCGT) enrichment [78]. Intriguingly, a subset of these CpG gene loci were also hypermethylated in tumors, with some showing a specificity for ER-positive tumors—the subtype that is more prevalent in older women—thus, drawing a connection between epigenetic changes that occur during aging and those that occur during tumor development [78]. Further investigation into the mechanisms by which epigenetic regulation influences cell identity within the mammary gland is needed. In particular, understanding how aging induces shifts in the epigenome and subsequent functional changes that contribute to altered cell fate decisions will be needed to determine how this can influence the heterogeneity of breast tumors.

Collectively, these observations put forth the possibility that as the breast ages, changes in tumor initiating/precursor cells might alter their differentiation potential and cell fate thereby priming the epigenetic landscape to impact tumor phenotype following neoplastic transformation. Additional studies and experimental evidence using models of aging will be necessary to directly support this hypothesis.

7. Conclusions

As discussed above, recent work has endeavored to understand the genetic and cellular changes to mammary epithelial cells that occur during mammary development and how they may be utilized during neoplastic transformation to promote breast tumor heterogeneity. Understanding the mechanisms that drive cell fate switching during the transformation process may be the key to improved therapies, and even disease prevention. It is too early to know whether this understanding will impact treatment but it may shed insights into how we may be able to prevent breast cancer from developing or progressing in the future.

Acknowledgments: This work was supported by the Raymond & Beverly Sackler Convergence Lab and grants from ArtBeCAUSE, the Breast Cancer Research Foundation, and the NIH/NCI CA125554 and CA092644.

Conflicts of Interest: The authors declare no conflict of interest.

References

1. Koboldt, D.C.; Fulton, R.S.; McLellan, M.D.; Schmidt, H.; Kalicki-Veizer, J.; McMichael, J.F.; Fulton, L.L.; Dooling, D.J.; Ding, L.; Mardis, E.R.; *et al.* Comprehensive molecular portraits of human breast tumours. *Nature* **2012**, *490*, 61–70. [CrossRef] [PubMed]
2. Bertos, N.R.; Park, M. Breast cancer—One term, many entities? *J. Clin. Investig.* **2011**, *121*, 3789–3796. [CrossRef] [PubMed]
3. Visvader, J.E. Keeping abreast of the mammary epithelial hierarchy and breast tumorigenesis. *Genes Dev.* **2009**, *23*, 2563–2577. [CrossRef] [PubMed]
4. Lehmann, B.D.; Bauer, J.A.; Chen, X.; Sanders, M.E.; Chakravarthy, A.B.; Shyr, Y.; Pietenpol, J.A. Identification of human triple-negative breast cancer subtypes and preclinical models for selection of targeted therapies. *J. Clin. Investig.* **2011**, *121*, 2750–2767. [CrossRef] [PubMed]
5. Perou, C.M.; Sørlie, T.; Eisen, M.B.; van de Rijn, M.; Jeffrey, S.S.; Rees, C.A.; Pollack, J.R.; Ross, D.T.; Johnsen, H.; Akslen, L.A.; *et al.* Molecular portraits of human breast tumors. *Nature* **2000**, *406*, 747–752. [CrossRef] [PubMed]
6. Prat, A.; Perou, C.M. Deconstructing the molecular portraits of breast cancer. *Mol. Oncol.* **2011**, *5*, 5–23. [CrossRef] [PubMed]
7. Sørlie, T.; Perou, C.M.; Tibshirani, R.; Aas, T.; Geisler, S.; Johnsen, H.; Hastie, T.; Eisen, M.B.; van de Rijn, M.; Jeffrey, S.S.; *et al.* Gene expression patterns of breast carcinomas distinguish tumor subclasses with clinical implications. *Proc. Natl. Acad. Sci. USA* **2001**, *98*, 10869–10874. [CrossRef] [PubMed]
8. Gupta, P.B.; Kuperwasser, C.; Brunet, J.P.; Ramaswamy, S.; Kuo, W.L.; Gray, J.W.; Naber, S.P.; Weinberg, R.A. The melanocyte differentiation program predisposes to metastasis after neoplastic transformation. *Nat. Genet.* **2005**, *37*, 1047–1054. [CrossRef] [PubMed]
9. Ince, T.A.; Richardson, A.L.; Bell, G.W.; Saitoh, M.; Godar, S.; Karnoub, A.E.; Iglehart, J.D.; Weinberg, R.A. Transformation of different human breast epithelial cell types leads to distinct tumor phenotypes. *Cancer Cell* **2007**, *12*, 160–170. [CrossRef] [PubMed]
10. Lim, E.; Vaillant, F.; Wu, D.; Forrest, N.C.; Pal, B.; Hart, A.H.; Asselin-Labat, M.L.; Gyorki, D.E.; Ward, T.; Partanen, A.; *et al.* Aberrant luminal progenitors as the candidate target population for basal tumor development in BRCA1 mutation carriers. *Nat. Med.* **2009**, *15*, 907–913. [CrossRef] [PubMed]

11. Raouf, A.; Zhao, Y.; To, K.; Stingl, J.; Delaney, A.; Barbara, M.; Iscove, N.; Jones, S.; McKinney, S.; Emerman, J.; *et al.* Transcriptome analysis of the normal human mammary cell commitment and differentiation process. *Cell Stem Cell* **2008**, *3*, 109–118. [CrossRef] [PubMed]
12. Prat, A.; Parker, J.S.; Karginova, O.; Fan, C.; Livasy, C.; Herschkowitz, J.I.; He, X.; Perou, C.M. Phenotypic and molecular characterization of the claudin-low intrinsic subtype of breast cancer. *Breast Cancer Res* **2010**, *12*, R68. [CrossRef] [PubMed]
13. Keller, P.J.; Lin, A.F.; Arendt, L.M.; Klebba, I.; Jones, A.D.; Rudnick, J.A.; DiMeo, T.A.; Gilmore, H.; Jefferson, D.M.; Graham, R.A.; *et al.* Mapping the cellular and molecular heterogeneity of normal and malignant breast tissues and cultured cell lines. *Breast Cancer Res* **2010**, *12*, R87. [CrossRef] [PubMed]
14. Proia, T.A.; Keller, P.J.; Gupta, P.B.; Klebba, I.; Jones, A.D.; Sedic, M.; Gilmore, H.; Tung, N.; Naber, S.P.; Schnitt, S.; *et al.* Genetic predisposition directs breast cancer phenotype by dictating progenitor cell fate. *Cell Stem Cell* **2011**, *8*, 149–163. [CrossRef] [PubMed]
15. Keller, P.J.; Arendt, L.M.; Skibinski, A.; Logvinenko, T.; Klebba, I.; Dong, S.; Smith, A.E.; Prat, A.; Perou, C.M.; Gilmore, H.; *et al.* Defining the cellular precursors to human breast cancer. *Proc. Natl. Acad. Sci. USA* **2012**, *109*, 2772–2777. [CrossRef] [PubMed]
16. Tao, L.; van Bragt, M.P.A.; Li, Z. A long-lived luminal subpopulation enriched with alveolar progenitors serves as cellular origin of heterogeneous mammary tumors. *Stem Cell Rep.* **2015**, *5*, 60–74. [CrossRef] [PubMed]
17. Molyneux, G.; Geyer, F.C.; Magnay, F.A.; McCarthy, A.; Kendrick, H.; Natrajan, R.; MacKay, A.; Grigoriadis, A.; Tutt, A.; Ashworth, A.; *et al.* BRCA1 basal-like breast cancers originate from luminal epithelial progenitors and not from basal stem cells. *Cell Stem Cell* **2010**, *7*, 403–417. [CrossRef] [PubMed]
18. Melchor, L.; Molyneux, G.; Mackay, A.; Magnay, F.A.; Atienza, M.; Kendrick, H.; Nava-Rodrigues, D.; López-García, M.Á.; Milanezi, F.; Greenow, K.; *et al.* Identification of cellular and genetic drivers of breast cancer heterogeneity in genetically engineered mouse tumour models: Breast cancer heterogeneity in mouse tumour models. *J. Pathol.* **2014**, *233*, 124–137. [CrossRef] [PubMed]
19. Meyer, D.S.; Brinkhaus, H.; Muller, U.; Muller, M.; Cardiff, R.D.; Bentires-Alj, M. Luminal expression of PIK3CA mutant H1047R in the mammary gland induces heterogeneous tumors. *Cancer Res.* **2011**, *71*, 4344–4351. [CrossRef] [PubMed]
20. Liu, P.; Cheng, H.; Santiago, S.; Raeder, M.; Zhang, F.; Isabella, A.; Yang, J.; Semaan, D.J.; Chen, C.; Fox, E.A.; *et al.* Oncogenic PIK3CA-driven mammary tumors frequently recur via PI3K pathway–dependent and PI3K pathway–independent mechanisms. *Nat. Med.* **2011**, *17*, 1116–1120. [CrossRef] [PubMed]
21. Meyer, D.S.; Koren, S.; Leroy, C.; Brinkhaus, H.; Müller, U.; Klebba, I.; Müller, M.; Cardiff, R.D.; Bentires-Alj, M. Expression of PIK3CA mutant E545K in the mammary gland induces heterogeneous tumors but is less potent than mutant H1047R. *Oncogenesis* **2013**, *2*, e74. [CrossRef] [PubMed]
22. Koren, S.; Reavie, L.; do Couto, J.P.; de Silva, D.; Stadler, M.B.; Roloff, T.; Britschgi, A.; Eichlisberger, T.; Kohler, H.; Aina, O.; *et al.* PIK3CAH1047R induces multipotency and multi-lineage mammary tumours. *Nature* **2015**, *525*, 114–118. [CrossRef] [PubMed]
23. Hein, S.M.; Haricharan, S.; Johnston, A.N.; Toneff, M.J.; Reddy, J.P.; Dong, J.; Bu, W.; Li, Y. Luminal epithelial cells within the mammary gland can produce basal cells upon oncogenic stress. *Oncogene* **2015**. [CrossRef] [PubMed]
24. Phillips, S.; Prat, A.; Sedic, M.; Proia, T.; Wronski, A.; Mazumdar, S.; Skibinski, A.; Shirley, S.H.; Perou, C.M.; Gill, G.; *et al.* Cell-state transitions regulated by SLUG are critical for tissue regeneration and tumor initiation. *Stem Cell Rep.* **2014**, *2*, 633–647. [CrossRef] [PubMed]
25. De la O, J.P.; Emerson, L.L.; Goodman, J.L.; Froebe, S.C.; Illum, B.E.; Curtis, A.B.; Murtaugh, L.C. Notch and Kras reprogram pancreatic acinar cells to ductal intraepithelial neoplasia. *Proc. Natl. Acad. Sci. USA* **2008**, *105*, 18907–18912. [CrossRef] [PubMed]
26. Habbe, N.; Shi, G.; Meguid, R.A.; Fendrich, V.; Esni, F.; Chen, H.; Feldmann, G.; Stoffers, D.A.; Konieczny, S.F.; Leach, S.D.; *et al.* Spontaneous induction of murine pancreatic intraepithelial neoplasia (mPanIN) by acinar cell targeting of oncogenic Kras in adult mice. *Proc. Natl. Acad. Sci. USA* **2008**, *105*, 18913–18918. [CrossRef] [PubMed]

27. Youssef, K.K.; Lapouge, G.; Bouvrée, K.; Rorive, S.; Brohée, S.; Appelstein, O.; Larsimont, J.C.; Sukumaran, V.; van de Sande, B.; Pucci, D.; *et al.* Adult interfollicular tumour-initiating cells are reprogrammed into an embryonic hair follicle progenitor-like fate during basal cell carcinoma initiation. *Nat. Cell Biol.* **2012**, *14*, 1282–1294. [CrossRef] [PubMed]
28. Beck, B.; Blanpain, C. Unravelling cancer stem cell potential. *Nat. Rev. Cancer* **2013**, *13*, 727–738. [CrossRef] [PubMed]
29. Visvader, J.E. Cells of origin in cancer. *Nature* **2011**, *469*, 314–322. [CrossRef] [PubMed]
30. Barker, N.; Ridgway, R.A.; van Es, J.H.; van de Wetering, M.; Begthel, H.; van den Born, M.; Danenberg, E.; Clarke, A.R.; Sansom, O.J.; Clevers, H. Crypt stem cells as the cells-of-origin of intestinal cancer. *Nature* **2009**, *457*, 608–611. [CrossRef] [PubMed]
31. Sangiorgi, E.; Capecchi, M.R. Bmi1 is expressed *in vivo* in intestinal stem cells. *Nat. Genet.* **2008**, *40*, 915–920. [CrossRef] [PubMed]
32. Schwitalla, S.; Fingerle, A.A.; Cammareri, P.; Nebelsiek, T.; Göktuna, S.I.; Ziegler, P.K.; Canli, O.; Heijmans, J.; Huels, D.J.; Moreaux, G.; *et al.* Intestinal tumorigenesis initiated by dedifferentiation and acquisition of stem-cell-like properties. *Cell* **2013**, *152*, 25–38. [CrossRef] [PubMed]
33. Sleeman, K.E.; Kendrick, H.; Robertson, D.; Isacke, C.M.; Ashworth, A.; Smalley, M.J. Dissociation of estrogen receptor expression and *in vivo* stem cell activity in the mammary gland. *J. Cell Biol.* **2007**, *176*, 19–26. [CrossRef] [PubMed]
34. Van Keymeulen, A.; Rocha, A.S.; Ousset, M.; Beck, B.; Bouvencourt, G.; Rock, J.; Sharma, N.; Dekoninck, S.; Blanpain, C. Distinct stem cells contribute to mammary gland development and maintenance. *Nature* **2011**, *479*, 189–193. [CrossRef] [PubMed]
35. Van Amerongen, R.; Bowman, A.N.; Nusse, R. Developmental stage and time dictate the fate of Wnt/β-catenin-responsive stem cells in the mammary gland. *Cell Stem Cell* **2012**, *11*, 387–400. [CrossRef] [PubMed]
36. Guo, W.; Keckesova, Z.; Donaher, J.L.; Shibue, T.; Tischler, V.; Reinhardt, F.; Itzkovitz, S.; Noske, A.; Zürrer-Härdi, U.; Bell, G.; *et al.* Slug and Sox9 cooperatively determine the mammary stem cell state. *Cell* **2012**, *148*, 1015–1028. [CrossRef] [PubMed]
37. Skibinski, A.; Breindel, J.L.; Prat, A.; Galván, P.; Smith, E.; Rolfs, A.; Gupta, P.B.; LaBaer, J.; Kuperwasser, C. The Hippo transducer TAZ interacts with the SWI/SNF complex to regulate breast epithelial lineage commitment. *Cell Rep.* **2014**, *6*, 1059–1072. [CrossRef] [PubMed]
38. Lindley, L.E.; Curtis, K.M.; Sanchez-Mejias, A.; Rieger, M.E.; Robbins, D.J.; Briegel, K.J. The Wnt-controlled transcriptional regulator LBH is required for mammary stem cell expansion and maintenance of the basal lineage. *Development* **2015**, *142*, 893–904. [CrossRef] [PubMed]
39. Lamb, R.; Ablett, M.P.; Spence, K.; Landberg, G.; Sims, A.H.; Clarke, R.B. Wnt pathway activity in breast cancer sub-types and stem-like cells. *PLoS ONE* **2013**, *8*, e67811. [CrossRef] [PubMed]
40. Rieger, M.E.; Sims, A.H.; Coats, E.R.; Clarke, R.B.; Briegel, K.J. The Embryonic transcription cofactor LBH is a direct target of the Wnt signaling pathway in epithelial development and in aggressive basal subtype breast cancers. *Mol. Cell. Biol.* **2010**, *30*, 4267–4279. [CrossRef] [PubMed]
41. Watanabe, K.; Villarreal-Ponce, A.; Sun, P.; Salmans, M.L.; Fallahi, M.; Andersen, B.; Dai, X. Mammary morphogenesis and regeneration require the inhibition of EMT at terminal end buds by Ovol2 transcriptional repressor. *Dev. Cell* **2014**, *29*, 59–74. [CrossRef] [PubMed]
42. Bloushtain-Qimron, N.; Yao, J.; Snyder, E.L.; Shipitsin, M.; Campbell, L.L.; Mani, S.A.; Hu, M.; Chen, H.; Ustyansky, V.; Antosiewicz, J.E.; *et al.* Cell type-specific DNA methylation patterns in the human breast. *Proc. Natl. Acad. Sci. USA* **2008**, *105*, 14076–14081. [CrossRef] [PubMed]
43. Gascard, P.; Bilenky, M.; Sigaroudinia, M.; Zhao, J.; Li, L.; Carles, A.; Delaney, A.; Tam, A.; Kamoh, B.; Cho, S.; *et al.* Epigenetic and transcriptional determinants of the human breast. *Nat. Commun.* **2015**, *6*, 6351. [CrossRef] [PubMed]
44. Pathania, R.; Ramachandran, S.; Elangovan, S.; Padia, R.; Yang, P.; Cinghu, S.; Veeranan-Karmegam, R.; Arjunan, P.; Gnana-Prakasam, J.P.; Sadanand, F.; *et al.* DNMT1 is essential for mammary and cancer stem cell maintenance and tumorigenesis. *Nat. Commun.* **2015**, *6*, 6910. [CrossRef] [PubMed]
45. Pal, B.; Bouras, T.; Shi, W.; Vaillant, F.; Sheridan, J.M.; Fu, N.; Breslin, K.; Jiang, K.; Ritchie, M.E.; Young, M.; *et al.* Global changes in the mammary epigenome are induced by hormonal cues and coordinated by Ezh2. *Cell Rep.* **2013**, *3*, 411–426. [CrossRef] [PubMed]

46. Gu, B.; Watanabe, K.; Sun, P.; Fallahi, M.; Dai, X. Chromatin effector Pygo2 mediates Wnt-Notch crosstalk to suppress luminal/alveolar potential of mammary stem and basal cells. *Cell Stem Cell* **2013**, *13*, 48–61. [CrossRef] [PubMed]
47. Maruyama, R.; Choudhury, S.; Kowalczyk, A.; Bessarabova, M.; Beresford-Smith, B.; Conway, T.; Kaspi, A.; Wu, Z.; Nikolskaya, T.; Merino, V.F.; *et al*. Epigenetic regulation of cell type-specific expression patterns in the human mammary epithelium. *PLoS Genet.* **2011**, *7*, e1001369. [CrossRef] [PubMed]
48. Chaffer, C.L.; Marjanovic, N.D.; Lee, T.; Bell, G.; Kleer, C.G.; Reinhardt, F.; D'Alessio, A.C.; Young, R.A.; Weinberg, R.A. Poised chromatin at the ZEB1 promoter enables breast cancer cell plasticity and enhances tumorigenicity. *Cell* **2013**, *154*, 61–74. [CrossRef] [PubMed]
49. Wagner, K.U.; Boulanger, C.A.; Henry, M.D.; Sgagias, M.; Hennighausen, L.; Smith, G.H. An adjunct mammary epithelial cell population in parous females: Its role in functional adaptation and tissue renewal. *Development* **2002**, *129*, 1377–1386. [PubMed]
50. Asselin-Labat, M.L.; Vaillant, F.; Sheridan, J.M.; Pal, B.; Wu, D.; Simpson, E.R.; Yasuda, H.; Smyth, G.K.; Martin, T.J.; Lindeman, G.J.; *et al*. Control of mammary stem cell function by steroid hormone signalling. *Nature* **2010**, *465*, 798–802. [CrossRef] [PubMed]
51. Kaanta, A.S.; Virtanen, C.; Selfors, L.M.; Brugge, J.S.; Neel, B.G. Evidence for a multipotent mammary progenitor with pregnancy-specific activity. *Breast Cancer Res* **2013**, *15*, R65. [CrossRef] [PubMed]
52. Chang, T.H.; Kunasegaran, K.; Tarulli, G.A.; de Silva, D.; Voorhoeve, P.M.; Pietersen, A.M. New insights into lineage restriction of mammary gland epithelium using parity-identified mammary epithelial cells. *Breast Cancer Res.* **2014**, *16*, R1. [CrossRef] [PubMed]
53. Rios, A.C.; Fu, N.Y.; Lindeman, G.J.; Visvader, J.E. *In situ* identification of bipotent stem cells in the mammary gland. *Nature* **2014**, *506*, 322–327. [CrossRef] [PubMed]
54. Visvader, J.E.; Stingl, J. Mammary stem cells and the differentiation hierarchy: Current status and perspectives. *Genes Dev.* **2014**, *28*, 1143–1158. [CrossRef] [PubMed]
55. Lambe, M.; Hsieh, C.; Trichopoulos, D.; Ekbom, A.; Pavia, M.; Adami, H.O. Transient increase in the risk of breast cancer after giving birth. *N. Engl. J. Med.* **1994**, *331*, 5–9. [CrossRef] [PubMed]
56. Lee, Y.; Kim, S.H.; Lee, Y.J.; Kang, E.S.; Lee, B.W.; Cha, B.S.; Kim, J.W.; Song, D.H.; Lee, H.C. Transcription factor Snail is a novel regulator of adipocyte differentiation via inhibiting the expression of peroxisome proliferator-activated receptor γ. *Cell. Mol. Life Sci.* **2013**, *70*, 3959–3971. [CrossRef] [PubMed]
57. Lombardi, S.; Honeth, G.; Ginestier, C.; Shinomiya, I.; Marlow, R.; Buchupalli, B.; Gazinska, P.; Brown, J.; Catchpole, S.; Liu, S.; *et al*. Growth hormone is secreted by normal breast epithelium upon progesterone stimulation and increases proliferation of stem/progenitor cells. *Stem Cell Rep.* **2014**, *2*, 780–793. [CrossRef]
58. Rajaram, R.D.; Buric, D.; Caikovski, M.; Ayyanan, A.; Rougemont, J.; Shan, J.; Vainio, S.J.; Yalcin-Ozuysal, O.; Brisken, C. Progesterone and Wnt4 control mammary stem cells via myoepithelial crosstalk. *EMBO J.* **2015**, *34*, 641–652. [CrossRef] [PubMed]
59. Badders, N.M.; Goel, S.; Clark, R.J.; Klos, K.S.; Kim, S.; Bafico, A.; Lindvall, C.; Williams, B.O.; Alexander, C.M. The Wnt receptor, Lrp5, is expressed by mouse mammary stem cells and is required to maintain the basal lineage. *PLoS ONE* **2009**, *4*, e6594. [CrossRef] [PubMed]
60. Shiah, Y.J.; Tharmapalan, P.; Casey, A.E.; Joshi, P.A.; McKee, T.D.; Jackson, H.W.; Beristain, A.G.; Chan-Seng-Yue, M.A.; Bader, G.D.; Lydon, J.P.; *et al*. A progesterone-CXCR4 axis controls mammary progenitor cell fate in the adult gland. *Stem Cell Rep.* **2015**, *4*, 313–322. [CrossRef]
61. Anonymous. Breast cancer and hormone replacement therapy: collaborative reanalysis of data from 51 epidemiological studies of 52,705 women with breast cancer and 108,411 women without breast cancer. *Lancet* **1997**, *350*, 1047–1059.
62. Chlebowski, R.T.; Hendrix, S.L.; Langer, R.D.; Stefanick, M.L.; Gass, M.; Lane, D.; Rodabough, R.J.; Gilligan, M.A.; Cyr, M.G.; Thomson, C.A.; *et al*. Influence of estrogen plus progestin on breast cancer and mammography in healthy postmenopausal women: The women's health initiative randomized trial. *JAMA* **2003**, *289*, 3243–3253. [CrossRef] [PubMed]
63. Ross, R.K.; Paganini-Hill, A.; Wan, P.C.; Pike, M.C. Effect of hormone replacement therapy on breast cancer risk: estrogen *versus* estrogen plus progestin. *J. Natl. Cancer Inst.* **2000**, *92*, 328–332. [CrossRef] [PubMed]
64. Russo, J.; Ao, X.; Grill, C.; Russo, I.H. Pattern of distribution of cells positive for estrogen receptor α and progesterone receptor in relation to proliferating cells in the mammary gland. *Breast Cancer Res. Treat.* **1999**, *53*, 217–227. [CrossRef] [PubMed]

65. Clarke, R.B.; Spence, K.; Anderson, E.; Howell, A.; Okano, H.; Potten, C.S. A putative human breast stem cell population is enriched for steroid receptor-positive cells. *Dev. Biol.* **2005**, *277*, 443–456. [CrossRef] [PubMed]
66. Schultz, J.R.; Petz, L.N.; Nardulli, A.M. Estrogen receptor α and Sp1 regulate progesterone receptor gene expression. *Mol. Cell. Endocrinol.* **2003**, *201*, 165–175. [CrossRef]
67. Santos, S.J.; Haslam, S.Z.; Conrad, S.E. Signal transducer and activator of transcription 5a mediates mammary ductal branching and proliferation in the nulliparous mouse. *Endocrinology* **2010**, *151*, 2876–2885. [CrossRef] [PubMed]
68. Jenkins, E.O.; Deal, A.M.; Anders, C.K.; Prat, A.; Perou, C.M.; Carey, L.A.; Muss, H.B. Age-specific changes in intrinsic breast cancer subtypes: A focus on older women. *Oncologist* **2014**, *19*, 1076–1083. [CrossRef] [PubMed]
69. Walker, K.J.; McClelland, R.A.; Candlish, W.; Blamey, R.W.; Nicholson, R.I. Hetergeneity of oestrogen receptor expression in normal and malignant breast tissue. *Eur. J. Cancer* **1992**, *28*, 34–37. [CrossRef]
70. Shoker, V.S.; Jarvis, C.; Clarke, R.B.; Anderson, E.; Hewlett, J.; Davies, M.P.A.; Sibson, D.R.; Sloane, J.P. Estrogen receptor-positive proliferating cells in the normal and precancerous breast. *Am. J. Pathol.* **1999**, *155*, 1811–1815. [CrossRef]
71. Shoker, B.S.; Jarvis, C.; Sibson, D.R.; Walker, C.; Sloane, J.P. Oestrogen receptor expression in the normal and pre-cancerous breast. *J. Pathol.* **1999**, *188*, 237–244. [CrossRef]
72. Walker, R.; Martin, C. The aged breast. *J. Pathol.* **2007**, *211*, 232–240. [CrossRef] [PubMed]
73. Garbe, J.C.; Pepin, F.; Pelissier, F.A.; Sputova, K.; Fridriksdottir, A.J.; Guo, D.E.; Villadsen, R.; Park, M.; Petersen, O.W.; Borowsky, A.D.; *et al.* Accumulation of multipotent progenitors with a basal differentiation bias during aging of human mammary epithelia. *Cancer Res.* **2012**, *72*, 3687–3701. [CrossRef] [PubMed]
74. Pelissier, F.A.; Garbe, J.C.; Ananthanarayanan, B.; Miyano, M.; Lin, C.; Jokela, T.; Kumar, S.; Stampfer, M.R.; Lorens, J.B.; LaBarge, M.A. Age-related dysfunction in mechanotransduction impairs differentiation of human mammary epithelial progenitors. *Cell Rep.* **2014**, *7*, 1926–1939. [CrossRef] [PubMed]
75. Kannan, N.; Huda, N.; Tu, L.; Droumeva, R.; Aubert, G.; Chavez, E.; Brinkman, R.R.; Lansdorp, P.; Emerman, J.; Abe, S.; *et al.* The luminal progenitor compartment of the normal human mammary gland constitutes a unique site of telomere dysfunction. *Stem Cell Rep.* **2013**, *1*, 28–37. [CrossRef] [PubMed]
76. Sputova, K.; Garbe, J.C.; Pelissier, F.A.; Chang, E.; Stampfer, M.R.; LaBarge, M.A. Aging phenotypes in cultured normal human mammary epithelial cells are correlated with decreased telomerase activity independent of telomere length. *Genome Integr.* **2013**, *4*. [CrossRef]
77. Pirone, J.R.; D'Arcy, M.; Stewart, D.A.; Hines, W.C.; Johnson, M.; Gould, M.N.; Yaswen, P.; Jerry, D.J.; Schneider, S.S.; Troester, M.A. Age-associated gene expression in normal breast tissue mirrors qualitative age-at-incidence patterns for breast cancer. *Cancer Epidemiol. Biomark. Prev.* **2012**, *21*, 1735–1744. [CrossRef] [PubMed]
78. Johnson, K.C.; Koestler, D.C.; Cheng, C.; Christensen, B.C. Age-related DNA methylation in normal breast tissue and its relationship with invasive breast tumor methylation. *Epigenetics* **2014**, *9*, 268–275. [CrossRef] [PubMed]

Permissions

All chapters in this book were first published in JDB, by MDPI; hereby published with permission under the Creative Commons Attribution License or equivalent. Every chapter published in this book has been scrutinized by our experts. Their significance has been extensively debated. The topics covered herein carry significant findings which will fuel the growth of the discipline. They may even be implemented as practical applications or may be referred to as a beginning point for another development.

The contributors of this book come from diverse backgrounds, making this book a truly international effort. This book will bring forth new frontiers with its revolutionizing research information and detailed analysis of the nascent developments around the world.

We would like to thank all the contributing authors for lending their expertise to make the book truly unique. They have played a crucial role in the development of this book. Without their invaluable contributions this book wouldn't have been possible. They have made vital efforts to compile up to date information on the varied aspects of this subject to make this book a valuable addition to the collection of many professionals and students.

This book was conceptualized with the vision of imparting up-to-date information and advanced data in this field. To ensure the same, a matchless editorial board was set up. Every individual on the board went through rigorous rounds of assessment to prove their worth. After which they invested a large part of their time researching and compiling the most relevant data for our readers.

The editorial board has been involved in producing this book since its inception. They have spent rigorous hours researching and exploring the diverse topics which have resulted in the successful publishing of this book. They have passed on their knowledge of decades through this book. To expedite this challenging task, the publisher supported the team at every step. A small team of assistant editors was also appointed to further simplify the editing procedure and attain best results for the readers.

Apart from the editorial board, the designing team has also invested a significant amount of their time in understanding the subject and creating the most relevant covers. They scrutinized every image to scout for the most suitable representation of the subject and create an appropriate cover for the book.

The publishing team has been an ardent support to the editorial, designing and production team. Their endless efforts to recruit the best for this project, has resulted in the accomplishment of this book. They are a veteran in the field of academics and their pool of knowledge is as vast as their experience in printing. Their expertise and guidance has proved useful at every step. Their uncompromising quality standards have made this book an exceptional effort. Their encouragement from time to time has been an inspiration for everyone.

The publisher and the editorial board hope that this book will prove to be a valuable piece of knowledge for researchers, students, practitioners and scholars across the globe.

List of Contributors

Anahid A. Birjandi
Centre for Craniofacial and Regenerative Biology, King's College London Dental Institute, Floor 27, Guy's Hospital, London SE1 9RT, UK

Maisa Seppala, Guilherme M. Xavier and Martyn T. Cobourne
Centre for Craniofacial and Regenerative Biology, King's College London Dental Institute, Floor 27, Guy's Hospital, London SE1 9RT, UK
Department of Orthodontics, King's College London Dental Institute, Floor 22, Guy's and St Thomas' NHS Foundation Trust, London SE1 9RT, UK

Gareth J. Fraser
Department of Animal and Plant Sciences, Alfred Denny Building, University of Sheffield, Sheffield S10 2TN, UK

Dagmara Korona, Stefan A. Koestler and Steven Russell
Department of Genetics, University of Cambridge, Downing Street, Cambridge CB2 3EH, UK

Romy Prühs and Reinhard Schröder
Department of Genetics, Biological Sciences, University of Rostock, Albert-Einstein-Str. 3, D 18059 Rostock, Germany

Anke Beermann
Animal Genetics, University of Tübingen, Auf der Morgenstelle 15, D 72076 Tübingen, Germany

Tapan A. Shah and Melissa B. Rogers
Rutgers—New Jersey Medical School, Microbiology, Biochemistry, & Molecular Genetics, Newark, NJ 07103, USA
Mireya Ruiz-Losada, David Blom-Dahl, Sergio Córdoba and Carlos Estella
Departamento de Biología Molecular and Centro de Biología Molecular Severo Ochoa, Universidad Autónoma de Madrid (UAM/CSIC), Nicolás Cabrera 1, 28049 Madrid, Spain;

Diana M. Juriloff and Muriel J. Harris
Department of Medical Genetics, University of British Columbia, 2350 Health Sciences Mall, Vancouver, BC V6T 1Z3, Canada

Joan T. Richtsmeier
Department of Anthropology, Pennsylvania State University, University Park, PA 16802, USA

Kevin Flaherty
Department of Anthropology, Pennsylvania State University, University Park, PA 16802, USA
Department of Biology, Augustana College, Rock Island, IL 61201, USA

Nikole R. Zuñiga and Esther T. Stoeckli
Department of Molecular Life Sciences and Neuroscience Center Zurich, University of Zurich, Winterthurerstrasse 190, 8057 Zurich, Switzerland

Simon A. Ramsbottom and Mary E. Pownall
Institute of Genetic Medicine, International Centre for Life, Newcastle University, NE1 3BZ Newcastle upon Tyne, UK
Biology Department, University of York, YO10 5YW York, UK

Adam Davis
Department of Biology and Physical Sciences, Gordon State College, Barnesville, GA 30204, USA

Michael C. Reubens
The Scripps Research Institute, 10550 N, Torrey Pines Road, MB3, La Jolla, CA 92037, USA

Edmund J. Stellwag
Department of Biology, Howell Science Complex, East Carolina University, Greenville, NC 27858, USA

Lucie Jeannotte, Florian Gotti and Kim Landry-Truchon
Centre de recherche sur le cancer de l' Université Laval; CRCHU de Québec, L' Hôtel-Dieu de Québec, QC G1R 3S3, Canada
Department of Molecular Biology, Medical Biochemistry and Pathology, Université Laval, QC G1V 0A6, Canada

Zunaira Afzal, T. C. Howton and Yali Sun
Department of Biology, University of Alabama at Birmingham, Birmingham, AL 35294, USA

M. Shahid Mukhtar
Department of Biology, University of Alabama at Birmingham, Birmingham, AL 35294, USA
Nutrition Obesity Research Center, University of Alabama at Birmingham, Birmingham, AL 35294, USA

Adam Skibinski and Sarah Phillips
Department of Developmental, Molecular and Chemical Biology, Sackler School of Graduate Biomedical Sciences, Tufts University School of Medicine, 136 Harrison Ave., Boston, M\A 02111, USA
Molecular Oncology Research Institute, Tufts Medical Center, 800Washington St., Boston, MA 02111, USA

Hong-Ming Zhou and Simon J. Conway
Herman B. Wells Center for Pediatric Research, Department of Pediatrics, Indiana University School of Medicine, Indianapolis, Indiana, IN 46202, USA

Kayla Gross, Ania Wronski and Charlotte Kuperwasser
Department of Developmental, Molecular and Chemical Biology, Sackler School of Graduate Biomedical Sciences, Tufts University School of Medicine, 136 Harrison Ave., Boston, MA 02111, USA
Raymond and Beverly Sackler Convergence Laboratory, Tufts University School of Medicine, 145 Harrison Ave., Boston, MA 02111, USA
Molecular Oncology Research Institute, Tufts Medical Center, 800Washington St., Boston, MA 02111, USA

Index

www.ingramcontent.com/pod-product-compliance
Lightning Source LLC
LaVergne TN
LVHW080957210926
848485LV00004B/840
9781639871568